KU-521-503

Elastic Waves in Solids I

Springer
Berlin
Heidelberg
New York
Barcelona
Hong Kong
London
Milan
Paris
Singapore
Tokyo

Advanced Texts in Physics

This program of advanced texts covers a broad spectrum of topics which are of current and emerging interest in physics. Each book provides a comprehensive and yet accessible introduction to a field at the forefront of modern research. As such, these texts are intended for senior undergraduate and graduate students at the MS and PhD level; however, research scientists seeking an introduction to particular areas of physics will also benefit from the titles in this collection.

Daniel Royer   Eugène Dieulesaint

Elastic Waves in Solids I

Free and Guided Propagation

Translated by David P. Morgan
With 203 Figures, Numerous Problems and Solutions

Springer

Professor Daniel Royer
Université Denis Diderot
Laboratoire Ondes et Acoustique
ESPCI
10 rue Vauquelin
F-75231 Paris Cedex 05, France
E-mail: Daniel.Royer@espci.fr

Professor Eugène Dieulesaint
Emeritus professor
Université Pierre et Marie Curie
École Supérieure de Physique et de Chimie
10 rue Vauquelin
F-75231 Paris Cedex 05, France

Translator:
David P. Morgan
23 Williton Close
Weston Favell
Northampton NN3 3BG, England

Title of the original French edition:
Ondes élastiques dans les solides.
Tome 1: Propagation libre et guidée.

Library of Congress Cataloging-in-Publication Data.
Royer, D. (Daniel) [Ondes élastiques dans les solides. English] Elastic waves in solids / Daniel Royer, Eugène Dieulesaint. p. cm. Includes bibliographical references and index. Contents: v. 1. Free and guided propagation – v. 2. Generation, acousto-optic interaction, applications. ISBN 3-540-65932-3 (v. 1 : hc. : alk. paper). – ISBN 3-540-65931-5 (v. 2 : hc. : alk. paper) 1. Acoustic surface waves. 2. Signal processing. 3. Elastic waves. I. Dieulesaint, E. II. Title. QC176.8.A3R6913 2000 531'.33–dc21 99-34758

ISSN 1439-2674

ISBN 3-540-65932-3 Springer-Verlag Berlin Heidelberg New York

Printed in Germany

Typesetting: Data conversion by Steingraeber Satztechnik GmbH, Heidelberg
Cover design: *design & production* GmbH, Heidelberg

Printed on acid-free paper SPIN 10633465 57/3144/di 5 4 3 2 1 0

Foreword

The story of elastic waves is fascinating, from the very early mechanical aspects – well covered in *The Theory of Sound* by Lord Rayleigh – right up to the quantum behaviour, initiated by Einstein and Debye early in the 20th century. There have been great advances such as the idea of second sound in helium, built up formally by Landau and explained physically by Feynman. Or such as the time-reversal mirror, recently introduced by M. Fink. And there are, of course, many applications. Here, at the Ecole Supérieure de Physique et de Chimie Industrielles, we zealously guard the tank with which Langevin made his first experiments on ultrasonic propagation in water. This was the birth of sonar, an instrument of major military and civilian significance. On the other hand, a central concern of seismologists and geophysicists was the propagation of surface waves. In our time, these very waves have an important role in electronics, where they provide a very useful technology exploiting the long delays obtainable.

Personally, I feel a little hesitant to introduce a book describing aspects of acoustics in such detail, since I have only a limited familiarity with this science. However, many years ago I assimilated Cagniard's theory showing how a subterranean explosion could generate surface waves. And later, I found that Landau and Feynman's idea of second sound could apply (to a degree) to physical situations very different to superfluid helium, namely smectic liquid crystals. Most recently, we have been investigating – with R. Kant and M. Hébert – how a stretched elastic strip contracts (with constant density) when released. In this problem, there are several surprises for novices such as us – the deformations propagate at a velocity which is not c_{T} (the transverse wave velocity), but rather $2c_{\mathrm{T}}$. I first found this result in the present book, where the classic theory on waves in a plate is described in such a clear and simple manner.

This example, in particular, leads me to appreciate the amount and the depth of the work contributed by Professors Dieulesaint and Royer. They are truly following in the steps of Langevin, in the spirit of our school of Physics and Chemistry, where they have been conducting their research for more than twenty years. Their book is a substantial work. I am sure that it will be of great service to a large audience of acoustic scientists, geophysicists and physicists, and I wish it every success.

Paris *P.-G. de Gennes*

Foreword

Preface

Elastic waves, also called acoustic waves, are mechanical vibrations. They propagate in gases, liquids and solids. We focus here on their propagation in solids. However, mention is made of liquids on several occasions. First, in Chap. 1, fundamental ideas are introduced. Later on we describe devices whose solid surface is in contact with a liquid.

The chief objective of this book is to study the propagation and generation of waves in crystals. A secondary topic is the application of elastic waves in a wide variety of fields. The examples chosen centre on signal processing and the sensing of physical quantities. The reason for this choice is the impressive development of telecommunications and instrumentation, which require many acousto-electronic filters and sensors.

To understand the principle of operation and the behaviour of such components requires a knowledge, on the one hand, of the modes of propagation of the waves and, on the other, of suitable methods for generating and exploiting them. These two major areas are covered separately in the two volumes of this textbook.

The purpose of this first volume, *Free and Guided Propagation*, is to describe the different types of waves that propagate in isotropic and anisotropic solids. Crystals are of primary importance since they support waves of several gigahertz. The role of piezoelectric crystals to describe the different type of waves that propagate in isotropic and is especially emphasized. The notions of crystalline structure and the tensorial relations needed for this study are reviewed so that the reader needs no more than a good knowledge of elementary mechanics and electricity.

The second volume, *Generation, Acousto-optic Interaction and Applications*, investigates the means of generating and detecting the various types of waves studied in the first volume, their interaction various types of waves studied in the first volume, their interaction with light waves, and the exploitation of their specific properties in the construction and functioning of filters and sensors.

As a whole, this book is addressed to advanced students, engineers and scientists working in telecommunications, and also in geophysics, internal nondestructive evaluation and medical ultrasonography.

A number of problems are included at the end of each chapter, complete with worked solutions. These will enable readers to consolidate the material and verify that they have understood the main concepts.

The authors express their thanks to the following for the help they have given during the course of the composition of this work: André Zarembowitch, Professor at the Université P. et M. Curie (Paris 6); Jacques Détaint, Engineer at the Centre National des Télécommunications (CNET); and members of the Waves and Acoustics laboratory, particularly Claire Prada, Olivier Casula and Laurent Lévin.

We also express our gratitude to Professor P.G. de Gennes, Director of the Ecole Supérieure de Physique et de Chimie Industrielles, who has done us the honour of providing a foreword for this work.

David Morgan (Impulse Consulting, G.B.) has translated this volume I from French into English. We are grateful to him for helpful comments.

Paris
September 1999

D. Royer
E. Dieulesaint

Contents

Contents of Volume II

Introduction

By way of introduction, we would like firstly to give a brief outline of the topics to be covered in the remainder of this book. Secondly, we will give a survey of the chief historical landmarks in the development of acoustics.

Outline of the Book

This first volume is divided into five chapters. Chapter 1 summarizes general ideas relating to elastic waves, which involve motion in physical materials, covering the wave equation, energy flow, reflection and refraction, and the radiation diagram of a source. These are introduced for the case of a fluid (i.e. for an acoustic wave) because the behaviour of a fluid can be described relatively simply in terms of scalar quantities, and also because a liquid is involved in some applications.

For high-frequency applications, such as signal processing above 100 MHz, crystalline solids are used because they generally give low attenuation of acoustic waves, owing to their ordered structure. Consequently, the first part of Chap. 2 covers the description of a crystal structure by the lattice and the unit cell, the symmetries of this structure, the allocation of crystals to the 32 classes belonging to the 7 crystal systems, and the structure of some common crystals. The second part is concerned with representation of physical properties of crystals by tensors. The number of independent components of each tensor depends on the crystal symmetry. The method for determining this number, for a given crystal and for any physical property, is explained and is illustrated by the example of the dielectric permittivity tensor.

The investigation of the elastic behaviour of a crystal, that is, its deformation (a second-rank tensor) in response to applied external forces which generate internal stress (another second-rank tensor) involves the use of a fourth-rank tensor, the stiffness tensor of the material. If the crystal is piezoelectric another tensor, of rank three, must be added. In Chap. 3 these tensors are defined, their independent components are enumerated, and the various forms of energy are expressed in terms of their components. The expression for the Poynting vector, introduced in Chap. 1, is derived here for a piezoelectric crystal. The values of the elastic and piezoelectric constants for the commonest crystals are given in two tables. The reader may of course postpone the detailed study of some derivations, for example those relating to the reduction of the number of independent components in the elastic and piezoelectric tensors by virtue of the symmetry, but it is important to appreciate the significance of the tables.

Chapter 4 begins by showing, using the model of a chain of atoms, that a crystal can be regarded as a continuous medium when considering elastic waves at practical frequencies (1 kHz to 10 GHz). The only crystalline property of relevance is its anisotropy. The three solutions of the wave equation are illustrated by three surfaces, such as the slowness surfaces (inverse of velocity). The slowness surface, analogous to the index surface used in optics, has the significance that its normal shows the direction of energy propagation. Its form is deduced for ordinary and piezoelectric crystals belonging to various symmetry classes. This surface also gives the directions of the reflected and refracted waves when a wave in one medium is incident on another. The amplitudes of the waves involved are calculated for cases where the media have high degrees of symmetry. This method is illustrated by three simple examples – the interfaces between solid and solid, solid and vacuum and solid and liquid – which are treated at the end.

The fifth, and last, chapter concerns propagation in bounded media, that is, waves simultaneously satisfying the wave equation and the boundary conditions. For a given crystal, this topic normally calls for numerical techniques. However, there are two particular cases in which the symmetry causes the equations to separate into two independent parts. The first section of this chapter describes these cases, illustrating the main types of waves guided along a plane surface, or along several parallel surfaces. The second section gives a general derivation of the power carried by a wave in a waveguide. This is followed by sections describing the important Rayleigh waves, guided along a plane surface, and several types of wave with transverse displacement – guided along a plate, guided by the surface of a piezoelectric solid (Bleustein–Gulyaev–Shimizu wave), and guided by a film on a substrate (Love wave). Later sections cover waves guided by an isotropic plate with displacements in the sagittal plane (Lamb waves), and finally waves guided by a cylinder.

Historical Survey

For many years, elastic waves were known only in the forms of audible waves (with frequencies of 50 Hz to 15 kHz) and seismic tremors with their disastrous effects. The latter were studied mainly by geophysicists who, with no access to the sources of the waves, attempted to explain the physics of propagation. At the end of the last century, it was understood that the vibrations in the earth propagate, in the volume, in the form of longitudinal waves and the (slower) transverse waves. Lord Rayleigh showed, in 1885, that they can also propagate partly in the form of surface waves, with a velocity less than that of the volume transverse waves. Subsequently, other forms of propagation were studied in particular structures – a layer on a substrate, a parallel-sided plate, and the interface between two solids. To Rayleigh waves were added the waves of other geophysicists – Love, Lamb and Stoneley.

The first controlled usage of elastic waves was that of Paul Langevin, during the first world war (1915), following C. Chilowsky's suggestion to use reflections of the waves for detection of submarines. Langevin had the idea of exploiting the piezoelectric effect, discovered in 1880 by the brothers Pierre and Jacques Curie, combined with the emerging electronics technology, to launch elastic waves into the sea (at 40 kHz) and to detect the fraction returned by an object in the beam. Using this echo principle, the first SONAR (SOund Navigation And Ranging) was realized. One of the authors of this book (E.D.) had the privilege of working with Marcel Tournier, a close collaborator of Langevin. Professor Tournier described the very first experiments carried out at ESPCI (Ecole Supérieure de Physique et de Chimie Industrielles, Paris), with a 'singing' condenser (in which the alternating electrostatic field caused the vibrations) as transmitter and a pendulum (sensing the pressure) or microphone as receiver. Later experiments, at Toulon, used a triplet transducer consisting of quartz crystals mounted between two steel plates, and a vacuum-tube amplifier (with triodes!). Indeed, the contribution of electronics to the development of acoustics (meaning acoustic waves at any frequency – infrasound, ultrasound, hypersound) was essential. Without an amplifier, the detection of weak signals reflected by a target several kilometres away would have been impossible.

Conversely, acoustics soon showed its gratitude to electronics. About 1920, A.M. Nicholson, W.G. Cady and G.W. Pierce introduced high-Q quartz resonators into electronic circuits, conferring a remarkable stability on oscillators for 'radiodiffusion' transmitters. In 1925, K.S. Van Dyke and D.W. Dye independently developed an equivalent circuit (electrical model) for the resonator, and its usage spread, particularly in electronic filters. These topics began a long dialogue between *electronics* and *acoustics*, whose fruits are summarized in the table below. Here we discuss some of the consequent applications.

After their discovery of piezoelectricity, more than a century ago, the brothers Curie exploited it in a quartz balance. However, the next application – the highly-significant Sonar – emerged only after a further thirty-five years. Between the two world wars, in the absence of enemy action, Sonar was used mainly for mapping the sea bed and locating shoals of fish. The first Langevin–Florisson probe was put into service in 1922, and by 1932 about 500 ships were equipped. Techniques for echo reception were regularly improved, permitting deeper and deeper sounding and improving the precision of position measurement for moving targets. Also, magnetostriction was partly replacing piezoelectricity. During the Second World war, Sonar played a major role in protecting the allied convoys.

Just before the second world war the Sonar principle was applied to electromagnetic waves to give RADAR (RAdio Detection And Ranging), exploiting the advance of electronics to higher frequencies (especially with the later British invention of the magnetron) and to improved time discrimination. Thus acoustic engineers preceded radio engineers in this context, though both were anticipated by nature (in bats and dolphins). The same principle is used in ultrasound scanning, as in present-day medicine (especially obstetrics). It was also applied to metallurgical non-destructive testing, though inventors such as Sokoloff realized, in the 1930s, that to be effective this application would need better time resolution and better transducers. In the laboratory, simple plates of quartz, already used as resonators in electronic equipment, would serve as high-frequency transducers (with harmonics for frequencies above 1 MHz), generating elastic waves in gases, liquids or solids in order to study phenomena such as wave absorption, coagulation of aerosols, cavitation, acceleration of chemical reactions, sterilization of liquids and biological effects. However, there was a need for better techniques for transforming electrical energy into acoustic energy, and for better piezoelectric materials. Following the Pierce oscillator in 1928, magnetostriction was found to be useful for low-frequency applications (below 100 kHz) such as ultrasonic drilling, and so could also be applied to Sonar. But piezoelectric materials, before 1940, were limited in practice to quartz and Seignette's salt (Rochelle salt), and the fragility of the latter restricted it to applications such as the pick-up head of record players.

The Second World war gave rise to intense research, of which the results were only partly divulged at the end of the conflict. Magnetostrictive cobalt alloys appeared, as did new piezoelectric crystals – mono-alkaline phosphates, practically replaced around 1950 by ceramics of barium titanate rendered piezoelectric by applying an electric field. These in turn were replaced, after 1955, by ceramics based on lead titanate and zirconate, called PZT (Piezoelectric lead Zirconium Titanate ceramics, Clevite Corporation). These ceramics, with electromechanical coefficients much larger than that of quartz, have considerably facilitated the generation of the waves at frequencies up to a few tens of MHz. They constitute the transmitters and receivers in equipment such as sonar and ultrasound scanning instruments for medicine and metallurgy. They are well suited to realization of sensitive transducers, which can be of large size and of varied geometry. To adapt the acoustic impedance to that of the wave propagation medium, it is possible to use them in the form of a matrix consisting of rods immersed in a resin, giving a composite transducer particularly suited to medical scanning. The number of applications, in both the professional and general public domains, grows steadily – they provide precise movement in advanced microscopes, they serve as a wedge of variable thickness for interferometers or mirrors, they act as lighters for gas cookers, ... However, intrinsic losses limit the use of ceramics to frequencies below 50 MHz, so it was necessary to pursue the fabrication of monocrystalline piezoelectrics. In the 1970s, this became possible with the production of single crystals of lithium niobate and tantalate. Nevertheless, single-crystal quartz, with its excellent temperature stability, remains irreplaceable in numerous situations, particularly as it has been manufactured industrially since the 1950s.

The elastic waves considered so far are called bulk acoustic waves (BAW) because they travel in the interior of the medium (the term 'bulk wave' is synonymous with 'volume wave', and the terms 'elastic' and 'acoustic' are often used interchangeably for solids or liquids). Since 1920, BAW's have been used in electronics in the form of stationary waves (in piezoelectric resonators), particularly, for several decades, to stabilize the operating frequency in professional equipment. Today, the effectiveness of the quartz resonator as a precise miniature timepiece is demonstrated by its ubiquitous presence in wrist-watches. Synchronization of functions in analogue or digital equipment is provided by quartz 'clocks'. The structure of these resonators, their performance (e.g. long-term stability) and their fabrication are continually evolving. Using ion milling, the thicknesses needed for fundamental mode operation above 200 MHz are readily obtainable. Coupled resonators, made on the same substrate, constitute monolithic filters usable for bandpass filtering.

The use of *travelling* bulk elastic waves appeared, in electronics, in response to the requirement for a signal memory in Radar systems. During the last war, the low velocity of these waves, typically 10^5 times smaller than that of electromagnetic waves, was exploited to construct delay lines using a liquid (water, mercury), and later a solid (silica), medium. Around 1960, the need for higher frequencies was met by depositing thin piezoelectric films on solid media in which the waves have low attenuation, such as sapphire (single-crystal alumina). The transducers, consisting of crystallographically oriented films such as zinc oxide, could generate the waves at frequencies above 1 GHz, and were also crucial in the development, by C.F. Quate and R.A. Lemons, of the acoustic microscope. For low frequencies, we must mention the development, in the 1980's, of sheets of piezoelectric polymer (PVF_2).

After the discovery of Rayleigh waves, or surface acoustic waves (SAWs), it was to be a long time, 80 years, before the advent of what is now the most common method for generation and detection. This was the use of interdigitated comb-shaped electrodes deposited on a piezoelectric material, demonstrated by R.M. White and F.W. Voltmer and also described in two patents filed in 1963, by W. Mortley in Britain and J. Rowen in the U.S.A. In effect, this technique provides a sequence of sources on the surface, with amplitudes and phases determined by the comb geometries (in contrast to a resonator-type transducer for volume waves, mentioned above, which represents only one well-localized source). It follows immediately that numerous functions can be synthesized. The exploitation of this technique quickly brought an elegant solution to the problem of extracting Radar signals from noise, using filters matched to these signals, and opened the way to many other advances. This success is partly due to the parallel development of microelectronic technology, which could be called on for electrode fabrication.

Among the main surface wave devices, we mention bandpass filters (which have many types, broadly classified into the two areas of travelling-wave devices and standing-wave devices), spectrum analysers and convolvers. These light, compact Rayleigh-wave filters, operating over a wide frequency range (50 MHz to 2 GHz) and suitable for 'batch' fabrication, have invaded the large consumer areas such as television and mobile telephones. Depending on the requirements, particularly temperature stability and cost, the substrate is usually a crystal of quartz, lithium niobate or lithium tantalate, or a glass plate supporting a piezoelectric film. In some cases the performance is improved by using a multi-strip coupler (F. Marshall and E. Paige), and reflective arrays can be used for radar pulse compression (R. Williamson and H. Smith). Several tens of laboratories world-wide have contributed to the development of surface wave devices, and a list of the main ones will be found in the introduction to reference [14] in the bibliography.

The interaction between *acoustics* and *optics* began with the diffraction of light by elastic waves, done experimentally by R. Lucas and P. Biquard in France (at ESPCI) and by P. Debye and F.W. Sears in the United States. This was in 1932, ten years after L. Brillouin had anticipated the effect. In the following years, C.V. Raman and N.S.N. Nath, in India, analysed the interaction under various conditions. It remained a laboratory curiosity until the advent, in 1960, of the laser, a powerful coherent light source which made experimentation easier and enabled the realization, around 1970, of components such as the acousto-optic modulator. This component became essential for instruments such as laser printers and optical probes for measurement of small mechanical movements. On the other hand, another promising interaction appeared – the direct excitation of elastic waves by a laser (using the thermoelastic effect) and their detection using an optical probe. This technique, which eliminates mechanical contact, now has its place in aeronautical non-destructive testing.

The development of acousto-electronic and acousto-optic components for signal processing demanded better understanding of theoretical topics such as acoustic propagation in crystals, the physics of transduction, wave generation by comb electrodes, and wave behaviour for different environmental conditions (variation of temperature or pressure). This resulted, *inter alia*, in the discovery of a new surface wave in a piezoelectric solid (the Bleustein–Gulyaev–Shimizu wave, 1968), the formulation of figures of merit for assessing new materials, new cuts of crystals, description of acoustic devices using electrical equivalent circuits, the possibility of exciting transverse waves using comb electrodes, and the development of sensors. The simplest example of the latter is a piezoelectric resonator using bulk acoustic waves, as might be used in an electronic filter. Near the cuts used for electronics, where the crystals are insensitive to temperature or acceleration, there are other cuts sensitive to these parameters. Consequently, resonators can play the role of sensors for temperature or pressure. In practice, it was quite early, after the Second World war, that the quartz resonator was used to measure the thick-

Chronology of Main Events in Acoustics

Date	Event	Applications
1880	Discovery of piezoelectricity. *P. and J. Curie*	Balance
1885	Surface acoustic waves (SAW). *Lord Rayleigh*	Seismology
1915 ↓	Ultrasound in water. *C. Chilowski and P. Langevin*	Sonar – target detection, sounding
1918	Triplet resonator, steel–quartz–steel (40 kHz)	
1920 ↓	Quartz resonators in electronic circuits. *A. Nicholson, W. Cady, G. Pierce*	Oscillators
1925	Equivalent circuit. *K. Van Dyke, D. Dye*	Filters
1928	Magnetostriction. *G. Pierce*	Sonar, Metallurgy
1932 ↓ 1935	Diffraction of light (foreseen in 1922 by *L. Brillouin*) $f < 30\,\text{MHz}$. *R. Lucas and P. Biquard, P. Debye and F. Sears.* Later studies by *C. Raman, N. Nath*	Measurement of elastic constants
1940	Delay lines (water, mercury, silica)	Radar
1950 ↓	Piezoelectric ceramics PZT ceramics (Clevite Corporation)	BAW transducers (LF) Sonar
1955	Synthetic quartz	
1960	Piezoelectric thin films Advent of laser	BAW delay lines (HF)
1965	Comb electrodes – SAW transducers. *R. White, F. Voltmer*	Dispersive filters – radar SAW delay line
1970	Lithium niobate and tantalate Multistrip coupler. *F. Marshall, E. Paige* Reflector array. *R. Williamson, H. Smith* Acousto-optic modulator (BAW) Ion milling SAW resonator. *E. Ash*	Bandpass filters for TV and telecommunications Radar, spectral analysis Instrumentation BAW filter ($f > 200\,\text{MHz}$) Bandpass filtering
1980	Focussing of BAW. *C. Quate, R. Lemons* Non-linear effect – convolution Acousto-optic probe Composite transducer Guided waves	Acoustic microscope Counter-measures Metrology Ultrasonic scanning SAW and BAW sensors
1990	SAW in-line coupled resonators Single phase unidirectional transducer Resonators using transverse surface waves Photothermal generation	Mobile telephone filters Filters to 3 GHz Non-destructive testing

ness of a thin film deposited in a vacuum – the resonant frequency changes with the mechanical loading on the faces. Other possibilities for sensors are delay lines using guided waves, such as waves in plates or cylinders or on a surface as in the case of Rayleigh waves; all of these have contact with the outside world and are potentially sensors under suitable conditions.

The table, taking account of the objectives of this book, concentrates on the developments in elastic waves and their main applications in the fields of electronic and optical signal processing (telecommunications) and instrumentation (sonar, radar, ultrasonic scanning). Of other disciplines, either deliberately using the waves or simply involving them, we mention two extreme examples – the reduction of noise by cancellation, and the explanation of sun spots (helioseismology).

The dates shown in the table should not be assumed accurate to one year, especially when referring to the appearance of a material. In the case of synthetic quartz it is generally accepted that, following work on crystal growth started by the Italian scientist Spezia in 1905, the methods were later developed by German, English and American researchers, particularly during the 1939–45 war when there was pressure to reduce the importation of natural quartz into their countries. Brazil was, at that time, the main supplier; American laboratories such as Brush Laboratories entered the market later, in 1953. The present annual production of synthetic quartz is assessed at thousands of tonnes [15].

1. Waves. Fluid as a Scalar Model

This first chapter consists of three sections. Section 1.1 summarizes expressions for propagating waves, plane waves and stationary waves, and the definition of group velocity. For convenience we assume here that the nature of these waves does not change during propagation, even on reflection at a surface.

Section 1.2 treats propagation of an elastic wave in a medium described by scalar relations, that is, in a fluid (gas or liquid). A wave propagating in such a medium is called an acoustic wave when the frequency is in the audible range, and an ultrasonic wave for higher frequencies. The scalar model serves as a simple introduction to some general notions needed for the study of propagation in isotropic solids, and also for crystals whose anisotropic behaviour is expressed by tensor relations. We describe the wave equation for plane waves, the acoustic impedance (analogous to that of an electrical transmission line), and the acoustic energy flux given by the Poynting vector. We also examine partial reflection, under normal or oblique incidence, of a wave at the interface of two media. The reflection and transmission coefficients, as functions of angle of incidence, follow directly, as do the impedance transformation given by a layer and the condition for total reflection.

Section 1.3 is concerned with the wave equation for a spherical wave. This equation is similar to that for a plane wave, but with the overpressure replaced by the product (overpressure $\times$ distance from source). The radiation from a planar source is studied, starting from the spherical wave emitted by an element of the surface. The objective is to distinguish the near-field and far-field regions and to deduce their characteristics. The results are applied particularly to transducers, usually piezoelectric, which are the sources of elastic waves in many devices. The idea of the diffraction impulse response is introduced at the end of the chapter.

1.1 Travelling, Stationary and Guided Waves

In general, a localized departure from equilibrium conditions can cause a perturbation which spreads out, and this is essentially a propagating wave. The description of a wave involves the parameters velocity, wavelength and

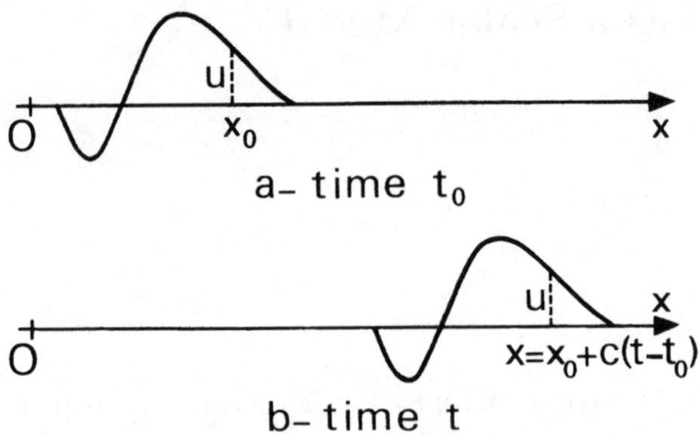

Fig. 1.1. Propagation of a disturbance. The disturbance propagates with velocity c, so that a feature with amplitude u at position x_0 and time t_0 is found at time t to be at position $x = x_0 + c(t - t_0)$

wavevector, whose definition does not depend on the nature of the disturbance. If the dimensions of the medium are finite, a travelling wave can be reflected at the faces which bound it, and the wave can then become a stationary or guided wave. In this section we clarify these concepts.

1.1.1 Expressions for a Plane Wave

We describe the propagation of a disturbance by considering, at some fixed point, the time variation of some characteristic quantity. At a position x_0, the disturbance is described by the value u of the quantity at time t_0 (Fig. 1.1). If the disturbance propagates without modifying its form, for example without attenuation, and is displaced with constant velocity c in the $+x$ direction, then the value u is the same at some other point x and at time t if

$$x = x_0 + c(t - t_0)$$

and hence

$$x - ct = x_0 - ct_0 \quad \text{or} \quad t - \frac{x}{c} = t_0 - \frac{x_0}{c}\,.$$

Any function $u(x,t)$, which describes a phenomenon propagating without distortion in the $+x$ direction, depends only on the expression $t - x/c$. It has the same value for all x and t if $t - x/c$ is constant, so that

$$\boxed{u(x,t) = F\left(t - \frac{x}{c}\right) = f(x - ct)}\,. \tag{1.1}$$

If the propagation is in the $-x$ direction, then

$$u(x,t) = G\left(t + \frac{x}{c}\right) = g(x + ct)\,.$$

If both of these waves propagate, in opposite directions, in the same medium, the total disturbance has the form

$$u(x,t) = F\left(t - \frac{x}{c}\right) + G\left(t + \frac{x}{c}\right) = f(x - ct) + g(x + ct)\,.$$

Among all the possible departures from equilibrium, *sinusoidal* oscillations about a mean value are of great importance since other perturbations can be expressed as sums of such oscillations. Consider for example a plane membrane vibrating in air with a sinusoidal displacement u_{M} with period T, so that

$$u_{\mathrm{M}} = A\cos\omega t \quad \text{with} \quad \omega = 2\pi/T\,.$$

The emitted acoustic wave has the form

$$u = A\cos\omega\left(t - \frac{x}{c}\right).$$

Here u denotes the displacement of the air in a plane parallel to the membrane, but it could alternatively represent the variation of pressure or density on such a plane. Equivalently, we can write

$$u = A\cos 2\pi\left(\frac{t}{T} - \frac{x}{\lambda}\right),$$

where the wavelength $\lambda = cT$ is the distance travelled by the disturbance in a time T. Since the phenomenon is unchanged after a time T, λ *represents, at a given instant, the distance between two identical states of the fluid* (air), for example two consecutive maxima of the amplitude u.

Another useful form is

$$u = A\cos\left(\omega t - \frac{\omega x}{c}\right) = A\cos(\omega t - kx)\,,$$

where $k = \omega/c = 2\pi/\lambda$ is the *wavenumber*. The term $-kx$ measures, at a given time, the *phase change* of the disturbance at point x, relative to the origin; in the present case the phase difference between the vibration at x and the vibration of the membrane at $x = 0$. The overall phase φ of the wave is defined by

$$\varphi = \omega t - kx\,.$$

The wavenumber k determines the variation of φ with distance x, at a given time, while the frequency ω gives the variation of φ with time at a given point. Thus,

$$k = -\left(\frac{\partial\varphi}{\partial x}\right)_t, \quad \omega = \left(\frac{\partial\varphi}{\partial t}\right)_x. \tag{1.2}$$

We therefore have the following correspondence between time and space variables:

$$\left.\begin{array}{ll}\text{Period} & T\\ \text{Frequency} & \omega\end{array}\right\}\ \omega = 2\pi/T \qquad \left.\begin{array}{ll}\text{Wavelength} & \lambda\\ \text{Wavenumber} & k\end{array}\right\}\ k = 2\pi/\lambda\,.$$

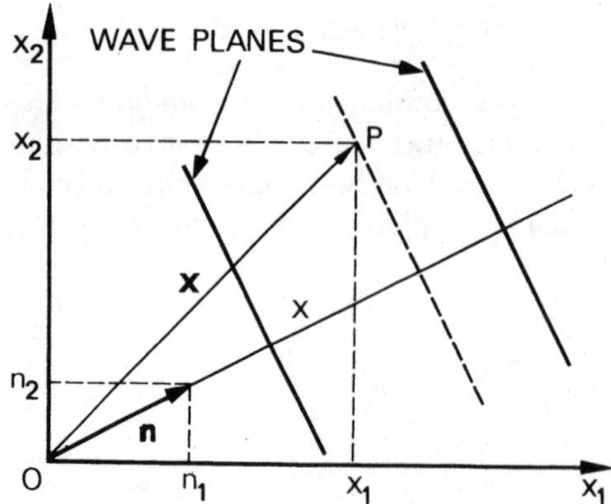

Fig. 1.2. At a typical point P denoted by the vector $\boldsymbol{x} = \boldsymbol{OP}$, the vibration due to the wave propagating with velocity c in the direction of the unit vector $\boldsymbol{n}$ is $u(\boldsymbol{x}, t) = F(t - \boldsymbol{n} \cdot \boldsymbol{x}/c)$

The propagation velocity $c = \omega/k$ is called the *phase velocity.* This is the velocity at which an observer would need to move in order to see the disturbance having the same phase at all times; the observed phase $\varphi = \omega(t - x/c)$ is constant if $x = ct + \text{constant}$. The wave appears stationary to this observer.

Wavefronts. In the above example the particles situated on a plane parallel to the membrane vibrate in phase, and the movement at all points on this 'wavefront' is determined if the distance x from the origin is given. In general, the wavefront is not perpendicular to a coordinate axis. The expression for the vibration at a point P, located by the vector $\boldsymbol{x} = \boldsymbol{OP}$ from the origin to this point, involves the unit vector $\boldsymbol{n}$ perpendicular to the wavefronts (Fig. 1.2). The 'abscissa' x of P is the projection of $\boldsymbol{x}$ onto $\boldsymbol{n}$, so that $x = \boldsymbol{x} \cdot \boldsymbol{n}$. Replacing x by this value in (1.1) gives

$$\boxed{u = F\left(t - \frac{\boldsymbol{n} \cdot \boldsymbol{x}}{c}\right)} . \tag{1.3}$$

We can express the scalar product in the usual system of orthogonal axes using the coordinates x_1, x_2, x_3 of the point (the components of the vector $\boldsymbol{x}$) and the direction cosines n_1, n_2, n_3 of the propagation direction (the components of the vector $\boldsymbol{n}$), so that

$$u = F\left(t - \frac{n_1 x_1 + n_2 x_2 + n_3 x_3}{c}\right) .$$

In the case of a sinusoidal plane wave, we have

$$u = A \cos \omega \left(t - \frac{\boldsymbol{n} \cdot \boldsymbol{x}}{c}\right) = A \cos(\omega t - \boldsymbol{k} \cdot \boldsymbol{x}) ,$$

where we have introduced the *wave vector*

$$\boldsymbol{k} = \frac{\omega}{c} \boldsymbol{n} = \frac{2\pi}{\lambda} \boldsymbol{n} = k \boldsymbol{n} .$$

In the above, the wave amplitude A was assumed to be constant. Clearly, if the propagation involves attenuation or amplification the amplitude will depend on the position of the observation point P, and therefore on x, so that

$$u = A(x)\cos(\omega t - \boldsymbol{k}\cdot\boldsymbol{x})\,.$$

Representation of sinusoidal functions by complex numbers. Although they are real, the amplitudes encountered in physics can often be represented by complex numbers. This usage, commonly applied to sinusoidal functions, rests on the fact that linear relations between complex numbers also apply to the real and imaginary parts separately, provided the coefficients are real. The equations of physics are often linear with real coefficients. For the sinusoidal case, the complex expression

$$u_\mathrm{c} = A\mathrm{e}^{\mathrm{i}(\omega t+\phi)}$$

can represent the real quantity

$$u = A\cos(\omega t+\phi) = \mathrm{Re}[u_\mathrm{c}]\,.$$

The significance of this follows from several properties of the exponential function:

- In a sum of sinusoids with the same frequency, the time-dependent term $\exp(\mathrm{i}\omega t)$ can be factored out, so that $\sum u_\mathrm{c} = \exp(\mathrm{i}\omega t)\sum A\exp(\mathrm{i}\phi)$.
- Differentiation with respect to time is equivalent to multiplication by $\mathrm{i}\omega$, so that $\mathrm{d}u_\mathrm{c}/\mathrm{d}t = \mathrm{i}\omega u_\mathrm{c}$.
- The square of the amplitude is $A^2 = u_\mathrm{c}u_\mathrm{c}^*$, where u_c^* is the complex conjugate of u_c.
- The energy (or mean power) transported by a wave, equal to the time-average of the product of two sinusoidal functions, can be expressed by a simple product. Taking the quantities

  $$u(t) = A\cos(\omega t+\phi) = \mathrm{Re}[u_\mathrm{c}] \quad\text{and}\quad v(t) = B\cos(\omega t+\psi) = \mathrm{Re}[v_\mathrm{c}],$$

  the time-averaged product P is

  $$P = \langle u(t), v(t)\rangle = \frac{1}{2}\mathrm{Re}[u_\mathrm{c}v_\mathrm{c}^*]\,, \tag{1.4}$$

  where the brackets $\langle\ \rangle$ indicate a time average. To show this, note that

  $$\langle u(t), v(t)\rangle = \frac{AB}{2}\langle\cos(2\omega t+\phi+\psi) + \cos(\phi-\psi)\rangle$$

  and hence, since the average of the first cosine is zero, we have

  $$\langle u(t), v(t)\rangle = \frac{AB}{2}\cos(\phi-\psi) = \frac{1}{2}\mathrm{Re}[u_\mathrm{c}v_\mathrm{c}^*]$$

  as required.

A sinusoidal plane wave can be expressed in complex notation as

$$u_\mathrm{c} = A\exp[\mathrm{i}(\omega t - \boldsymbol{k}\cdot\boldsymbol{x})]\,.$$

1.1.2 Total Reflection

If the propagation medium is not infinite, particularly if its properties vary rapidly on the scale of the wavelength, the wave is either partially or totally reflected. This is the case at the boundary between two media with different properties. The *boundary conditions* at the interface depend on the nature of the wave. They are simplest if the disturbance cannot propagate in one medium, and if a variable characterizing the disturbance is zero on one side of the boundary and must also vanish on the other side for reasons of continuity. This case, assumed here for illustration, is exemplified by the particle velocity in an acoustic wave reflected at a rigid wall, or the stresses of a transverse wave reflected at the free surface of its propagation medium (see Sect. 4.4.2.1).

1.1.2.1 Normal Incidence – Stationary Waves. An incident wave

$$u_{\mathrm{I}}(x,t) = F_{\mathrm{I}}\left(t - \frac{x}{c}\right)$$

arriving perpendicularly at the plane surface separating two media gives rise to a reflected wave propagating in the reverse direction, given by

$$u_{\mathrm{R}}(x,t) = F_{\mathrm{R}}\left(t + \frac{x}{c}\right).$$

We assume a perfectly-reflecting boundary at which the amplitude is zero at all times. Taking the origin to be at the boundary, we have

$$u_{\mathrm{I}}(0,t) + u_{\mathrm{R}}(0,t) = 0 \quad \text{for all } t\text{, giving} \quad F_{\mathrm{R}}(t) = -F_{\mathrm{I}}(t).$$

It follows that the reflected wave is

$$u_{\mathrm{R}}(x,t) = -F_{\mathrm{I}}\left(t + \frac{x}{c}\right).$$

The reflected and incident waves thus have an antisymmetric relation when referred to the origin, as shown in Fig. 1.3.

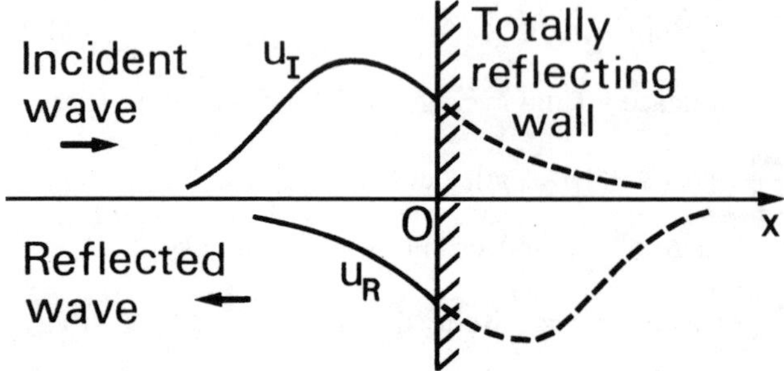

Fig. 1.3. Total reflection of a propagating wave. The reflected disturbance is a symmetric version of the incident disturbance with respect to the origin, which is taken to be on the reflecting plane

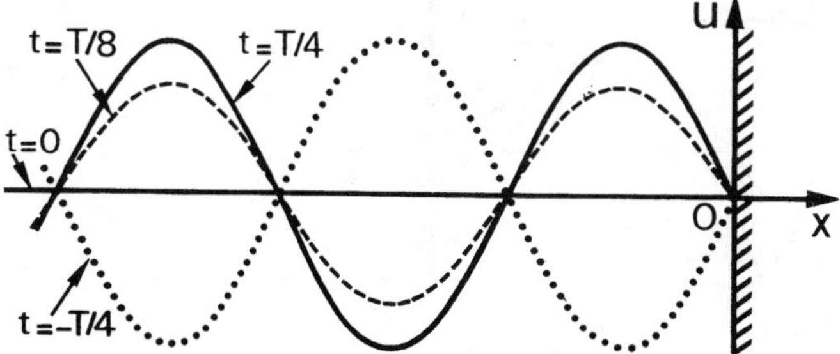

Fig. 1.4. Stationary wave. The total reflection of a propagating sinusoidal wave creates a stationary wave

As a result of the reflection, the total disturbance is

$$u = u_\mathrm{I} + u_\mathrm{R} = F_\mathrm{I}\left(t - \frac{x}{c}\right) - F_\mathrm{I}\left(t + \frac{x}{c}\right).$$

For the *sinusoidal* case the disturbance

$$u = A\left[\cos\omega\left(t - \frac{x}{c}\right) - \cos\omega\left(t + \frac{x}{c}\right)\right] = 2A\sin\omega t\sin\frac{\omega x}{c} \tag{1.5}$$

is the *product of two distinct functions*, one of time and the other of position. All semblance of propagation has disappeared. The result is a *stationary wave* (whose energy remains localized), with the characteristic feature that the phase is the same at all points but the amplitude varies with x. This is shown in Fig. 1.4, which depicts the vibration at several successive times. The x-values given by

$$\frac{\omega x}{c} = -n\pi \quad n = 0, 1, 2, 3, \ldots$$

are points where the vibration is zero for all t, and are called *nodes*. Midway between the nodes are the *anti-nodes*, where the amplitude is maximized. The interval between successive nodes or anti-nodes is equal to half the wavelength.

Standing-wave ratio. If the reflection is not total, the amplitude A_R of the reflected wave is smaller than the amplitude A_I of the incident wave, since part of the latter has been transmitted into the second medium. The resulting wave is

$$u = A_\mathrm{I}\cos\omega\left(t - \frac{x}{c}\right) - A_\mathrm{R}\cos\omega\left(t + \frac{x}{c}\right)$$

$$u = (A_\mathrm{I} - A_\mathrm{R})\cos\omega\left(t - \frac{x}{c}\right) + A_\mathrm{R}\left[\cos\omega\left(t - \frac{x}{c}\right) - \cos\omega\left(t + \frac{x}{c}\right)\right]$$

and this can be expressed as the sum of a travelling wave and a stationary wave, so that

$$u = (A_\mathrm{I} - A_\mathrm{R})\cos\omega\left(t - \frac{x}{c}\right) + 2A_\mathrm{R}\sin\omega t\sin\frac{\omega x}{c}.$$

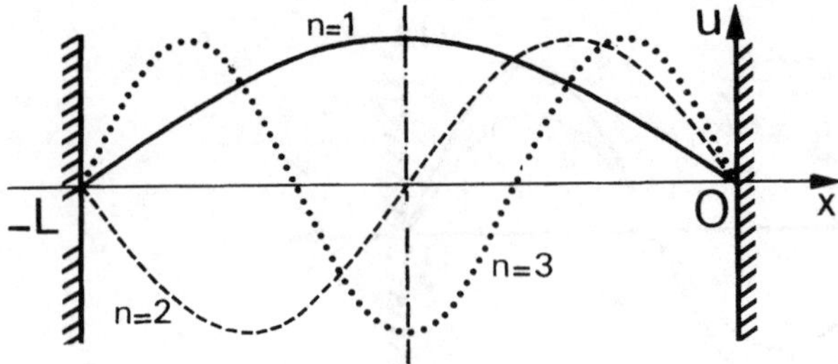

Fig. 1.5. One-dimensional resonator. The first few modes of a resonator comprising two parallel perfectly reflecting walls

The ratio of the maximum and minimum amplitudes is, by definition, the *standing-wave ratio*, (*SWR*), so that

$$SWR = \frac{A_{\mathrm{I}} + A_{\mathrm{R}}}{A_{\mathrm{I}} - A_{\mathrm{R}}} . \tag{1.6}$$

The *SWR* is infinite for a stationary wave, and unity for a travelling wave. More generally, it is determined by the reflection coefficient $r = A_{\mathrm{R}}/A_{\mathrm{I}}$ and given by (see Sect. 1.2.4.2)

$$SWR = \frac{1+r}{1-r} .$$

One-dimensional resonator. We now consider a wave confined to travel between two perfectly reflecting walls, at $x = 0$ and $x = -L$. For the stationary wave (1.5) to be zero at the boundary $x = -L$, as well as at $x = 0$, we must have

$$\frac{\omega L}{c} = n\pi \quad \text{or} \quad f = n\frac{c}{2L}, \quad \text{with } n \text{ integer} . \tag{1.7}$$

Only these stationary waves, whose frequency f is a multiple of $c/2L$, are allowed to exist. The first few of these *eigen modes* of the resonator are shown in Fig. 1.5. The modes with even (odd) order are antisymmetric (symmetric) with respect to the median plane.

1.1.2.2 Oblique Incidence. Guided Waves. Suppose that the wave vector $\boldsymbol{k}$ is no longer normal to the wall. In the chosen system of axes (Fig. 1.6) the incident wave is expressed as

$$u_{\mathrm{I}} = A\cos\left(\omega t - k_1 x_1 - k_2 x_2\right) .$$

Taking $\boldsymbol{k}'$ as the wave vector of the reflected wave (reflected totally at the plane $x_1 = 0$), we have

$$u_{\mathrm{R}} = -A\cos\left(\omega t - k_1' x_1 - k_2' x_2\right) .$$

The total disturbance is

$$u = 2A\sin\left(\omega t - \frac{k_1 + k_1'}{2}x_1 - \frac{k_2 + k_2'}{2}x_2\right)\sin\left(\frac{k_1 - k_1'}{2}x_1 + \frac{k_2 - k_2'}{2}x_2\right)$$

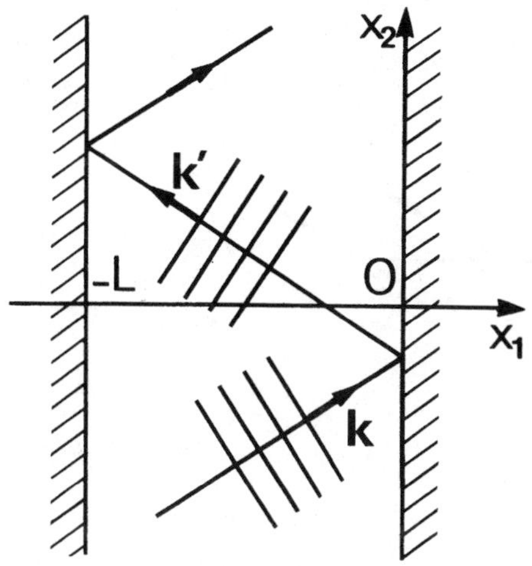

Fig. 1.6. Reflections at oblique incidence. The wave propagates in the Ox_2 direction by successive reflections from the two parallel walls

and for this to be zero at all points on the plane $x_1 = 0$ we need $k'_2 = k_2$. Since the wavenumbers are the same ($k = k' = \omega/c$) we find $k'_1 = \pm k_1$, and only the case $k'_1 = -k_1$ gives a non-zero solution, which is

$$u = 2A \sin(\omega t - k_2 x_2) \sin k_1 x_1 \,.$$

Thus the wave vectors of the incident and reflected waves are symmetric with respect to the reflecting surface. In addition, for u to be zero at the second boundary $x_1 = -L$ we need

$$k_1 = n\frac{\pi}{L} \quad n = 1, 2, 3 \ldots .$$

The final expression,

$$u = 2A \sin(\omega t - k_2 x_2) \sin\left(\frac{n\pi}{L} x_1\right) \tag{1.8}$$

is that of a wave propagating along Ox_2 inside the guide defined by the walls at $x_1 = 0$ and $x_1 = -L$, which are nodal planes. This is a *guided wave*. The phase velocity $c_\varphi = \omega/k_2$ is higher than that of an unguided plane wave, $c = \omega/k$. In essence, the point where a wavefront intercepts the guide axis has velocity greater than c because the wavefront itself travels obliquely. For each value of the integer n the solution is a *mode*; between the walls, the n-th mode has $n - 1$ nodal planes, equidistant and parallel to them (Fig. 1.7).

The wavenumber $\beta = k_2$ of the guided wave is related to that of an unguided wave, k, by

$$\beta^2 = k^2 - k_1^2 = k^2 - \left(\frac{n\pi}{L}\right)^2 .$$

The smallest possible value for k, obtained for $n = 1$, is $k_c = \pi/L$. This corresponds to a maximum wavelength $\lambda_c = 2L$, called the cut-off wavelength. Only waves with frequencies higher than the *cut-off frequency* $f_c = c/2L$ can

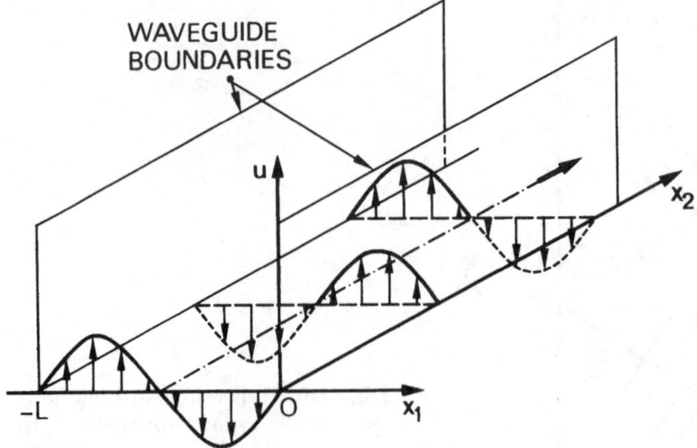

Fig. 1.7. Propagation of a guided wave. A guided wave has nodal planes at the guide walls and also, if the order n exceeds 1, inside the guide. Here $n = 2$ (the first anti-symmetric mode)

propagate in the guide. Indeed, for $f < f_c$ the wavenumber β is imaginary, and the wave attenuates once it has entered the guide. For $n > 1$ the cut-off frequencies are at $f = nf_c$.

A waveguide is an example of a *dispersive structure*, in which the phase velocity of the wave depends on its frequency; the frequency ω is not proportional to the wavenumber β. The dispersion is determined by the guide geometry. The relation between ω and β, called the dispersion relation, is given by

$$\omega = ck = c\left[\beta^2 + \left(\frac{n\pi}{L}\right)^2\right]^{1/2} . \tag{1.9}$$

Figure 1.8 shows this, in normalized coordinates, for the first few modes. Here,

$$\frac{\omega}{\omega_c} = \left(\frac{\beta^2 L^2}{\pi^2} + n^2\right)^{1/2} \quad \text{defining} \quad \omega_c = \frac{\pi c}{L} .$$

The guide is very dispersive near the cut-off frequency and its multiples.

Two-dimensional resonator. If two further reflecting planes are introduced at $x_2 = 0$ and $x_2 = -L_2$, a new situation results. In addition to the incident wave, given by (1.8) (with L replaced by L_1 for convenience), there is also a wave reflected at the plane $x_2 = 0$, given by

$$u_R = -2A \sin(\omega t + k_2 x_2) \sin\left(\frac{n_1 \pi}{L_1} x_1\right) .$$

The stationary wave

$$u = u_I + u_R = -4A \sin k_2 x_2 \sin\left(\frac{n_1 \pi}{L_1} x_1\right) \cos \omega t$$

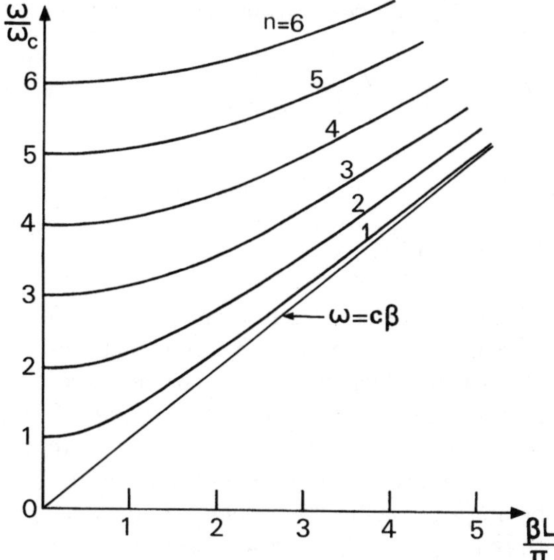

Fig. 1.8. Dispersion curves for a waveguide. At an angular frequency ω, the normalized phase velocity c_φ/c of the n-th mode is the slope of the line joining the corresponding point on the curve to the origin

needs to be zero at $x_2 = -L_2$, and this requires

$$k_2 = \frac{n_2\pi}{L_2} \quad \text{with} \quad n_2 = 1, 2, 3, \ldots .$$

The disturbance is

$$u = -4A \sin\left(\frac{n_1\pi}{L_1}x_1\right) \sin\left(\frac{n_2\pi}{L_2}x_2\right) \cos\omega t$$

and the allowed frequencies are given by

$$\frac{\omega^2}{c^2} = k^2 = k_1^2 + k_2^2\,,$$

so that

$$f = \frac{c}{2}\left[\left(\frac{n_1}{L_1}\right)^2 + \left(\frac{n_2}{L_2}\right)^2\right]^{1/2} . \tag{1.10}$$

In this two-dimensional cavity, the only solutions are modes for which the frequency is a *resonant frequency* (eigenfrequency) given by (1.10).

Three-dimensional resonator. To extend this to the three-dimensional case, it is sufficient to allow the wave to vary in the Ox_3 direction, perpendicular to the x_1Ox_2 plane. If there is a reflecting wall at $x_3 = 0$, the interference between the incident wave u_I and the corresponding reflected wave u_R gives rise to a stationary wave, at any frequency. However, with the addition of another wall at $x_3 = -L_3$, the cavity thus formed only allows solutions of the form

$$u = 8A \sin\left(\frac{n_1\pi}{L_1}x_1\right) \sin\left(\frac{n_2\pi}{L_2}x_2\right) \sin\left(\frac{n_3\pi}{L_3}x_3\right) \sin\omega t$$

with specific allowed frequencies

$$f = \frac{c}{2}\left[\left(\frac{n_1}{L_1}\right)^2 + \left(\frac{n_2}{L_2}\right)^2 + \left(\frac{n_3}{L_3}\right)^2\right]^{1/2} .$$

Comment. As stated at the beginning of this section, these simple formulae for waveguides and resonators assume that there is only one reflected wave, and that it has the same type as the incident wave. They are applicable to elastic waves propagating in a fluid filling the space between the walls. They are not generally valid for elastic waves in a solid, because in this case reflection at a plane surface can give several waves with different polarizations (i.e. directions of vibration), each having a characteristic velocity. The reader will appreciate this better after studying Sect. 4.4. However, there is an exception – a 'transverse' elastic wave with its mechanical vibration perpendicular to the wave vector and parallel to the reflecting planes. Equations (1.7) and (1.8) and the curves in Fig. 1.8 are applicable to this wave, which retains its transverse nature on reflection (it has no equivalent in electromagnetism).

1.1.3 Velocity of a Wave Packet

Up to this point we have only considered monochromatic waves, that is, sinusoidal vibrations of unlimited duration, with amplitude and frequency independent of time. In themselves, these pure waves are of little practical interest – an observer seeing the progress of a monochromatic wave gains no more information from its continuous propagation than he would from the regular flow of a homogeneous fluid. The transfer of information demands an 'anomaly', a change in some characteristic quantity. For example, a river with constant flow might carry a message in the form of a floating twig. Similarly, the transfer of information by a wave requires the variation of one or more of the parameters – amplitude or phase. This complex wave is no longer monochromatic, since its amplitude or frequency is modulated according to the signal transmitted. However, it can be expressed as the superposition of an infinite number of monochromatic waves with different amplitudes and frequencies. This superposition can be such as to form a finite-length group, or pulse, of waves, often called a *wave packet*. Taking the wavenumber k as the variable, a wave packet can be written as

$$u(x,t) = \frac{1}{2\pi}\int_{-\infty}^{+\infty} A(k)\mathrm{e}^{\mathrm{i}(\omega t - kx)}\mathrm{d}k\,, \tag{1.11}$$

where $A(k)$ is the amplitude 'density' and ω is determined by k.

Two cases need to be distinguished – the medium may or may not be dispersive. For a path length x_0 in a *non-dispersive* medium, where the phase velocity c is independent of frequency, each component of the group is delayed by the same amount $t_0 = x_0/c$. After this time t_0, the wave group is found to be displaced but unchanged in form. Thus, in a non-dispersive medium with no attenuation, a complex wave propagates without distortion. This property can be expressed in several ways:

- the phase velocity is independent of frequency;
- the phase change $\phi = -\omega x_0/c$ for a path length x_0 is proportional to frequency;
- the group delay, defined as $\tau_{\mathrm{g}} = -\mathrm{d}\phi/\mathrm{d}\omega$, is equal to x_0/c and is independent of frequency.

Consider now the propagation of a wave packet in a *dispersive* medium. In practice, the frequency of the carrier wave is often much greater than the spread of frequencies in the signal, so that the fractional bandwidth is small. Thus the amplitude density $A(k)$ is significant only for a small region around the value $k_0 = \omega_0/c(\omega_0)$. In these conditions, it is sufficient to expand the dispersion relation to first order about the value k_0, giving

$$\omega(k) \cong \omega(k_0) + \left(\frac{\mathrm{d}\omega}{\mathrm{d}k}\right)_{k_0} (k - k_0) .$$

The quantity $c_{\mathrm{g}} = (\mathrm{d}\omega/\mathrm{d}k)_{k_0}$ has the dimensions of velocity. Substituting into (1.11) gives

$$u(x,t) = \frac{1}{2\pi} \mathrm{e}^{\mathrm{i}[\omega(k_0) - k_0 c_{\mathrm{g}}]t} \int_{-\infty}^{+\infty} A(k) \mathrm{e}^{-\mathrm{i}k(x - c_{\mathrm{g}} t)} \mathrm{d}k$$

and further, on taking $u(x,0) = f(x)$ and $\Omega_0 = \omega(k_0) - k_0 c_{\mathrm{g}}$, this becomes

$$u(x,t) = \mathrm{e}^{\mathrm{i}\Omega_0 t} f(x - c_{\mathrm{g}} t) .$$

This equation shows that, during time t, the wave group is displaced by an amount $x = c_{\mathrm{g}} t$. Hence, for a dispersive medium a wave packet centred at wavenumber k_0 propagates with a velocity

$$\boxed{c_{\mathrm{g}} = \left(\frac{\mathrm{d}\omega}{\mathrm{d}k}\right)_{k_0}} , \tag{1.12}$$

which is called the *group velocity*.

The above result uses the first-order expansion of the dispersion relation (equivalent to replacing the curve $\omega(k)$ by its tangent), and consequently fails to show that the wavepacket can be distorted during propagation. For example, some component waves formed at the front of the group might start early and travel slowly, and could be overtaken by other waves leaving later but travelling faster.

The peak amplitude of the wave packet, in which the relative positions of the component waves are always changing, is given at time t by the point

x for which the amplitudes of the larger components interfere constructively. This requires the components with wavenumbers near k_0 to be in phase, a condition expressed by writing

$$\varphi(k) = \omega(k)t - kx = \text{ constant for } \quad k \approx k_0$$

or alternatively,

$$\left(\frac{\mathrm{d}\varphi}{\mathrm{d}k}\right)_{k_0} = \left(\frac{\mathrm{d}\omega}{\mathrm{d}k}\right)_{k_0} t - x = 0\,.$$

This method, called the *stationary phase method*, is another way of showing that the peak of the wavepacket travels with the group velocity c_{g}, given by the (1.12). If there is no attenuation, this is also the velocity of energy transported by the wave packet (Sect. 5.2.3).

As an example, consider the waveguide of Sect. 1.1.2.2. Equation (1.9) gives directly the group velocity for guided waves as

$$c_{\mathrm{g}} = \frac{\mathrm{d}\omega}{\mathrm{d}\beta} = c\beta\left(\beta^2 + \frac{n^2\pi^2}{L^2}\right)^{-1/2}\,. \tag{1.13}$$

Since the phase velocity is

$$c_{\varphi} = \frac{\omega}{\beta} = \frac{c}{\beta}\left(\beta^2 + \frac{n^2\pi^2}{L^2}\right)^{1/2} \tag{1.14}$$

we see that the product

$$c_{\mathrm{g}}c_{\varphi} = c^2 \tag{1.15}$$

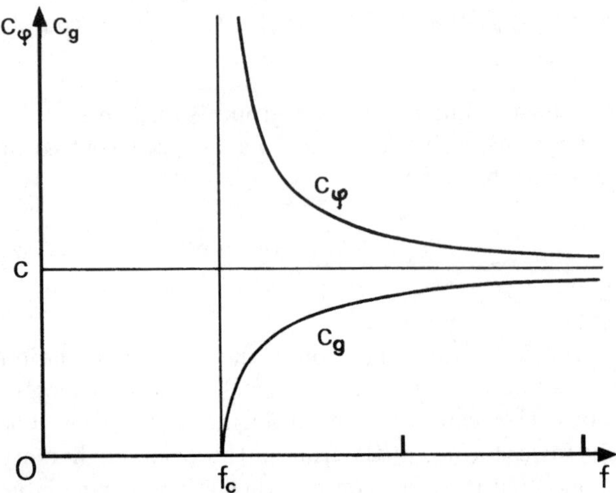

Fig. 1.9. In a waveguide, the phase velocity c_{φ} is higher than the velocity c of unguided waves, and the group velocity c_{g} is lower

is a constant, equal to the square of the phase velocity of unguided waves. The frequency variations of these velocities are shown in Fig. 1.9.

From the relation $\omega = kc_\varphi(k)$ we obtain the useful formulae

$$\frac{d\omega}{dk} = c_\varphi + k\frac{dc_\varphi}{dk} \quad \Rightarrow \quad c_g = c_\varphi - \lambda\frac{dc_\varphi}{d\lambda} .$$

1.2 Acoustic Waves

The above ideas apply to waves in general, whatever the nature of the disturbance or the mechanism of propagation, though they do assume that a discontinuity in the medium gives only one reflected wave, of the same type as the incident wave. These ideas are illustrated here by the physical example of an acoustic wave. The variables involved in reflections can now be specified physically.

The oscillatory movement of a loudspeaker membrane in air generates a propagating sound, that is, a sound wave. For frequencies in the audible region (20 Hz to 20 kHz), this is an acoustic wave. It is the result of the displacement of air molecules, subject to Newton's laws. Such *matter waves* (which do not propagate in a vacuum, unlike electromagnetic waves governed by Maxwell's equations) also propagate in liquids and solids. In these media the waves are often inaudible – the frequency can extend to 5 GHz in crystals. For such frequencies the term 'acoustics' is replaced by 'ultrasonics' or 'hypersonics'. All these 'sounds' are also called elastic waves. Compared with a solid, the physics of propagation in a compressible fluid, such as air, is simpler because the amplitude can be represented by a *scalar* – the local pressure. Here we highlight the main aspects of the physics – the wave equation and its solutions; the characterization of a medium by an acoustic impedance; energy transport and the Poynting vector; and reflection and refraction at the interface between two media with different impedances.

1.2.1 Wave Equation for a Plane Wave

We assume here a continuous isotropic homogeneous fluid which is perfectly compressible. We also assume no viscosity, thus excluding any dissipation.

Lagrange description and Euler description. In the mechanics of continuous media, the motion described by Lagrangian variables is associated with a material point, while that of Eulerian variables refers to a geometrical point in the coordinate system. In the one-dimensional case, the Lagrangian independent variables (used in the 'material' description) are the initial or equilibrium position a of a particle, and the time t. Any physical quantity g can be expressed as a function $G(a, t)$. The Eulerian independent variables (used in the 'spatial' description) are the abscissa x of a geometric point on the axis, and the time t. The quantity g is defined by a function $g(x, t)$.

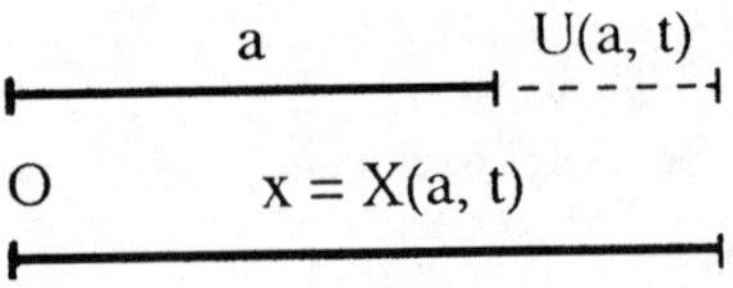

Fig. 1.10.

The relation between the two functions $G(a,t)$ and $g(x,t)$ is obtained on specifying the position x, at time t, of the particle with initial position a; this is a function X of a and t, so that

$$x = X(a,t), \quad \text{and hence} \quad G(a,t) = g[X(a,t),t]\,.$$

In particular, from Fig. 1.10, the displacement of the particle of fluid with initial position a is given by

$$U(a,t) = X(a,t) - a = u[X(a,t),t]\,. \tag{1.16}$$

The velocity of a particle, following its motion, is $\mathrm{d}U/\mathrm{d}t$. This is called the particle velocity, and is denoted by

$$V_\mathrm{p} = V_\mathrm{p}(a,t) = \frac{\mathrm{D}U}{\mathrm{D}t}\,.$$

This is distinct from the local velocity at the geometrical point x, which is

$$v = v(x,t) = \left(\frac{\partial u}{\partial t}\right)_x\,.$$

Taking account of (1.16), this leads to

$$V_\mathrm{p} = \left(\frac{\partial u}{\partial t}\right)_x + \left(\frac{\partial u}{\partial x}\right)_t\left(\frac{\mathrm{d}X}{\mathrm{d}t}\right) = v + V_\mathrm{p}\left(\frac{\partial u}{\partial x}\right)_t \tag{1.17}$$

since $\mathrm{d}X/\mathrm{d}t = \mathrm{D}U/\mathrm{D}t = V_\mathrm{p}$. Here V_p is the velocity of a particle identified by its initial or equilibrium position a and v is the local velocity of whichever particle passes coordinate point x at time t (different particles for different t-values). The third term, the difference between V_p and v, is called the convective velocity.

Equation (1.17) can be applied to any variable. The time derivative of a variable g attached to a particle, following its motion, is called the material derivative, or total derivative, and is denoted by $\mathrm{d}g/\mathrm{d}t$ or $\mathrm{D}G/\mathrm{D}t$. In contrast, the derivative at a fixed geometric point x is the local derivative (or partial derivative), denoted $\partial g/\partial t$. Thus

$$\frac{\mathrm{D}G}{\mathrm{D}t} = \left(\frac{\partial g}{\partial t}\right)_x + V_\mathrm{p}\left(\frac{\partial g}{\partial x}\right)_t\,.$$

For example, the acceleration of a particle of fluid is

$$\Gamma = \frac{\mathrm{D}V}{\mathrm{D}t} = \left(\frac{\partial v}{\partial t}\right)_x + V_\mathrm{p}\left(\frac{\partial v}{\partial x}\right)_t\,. \tag{1.18}$$

Linear acoustics is the regime in which the amplitudes are small enough for the second-order terms in the above equations to be neglected. Thus

$$\frac{\partial u}{\partial x} \ll 1 \quad \text{giving} \quad V_{\mathrm{p}} \cong \left(\frac{\partial u}{\partial t}\right)_x = v \quad \text{and} \quad \Gamma \cong \left(\frac{\partial v}{\partial t}\right)_x = \frac{\partial^2 u}{\partial t^2} = \gamma \,. \tag{1.19}$$

In this case the distinction between the above two descriptions disappears.

For a plane acoustic wave, we have the following variables:

- $p(x,t)$ is the instantaneous pressure at point x;
- $\delta p(x,t)$ is the *overpressure*, i.e. local pressure relative to the equilibrium pressure p_0, so that $\delta p = p - p_0$;
- S is the local dilatation corresponding to the overpressure δp, i.e. the fractional change of volume $\delta(\mathrm{d}V)/\mathrm{d}V$ of a particle of fluid. The term *particle of fluid* refers to a volume sufficiently large for the molecular aspect to be ignored, but much smaller than the wavelength so that physical quantities can be considered constant within it.

1.2.1.1 One-Dimensional Case. Consider a plane membrane immersed in a fluid such as air. When at rest, the pressure in the fluid is uniform and equal to p_0. When displaced, for example in the $+x$ direction, the membrane compresses an adjacent layer of air OA (Fig. 1.11). This situation is unstable – the fluid here relaxes as it compresses, in turn, the neighbouring layer AB. Thus the wave propagates step by step, in a succession of compressions and dilatations. The pressure $p(x,t)$ is a function of position x and time t. The forces acting on a layer such as MN, with boundaries at x and $x + \mathrm{d}x$, are no longer in equilibrium, but cause it to move to $M'N'$.

Let $u(x,t)$ be the displacement at time t of the plane M originally at x, and $u + \mathrm{d}u$ the displacement of plane N at the same time. We have

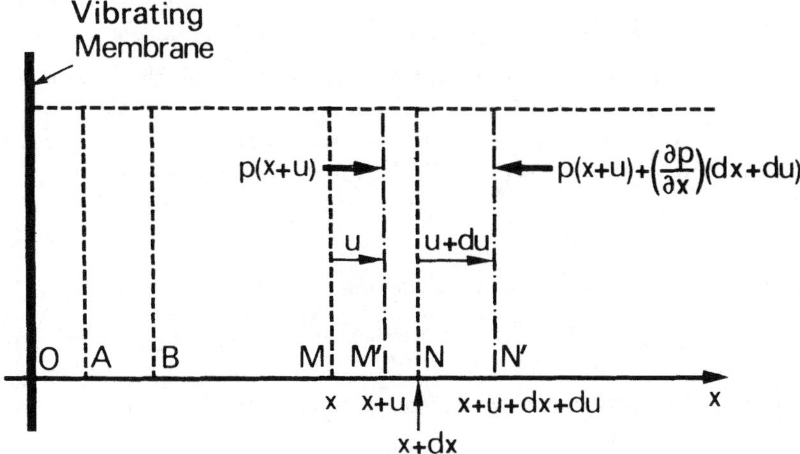

Fig. 1.11. The vibration of a membrane in a fluid creates a longitudinal wave which propagates by compressing and releasing layers of fluid OA, AB, ... , MN

$$\mathrm{d}u = \left(\frac{\partial u}{\partial x}\right) \mathrm{d}x \,.$$

If s is the surface area of the section considered, the force exerted by the fluid on the left when a wavefront passes M' is $F_{M'} = sp(x+u,t)$. The force exerted by the fluid on the right of N' is

$$F_{N'} = -sp(x+u+\mathrm{d}x+\mathrm{d}u,t) = -s\left[p(x+u,t) + \frac{\partial p}{\partial x}(\mathrm{d}x+\mathrm{d}u)\right] .$$

The resulting force acting on the layer $M'N'$ of fluid is

$$\mathrm{d}F = F_{M'} + F_{N'} = -s\frac{\partial p}{\partial x}(\mathrm{d}x+\mathrm{d}u) = -s\frac{\partial p}{\partial x}\left(1+\frac{\partial u}{\partial x}\right)\mathrm{d}x$$

and this causes an acceleration γ of the mass $dm = \rho_0 s\,\mathrm{d}x$ of the layer originally at MN. Here ρ_0 is the density of the fluid when at rest, with pressure p_0. Using Newton's law $\mathrm{d}F = \gamma\,dm$ we have, with the approximation in (1.19),

$$\rho_0\frac{\partial^2 u}{\partial t^2} = -\frac{\partial p}{\partial x}\left(1+\frac{\partial u}{\partial x}\right) . \tag{1.20}$$

This equation relates the pressure p to the material displacement u. These two quantities are not independent in a compressible fluid. At the position $x+u$ of the layer $M'N'$, the pressure $p(x+u,t)$ can be written as

$$p(x+u,t) = p_0 + \delta p \,.$$

The deviation δp of the pressure from its equilibrium value p_0 (i.e. the overpressure) is related to the volume change $\delta(\mathrm{d}V)$ by the *compressibility coefficient* χ of the fluid, defined by

$$\boxed{\chi = -\frac{1}{\mathrm{d}V}\frac{\delta(\mathrm{d}V)}{\delta p}} . \tag{1.21}$$

This coefficient, always positive, is sufficient to characterize the *elastic behaviour* of a non-viscous fluid. For the original layer MN the volume is $\mathrm{d}V = s\,\mathrm{d}x$ and the pressure is p_0, and this becomes $\mathrm{d}V + \delta(\mathrm{d}V) = s(1+\partial u/\partial x)\mathrm{d}x$ under pressure $p_0 + \delta p$ (at position $M'N'$). The local dilatation is

$$S(x,t) = \frac{\delta(\mathrm{d}V)}{\mathrm{d}V} = \frac{\partial u}{\partial x} . \tag{1.22}$$

The overpressure is linearly related to the dilatation S by

$$\delta p = -\frac{S}{\chi} = -\frac{1}{\chi}\cdot\frac{\partial u}{\partial x} \tag{1.23}$$

provided S is small, so that $S = \partial u/\partial x \ll 1$. With this approximation to *linear acoustics*, (1.20) becomes

$$\frac{\partial^2 u}{\partial t^2} = -\frac{1}{\rho_0}\frac{\partial(\delta p)}{\partial x} . \tag{1.24}$$

On substituting for δp using (1.23), we obtain the wave equation

$$\boxed{\frac{\partial^2 u}{\partial t^2} = c^2 \frac{\partial^2 u}{\partial x^2} \quad \text{with} \quad c = \frac{1}{\sqrt{\rho_0 \chi}}}\,. \tag{1.25}$$

This partial differential equation, obtained by considering the progressive action of a layer of fluid on its neighbours, governs the spatial and temporal evolution of the displacement in a fluid. Any function $u(x,t)$ of the form

$$u(x,t) = F\left(t - \frac{x}{c}\right) + G\left(t + \frac{x}{c}\right) \tag{1.26}$$

is a solution. The acoustic disturbance propagates with a *velocity* $c = 1/(\rho_0\chi)^{1/2}$.

The other variables (v, S, δp) also satisfy the same wave equation. To verify this, we simply note that they are related to u by the equations

$$\boxed{v = \frac{\partial u}{\partial t}, \quad S = \frac{\partial u}{\partial x}, \quad \delta p = -\rho_0 c^2 S}\,. \tag{1.27}$$

Since $\rho V = \rho_0 V_0 =$ constant, the compressibility coefficient and the sound velocity can be put in the forms

$$\chi = -\frac{1}{V_0}\frac{\partial V}{\partial p} = \frac{1}{\rho_0}\frac{\partial \rho}{\partial p} \quad \text{and hence} \quad c = \sqrt{\frac{\partial p}{\partial \rho}}\,.$$

From the relations (1.23) and (1.24), *the particle velocity v and the overpressure δp* are also solutions of the two first-order differential equations

$$\boxed{\frac{\partial v}{\partial t} + \frac{1}{\rho_0}\cdot\frac{\partial(\delta p)}{\partial x} = 0, \qquad \frac{\partial(\delta p)}{\partial t} + \frac{1}{\chi}\cdot\frac{\partial v}{\partial x} = 0}\,, \tag{1.28}$$

which are equivalent to the wave equation.

Introducing the derivatives $f = F'$ and $g = G'$ of the functions F and G in the general solution (1.26), the particle velocity can be expressed as

$$v = f\left(t - \frac{x}{c}\right) + g\left(t + \frac{x}{c}\right)\,. \tag{1.29}$$

The overpressure δp is determined by (1.23), which can be written $\delta p = -\rho_0 c^2 \partial u/\partial x$, giving

$$\delta p = \rho_0 c\left[f\left(t - \frac{x}{c}\right) - g\left(t + \frac{x}{c}\right)\right]\,. \tag{1.30}$$

We note that *the overpressure is proportional to the particle velocity*, since $\delta p = \pm\rho_0 c v$ for waves travelling in the $\pm x$ directions.

Velocity of sound. For an ideal gas of molecular weight M, the equation of state for n moles at equilibrium is

$$pV = nRT \quad \text{or} \quad p = \frac{RT}{M}\rho\,,$$

where T is the absolute temperature and $R = 8.314\,\mathrm{J/mole\cdot K}$ is the gas constant for an ideal gas. The compressibility χ depends on thermodynamic effects produced by the compression and dilatation associated with the wave. Experiments show that this process is adiabatic, i.e. it involves no heat flow and so is not isothermal. It gives the relation pV^γ = constant, where γ is the ratio of the specific heats at constant pressure and constant volume. This gives

$$\frac{\mathrm{d}p}{p} + \gamma\frac{\mathrm{d}V}{V} = 0 \quad \text{and hence} \quad \chi = -\frac{1}{V}\frac{\partial V}{\partial p} = \frac{1}{\gamma p_0}\,.$$

The sound velocity is therefore

$$c = \sqrt{\gamma\frac{p_0}{\rho_0}} = \sqrt{\gamma\frac{RT}{M}}\,. \tag{1.31}$$

The value calculated for *air* with $\gamma = 1.4$ (appropriate for a diatomic gas) is 343 m/s at 20 °C, in good agreement with experiment.

It is instructive to compare the phase velocity of the wave, c (the velocity at which the acoustic wavefronts travel), with the *thermal* velocity of the gas molecules. From the kinetic theory of gases, the root-mean-square velocity for molecules of mass m is

$$v_{\mathrm{rms}} = \sqrt{3\frac{kT}{m}} = \sqrt{3\frac{RT}{M}}\,, \quad \text{where } k = R/N \text{ is the Boltzmann's constant.}$$

Since γ is between $5/3 = 1.67$ (for a monatomic gas) and $7/5 = 1.4$ (for a diatomic gas), we see that c and v_{rms} have the same order of magnitude. In a gas with low pressure (an almost ideal gas), the interaction between molecules is mainly due to collisions, and this is also the mechanism for the propagation of acoustic waves.

Thermal motion imposes an upper limit to the *frequency* of acoustic waves. The wave can only propagate if the distance between the maximum and minimum of pressure or density is large compared with the mean free path l_{m} of the molecules (the mean path between collisions). Otherwise, the molecules simply diffuse into regions of low density, and the perturbation carrying the wave disappears. This imposes the condition

$$\lambda \gg l_{\mathrm{m}} \quad \text{giving} \quad f \ll c/l_{\mathrm{m}}\,.$$

Numerical example. Oxygen has $c = 315\,\mathrm{m/s}$ at 0 °C, and for a pressure of 0.1 Pa ($= 0.1\,\mathrm{N/m^2}$) we have $l_{\mathrm{m}} = 4.9\,\mathrm{cm}$, giving a frequency limit of 6400 Hz.

The particle velocity v in the fluid is proportional to the overpressure (since $v = \delta p/\rho c$) and therefore depends on the acoustic power (Sect. 1.2.2 below). For the sinusoidal case, the particle oscillates about its mean position x_0, and v has the form

$$v = v_{\mathrm{m}} \cos\omega(t - x_0/c) \quad \text{with} \quad v_{\mathrm{m}} = \delta p/\rho_0 c\,.$$

Taking $\delta p \approx 30\,\mathrm{Pa}$ (loud enough to be painful) and with $\rho = 1.29\,\mathrm{kg/m^3}$ and $c = 340\,\mathrm{m/s}$, we find $v_\mathrm{m} = 0.07\,\mathrm{m/s}$. Thus the particle velocity is much smaller than the phase velocity c.

In a *liquid*, the formula for the sound wave velocity, similar to (1.31), is

$$c = \sqrt{\frac{A}{\rho_0}} \quad \text{with} \quad A = \frac{1}{\chi}\,.$$

Here A is the bulk elastic modulus (reciprocal of χ) of the liquid. For *water*, $c = 1480\,\mathrm{m/s}$ at $20\,°\mathrm{C}$.

In a *solid* the velocity of longitudinal elastic waves, which have the particle displacement parallel to the wave vector as in a fluid, is of the order of 5000 m/s. It is less than 1000 m/s for soft materials such as lead or indium, and above 10,000 m/s for hard materials such as sapphire (crystalline alumina) or beryllium.

1.2.1.2 Electrical Analogy. Acoustic Impedance. The characteristic parameters of an electrical transmission line, Fig. 1.12a, are its inductance L and capacitance C per unit length. At a position x, each of the two conductors carries a current with the same magnitude $I(x,t)$, but with opposite directions. Let $U(x,t)$ be the voltage across the two conductors.

An element of the line, with length Δx, is equivalent to a circuit with lumped components – an inductor $L\Delta x$ and a capacitor $C\Delta x$. Between the points x and $x + \Delta x$ the voltage changes by an amount ΔU, due to the time-variation of current in the inductor, so that

$$\Delta U = \frac{\partial U}{\partial x}\Delta x = -L\Delta x\frac{\partial I}{\partial t} \quad \text{giving} \quad \frac{\partial U}{\partial x} + L\frac{\partial I}{\partial t} = 0\,. \tag{1.32}$$

The change of current, ΔI, is obtained from the charge $\Delta Q = C\,\Delta x\,U$, so that

$$\Delta I = \frac{\partial I}{\partial x}\Delta x = \frac{\partial(\Delta Q)}{\partial t} = -C\Delta x\frac{\partial U}{\partial t} \quad \text{giving} \quad \frac{\partial I}{\partial x} + C\frac{\partial U}{\partial t} = 0\,. \tag{1.33}$$

The variables U and I can be separated by differentiating these equations, one with respect to x and the other with respect to t, giving

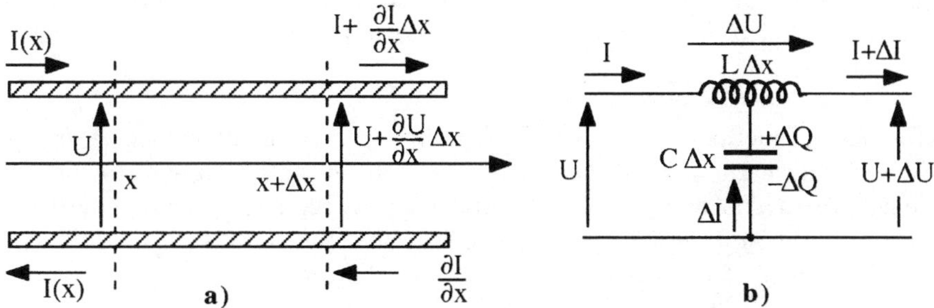

Fig. 1.12. (**a**) Element of transmission line, of length Δx. (**b**) Equivalent circuit. L and C are the inductance and capacitance per unit length

$$\frac{\partial^2 U}{\partial x^2} = LC\frac{\partial^2 U}{\partial t^2} \quad \text{and} \quad \frac{\partial^2 I}{\partial x^2} = LC\frac{\partial^2 I}{\partial t^2}\,. \tag{1.34}$$

Thus, in a transmission line the voltage and current propagate with velocity $c = 1/(LC)^{1/2}$. The solutions have the form of (1.26), so

$$U(x,t) = f(x - ct) + g(x + ct)\,.$$

Substituting into (1.32) and integrating, we also find

$$I(x,t) = \sqrt{\frac{C}{L}}\,[f(x-ct) - g(x+ct)]\,.$$

At each point and at every instant, the voltage and current are proportional. Depending on the direction of propagation we have

$$\frac{U}{I} = Z_\mathrm{c} = \sqrt{\frac{L}{C}} \quad \text{or} \quad \frac{U}{I} = -Z_\mathrm{c}\,. \tag{1.35}$$

Comparing (1.32) and (1.33) with (1.28) for plane acoustic waves, we see a correspondence between the overpressure δp and voltage U, and between particle velocity v and current I. By analogy with the electrical impedance U/I, the *acoustic impedance* is defined by

$$\boxed{Z = \frac{\delta p}{v}}\,. \tag{1.36}$$

The relations between overpressure and velocity,

$$\delta p_+ = \rho_0 c\, v_+ \quad \text{or} \quad \delta p_- = -\rho_0 c\, v_-$$

respectively for a wave propagating in the $\pm x$ direction, show that the acoustic impedance is a real constant given by

$$Z_\mathrm{c} = \rho_0\, c = \sqrt{\frac{\rho_0}{\chi}}\,. \tag{1.37}$$

This is the *characteristic impedance* of the medium (in units $\mathrm{kg\,m^{-2}\,s^{-1}}$).

1.2.1.3 Three-Dimensional Case. The displacement of the particles at a position given by the vector $\boldsymbol{r}$ is a vector $\boldsymbol{u}(\boldsymbol{r},t)$, with three components u_x, u_y, u_z. The local overpressure due to an acoustic wave,

$$\delta p(\boldsymbol{r},t) = p(\boldsymbol{r},t) - p_0(\boldsymbol{r})$$

is a *scalar* function of $\boldsymbol{r}$ and t. It is therefore convenient to choose δp as the variable characterizing the disturbance. Under the action of δp, a surface element of area $\mathrm{d}s$ is displaced by an amount $\boldsymbol{u}$, sweeping out a volume $\boldsymbol{u}\cdot\boldsymbol{l}\,\mathrm{d}s$, where $\boldsymbol{l}$ is the unit vector normal to the surface (Fig. 1.13).

The change of volume V occupied by the fluid is

$$\Delta V = \int_s \boldsymbol{u}\cdot\boldsymbol{l}\,\mathrm{d}s$$

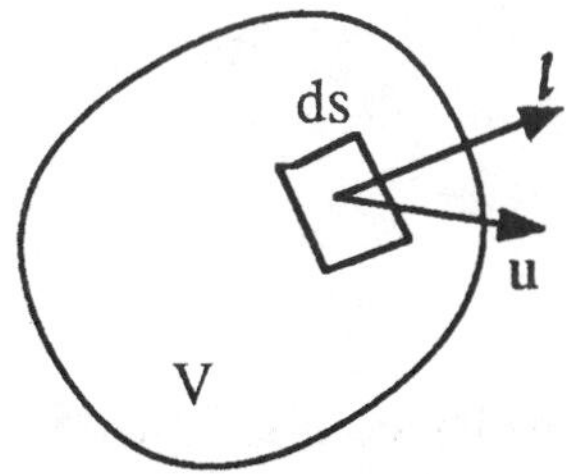

Fig. 1.13. A surface element with area ds and displacement $\boldsymbol{u}$ sweeps out a volume $\boldsymbol{u}\cdot\boldsymbol{l}\,\mathrm{d}s$

and, on applying Green's theorem (Appendix C), this becomes

$$\Delta V = \int_V (\nabla\cdot\boldsymbol{u})\mathrm{d}V = \int_V S(\boldsymbol{r})\mathrm{d}V\,,$$

where $S(\boldsymbol{r}) = \delta(\mathrm{d}V)/\mathrm{d}V$ is the local dilatation and $\nabla\cdot\boldsymbol{u} \equiv \operatorname{div}\boldsymbol{u}$. Thus the dilatation is

$$\boxed{\mathrm{S}(\boldsymbol{r}) = \boldsymbol{\nabla}\cdot\boldsymbol{u} = \frac{\partial u_x}{\partial x} + \frac{\partial u_y}{\partial y} + \frac{\partial u_z}{\partial z}}\,, \tag{1.38}$$

and is therefore equal to the divergence of the displacement vector.

Considering an elementary cube of volume dxdydz, as in Fig. 1.14, the applied force, in the x-direction, is

$$F_x = p\,\mathrm{d}y\,\mathrm{d}z - \left(p + \frac{\partial p}{\partial x}\mathrm{d}x\right)\mathrm{d}y\,\mathrm{d}z = -\frac{\partial p}{\partial x}\mathrm{d}x\,\mathrm{d}y\,\mathrm{d}z\,.$$

Applying Newton's law we have

$$\rho_0\,\mathrm{d}x\,\mathrm{d}y\,\mathrm{d}z\frac{\partial^2 u_x}{\partial t^2} = -\frac{\partial p}{\partial x}\mathrm{d}x\,\mathrm{d}y\,\mathrm{d}z \quad \text{giving} \quad \rho_0\frac{\partial^2 u_x}{\partial t^2} = -\frac{\partial p}{\partial x}\,. \tag{1.39}$$

Taking account of similar equations for the y and z axes, Newton's law can be written in the vectorial form

$$\rho_0\frac{\partial^2 \boldsymbol{u}}{\partial t^2} = -\nabla(\delta p) = -\operatorname{grad}\delta p\,. \tag{1.40}$$

Taking the divergence of each side gives

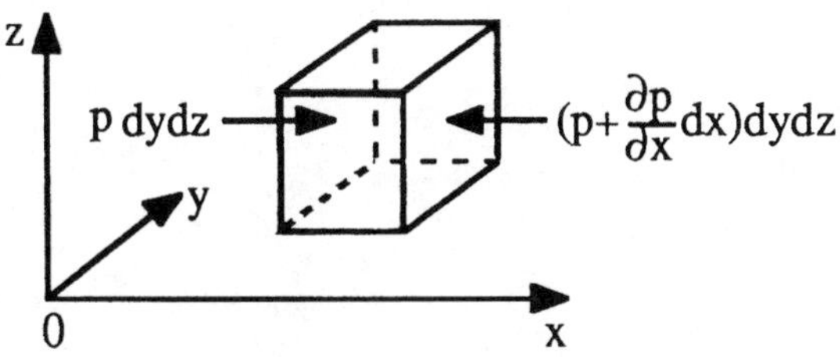

Fig. 1.14. Forces exerted, in the x-direction, on an elementary cube of volume dx dy dz

$$\rho_0 \frac{\partial^2 S}{\partial t^2} = -\Delta(\delta p)\,, \tag{1.41}$$

where the operator Δ is the Laplacian

$$\Delta = \nabla \cdot \nabla = \frac{\partial^2}{\partial x^2} + \frac{\partial^2}{\partial y^2} + \frac{\partial^2}{\partial z^2}\,.$$

For a *compressible* fluid with *linear* behaviour, the dilatation is related to the overpressure by $S = -\chi \delta p$. The overpressure obeys the three-dimensional wave equation

$$\boxed{\Delta(\delta p) = \frac{1}{c^2}\frac{\partial^2(\delta p)}{\partial t^2} \quad \text{with} \quad c = \frac{1}{\sqrt{\rho_0 \chi}}}\,, \tag{1.42}$$

where the velocity c is the same as that found previously for the one-dimensional case.

If there is an applied force with density $\boldsymbol{f}$ per unit volume, due for example to gravity ($\boldsymbol{f} = \rho \boldsymbol{g}$) or external forces, the basic dynamical equation (1.40) takes on an additional term, giving

$$\boxed{\rho_0 \frac{\mathrm{d}\boldsymbol{v}}{\mathrm{d}t} = -\nabla(\delta p) + \boldsymbol{f}}\,. \tag{1.43}$$

1.2.2 Power Flow. Poynting Vector

The progression of an acoustic disturbance in a fluid involves the transport of energy. We consider an elementary volume $\mathrm{d}V$ inside the fluid, as in Fig. 1.15. The work done by the applied forces, with density $\boldsymbol{f}$ per unit volume, is

$$\mathrm{d}w = \boldsymbol{f} \cdot \mathrm{d}\boldsymbol{u} = (\boldsymbol{f} \cdot \boldsymbol{v})\mathrm{d}t \tag{1.44}$$

where $\boldsymbol{f} \cdot \boldsymbol{v}$ is the instantaneous power provided per unit volume.

To express this work as a function of the variables $\boldsymbol{v}$, δp and S characterizing the disturbance, we replace $\boldsymbol{f}$ using the equation of motion (1.43), giving

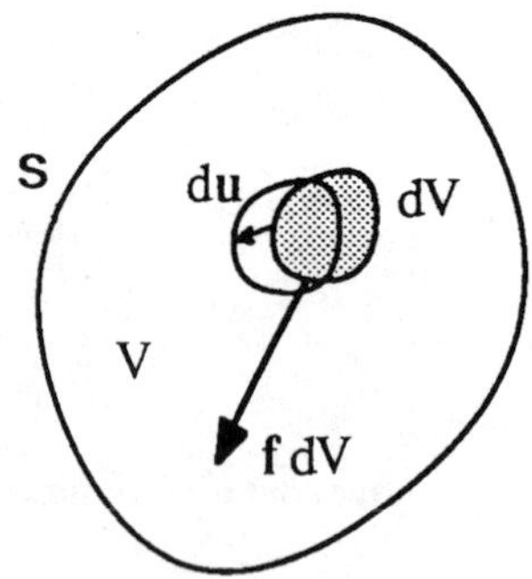

Fig. 1.15. The force acting on a volume $\mathrm{d}V$ is $\boldsymbol{f}\,\mathrm{d}V$, and the work done by this force per unit volume is $\boldsymbol{f} \cdot \mathrm{d}\boldsymbol{u}$

$$\mathrm{d}w = \rho_0\, \boldsymbol{v} \cdot \mathrm{d}\boldsymbol{v} + \nabla(\delta p) \cdot \mathrm{d}\boldsymbol{u}\,.$$

Now, since

$$\nabla(\delta p) \cdot \mathrm{d}\boldsymbol{u} = \nabla \cdot (\delta p\, \mathrm{d}\boldsymbol{u}) - \delta p\, \nabla \cdot \mathrm{d}\boldsymbol{u} \quad \text{and} \quad \nabla \cdot \mathrm{d}\boldsymbol{u} = \mathrm{d}S$$

the work done per unit volume becomes

$$\mathrm{d}w = \rho_0\, \boldsymbol{v} \cdot \mathrm{d}\boldsymbol{v} - \delta p\, \mathrm{d}S + \nabla \cdot (\delta p\, \mathrm{d}\boldsymbol{u})$$

and can be put in the form

$$\mathrm{d}w = (\boldsymbol{f} \cdot \boldsymbol{v})\mathrm{d}t = \mathrm{d}\left(\frac{1}{2}\rho_0\, \boldsymbol{v}^2\right) - \delta p\, \mathrm{d}S + \nabla \cdot (\delta p\, \mathrm{d}\boldsymbol{u})\,. \tag{1.45}$$

Here the first term on the right gives the change of the *kinetic energy* per unit volume, defined by

$$\boxed{e_\mathrm{k} = \frac{1}{2}\rho_0\, \boldsymbol{v}^2}\,. \tag{1.46}$$

The second term, $\mathrm{d}e_\mathrm{p} = -\delta p\, \mathrm{d}S$, gives the change of *potential energy* per unit volume for a change $\mathrm{d}S$ in the dilatation. Indeed, for a finite change of dilatation from 0 to S, and with $\delta p = -S/\chi$, we have

$$e_\mathrm{p} = -\int_0^S \delta p\, \mathrm{d}S = \int_0^S \frac{S}{\chi}\mathrm{d}S \quad \text{giving} \quad \boxed{e_\mathrm{p} = \frac{1}{2}\frac{S^2}{\chi} = \frac{1}{2}\chi(\delta p)^2}\,. \tag{1.47}$$

Calculating the work done on a mass of fluid passing from the state V_0, p_0 to the state V, $p_0 + \delta p$ under the influence of an acoustic wave, we have

$$\mathrm{d}W = -\delta p\, \mathrm{d}V \text{ giving } W = -\int_0^{\delta p} \delta p\, \mathrm{d}V = \chi V_0 \int_0^{\delta p} \delta p\, \mathrm{d}(\delta p) = \frac{1}{2}\chi V_0 (\delta p)^2\,.$$

This shows that e_p is the change in potential energy per unit volume.

We now introduce the *Poynting vector*, defined by

$$\boxed{\boldsymbol{P} = \delta p\, \boldsymbol{v}}\,. \tag{1.48}$$

Using this, (1.45) becomes, after division by $\mathrm{d}t$,

$$\frac{\mathrm{d}w}{\mathrm{d}t} = \boldsymbol{f} \cdot \boldsymbol{v} = \frac{\mathrm{d}}{\mathrm{d}t}(e_\mathrm{k} + e_\mathrm{p}) + \nabla \cdot \boldsymbol{P}\,. \tag{1.49}$$

We designate respectively

$$W = \int_V w\, \mathrm{d}V, \quad E_\mathrm{k} = \int_V e_\mathrm{k}\, \mathrm{d}V \quad \text{and} \quad E_\mathrm{p} = \int_V e_\mathrm{p}\, \mathrm{d}V$$

as the work done by the sources, the kinetic energy and the potential energy contained in a volume V of fluid. Integration of (1.49) shows that the power $\mathrm{d}W/\mathrm{d}t$ produced by the sources acting on this volume is given by

$$\boxed{\frac{\mathrm{d}W}{\mathrm{d}t} = \frac{\mathrm{d}}{\mathrm{d}t}(E_\mathrm{k} + E_\mathrm{p}) + \int_s \boldsymbol{P} \cdot \boldsymbol{l}\, \mathrm{d}s}\,. \tag{1.50}$$

This relation expresses, at each instant, the law of *energy conservation*:

The power produced by the internal sources in a volume V is accounted for partly as energy stored in the form of kinetic and potential energy, with the remainder as energy radiated outwards. The radiated power is equal to the flux of the Poynting vector crossing the surface s that delimits the volume V.

Thus the Poynting vector represents, in magnitude and direction, the instantaneous power per unit area transported by the wave. Its time-average

$$I = \langle P(t) \rangle = \lim_{T\to\infty} \frac{1}{T} \int_0^T \delta p(t)\, v(t)\, \mathrm{d}t \tag{1.51}$$

is, by definition, the acoustic *intensity*. This is the amount of energy crossing unit area in unit time, expressed in units of $\mathrm{W/m^2}$.

Plane wave. For this case, the material displacement, the particle velocity and the overpressure are given by

$$u = F\left(t - \frac{x}{c}\right), \quad v = F'\left(t - \frac{x}{c}\right) \quad \delta p = \rho_0 c v = \rho_0 c\, F'\left(t - \frac{x}{c}\right),$$

and the densities of potential and kinetic energy are, from (1.46) and (1.47),

$$e_{\mathrm{p}} = \frac{1}{2}\rho_0^2 c^2 \chi\, F'^2 = \frac{1}{2}\rho_0\, F'^2 = e_{\mathrm{k}}\,. \tag{1.52}$$

Thus, for a propagating wave, *the densities of potential and kinetic energy are equal.* The power flow per unit area is the magnitude of the Poynting vector, given by

$$P = \delta p\, v = \rho_0\, c F'^2 = c(e_{\mathrm{p}} + e_{\mathrm{k}})\,. \tag{1.53}$$

This is seen to be the total energy per unit volume multiplied by the phase velocity. Hence this is also the *velocity of energy transfer.* This property is always true for a plane wave if the medium is isotropic and non-dispersive.

For the *sinusoidal* case the displacement has the form

$$u = u_{\mathrm{m}} \sin\left[\omega\left(t - \frac{x}{c}\right)\right] \quad \text{giving} \quad F'\left(t - \frac{x}{c}\right) = \omega u_{\mathrm{m}} \cos\left[\omega\left(t - \frac{x}{c}\right)\right],$$

the power flow per unit area is

$$P = Z_{\mathrm{c}}\omega^2\, u_{\mathrm{m}}^2 \cos^2\left[\omega\left(t - \frac{x}{c}\right)\right] \quad \text{with} \quad Z_{\mathrm{c}} = \rho_0\, c\,,$$

and the *acoustic intensity* $I = \langle P \rangle$ for a beam of plane waves is

$$\boxed{I = \frac{1}{2} Z_{\mathrm{c}}\omega^2 (u_{\mathrm{m}})^2 = \frac{1}{2} Z_{\mathrm{c}} (v_{\mathrm{m}})^2 = \frac{1}{2Z_{\mathrm{c}}} (\delta p_{\mathrm{m}})^2}\,. \tag{1.54}$$

We note that frequency does not occur in expressions written in terms of velocity or overpressure. These formulae are identical to those giving the electrical power, on using the following correspondence established in Sect. 1.2.1.2:

overpressure δp	$\to$	electrical voltage U
particle velocity v	$\to$	electrical current I
acoustic impedance $Z = \delta p/v$	$\to$	electrical impedance $Z = U/I$.

More generally (for example for non-plane waves or superimposed waves propagating in opposite directions), it is convenient to use the complex form

$$\boxed{I = \frac{1}{2}\mathrm{Re}\left[(\delta p)\, v^*\right]}\,. \tag{1.55}$$

1.2.3 Attenuation

Up to here we have taken the fluid to be ideal, responding instantaneously to any excitation. Thus a local overpressure δp immediately causes a reduction of volume (i.e. a dilatation $S = \partial u/\partial x < 0$) and an increase of mass density ($\delta\rho = -\rho_0 S$), expressed by the relations

$$\delta p = -\frac{S}{\chi} = -\rho_0 c^2 S\,, \quad \delta p = c^2 \delta\rho\,.$$

This assumption is, a priori, acceptable provided the phenomena act slowly, but the time-scale involved needs to be quantified.

One way to take account of the fact that the response $(S, \delta\rho)$ of the fluid to an overpressure δp is not instantaneous is to introduce a term proportional to the rate of change of the observed variable. We can write, for example

$$\delta p = -\frac{S}{\chi} - b\frac{\partial S}{\partial t}\,.$$

This relation describes a first-order process [1.1]. It involves a time constant (in the language of systems analysis), or *relaxation time* (in the language of physics). This characteristic time τ appears when we write

$$\tau\frac{\partial S}{\partial t} + S(t) = K\delta p \quad \text{with} \quad \tau = b\chi \quad \text{and} \quad K = -\chi\,.$$

The response of the system tends to drag behind the excitation. For example, a pressure change in the form of a step function $[\delta p = \Gamma(t)]$ gives rise to a dilatation

$$S = K(1 - \mathrm{e}^{-t/\tau})\Gamma(t)\,.$$

The relaxation time τ is the time at which the dilatation reaches a fraction $1 - e^{-1} \approx 0.63$ of its final value $S(t = \infty) = K$. At time 3τ the dilatation S is $0.95\,K$, which is 5% below its final value. This time 3τ is called the *5 % response time.*

Substituting the new expression for δp into (1.24) and (1.40) gives the one- and three-dimensional wave equations

$$\rho_0\frac{\partial^2 u}{\partial t^2} = \frac{1}{\chi}\frac{\partial^2 u}{\partial x^2} + b\frac{\partial^2}{\partial x^2}\left(\frac{\partial u}{\partial t}\right) \tag{1.56a}$$

and

$$\rho_0\frac{\partial^2 \boldsymbol{u}}{\partial t^2} = \frac{1}{\chi}\nabla^2\boldsymbol{u} + b\nabla^2\left(\frac{\partial \boldsymbol{u}}{\partial t}\right)\,. \tag{1.56b}$$

To illustrate the effects of relaxation, it is sufficient to consider a sinusoidal solution of the first equation. The solution has the form

$$u = u_0\,\mathrm{e}^{-\alpha x}\mathrm{e}^{\mathrm{i}(\omega t - kx)} = u_0\,\mathrm{e}^{\mathrm{i}[\omega t - (k - \mathrm{i}\alpha)x]}\,,$$

where α is the attenuation coefficient for the wave, which propagates along x. Using this in (1.56a), with $\partial u/\partial t$ replaced by $\mathrm{i}\omega u$ on the right, we obtain

$$\frac{\partial^2 u}{\partial t^2} = c^2(1 + \mathrm{i}\omega\tau)\frac{\partial^2 u}{\partial x^2}$$

and hence

$$\omega^2 = c^2(1 + \mathrm{i}\omega\tau)(k - \mathrm{i}\alpha)^2\,.$$

The real part and magnitude of $(k - \mathrm{i}\alpha)^2$ are respectively given by

$$k^2 - \alpha^2 = \frac{\omega^2}{c^2(1 + \omega^2\tau^2)} \quad \text{and} \quad k^2 + \alpha^2 = \frac{\omega^2}{c^2\sqrt{1 + \omega^2\tau^2}}$$

from which we obtain

$$\alpha^2 = \frac{\omega^2}{2c^2}\left(\frac{1}{\sqrt{1 + \omega^2\tau^2}} - \frac{1}{1 + \omega^2\tau^2}\right)$$

and

$$k^2 = \frac{\omega^2}{2c^2}\left(\frac{1}{\sqrt{1 + \omega^2\tau^2}} + \frac{1}{1 + \omega^2\tau^2}\right)\,.$$

Hence the phase velocity of the wave, denoted by c_φ, is given by

$$c_\varphi^2 = \frac{\omega^2}{k^2} = c^2\frac{2(1 + \omega^2\tau^2)}{1 + \sqrt{1 + \omega^2\tau^2}}\,.$$

Thus the relaxation process affects the phase velocity of the wave as well as its amplitude. Moreover, both of these quantities have become functions of frequency.

If the relaxation time τ of the fluid is much less than the period $T = 2\pi/\omega$ of the wave, so that $\omega\tau \ll 1$, we have

$$\alpha^2 \cong \frac{\omega^4\tau^2}{4c^2} \quad \text{giving} \quad \alpha \cong \frac{\omega^2\tau}{2c} = 2\pi^2\frac{\tau}{c}f^2 \tag{1.57}$$

and

$$k^2 \cong \frac{\omega^2}{c^2}\left(1 - \frac{3}{4}\omega^2\tau^2\right) \quad \text{giving} \quad k \cong \frac{\omega}{c}\left(1 - \frac{3}{8}\omega^2\tau^2\right)$$

and

$$c_\varphi \cong c\left(1 + \frac{3}{8}\omega^2\tau^2\right)\,.$$

Thus, assuming τ to be independent of frequency, the condition $\omega\tau \ll 1$ makes the attenuation coefficient α *proportional to frequency squared*, while the phase velocity c_φ is little different from c. The ratio r_{12} of the wave amplitudes at two points x_1, x_2 (with $x_2 > x_1$) is

$$r_{12} = \mathrm{e}^{-\alpha(x_1 - x_2)}\,.$$

For practical purposes it is usual to express the attenuation either in Nepers or in decibels (dB).

(a) In Nepers, by Log(r_{12}), where Log( ) denotes the natural logarithm (i.e. to base e). This is a non-dimensional quantity; the logarithm is used for convenience to contract the range of numerical values. Thus,

$$att(\text{Neper}) = \text{Log}(r_{12}) = \alpha(x_2 - x_1)\text{Nepers}\,.$$

Similarly, the coefficient α can be expressed in Neper/m.

(b) In dB, the expression is

$$att(\text{decibels}) = 10\log(r_{12})^2 = 20\log(e)\alpha(x_2 - x_1)\text{dB}\,,$$

where log( ) denotes the logarithm to base 10.

The two forms are related by

$$\alpha(\text{dB/m}) = 20\text{log(e)} \times \alpha(\text{Neper/m}) \approx 8.7\alpha(\text{Neper/m})\,.$$

The power dissipated in the path from x_1 to x_2 is proportional to $(r_{12})^2$. The power has the form

$$P_\mathrm{d} \propto u_0^2 \mathrm{e}^{-2\alpha x} \quad \text{and hence} \quad \frac{1}{P_\mathrm{d}}\frac{\mathrm{d}P_\mathrm{d}}{\mathrm{d}x} = -2\alpha = -\frac{\omega^2\tau}{c}\,.$$

The coefficient α can be expressed in dB/m as above or, since the velocity is known, in dB/µs. The attenuation per wavelength, $\alpha\lambda = \pi\omega\tau$, is proportional to τ and to ω.

If the amplitude of the wave diminishes as it propagates, the fluid must be absorbing part of the energy transported by the wave. There are several mechanisms for this. Viscosity arises from the relative motion of groups of particles, and thermal conduction occurs because the temperature rises (falls) in an area of compression (dilatation). Thus the adiabatic condition assumed earlier does not apply exactly, and the conversion of some mechanical energy to heat causes loss. For a molecular fluid, there is also the possibility of energy exchange between the wave and the internal motions (translation, vibration, rotation) of the molecules. In a liquid, chemical or structural changes can occur. These phenomena are made all the more complex by their dependence on the ambient temperature and pressure, the presence of impurities in the fluid, the humidity (in the case of air), and also on the wave amplitude (overpressure) and frequency. Consequently, it is appropriate to attribute to them different relaxation times. However, depending on the nature of the fluid – gas or liquid – and the circumstances (particularly the frequency), it is usually found that only one or two processes dominate. The fluid can thus be characterized by two values, or one average value, of the relaxation time.

The attenuation of a wave in a fluid has been introduced here because, on the one hand, the idea of a relaxation time with its consequences (e.g. the f^2 dependence of α) is readily transferred to the study of solids (Sect. 4.2.6), and on the other hand a fluid is present in many components (sensors) and instruments (acoustic microscopes).

Typical values. At room temperature, the attenuation coefficient α for air is typically 1 dB/m at 0.1 MHz [1.2], giving $\tau \approx 0.2$ ns. For water it is typically 0.25 dB/m at 1 MHz [1.3], giving $\tau \approx 2$ ps.

1.2.4 Reflection and Refraction

Waves are generally reflected at an abrupt discontinuity (abrupt on the scale of the wavelength), such as the interface between two media. This was introduced in Sect. 1.1.2 above assuming that the wave remains of the same type and is totally reflected, though the wave type was not specified. In this Section we consider an acoustic wave, taking the characteristic quantities to be the over-pressure and the particle displacement.

1.2.4.1 Boundary Conditions. Consider a surface Σ, at $x = 0$, separating two non-viscous fluids designated as fluids 1 and 2 (Fig. 1.16). Since the wave equation is of second order, two boundary conditions are needed.

(a) At the surface, the *normal components* of the particle displacements, $u_{1\mathrm{N}}$ and $u_{2\mathrm{N}}$, must be equal at all times, so

$$u_{1\mathrm{N}}(x = 0, t) = u_{2\mathrm{N}}(x = 0, t) . \tag{1.58}$$

This condition expresses the fact that neither fluid penetrates into the other. It also shows that the particle velocity is continuous.

(b) The acoustic overpressures δp_1 and δp_2 must be equal on the two sides of the boundary, so that

$$\delta p_1(x = 0, t) = \delta p_2(x = 0, t) \quad \text{for all } t . \tag{1.59}$$

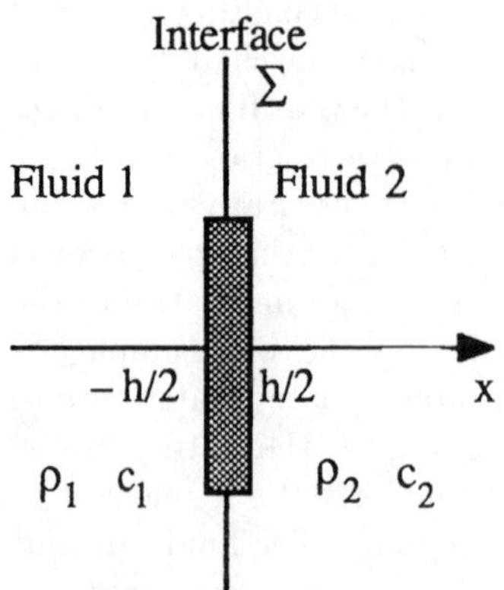

Fig. 1.16. The overpressure and the normal components of particle velocity are equal on either side of the surface Σ separating two fluids.

This *continuity of overpressure* can be seen from the equilibrium of a disc of thickness h and surface area s situated astride the interface, as in Fig. 1.16. Newton's law can be written

$$\left(\frac{\rho_1+\rho_2}{2}\right) hs\frac{\mathrm{d}^2 u}{\mathrm{d}t^2} = s\left[\delta p_1\left(x=-\frac{h}{2},t\right) - \delta p_2\left(x=\frac{h}{2},t\right)\right] .$$

Since the acceleration must be finite, the left side becomes zero as the thickness approaches zero, giving $\delta p_1(0,t) = \delta p_2(0,t)$. This expresses the continuity of the forces at the interface.

1.2.4.2 Normal Incidence. Reflection and Transmission Coefficients. If the interface is a *rigid* plane (a wall), the particle displacement must be zero at this surface. For normal incidence, this condition is only satisfied if the incident wave gives rise to another wave, symmetric about the origin (the wall location), and propagating in the opposite direction, as in Fig. 1.3. We then have total reflection.

If the interface is not rigid, part of the incident wave is transmitted into the second medium. The reflection and transmission coefficients depend on the characteristic impedances of the two media. Suppose that

$$v_\mathrm{I} = F_\mathrm{I}\left(t-\frac{x}{c_1}\right) \quad \text{and} \quad \delta p_\mathrm{I} = Z_1 v_\mathrm{I} \quad \text{for} \quad -\infty < x < 0$$

are the particle velocity and overpressure of the incident wave in medium 1, with acoustic impedance $Z_1 = \rho_1 c_1$, and that

$$v_\mathrm{T} = F_\mathrm{T}\left(t-\frac{x}{c_2}\right) \quad \text{and} \quad \delta p_\mathrm{T} = Z_2 v_\mathrm{T} \quad \text{for} \quad 0 < x < \infty$$

are the same for the wave transmitted into medium 2, with acoustic impedance $Z_2 = \rho_2 c_2$. If $Z_1 \neq Z_2$, these two waves are not sufficient to satisfy the boundary conditions. We need to add a third wave, reflected at the interface and propagating in the $-x$ direction, given by

$$v_\mathrm{R} = F_\mathrm{R}\left(t+\frac{x}{c_1}\right) \quad \text{and} \quad \delta p_\mathrm{R} = -Z_1 v_\mathrm{R} \quad \text{for} \quad -\infty < x < 0 .$$

The boundary conditions become

$$\begin{cases} v_\mathrm{I} + v_\mathrm{R} = v_\mathrm{T} & \text{giving} \quad F_\mathrm{I}(t) + F_\mathrm{R}(t) = F_\mathrm{T}(t) \\ \delta p_\mathrm{I} + \delta p_\mathrm{R} = \delta p_\mathrm{T} & \text{giving} \quad Z_1 F_\mathrm{I}(t) - Z_1 F_\mathrm{R}(t) = Z_2 F_\mathrm{T}(t) . \end{cases} \tag{1.60}$$

The forms of the transmitted and reflected waves resemble that of the incident wave. The *transmission coefficients* for particle velocity, t_v, and for overpressure, t_p, are

$$t_\mathrm{v} = \frac{F_\mathrm{T}(t)}{F_\mathrm{I}(t)} = \frac{2Z_1}{Z_1+Z_2}, \quad t_\mathrm{p} = \frac{Z_2}{Z_1} t_\mathrm{v} = \frac{2Z_2}{Z_1+Z_2} \tag{1.61}$$

and the *reflection coefficients* are

$$r_\mathrm{v} = \frac{F_\mathrm{R}(t)}{F_\mathrm{I}(t)} = \frac{Z_1-Z_2}{Z_1+Z_2} \quad \text{and} \quad r_\mathrm{p} = -r_\mathrm{v} = \frac{Z_2-Z_1}{Z_1+Z_2} . \tag{1.62}$$

If $Z_1 = Z_2$ there is total transmission ($t = 1$) and no reflection ($r = 0$).

Reflected and transmitted power. The power per unit area carried by each wave is given by the magnitude of the Poynting vector, so for the incident wave

$$|P_\mathrm{I}| = P_\mathrm{I} = \delta p_\mathrm{I} v_\mathrm{I} = Z_1 \left[F_\mathrm{I}\left(t - \frac{x}{c_1}\right)\right]^2 .$$

For the reflected wave, propagating in the $-x$ direction,

$$|P_\mathrm{R}| = -P_\mathrm{R} = -\delta p_\mathrm{R} v_\mathrm{R} = Z_1 \left[F_\mathrm{R}\left(t + \frac{x}{c_1}\right)\right]^2 .$$

Using (1.61) and (1.62), the *power reflection coefficient* R is

$$\boxed{R = \frac{|P_\mathrm{R}|}{|P_\mathrm{I}|} = \left(\frac{Z_1 - Z_2}{Z_1 + Z_2}\right)^2} . \tag{1.63}$$

Using the boundary conditions of (1.58) and the relation $\delta p_\mathrm{R} v_\mathrm{I} = -Z_1 v_\mathrm{R} v_\mathrm{I} = -\delta p_\mathrm{I} v_\mathrm{R}$, the transmitted power density is written as

$$|P_\mathrm{T}| = \delta p_\mathrm{T} v_\mathrm{T} = (\delta p_\mathrm{I} + \delta p_\mathrm{R})(v_\mathrm{I} + v_\mathrm{R}) = \delta p_\mathrm{I} v_\mathrm{I} + \delta p_\mathrm{R} v_\mathrm{R}$$

giving

$$|P_\mathrm{T}| = |P_\mathrm{I}| - |P_\mathrm{R}| .$$

We thus have *energy conservation.* The acoustic *power transmission coefficient*, T, is thus

$$\boxed{T = \frac{|P_\mathrm{T}|}{|P_\mathrm{I}|} = 1 - R = \frac{4Z_1 Z_2}{(Z_1 + Z_2)^2}} . \tag{1.64}$$

Impedance transformation by a layer. We consider the acoustic impedance in an intermediate medium with characteristic impedance $Z_\mathrm{i} = \rho_\mathrm{i} c_\mathrm{i}$ having a boundary at $x = L$ with another medium of impedance Z_L (Fig. 1.17). Initially, the medium to the left of the boundary at $x = 0$ in the figure is ignored. In the intermediate medium, assuming *sinusoidal* variations, (1.29) and (1.30) give the particle velocity as

$$v = A\mathrm{e}^{\mathrm{i}(\omega t - kx)} + B\mathrm{e}^{\mathrm{i}(\omega t + kx)} \quad \text{with} \quad k = \omega / c_\mathrm{i} ,$$

where A and B are constants. The corresponding overpressure is

$$\delta p = Z_\mathrm{i}\left(A\mathrm{e}^{-\mathrm{i}kx} - B\mathrm{e}^{\mathrm{i}kx}\right)\mathrm{e}^{\mathrm{i}\omega t} \quad \text{with} \quad Z_\mathrm{i} = \rho_\mathrm{i} c_\mathrm{i} .$$

The acoustic impedance, *at point* x, is

$$Z(x) = \frac{\delta p}{v} = Z_\mathrm{i} \frac{A\mathrm{e}^{-\mathrm{i}kx} - B\mathrm{e}^{\mathrm{i}kx}}{A\mathrm{e}^{-\mathrm{i}kx} + B\mathrm{e}^{\mathrm{i}kx}} \tag{1.65}$$

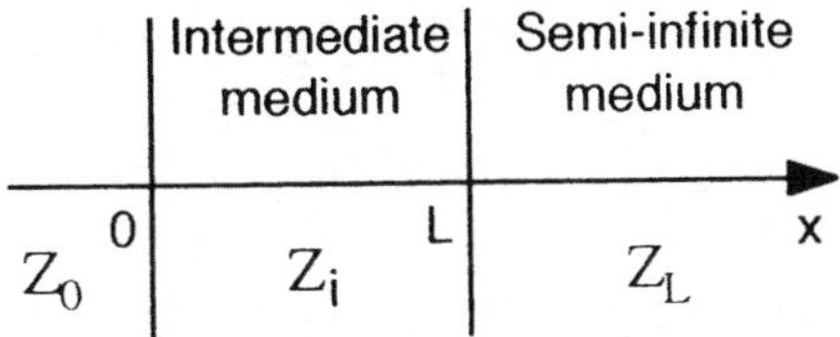

Fig. 1.17. Three successive media with impedances Z_0, Z_i, Z_L

and this is independent of time. The coefficients A and B are determined by the boundary conditions at $x = L$. Since the medium at $x > L$ is infinite, the impedance at $x = L$ is the characteristic impedance Z_L. We can write

$$Z(x) = Z_i \frac{a e^{ik(L-x)} - b e^{-ik(L-x)}}{a e^{ik(L-x)} + b e^{-ik(L-x)}}$$

on defining $A = a e^{ikL}$ and $B = b e^{-ikL}$ and for $x = L$ we find

$$Z(L) = Z_L = Z_i \frac{a-b}{a+b} \quad \text{giving} \quad \frac{a}{b} = \frac{Z_i + Z_L}{Z_i - Z_L} .$$

At any point $x < L$ the impedance is

$$Z(x) = Z_i \frac{(Z_i + Z_L) e^{ik(L-x)} - (Z_i - Z_L) e^{-ik(L-x)}}{(Z_i + Z_L) e^{ik(L-x)} + (Z_i - Z_L) e^{-ik(L-x)}}$$

from which we have

$$\boxed{Z(x) = Z_i \frac{Z_L \cos k(L-x) + iZ_i \sin k(L-x)}{Z_i \cos k(L-x) + iZ_L \sin k(L-x)}} . \tag{1.66}$$

This is the *transformed impedance* at a distance $d = L - x$ from the point where the impedance becomes Z_L; its form is as for an electrical transmission line with load Z_L. In general $Z(x)$ is complex, but not for the following special cases:

- $k(L-x) = 0 \pmod{\pi}$ giving $Z(x) = Z_L$;
- $k(L-x) = \frac{\pi}{2} \pmod{\pi}$ giving $\boxed{Z(x) = Z_i^2/Z_L}$. (1.67)

Quarter-wave layer. The last case leads to a method for cancelling the reflections between two media with different impedances Z_0 and Z_L, as in Fig. 1.17. It consists of adding an intermediate layer with impedance $Z_i = (Z_0 Z_L)^{1/2}$ and thickness d such that $k_i d = \pi/2 + n\pi$, so that d is related to the wavelength λ_i in this medium by

$$d = \frac{\lambda_i}{4} + n\frac{\lambda_i}{2} .$$

The impedance at the front of the intermediate medium is thus Z_0; there is no reflection, and all the energy of a wave incident from the left is transmitted into the medium on the right. However, the *impedance transformation* provided by this *quarter-wave layer* is correct only at a specific frequency $f_0 = c_i/4d$, as shown in Problem 1.6.

1.2.4.3 Oblique Incidence. If the propagation direction of the incident wave makes an angle with the normal to the interface plane, this angle naturally affects the directions and amplitudes of the reflected and transmitted waves.

Snell–Descartes law. The equation for the interface plane Σ in Fig. 1.18 is $\boldsymbol{l} \cdot \boldsymbol{r} = 0$, where $\boldsymbol{l}$ is the unit vector normal to the plane. The incident wave, plane and monochromatic and with wave vector $\boldsymbol{k}_\mathrm{I}$, causes an overpressure

$$\delta p_\mathrm{I} = A_\mathrm{I} \mathrm{e}^{\mathrm{i}(\omega_\mathrm{I} t - \boldsymbol{k}_\mathrm{I} \cdot \boldsymbol{r})} \,.$$

Using subscripts R and T to denote the reflected and transmitted waves, equality of the overpressure on the two sides of the boundary gives

$$A_\mathrm{I} \mathrm{e}^{\mathrm{i}(\omega_\mathrm{I} t - \boldsymbol{k}_\mathrm{I} \cdot \boldsymbol{r})} + A_\mathrm{R} \mathrm{e}^{\mathrm{i}(\omega_\mathrm{R} t - \boldsymbol{k}_\mathrm{R} \cdot \boldsymbol{r})} = A_\mathrm{T} \mathrm{e}^{\mathrm{i}(\omega_\mathrm{T} t - \boldsymbol{k}_\mathrm{T} \cdot \boldsymbol{r})} \,. \tag{1.68}$$

This implies two conditions:

(a) At each instant, $\omega_\mathrm{R} = \omega_\mathrm{T} = \omega_\mathrm{I}$, as for normal incidence. That is, *reflection and transmission at a stationary surface do not cause a change of frequency.*

(b) At each point $\boldsymbol{r}$ on the surface,

$$\boldsymbol{k}_\mathrm{R} \cdot \boldsymbol{r} = \boldsymbol{k}_\mathrm{T} \cdot \boldsymbol{r} = \boldsymbol{k}_\mathrm{I} \cdot \boldsymbol{r}$$

giving

$$(\boldsymbol{k}_\mathrm{R} - \boldsymbol{k}_\mathrm{I}) \cdot \boldsymbol{r} = 0 \quad \text{and} \quad (\boldsymbol{k}_\mathrm{T} - \boldsymbol{k}_\mathrm{I}) \cdot \boldsymbol{r} = 0 \,. \tag{1.69}$$

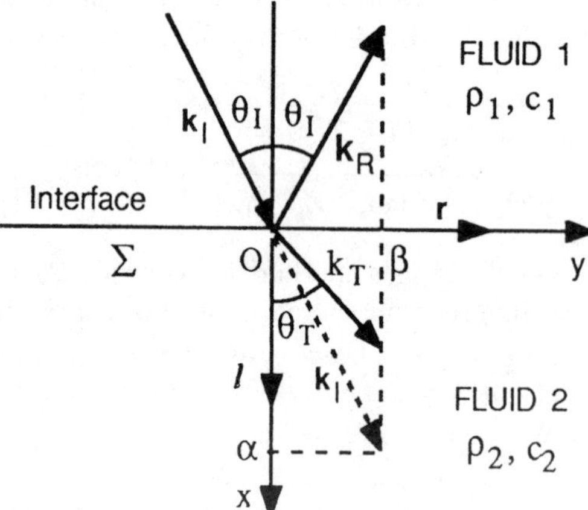

Fig. 1.18. The Snell–Descartes law. The projections onto the interface of the wave vectors $\boldsymbol{k}_\mathrm{R}$ and $\boldsymbol{k}_\mathrm{T}$ of the reflected and transmitted waves are equal to that of the wave vector of the incident wave

Comparison with the equation of the surface, $\boldsymbol{l} \cdot \boldsymbol{r} = 0$, shows that the vectors $(\boldsymbol{k}_\mathrm{R} - \boldsymbol{k}_\mathrm{I})$ and $(\boldsymbol{k}_\mathrm{T} - \boldsymbol{k}_\mathrm{I})$ are normal to the surface. Consequently,

(1) the wave vectors of the reflected and refracted waves are in the *plane of incidence*, defined by the normal vector $\boldsymbol{l}$ and the incident wave vector $\boldsymbol{k}_\mathrm{I}$;
(2) the projections of these vectors on to the interface are equal to that of the incident wave vector.

The latter condition is a form of the Snell–Descartes law, which can be written as (Fig. 1.18)

$$k_\mathrm{R} \sin\theta_\mathrm{R} = k_\mathrm{T} \sin\theta_\mathrm{T} = k_\mathrm{I} \sin\theta_\mathrm{I} \,. \tag{1.70}$$

Here $\theta_\mathrm{I}, \theta_\mathrm{R}$ and θ_T denote the angles of incidence, reflection and refraction. Since the frequencies ω_R and ω_I are the same we have $k_\mathrm{R} = k_\mathrm{I}$, giving

$$\theta_\mathrm{R} = \theta_\mathrm{I} \quad \text{and} \quad \boxed{\frac{\sin\theta_\mathrm{T}}{c_2} = \frac{\sin\theta_\mathrm{I}}{c_1}} \,. \tag{1.71}$$

The ends of the vectors $\boldsymbol{k}_\mathrm{R}$ and $\boldsymbol{k}_\mathrm{T}$ are located on semi-circles with radii proportional to $1/c_1$ and $1/c_2$, drawn on either side of the interface. This construction, due to Huygens, is useful in the case of solid media (Sect. 4.4.1.1).

Critical angle. Evanescent waves. If $c_2 < c_1$, (1.71) has a solution for all angles of incidence, i.e. for $0 < \theta_\mathrm{I} < \pi/2$. In the opposite case $(c_2 > c_1)$ there is a *critical angle* of incidence θ_c given by

$$\sin\theta_\mathrm{c} = \frac{c_1}{c_2} \tag{1.72}$$

such that for $\theta_\mathrm{I} > \theta_\mathrm{c}$ there is no wave transmitted straightforwardly into the second medium. We then have *total reflection*. The wavenumber of the 'transmitted' wave is related to its components α and β by

$$k_\mathrm{T}^2 = \alpha^2 + \beta^2 = \omega^2/c_2^2 \,,$$

where $\beta = k_\mathrm{I} \sin\theta_\mathrm{I}$ is the projection of $\boldsymbol{k}_\mathrm{T}$ onto the surface. The normal component α of the refracted wave, given by

$$\alpha^2 = \frac{\omega^2}{c_1^2}\left[\left(\frac{c_1}{c_2}\right)^2 - \sin^2\theta_\mathrm{I}\right]$$

is purely imaginary if $\theta_\mathrm{I} > \theta_\mathrm{c}$, since then we have $\sin\,\theta_\mathrm{I} > c_1/c_2$. The root

$$\alpha = -\mathrm{i}\frac{\omega}{c_1}\sqrt{\sin^2\theta_\mathrm{I} - \left(\frac{c_1}{c_2}\right)^2} = -\mathrm{i}|\alpha|$$

corresponds to a transmitted wave whose amplitude in the second medium $(x > 0)$ is

$$\delta p_\mathrm{T} = A_\mathrm{T} \mathrm{e}^{\mathrm{i}(\omega t - \alpha x - \beta y)} = A_\mathrm{T} \mathrm{e}^{-|\alpha| x} \mathrm{e}^{\mathrm{i}(\omega t - \beta y)} \,.$$

This wave, decreasing exponentially, is an *evanescent wave* which carries no power. The other root, $\alpha = +\mathrm{i}|\alpha|$, has its amplitude increasing exponentially in the second medium, and is not a physical solution.

Transmission and reflection coefficients. Angular dependence. The equality of the frequencies and the phases $\boldsymbol{k} \cdot \boldsymbol{r}$ in (1.68) implies that the coefficients giving the overpressure are related by

$$A_{\mathrm{I}} + A_{\mathrm{R}} = A_{\mathrm{T}} . \tag{1.73}$$

The continuity of the normal component of particle velocity is written in the same manner, so

$$v_{\mathrm{I}} \cos\theta_{\mathrm{I}} + v_{\mathrm{R}} \cos\theta_{\mathrm{R}} = v_{\mathrm{T}} \cos\theta_{\mathrm{T}} .$$

Substituting $v_{\mathrm{I}} = \delta p_{\mathrm{I}}/Z_1$, $v_{\mathrm{T}} = \delta p_{\mathrm{T}}/Z_2$ and $v_{\mathrm{R}} = -\delta p_{\mathrm{R}}/Z_1$, and using $\theta_{\mathrm{I}} = \theta_{\mathrm{R}}$, this condition becomes

$$\frac{A_{\mathrm{I}}}{Z_1}\cos\theta_{\mathrm{I}} - \frac{A_{\mathrm{R}}}{Z_1}\cos\theta_{\mathrm{I}} = \frac{A_{\mathrm{T}}}{Z_2}\cos\theta_{\mathrm{T}} .$$

Combining this equation with (1.73), and eliminating A_{T}, the *reflection coefficient* for the overpressure is

$$\boxed{r_{\mathrm{p}} = \frac{A_{\mathrm{R}}}{A_{\mathrm{I}}} = \frac{Z_2\cos\theta_{\mathrm{I}} - Z_1\cos\theta_{\mathrm{T}}}{Z_2\cos\theta_{\mathrm{I}} + Z_1\cos\theta_{\mathrm{T}}}} , \tag{1.74}$$

from which the *transmission* coefficient

$$t_{\mathrm{p}} = \frac{A_{\mathrm{T}}}{A_{\mathrm{I}}} = 1 + \frac{A_{\mathrm{R}}}{A_{\mathrm{I}}} = 1 + r_{\mathrm{p}}$$

is given by

$$t_{\mathrm{p}} = \frac{2Z_2\cos\theta_{\mathrm{I}}}{Z_2\cos\theta_{\mathrm{I}} + Z_1\cos\theta_{\mathrm{T}}} , \tag{1.75}$$

where θ_{T} is related to θ_{I} by the Snell–Descartes law.

The manner in which these reflection and transmission coefficients vary with the angle of incidence θ_{I} depends on the acoustic impedance ratio Z_2/Z_1 and the wave velocity ratio c_2/c_1. If c_2 is less than c_1 then θ_{T} is less than θ_{I}, and there is no critical angle for the transmitted wave. If in addition $Z_2 > Z_1$, the reflection coefficient r_{p} is zero for an angle of incidence θ_0 such that

$$\frac{\cos\theta_0}{\cos\theta_{\mathrm{T}}} = \frac{Z_1}{Z_2} \quad \text{giving} \quad \tan^2\theta_0 = \frac{(Z_2/Z_1)^2 - 1}{1 - (c_2/c_1)^2} .$$

The transmission coefficient is of course unity in this case, as shown in Fig. 1.19.

If $c_2 > c_1$, the transmitted wave is evanescent when $\theta_{\mathrm{I}} > \theta_{\mathrm{c}}$. Introducing the normal components $\alpha = k\cos\theta$ of the wave vectors ($k = \omega/c$), (1.74) becomes

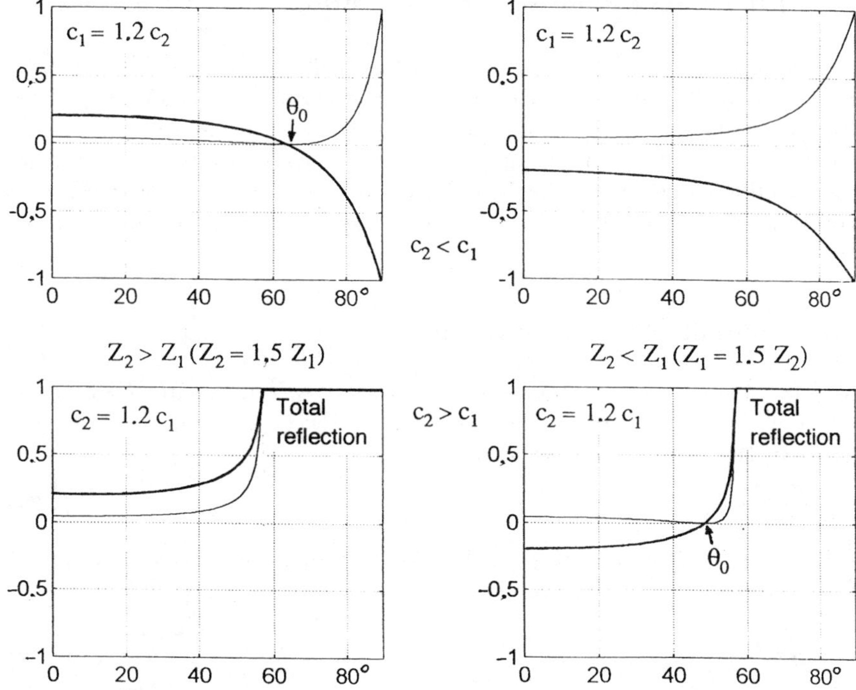

Fig. 1.19. Reflection coefficients for overpressure (*thick lines*) and acoustic power (*thin lines*) as functions of angle of incidence, for several ratios of velocity and characteristic impedance

$$r_\mathrm{p} = \frac{\dfrac{\omega\cos\theta_\mathrm{I}}{\rho_1 c_1} - \dfrac{\omega\cos\theta_\mathrm{T}}{\rho_2 c_2}}{\dfrac{\omega\cos\theta_\mathrm{I}}{\rho_1 c_1} + \dfrac{\omega\cos\theta_\mathrm{T}}{\rho_2 c_2}} = \frac{\dfrac{\alpha_\mathrm{I}}{\rho_1} - \dfrac{\alpha_\mathrm{T}}{\rho_2}}{\dfrac{\alpha_\mathrm{I}}{\rho_1} + \dfrac{\alpha_\mathrm{T}}{\rho_2}}\,. \tag{1.76}$$

Since α_T is purely imaginary ($\alpha_\mathrm{T} = -\mathrm{i}|\alpha|$), the reflection coefficient r_p is a complex number having the form $(a + \mathrm{i}b)/(a - \mathrm{i}b)$, with magnitude unity. Thus the *total reflection* of a wave with angle of incidence greater than the critical angle is accompanied by a phase change. The ratio $A_\mathrm{T}/A_\mathrm{I}$ of the overpressure at the surface no longer represents the transmission coefficient, since the amplitude of the transmitted wave is not constant – it decreases exponentially in the second medium.

While $R = r_\mathrm{p}^2$ gives correctly the fraction of energy reflected, t_p^2 does not give the transmission of energy because the impedances Z_1 and Z_2 are different, and also because the transmitted beam has its width different to that of the incident beam. Given the conservation of acoustic energy, the energy transmission factor T must be

$$T = 1 - R\,. \tag{1.77}$$

This is valid for all cases (irrespective of whether $\theta_\mathrm{I} > \theta_\mathrm{c}$).

1.3 Spherical Acoustic Waves. Radiation

In practice, plane waves are not physically realizable because real sources must have finite dimensions. Acoustic waves generated by a real source must diverge, so that the overpressure and the power density decrease with distance from the source. Waves with spherical symmetry are a particularly important case, leading to understanding of radiation from sources and hence from the 'transducers' which generate the waves in acousto-electric components and systems.

1.3.1 Solution of the Wave Equation

For a non-planar wave, it is convenient to associate a scalar potential function $\phi(\boldsymbol{r},t)$ with the fields of the displacement vector $\boldsymbol{u}$ or the particle velocity vector $\boldsymbol{v}$. Here we define ϕ such that

$$\boxed{\boldsymbol{v} = -\nabla\phi} \tag{1.78}$$

so that ϕ is the potential for particle velocity. Using the relation (1.43), with $\boldsymbol{f} = 0$, the overpressure is expressed as

$$\delta p = \rho_0 \frac{\partial\phi}{\partial t} \,. \tag{1.79}$$

The potential for velocities (or for displacements) thus satisfies the wave equation (1.42), so that

$$\boxed{\frac{1}{c^2}\frac{\partial^2\phi}{\partial t^2} = \Delta\phi} \,. \tag{1.80}$$

For a disturbance with spherical symmetry about the origin O, that is, a spherical wave, the potential ϕ at some point M depends only on the distance $r = OM$. Bearing in mind the expression for the Laplacian Δ in spherical coordinates (Appendix A), the wave equation becomes

$$\frac{1}{c^2}\frac{\partial^2\phi}{\partial t^2} = \frac{2}{r}\frac{\partial\phi}{\partial r} + \frac{\partial^2\phi}{\partial r^2} \,.$$

Here the right side is equal to $(1/r)\partial^2(r\phi)/\partial r^2$. Consequently, the *product* $r\phi$ obeys the wave equation

$$\boxed{\frac{1}{c^2}\frac{\partial^2(r\phi)}{\partial t^2} = \frac{\partial^2(r\phi)}{\partial r^2}} \tag{1.81}$$

which has the same form as that for plane waves. The *general solution* is

$$\boxed{\phi(r,t) = \frac{1}{r}F\left(t - \frac{r}{c}\right) + \frac{1}{r}G\left(t + \frac{r}{c}\right)} \tag{1.82}$$

that is, the sum of a *divergent* wave (receding from the origin with velocity c) represented by the first term, and a *convergent* wave approaching the origin. Using (1.79), the *divergent wave* has overpressure

$$\delta p\,(r,t) = \frac{\rho_0}{r} F'\left(t - \frac{r}{c}\right) \tag{1.83}$$

and thus propagates without distortion except that its amplitude decreases with the distance from the origin. This decrease, as $1/r$, expresses conservation of energy. The power density is proportional to amplitude squared, and the total power flowing across the surface of a sphere of radius r must be independent of r. Thus,

$$4\pi r^2 A^2 = \text{ constant, giving } A = A_0/r\,.$$

At any point M, the particle velocity $\boldsymbol{v}$ has the direction of the vector $\boldsymbol{r} = \boldsymbol{OM}$. The *radial particle velocity* v of the wave, obtained from (1.78), is

$$v = -\frac{\partial \phi}{\partial r} = \frac{1}{rc} F'\left(t - \frac{r}{c}\right) + \frac{1}{r^2} F\left(t - \frac{r}{c}\right) \tag{1.84}$$

and this is a sum of two terms. The second term is dominant near the origin, where the source is, while the first dominates at points far from the origin. In passing from one region to the other, the form of the velocity wave changes.

The form of a *convergent* velocity wave

$$v\,(r,t) = -\frac{1}{rc} G'\left(t + \frac{r}{c}\right) + \frac{1}{r^2} G\left(t + \frac{r}{c}\right)$$

shows the same phenomenon. This change of form of the velocity wave, in comparison with the pressure wave, is expressed for the sinusoidal case by a change of phase on passing through a focus.

Sinusoidal case. Transit through a focus. Consider a *convergent* sinusoidal pressure wave, with wavenumber $k = \omega/c$, given by

$$\delta p\,(r,t) = \frac{B}{r} \mathrm{e}^{\mathrm{i}(\omega t + kr)}\,.$$

The relation between the *radial particle velocity* and the overpressure, deduced from (1.78) and (1.79), is

$$\rho_0 \frac{\partial v}{\partial t} = -\frac{\partial\,(\delta p)}{\partial r} \tag{1.85}$$

showing that

$$\rho_0 \mathrm{i}\omega v = -\left(\mathrm{i}k - \frac{1}{r}\right) \frac{B}{r} \mathrm{e}^{\mathrm{i}(\omega t + kr)} \quad \text{giving} \quad v = -\left(1 + \frac{\mathrm{i}}{kr}\right) \frac{\delta p}{\rho_0 c}\,. \tag{1.86}$$

Far from the origin (for $kr \gg 1$, i.e. $r \gg \lambda/2\pi$), the velocity is in antiphase with the overpressure, with phase difference $-\pi$. Near the origin it is in phase quadrature (relative phase $-\pi/2$) with the overpressure.

For a *divergent* wave, the particle velocity

$$v = \frac{B}{\rho_0 rc}\left(1 - \frac{\mathrm{i}}{kr}\right)\mathrm{e}^{\mathrm{i}(\omega t - kr)} = \left(1 - \frac{\mathrm{i}}{kr}\right)\frac{\delta p}{\rho_0 c} \tag{1.87}$$

is in phase with the overpressure at points far from the origin ($kr \gg 1$) and in quadrature, with relative phase $-\pi/2$, when $kr \ll 1$.

For a wave that converges towards a focus and then diverges, the phase of the particle velocity, relative to the overpressure, varies from $-\pi$ to 0, passing through $-\pi/2$ in the region of the focus ($r \ll \lambda$). Thus, relative to the overpressure, the velocity wave undergoes a *phase advance of* π *on transit through a focus.* This behaviour is also found in optics.

1.3.2 Radiation

The objective of this section is to derive firstly the velocity potential for a sphere vibrating radially, and secondly the acoustic pressure radiated by a plane disc. The first part gives the basis for calculating the second, that is, the acoustic potential created by a harmonic source comprising a surface element with diameter $d \ll \lambda$. The example of a disc, with vibration parallel to the axis, is of great practical importance. When all the points vibrate in phase and with the same amplitude it becomes a circular *piston*, serving as a model for a loudspeaker membrane or a piezoelectric ceramic disc used to generate waves in, for example, a Sonar system. If this disc (or transducer) is not excited sinusoidally, its behaviour can be characterized by a time-domain description of diffraction, i.e. an impulse response involving diffraction.

1.3.2.1 Radiation by a Sphere. Consider a sphere whose radius varies sinusoidally with time about a mean value a, as shown in Fig. 1.20a. The surface of the sphere moves in the radial direction with velocity v_s, which is taken to have amplitude V_n, so that

$$v_\mathrm{s} = V_\mathrm{n}\cos\omega t\,.$$

At a distance $r > a$ from the centre of the sphere, the potential of the divergent spherical wave emitted has the form

$$\phi(r,t) = \frac{A}{r}\cos(\omega t - kr)\,. \tag{1.88}$$

The radial velocity v of a particle of fluid is given by (1.84), so that

$$v(r,t) = \frac{A}{r^2}\left[\cos(\omega t - kr) - kr\sin(\omega t - kr)\right]\,. \tag{1.89}$$

The constant A is determined by equating the velocity in the fluid $v(r,t)$, at $r = a$, with the velocity v_s of the sphere surface, invoking continuity of velocity. We assume that the radius a is much less than the wavelength, so that the first term in $v(r = a, t)$ is dominant, giving the result $A = V_\mathrm{n}a^2$. Thus, the potential generated at $r > a$ by a vibrating sphere with radius $a \ll \lambda$ and with radial velocity $v_\mathrm{s} = V_\mathrm{n}\cos\omega t$ is

$$\phi(\boldsymbol{r},t) = a^2\frac{V_\mathrm{n}}{r}\cos(\omega t - kr)\,. \tag{1.90}$$

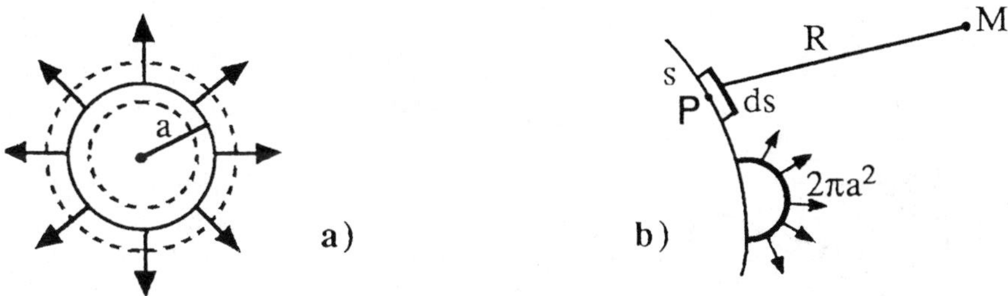

Fig. 1.20. Radiation, (**a**) by a sphere vibrating radially, and (**b**) by an element of a surface.

Radiation by a surface element. The above formula shows that the potential is proportional to the product $a^2 V_n$, and therefore to the volume of fluid displaced by the sphere in unit time. This result can be used to find the wave radiated by a source of any form, by decomposing it into elements of area $\mathrm{d}s$ with dimensions much smaller than the wavelength, as in Fig. 1.20b. Each element $\mathrm{d}s$ can be considered as a *point* source, emitting a spherical wave with amplitude $\mathrm{d}A$. The resulting potential at some point M is the sum

$$\phi(\boldsymbol{r},t) = \int_s \frac{\mathrm{d}A}{R} \exp\left[\mathrm{i}\left(\omega t - kR\right)\right] , \tag{1.91}$$

where R is the distance of the element $\mathrm{d}s$ from the point M. The amplitude $\mathrm{d}A$ is equal to that generated by a hemisphere displacing the same volume of fluid, that is, with radius a such that $2\pi a^2 V_\mathrm{n} = V_\mathrm{n}\mathrm{d}s$ (the waves emitted by the other half of the sphere do not reach the point M). Using (1.90), the amplitude of the potential is

$$\mathrm{d}A = \frac{\mathrm{d}s}{2\pi} V_\mathrm{n}(P) .$$

This is proportional to the area $\mathrm{d}s$ of the element and to the amplitude $V_\mathrm{n}(P)$ of the normal velocity at the point P where the element is located.

The total potential is given by the *Rayleigh integral*

$$\boxed{\phi(\boldsymbol{r},t) = \mathrm{e}^{\mathrm{i}\omega t} \int_s \frac{\mathrm{e}^{-\mathrm{i}kR}}{2\pi R} V_\mathrm{n}(P)\, \mathrm{d}s} . \tag{1.92}$$

The radiation from a vibrating disc can be calculated from this formula, which expresses Huygen's principle: the wave generated can be calculated by assuming that all points on the surface generate hemispherical waves, which are added using superposition.

1.3.2.2 Radiation by a Disc. Consider a plane disc with radius a on which all points vibrate in phase with the same frequency ω and with a uniform normal velocity of amplitude V_n (i.e. a circular piston). The beam of waves

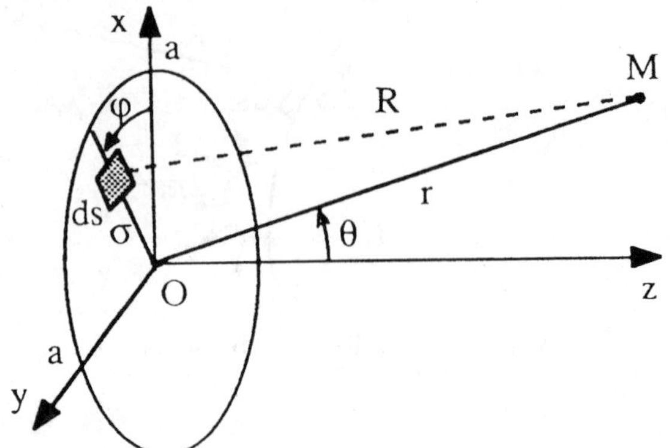

Fig. 1.21. The circular piston has a uniform velocity with amplitude V_n. The point M is in the xz plane

generated is symmetrical about the disc axis Oz. To find the potential at any point, it is sufficient to consider a point M at a distance r from the disc centre O and with its radius vector $\boldsymbol{OM}$ making an angle θ with Oz, as in Fig. 1.21.

Using polar coordinates σ and φ to define the position of a radiating element, which has area $\mathrm{d}s = \sigma\,\mathrm{d}\sigma\,\mathrm{d}\phi$, the potential is given by the Rayleigh integral as

$$\phi(r,t) = V_\mathrm{n}\mathrm{e}^{\mathrm{i}\omega t}\int_0^{2\pi}\frac{\mathrm{d}\varphi}{2\pi}\int_0^a \frac{\sigma}{R}\exp(-\mathrm{i}kR)\,\mathrm{d}\sigma\,. \tag{1.93}$$

Here R is the distance between the source and the point M, given by

$$R = \left(r^2+\sigma^2-2r\sigma\sin\theta\cos\varphi\right)^{1/2}\,. \tag{1.94}$$

In general the integral is not amenable, so here we consider two specific cases:

(a) the potential at any point on the axis, and the overpressure on the axis and near the disc. This is analogous to the amplitude of a light wave in Fresnel's theory of diffraction by a circular aperture;

(b) the potential at a large distance ($r \gg a$) in any direction, and the radiation diagram, i.e. the angular variation of acoustic intensity. This is analogous to the diffraction pattern at infinity produced by light passing through a circular aperture (Fraunhofer diffraction).

Pressure on the axis. On the axis we have $\theta = 0$, and R is given by

$$R^2 = r^2+\sigma^2 \quad \text{giving} \quad \sigma\,\mathrm{d}\sigma = R\,\mathrm{d}R\,.$$

The calculation is simplified because R no longer depends on φ. The integral (1.93) becomes, after changing the variable of integration from σ to R,

$$\phi(z,t) = V_\mathrm{n}\mathrm{e}^{\mathrm{i}\omega t}\int_z^{\sqrt{z^2+a^2}}\exp(-\mathrm{i}kR)\,\mathrm{d}R \tag{1.95}$$

giving

$$\phi(z,t) = \mathrm{i}\frac{V_\mathrm{n}}{k}\mathrm{e}^{\mathrm{i}\omega t}\left[\exp\left(-\mathrm{i}k\sqrt{z^2+a^2}\right) - \exp(-\mathrm{i}kz)\right] .$$

To find the overpressure on the axis, we take the time differential as in (1.79), and separating out the propagation phase term gives

$$\delta p(z,t) = \rho_0 c V_\mathrm{n}\left(1-\mathrm{e}^{-\mathrm{i}k\beta}\right)\exp\mathrm{i}(\omega t - kz) \quad \text{with } \beta = \sqrt{z^2+a^2} - z .$$

The magnitude of δp is

$$|\delta p| = \rho_0 c V_\mathrm{n}\left|1-\mathrm{e}^{-\mathrm{i}k\beta}\right| = \rho_0 c V_\mathrm{n}\sqrt{2-2\cos k\beta}$$

and this can be written

$$|\delta p| = 2\delta p_0\left|\sin\frac{k\beta}{2}\right| = 2\delta p_0\left|\sin\frac{\pi\beta}{\lambda}\right| , \tag{1.96}$$

where we define $\delta p_0 = \rho_0 c V_\mathrm{n}$, as the amplitude of the overpressure for a plane wave of frequency ω, with particle velocity equal to the velocity V_n of the piston.

The overpressure is zero whenever $\beta = n\lambda$. Since β decreases from a to zero as z varies from zero to infinity, zeros occur only if $a > \lambda$. Their locations are such that

$$\beta = \sqrt{z^2+a^2} - z = n\lambda \quad \text{giving} \quad z_n = \frac{a^2}{2n\lambda} - n\frac{\lambda}{2}, \quad n \text{ integer} . \tag{1.97}$$

Figure 1.22 shows that, for this case ($a > \lambda$), *the overpressure on the axis varies very rapidly in the region near the piston.* It has a series of zeros, and a series of maxima with equal magnitude. The first zero is at the origin if a is a multiple of λ $(a = p\lambda)$, and the position z_1 of the last zero is obtained by using $n = 1$ above, so

$$z_1 = \frac{a^2}{2\lambda} - \frac{\lambda}{2} \cong \frac{a^2}{2\lambda} \quad \text{if} \quad \mathrm{a} \gg \lambda .$$

The position of the last maximum, corresponding to $\beta = \lambda/2$, is obtained by setting $n = 1/2$, giving

$$z_m = \frac{a^2}{\lambda} - \frac{\lambda}{4} \cong \frac{a^2}{\lambda} \quad \text{if} \quad \mathrm{a} \gg \lambda .$$

Fresnel zones. At some points on the axis the overpressure is zero, and at others it reaches a maximum magnitude of $2\delta p_0$. To clarify this phenomenon, we divide the disc into zones defined by the intersection of the disc surface with spheres centred at M, with radii

$$R_n = z + n\frac{\lambda}{2} = \sqrt{\sigma_n^2 + z^2} . \tag{1.98}$$

The circular annulus between neighbouring circles with radii σ_n and σ_{n+1}, shown in Fig. 1.23, is called a Fresnel zone. The integral (1.95) can be arranged as a sum of component integrals, one for each of the Fresnel zones, so that

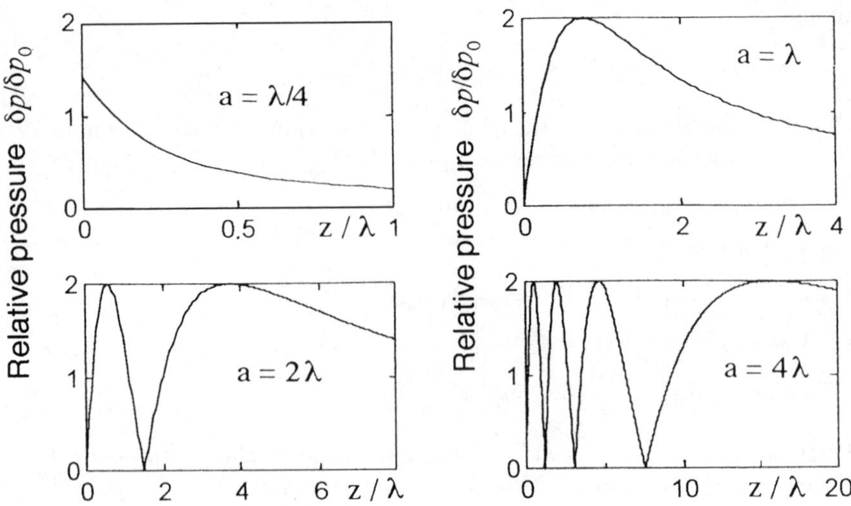

Fig. 1.22. Amplitude variation of the relative overpressure $|\delta p|/\delta p_0$ on the disc axis, for several values of the ratio a/λ

$$\phi(z,t) = V_{\mathrm{n}} \mathrm{e}^{\mathrm{i}\omega t} \sum_{n=0}^{N} I_n \quad \text{with} \quad I_n = \int_{R_n}^{R_{n+1}} \mathrm{e}^{-\mathrm{i}kR} \mathrm{d}R \,.$$

Since $R_{n+1} = R_n + \lambda/2$ and $k\lambda/2 = \pi$, this gives $I_n = -(2\mathrm{i}/k)\exp(-\mathrm{i}kR_n)$. The potential is written as

$$\phi = -\frac{2\mathrm{i}}{k} V_{\mathrm{n}} \mathrm{e}^{\mathrm{i}\omega t} \sum_{n=0}^{N} \mathrm{e}^{-\mathrm{i}kR_n} \quad \text{with} \quad kR_n = kz + n\pi$$

and this gives the overpressure as (with the propagation phase separated out)

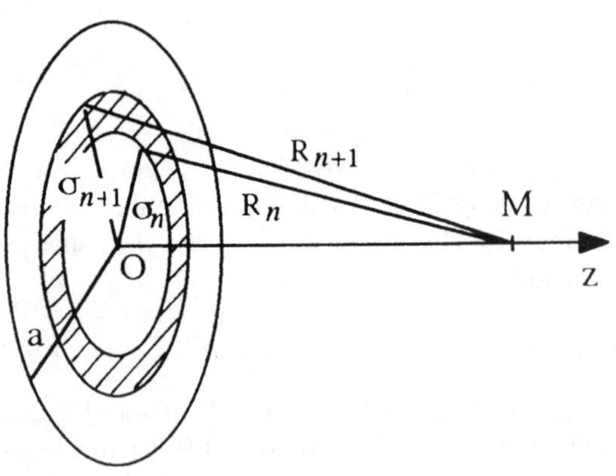

Fig. 1.23. Fresnel zones

$$\delta p = 2\delta p_0 \mathrm{e}^{\mathrm{i}(\omega t - kz)} \sum_{n=0}^{N} \mathrm{e}^{-\mathrm{i}n\pi} . \tag{1.99}$$

In the summation, the terms with n even give $+1$, and those with n odd give -1. Thus, *the contributions from two successive Fresnel zones cancel each other*. If the radius a and the distance z are such that the disc contains an exact number N of zone pairs, the amplitude of the overpressure at point P is zero. If the disc contains an odd number of Fresnel zones, the sum has absolute value unity and the overpressure has its maximum value of $2\delta p_0$. For general values of a and z the disc does not have an integer number of Fresnel zones, and the overpressure has an intermediate value.

Fresnel's approximation. The above applies near the 'transducer', i.e. in the *near field* region. At larger distances the overpressure decreases as $1/z$. To see this, we write

$$\frac{\beta}{\lambda} = \frac{z}{\lambda}\left(\sqrt{1+\frac{a^2}{z^2}} - 1\right) \cong \frac{a^2}{2z\lambda} \ll 1 , \quad \text{if} \quad z \gg \frac{a^2}{\lambda} .$$

Then (1.96) gives

$$\delta p \cong 2\delta p_0 \pi \frac{\beta}{\lambda} = \delta p_0 \frac{ka^2}{2z} .$$

This decrease corresponds to the angular spreading of the beam.

Variation at infinity. Radiation diagram. At a large distance from the disc ($r \gg a^2/\lambda$), the distance R between the element $\mathrm{d}s$ and the observation point M is almost independent of the position of the element. In the denominator of the integral (1.93) for the overpressure, we may replace R by r to a first approximation. The term kR in the exponential gives the phase as a function of the source (i.e. element) position. Here it is sufficient to express this as a first-order function of σ, the distance from the disc centre ($\sigma < a \ll r$). From (1.94),

$$R = r\left(1 + \frac{\sigma^2}{r^2} - 2\frac{\sigma}{r}\sin\theta\cos\varphi\right)^{1/2} \cong r - \sigma\sin\theta\cos\varphi .$$

Thus the potential can be put in the form

$$\phi(r,\theta,t) = \frac{V_{\mathrm{n}}\mathrm{e}^{\mathrm{i}\omega t}}{2\pi r}\left(\int_0^a \sigma\,\mathrm{d}\sigma \int_0^{2\pi} \mathrm{e}^{\mathrm{i}k\sigma\sin\theta\cos\varphi}\mathrm{d}\varphi\right)\mathrm{e}^{-\mathrm{i}kr} .$$

The integral over the angle φ involves the Bessel function $J_0(k\sigma\sin\theta)$, because this is defined by

$$J_0(x) = \frac{1}{2\pi}\int_0^{2\pi} \mathrm{e}^{\mathrm{i}x\cos\varphi}\mathrm{d}\varphi \quad \text{with} \quad x = k\sigma\sin\theta .$$

Thus the potential is

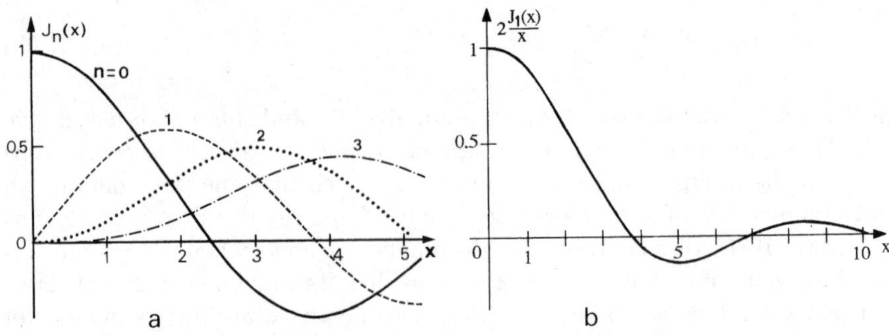

Fig. 1.24. (**a**) Bessel functions $J_0(x)$, $J_1(x)$, $J_2(x)$.... (**b**) The function $(2/x)J_1(x)$

$$\phi(r,\theta,t) = \frac{V_n}{r}\left[\int_0^a J_0(k\sigma\sin\theta)\,\sigma\,\mathrm{d}\sigma\right]\mathrm{e}^{\mathrm{i}(\omega t-kr)}\,.$$

To calculate this integral, we use the standard relation $\int xJ_0(x)\mathrm{d}x = xJ_1(x)$ between Bessel functions of order 0 and 1, giving

$$\int_0^a J_0(k\,\sigma\sin\theta)\,\sigma\,\mathrm{d}\sigma = \left[\frac{\sigma J_1(k\,\sigma\sin\theta)}{k\sin\theta}\right]_0^a = a^2\frac{J_1(k\,a\sin\theta)}{k\,a\sin\theta}$$

and hence

$$\phi(r,\theta,t) = \frac{a^2V_n}{2r}\left[\frac{2J_1(k\,a\sin\theta)}{k\,a\sin\theta}\right]\mathrm{e}^{\mathrm{i}(\omega t-kr)}\,. \tag{1.100}$$

The angular dependence of the overpressure thus has the form $2J_1(x)/x$, with $x = ka\sin\theta$. This function has its maximum value of unity when $x = 0$, and is zero at the zeros of $J_1(x)$, for example at $x_1 = 3.83$, $x_2 = 7.02$, $x_3 = 10.15$, as shown in Fig. 1.24. At a constant distance r, the overpressure decreases as θ moves from zero, and its first zero is at $\theta = \theta_1$, given by

$$\sin\theta_1 = \frac{3.83}{ka} = 0.61\frac{\lambda}{a}\,.$$

Figure 1.25 shows the radiation diagram for a disc with radius $a = 3\lambda$, and also shows that most of the *acoustic intensity*, given by (1.54), is in the main lobe between the angles $-\theta_1$ and $+\theta_1$. The energy in this lobe is radiated in a cone with vertex angle $\alpha = 2\theta_1$. This *'beam divergence angle'* is

$$\boxed{\alpha = 1.22\frac{\lambda}{a} \quad \text{if} \quad a \gg \lambda}\,. \tag{1.101}$$

For example, if $\lambda = 1\,\mathrm{mm}$ and $a = 10\,\mathrm{mm}$, then $\alpha = 0.122\,\mathrm{rad} \approx 7°$. The amplitude of the first secondary lobe is 0.133 times that of the main lobe, i.e. it is 17.5 dB below it. The acoustic beam becomes more directional as the diameter of the disc, in wavelengths, is increased.

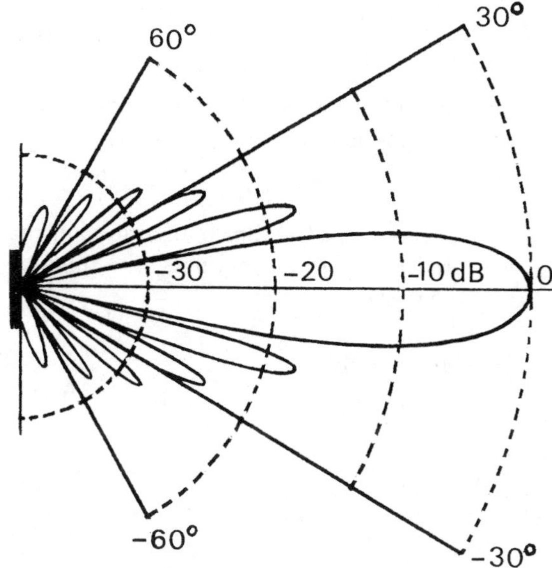

Fig. 1.25. Radiation diagram showing the acoustic intensity emitted by a circular piston with radius $a = 3\lambda$

1.3.2.3 Impulse Response. In areas such as medical scanning and non-destructive evaluation of materials, the transducers generating the waves are excited by short pulses, so the above analysis, referring to the harmonic case, needs to be generalized. We start from the Rayleigh integral, given for the sinusoidal case by (1.92), and replace the wavenumber k by $2\pi f/c$. This gives the velocity potential at point M and at frequency f as

$$\Phi(M,f) = \int_s \frac{V_\mathrm{n}(P,f)}{2\pi R} \mathrm{e}^{-\mathrm{i}2\pi f R/c} \mathrm{d}s\,. \tag{1.102}$$

Here $\Phi(M,f)$ is the spectral component, at frequency f, of the instantaneous potential $\phi(M,t)$ at point M. Similarly, $V_\mathrm{n}(P,f)$ is the spectral component of the normal velocity $v_\mathrm{n}(P,t)$ at point P on the emitting surface. Using the Fourier integral

$$\phi(M,t) = \int_{-\infty}^{+\infty} \Phi(M,f)\,\mathrm{e}^{\mathrm{i}2\pi f t}\mathrm{d}f$$

gives, on changing the order of integration

$$\phi(M,t) = \int_s \frac{\mathrm{d}s}{2\pi R}\left[\int_{-\infty}^{+\infty} V_\mathrm{n}(P,f) \exp\left[\mathrm{i}2\pi f\,(t - R/c)\right] \mathrm{d}f\right].$$

Here the integral in brackets represents the normal component of velocity at point P at time $t - R/c$, that is, $v_\mathrm{n}(P, t - R/c)$. Thus the Rayleigh integral can be expressed in the time domain as

$$\boxed{\phi(M,t) = \int_s \frac{v_\mathrm{n}(P, t - R/c)}{2\pi R} \mathrm{d}s} . \qquad (1.103)$$

Piston mode. In this case the points on the source all vibrate with *normal* velocity proportional to the same time function $v(t)$, so we can write

$$v_\mathrm{n}(P,t) = V_\mathrm{n}(P)\, v(t) ,$$

where $V_\mathrm{n}(P)$ is the amplitude profile over the piston. Since any function can be decomposed into a sum of weighted and delayed impulses, the function $v(t - R/c)$ needed for (1.103) can be written

$$v(t - R/c) = \int_{-\infty}^{+\infty} v(\tau)\, \delta(t - R/c - \tau)\, \mathrm{d}\tau . \qquad (1.104)$$

We define

$$\boxed{h(M,t) = \int_s \frac{V_\mathrm{n}(P)\, \delta(t - R/c)}{2\pi R} \mathrm{d}s} \qquad (1.105)$$

and this gives

$$\phi(M,t) = \int_{-\infty}^{+\infty} v(\tau)\, h(M, t - \tau)\, \mathrm{d}\tau = v(t) \otimes h(M,t) . \qquad (1.106)$$

This equation also defines the convolution integral, symbolically represented by a cross within a circle. Thus the velocity potential at point M is the convolution of the velocity $v(t)$ at the piston surface with the time- and position-dependent function $h(M,t)$, which is called the *diffraction impulse response.* This result expresses the assumption in the analysis of linearity and time invariance. The notion of an impulse response, essentially the response of the system to a pulse shorter than any characteristic time of the system, is frequently used in signal processing.

In the sinusoidal regime, where $v(t) = v_0 \exp(\mathrm{i}2\pi f t)$, the velocity potential (1.103) becomes

$$\phi(M,t) = \int_s \frac{V_\mathrm{n}(P)\, v_0}{2\pi R} \mathrm{e}^{\mathrm{i}2\pi f (t - R/c)} \mathrm{d}s = H(M,f)\, v_0 \mathrm{e}^{\mathrm{i}2\pi f t} ,$$

which is proportional to the frequency-domain diffraction response

$$H(M,f) = \int_s \frac{V_\mathrm{n}(P)}{2\pi R} \mathrm{e}^{-\mathrm{i}2\pi f R/c} \mathrm{d}s . \qquad (1.107)$$

This is the Fourier transform of $h(M,t)$.

Plane uniform transducer (piston). Here the vibration is uniform, so we can take $V_\mathrm{n}(P) = 1$. To calculate the diffraction impulse response for this case we note that, because of the term $\delta(t - R/c)$ in the integral (1.105), contributions to the field at M at time t arise only from points at a distance $R = ct$ from M (Fig. 1.26). These points are situated on a circular arc AB

centred at M_0, which is the projection of the point M on to the plane of the disc. The radius ρ of this arc is given by

$$\rho^2 + z^2 = R^2 \quad \text{so that} \quad \rho\,\mathrm{d}\rho = R\,\mathrm{d}R\,.$$

The surface element has area $\mathrm{d}s = L(R)\mathrm{d}\rho$, where $L(R)$ is the length of the arc. Using $\mathrm{d}\rho = \mathrm{d}R/\sin\theta(R)$, we obtain the impulse response in the following form, also given by Stephanishen [1.4]:

$$h\,(M,t) = \int_{R_1}^{R_2} \frac{\delta\,(t - R/c)}{2\pi R} \cdot \frac{L\,(R)}{\sin\theta\,(R)} \mathrm{d}R\,, \tag{1.108}$$

where $R_1 = ct_1$ and $R_2 = ct_2$ designate respectively the minimum and maximum distances from M to points on the disc, corresponding to points P_1 and P_2. Since $\delta(t - R/c) = c\delta(R - ct)$, we have

$$\begin{cases} h\,(M,t) = \dfrac{L\,(ct)}{2\pi t \sin\theta\,(ct)} & \text{for} \quad t_1 < t < t_2 \\ h\,(M,t) = 0 \quad \text{otherwise.} \end{cases} \tag{1.109}$$

An alternative and simpler expression is obtained by introducing the angle $\Omega(R = ct)$ which the arc AB subtends at its centre M_0, as shown in Fig. 1.26. The arc length is written

$$L(ct) = \Omega(ct)\rho(ct) = \Omega(ct)ct\sin\theta(ct)\,.$$

The impulse response (1.109) can thus be put in the equivalent form (using $R = ct$)

$$\begin{cases} h\,(M,t) = \dfrac{c}{2\pi}\Omega(ct) & \text{for} \quad t_1 < t < t_2 \\ h\,(M,t) = 0 \quad \text{otherwise.} \end{cases} \tag{1.110}$$

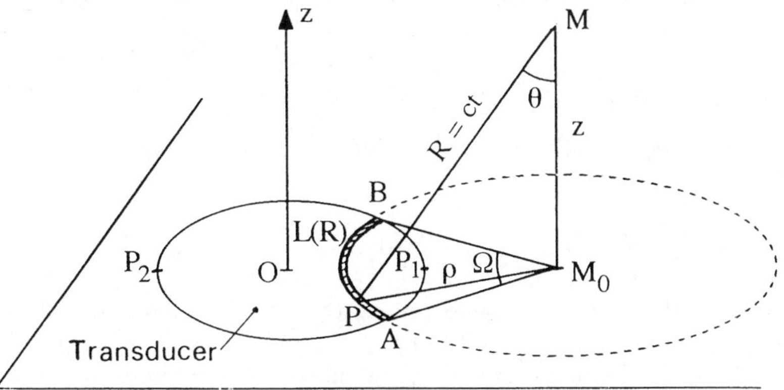

Fig. 1.26. At time t, contributions to the field at M originate from sources at points such as P, on the circular arc AB where the transducer plane intersects a sphere of radius $R = ct$ centred at M

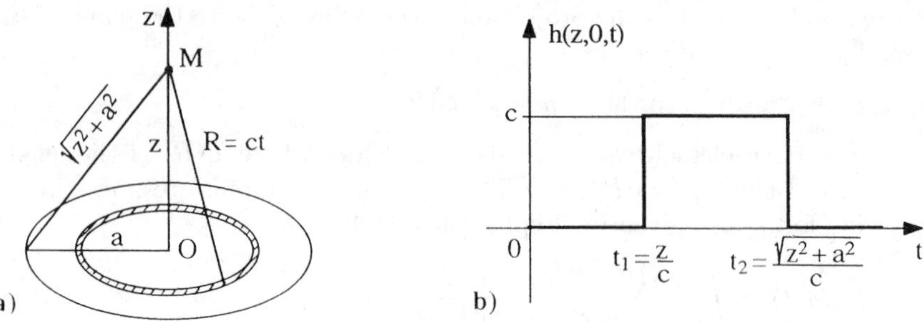

Fig. 1.27. Uniform disc with radius a. For a point M on the z axis, as in (**a**), the diffraction impulse response (**b**) is a rectangle with height c, extending from $t_1 = z/c$ to $t_2 = (z^2 + a^2)^{1/2}/c$

In the case of a *uniform disc* with radius a, the field depends only on the variables r, z and t, where r is the radial coordinate measured normal to the disc axis. For points on the axis ($r = 0$) and for times such that

$$t_1 = z/c < t < t_2 = \left(z^2 + a^2\right)^{1/2}/c \tag{1.111}$$

the arcs of points equidistant from M become complete circles, for which $\Omega(ct) = 2\pi$, and so

$$h\left(z, r = 0, t\right) = c \quad \text{for} \quad t_1 < t < t_2 \,. \tag{1.112}$$

Since it is zero for other times, the diffraction impulse response is a *rectangular pulse*, as in Fig. 1.27b. Its duration

$$\Theta\left(z\right) = \left[\left(z^2 + a^2\right)^{1/2} - z\right]/c$$

decreases as z increases.

When Fresnel's approximation is valid, the duration of the impulse response is inversely proportional to z, since

$$\left(z^2 + a^2\right)^{1/2} \cong z + \frac{a^2}{2z} \quad \Rightarrow \quad \Theta\left(z\right) \cong \frac{a^2}{2cz} \,. \tag{1.113}$$

The frequency-domain diffraction response $H(M, f)$ is the Fourier transform of $h(M, t)$, so that

$$H\left(z, 0, f\right) = \int_{-\infty}^{+\infty} h\left(z, 0, t\right) \mathrm{e}^{-\mathrm{i}2\pi f t} \mathrm{d}t \,.$$

As z increases this function expands, along the frequency axis, by a factor which is proportional to z when $z^2 \gg a^2$.

In the *far-field* (Fraunhofer) region, the diffraction impulse response approaches a Dirac delta function, so that

$$h\left(z, 0, t\right) = c\,\Theta\left(z\right) \delta\left(t\right) \cong \frac{a^2}{2z} \delta\left(t\right) \,, \tag{1.114}$$

bearing in mind that the delta function has unit area. Using (1.106), the velocity potential becomes

$$\phi(z,0,t) \cong \frac{a^2}{2z} v(t) ,$$

and this resembles the velocity $v(t)$ of points on the source disc.

The *overpressure* generated by the disc, which has surface velocity $v(t)$, follows from the time differential of the potential, as in (1.79), and the convolution (1.106), giving

$$\delta p = \rho_0 v(t) \otimes \frac{\mathrm{d}h}{\mathrm{d}t} = \rho_0 h(t) \otimes \frac{\mathrm{d}v}{\mathrm{d}t} . \tag{1.115}$$

On the axis of the piston, the differential of the impulse response, Fig. 1.27b, is two delta-functions at times t_1 and t_2, so that $h'(t) = c[\delta(t-t_1) - \delta(t-t_2)]$; this gives

$$\delta p = \rho_0 c\left[v(t-t_1) - v(t-t_2)\right] = \rho_0 c\left[v\left(t - \frac{z}{c}\right) - v\left(t - \frac{\sqrt{z^2+a^2}}{c}\right)\right] .$$

This function is shown in Fig. 1.28. Near the disc ($z \ll a$, Fig. 1.28a), the two terms of the overpressure are distinct, and both vary with time as the surface velocity $v(t)$. The first term is as for a plane wave arriving at a time z/c, and the second, which is inverted, arrives after a time corresponding to the distance to the edge. This *edge wave* expresses diffraction effects associated with the finite source radius a. At a larger distance ($z = 2a$), the two terms begin to overlap, as in Fig. 1.28b. Far from the source ($z \gg a$), we can use the diffraction impulse response in (1.114) to obtain

$$\delta p = \rho_0 \frac{a^2}{2z} v'(t) = \rho_0 c\,\Theta(z)\, v'(t) .$$

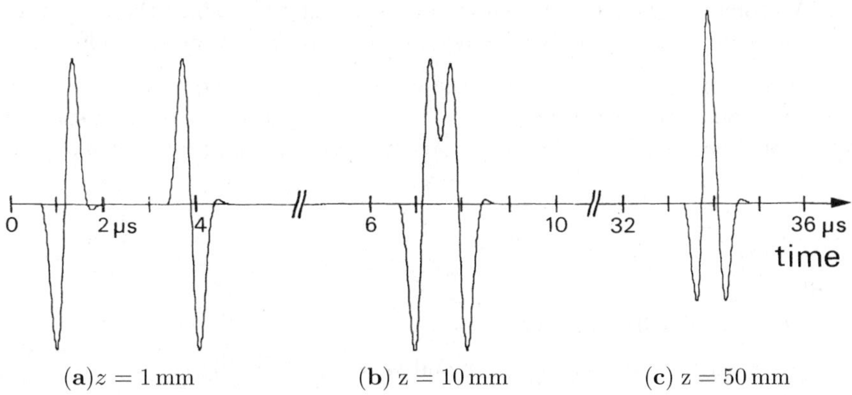

Fig. 1.28. Time-domain overpressure, on the z axis, produced by a plane circular transducer with radius $a = 5\,\mathrm{mm}$ and centre frequency 1 MHz. In the far field (**c**) this function has the form of the differential of the surface velocity on the transducer

This has the form of the differential $v'(t)$ of the surface velocity, as in Fig. 1.28c.

The interference phenomena causing the pressure oscillations seen on the axis in the sinusoidal case (Fig. 1.22) are absent when the transducer is impulsed, because then the acoustic field received at one point and at one instant is generated at a well-localized zone of the transducer surface.

Problems

1.1. What is the nature (plane, propagating, etc.) of waves represented by the following expressions:

(a) $\cos(\omega t - kx) + \sin(\omega t + kx)$
(b) $a\cos^2(\omega t - kx) + b\sin(\omega t - kx)$
(c) $a\cos\omega t\cos kx + b\cos\omega t\sin kx$
(d) $\sin\omega t\cos kx - \cos\omega t\ sinkx$
(e) $a\sin(\omega t - kx) + b\cos(-2\omega t + 2kx)$
(f) $a\exp(-x^2 - y^2)\cos(\omega t - kz)$?

Solution.

(a) The sum of a plane wave propagating along $+x$ and a plane wave of the same amplitude propagating along $-x$, giving a stationary wave.
(b) The first term is equal to $a[1 + \cos 2(\omega t - kx)]/2$, so the wave is plane but not monochromatic; it has frequencies ω, 2ω. Propagation is along $+x$.
(c) A plane stationary wave, $(a\cos kx + b\sin kx)\cos\omega t$.
(d) $\sin(\omega t - kx)$, a plane wave propagating along $+x$.
(e) $\cos(-2\omega t + 2kx) = \cos 2(\omega t - kx)$, so this is a plane propagating wave, but not monochromatic.
(f) A propagating monochromatic wave, but not plane since the amplitude varies in the xy plane perpendicular to the propagation direction $+z$.

Comment. The general solution to the wave equation (1.25) is the sum of two counter-propagating waves: $u(x,t) = f(x - ct) + g(x + ct)$. For a stationary wave we have $u(x,t) = F(x)\cdot G(t)$, and the equation can be split into two, so that

$$\frac{1}{F(x)}\frac{\partial^2 F}{\partial x^2} = \frac{1}{c^2}\cdot\frac{1}{G(t)}\frac{\partial^2 G}{\partial t^2} = \text{constant} = -k^2,$$

giving:

$$F'' + k^2 F = 0 \quad \text{and} \quad G'' + \omega^2 G = 0\,,$$

where $\omega = ck$, $F = \cos kx$ or $\sin kx$ and $G = \cos\omega t$ or $\sin\omega t$.

1.2. What is the complex representation $P(x,t)$ for an acoustic wave with overpressure $p(x,t) = a\cos(\alpha t + \beta x) + b\sin(\alpha t + \beta x)$?

Answer: $P(x,t) = a\mathrm{e}^{\mathrm{i}(\alpha t + \beta x)} + b\,\mathrm{e}^{\mathrm{i}(\alpha t + \beta x - \pi/2)} = (a - \mathrm{i}b)\mathrm{e}^{\mathrm{i}(\alpha t + \beta x)}\,.$

1.3. Write the expression for a plane acoustic wave propagating in the direction $Ox + 30°$ in the xOy plane.

Solution: $F[t - (x\cos\theta + y\sin\theta)/c]$ with $\theta = 30°$

giving:

$$F\left[t - \left(x\sqrt{3} + y\right)/2c\right] .$$

1.4. What is the complex representation of a monochromatic wave $P(z,t)$ propagating in the $-z$ direction, such that at the point $z = L$ its amplitude is $P(L,t) = A\exp(\mathrm{i}\omega t)$?

Solution: $P(z,t) = A\mathrm{e}^{\mathrm{i}(\omega t + kz + \phi)}$, $\phi = -kL$

giving:

$$P(z,t) = A\mathrm{e}^{\mathrm{i}[\omega t + k(z-L)]} .$$

1.5 Waveguides. A sinusoidal wave, with pressure $p(x,y,t) = f(y) \times \exp\mathrm{i}(\omega t - kx)$ for $|y| < h$ and $p(x,y,t) = g(y)\exp[\mathrm{i}(\omega t - kx)]$ for $|y| > h$, propagates along x in the two-dimensional structure of Fig. 1.29. Write the differential equations giving $f(y)$ and $g(y)$ as functions of y. The phase velocity c of this wave is between c_1 and c_2. Defining $\alpha^2 = (\omega/c_1)^2 - k^2$ and $\beta^2 = k^2 - (\omega/c_2)^2$, write the general solution in fluid 1 and fluid 2 such that the pressure p is zero for $|y| = \infty$. Express p in the two fluids for a mode that is (1) symmetric, and (2) antisymmetric, about the x axis. Show that the continuity of the normal component of the fluid particle velocity is equivalent to continuity of the component $\partial p/\partial y$ of the pressure gradient. For the symmetric mode, derive expressions for the overpressure p in the different regions and the relation between α, β and h. Sketch the curve showing the variation of p with y when $\alpha h = \pi/3$, for $-\infty < y < \infty$. Derive the same relations for the antisymmetric mode, and sketch the function $p(y)$ for $\alpha h = 2\pi/3$. How do the continuity conditions constrain these curves?

Solution. Substituting the expressions for p into the wave equation, we find

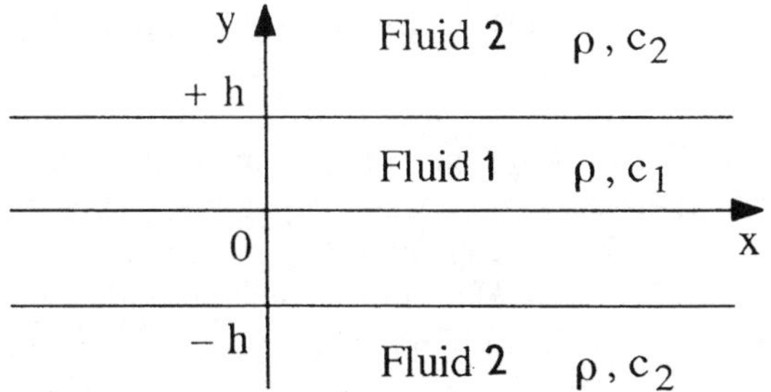

Fig. 1.29. Waveguide comprising two fluids with the same mass density but with different acoustic wave velocities, $c_1 < c_2$

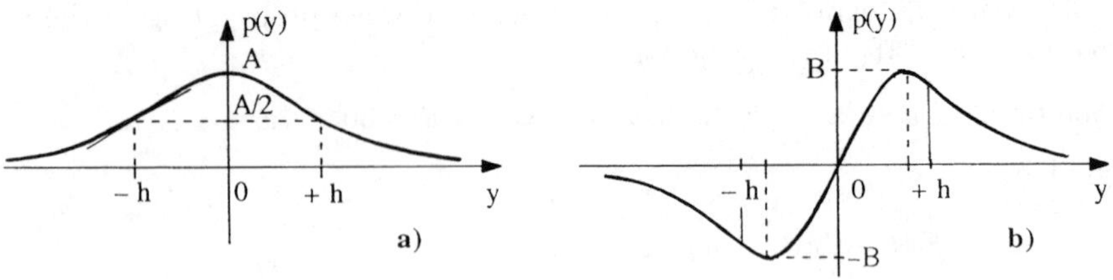

Fig. 1.30. Variation of pressure $p(y)$ for (**a**) a symmetric mode and (**b**) an antisymmetric mode

$$\frac{d^2 f}{dy^2} + \alpha^2 f(y) = 0 \quad \text{and} \quad \frac{d^2 g}{dy^2} - \beta^2 g(y) = 0, \quad \text{with } \alpha^2 > 0 \text{ and } \beta^2 > 0 .$$

In fluid (1), $f(y) = A\cos\alpha y + B\sin\alpha y$.

In fluid (2), for $y > h$, $g(y) = C\exp(-\beta y)$,

and for $y < -h$, $g(y) = D\exp(\beta y)$.

- The symmetric mode gives $B = 0$ and $C = D$, and the antisymmetric mode gives $A = 0$ and $D = -C$.
- Using Newton's law (1.40) for the sinusoidal case, $\mathrm{i}\omega\rho v_y = -\partial p/\partial y$. Thus continuity of v_y implies continuity of $\partial p/\partial y$.
- For the symmetric mode, continuity of p gives A $\cos\alpha h = C\exp(-\beta h)$, and continuity of $\partial p/\partial y$ gives $-\alpha A\sin\alpha h = -\beta C\exp(-\beta h)$; hence $\beta = \alpha\tan\alpha h$. For $|y| < h$, $f(y) = A\cos\alpha y$. For $|y| > h$, $g(y) = A\cos\alpha h\exp[\beta(h - |y|)]$.
- If $\alpha h = \pi/3$, then $f(h) = g(h) = A/2$.
- For the antisymmetric mode, continuity of p gives $B\sin\alpha h = C\exp(-\beta h)$, and continuity of $\partial p/\partial y$ gives $\alpha B\cos\alpha h = -\beta C\exp(-\beta h)$, and hence $\beta = -\alpha\,\mathrm{cotan}\,\alpha h$. For $|y| < h$, $f(y) = B\ \sin\alpha y$. For $|y| > h$, $g(y) = \pm B\sin\alpha h\exp\left[\beta(h - |y|)\right]$.
- If $\alpha h = 2\pi/3$, then $f(h) = g(h) = B\sqrt{3}/2$, and $f(y)$ is a maximum for $\alpha y = \pi/2$, so $y = 3h/4$.
- The continuity conditions require continuity of the curve $p(y)$ and its slope at the points $y = \pm h$ (Fig. 1.30).

1.6 Quarter-wave layer. An acoustic wave emerging from solid 1, which has impedance $Z_1 = 24\,\mathrm{MRayl.}$, traverses solid 2 and propagates into water (Fig. 1.31). Find the impedance Z_2 and the thickness L of medium 2 (with $c_2 = 2000\,\mathrm{m/s}$) for which the energy is totally transferred into the water (with impedance Z_3) at frequency $f_0 = 2\,\mathrm{MHz}$. For this purpose, express the reflection and transmission factors, R and T, as functions of $\cos kL$. Sketch the curve of T as a function of frequency for $0 < f < 2f_0$. For what frequencies f_1 and f_2 is half of the power transferred to the water?

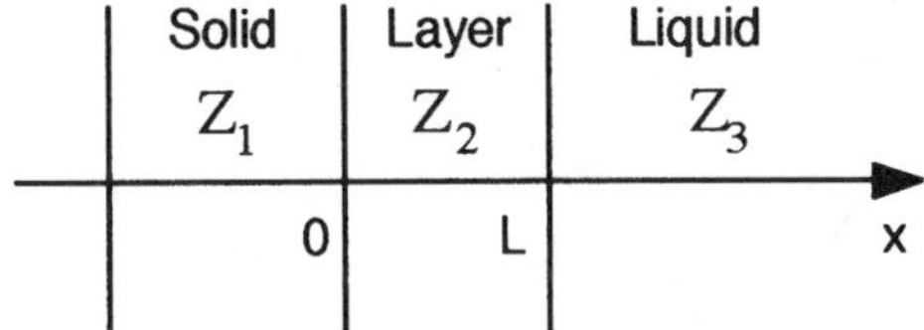

Fig. 1.31. Transmission of energy across an impedance-matching layer, from a solid into a liquid (water)

Solution. The transformed impedance at $x = 0$ is

$$Z_{\mathrm{R}}(0) = Z_2 \frac{Z_3 \cos kL + \mathrm{i} Z_2 \sin kL}{Z_2 \cos kL + \mathrm{i} Z_3 \sin kL} .$$

The reflection coefficient and the reflection factor are given by

$$r = \frac{Z_{\mathrm{R}}(0) - Z_1}{Z_{\mathrm{R}}(0) + Z_1}$$

and

$$R = r^2 = \frac{Z_2^2 (Z_3 - Z_1)^2 \cos^2 kL + (Z_2^2 - Z_1 Z_3)^2 \sin^2 kL}{Z_2^2 (Z_3 + Z_1)^2 \cos^2 kL + (Z_2^2 + Z_1 Z_3)^2 \sin^2 kL} . \tag{1.116}$$

$R = 0$ requires $\cos kL = 0$, giving $kL = \pi/2 + n\pi$. And hence $L = c_2/4f_0$ and $Z_2 = \sqrt{Z_1 Z_3}$. With this value we have

$$T = 1 - R = \frac{4Z_1 Z_3}{4Z_1 Z_3 + (Z_3 - Z_1)^2 \cos^2 kL} = \frac{4}{4 + a^2 \cos^2 kL} ,$$

where $a = \sqrt{\frac{Z_3}{Z_1}} - \sqrt{\frac{Z_1}{Z_3}}$. Here $\frac{Z_1}{Z_3} = \frac{24}{1.5} = 16$ giving $a = 3.75$.

$T = 1$ for $kL = \pi/2$. For $T = 1/2$ we have $\cos kL = \pm 2/a$, giving $kL = 1\,\mathrm{rad}$ ($f_1 = 2f_0/\pi$), or $kL = \pi - 1$, giving $f_2 = 2(\pi - 1)f_0/\pi$. From this $\Delta f = f_2 - f_1 = 0.72 f_0$. Figure 1.32 shows the transmission as a function of frequency.

Note. Without the impedance transforming layer, only 22 % of the power is transmitted into the water, since $R = (Z_1 - Z_3)^2/(Z_1 + Z_3)^2 = 0.78$, and $T = 0.22$.

1.7 Array of sources. Consider N plane waves with equal amplitude and frequency and with parallel wavefronts, emitted by sources in a line separated by the same distance d, and linearly superimposed. What is the total amplitude A of this disturbance?

Solution. The phase difference between one wave and the previous one is the constant $\phi = \omega d/c$, so the sum has the form

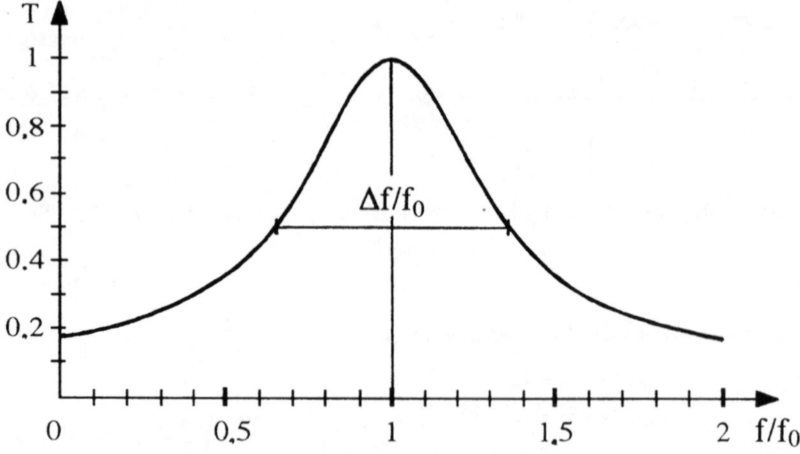

Fig. 1.32. Variation of transmission factor T with frequency. The 3 dB pass-band width ($T/T\,\mathrm{max} = 1/2$) is 0.72 f_0

$$\sum_{n=0}^{n=N-1} a\,\mathrm{e}^{-in\phi} \quad \text{giving} \quad A = Na\frac{\sin N\phi/2}{N\sin\phi/2}\,.$$

This classic result applies for an optical grating, and also appears in the simple analysis of an interdigital transducer (Chap. 2 of Vol. 2). A limiting form $(\sin x)/x$ appears when the number of sources is infinite and the amplitudes and phase changes are infinitesimal, but the total length of the source region remains finite. This corresponds to optical diffraction by a slit, observed at infinity.

1.8 Mirage effect. A variation of velocity in a fluid (due for example to a temperature gradient) causes curvature of the acoustic rays. Using the Snell–Descartes law, express this curvature as a function of the velocity gradient. In air, the velocity decreases with height above the ground. What form does the acoustic ray take if the velocity is a linear function of altitude?

Assuming the air to be a perfect gas, express the curvature as a function of temperature gradient. Use the information that, in clear weather, the air temperature decreases with height by $\Delta T = 1\,K$ for $\Delta z = 100\,\mathrm{m}$, taking the surface temperature as $T_0 \approx 300\,\mathrm{K}$ and $\theta_0 = \pi/2$. Calculate the radius of curvature. What height h does the ray reach to cover a distance $l = 1000\,\mathrm{m}$? How is this mirage effect expressed in optics?

Solution. As shown in Fig. 1.33a, the continuous refraction of an acoustic ray is given by

$$\frac{\sin\theta(z)}{c(z)} = \frac{\sin\theta_0}{c_0} = \text{constant} \quad \Rightarrow \quad \cos\theta\cdot\frac{\mathrm{d}\theta}{\mathrm{d}z} = \frac{\sin\theta_0}{c_0}\cdot\frac{\mathrm{d}c}{\mathrm{d}z}\,.$$

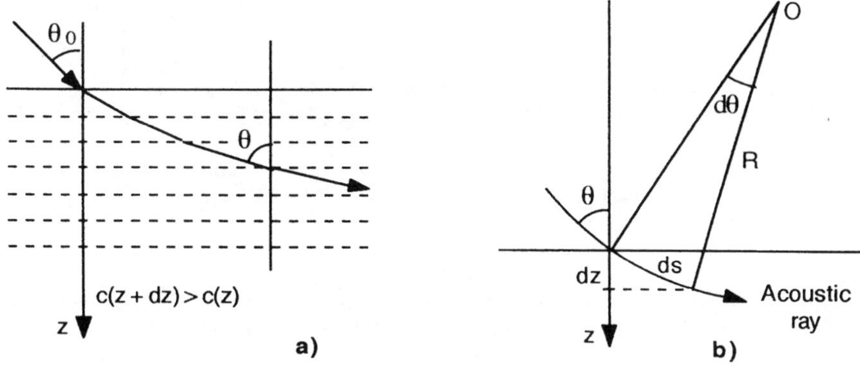

Fig. 1.33. A continuous velocity variation (**a**) causes a curvature of the acoustic rays (**b**)

From Fig. 1.33b the radius of curvature R of the ray is given by

$$\cos\theta = \frac{\mathrm{d}z}{\mathrm{d}s} = \frac{\mathrm{d}z}{R\mathrm{d}\theta} \quad \Rightarrow \quad \cos\theta \cdot \frac{\mathrm{d}\theta}{\mathrm{d}z} = \frac{1}{R}$$

and hence

$$\frac{1}{R} = \frac{\sin\theta_0}{c_0} \cdot \frac{\mathrm{d}c}{\mathrm{d}z} = \frac{\sin\theta(z)}{c(z)} \cdot \frac{\mathrm{d}c}{\mathrm{d}z} . \tag{1.117}$$

The curvature is proportional to the relative gradient of velocity, $G = (1/c)dc/dz$. For a linear variation of velocity with altitude we write $c = c_0(1 + az)$, so

$$\frac{\mathrm{d}c}{\mathrm{d}z} = c_0 a \quad \Rightarrow \quad \frac{1}{R} = a\sin\theta_0 = \text{constant} .$$

Thus the acoustic ray is a circular arc. For a temperature gradient in the air, taken as a perfect gas,

$$c = \sqrt{\gamma\frac{RT}{M}} .$$

This gives

$$\frac{1}{c} \cdot \frac{\mathrm{d}c}{\mathrm{d}z} = \frac{1}{2T} \cdot \frac{\mathrm{d}T}{\mathrm{d}z}, \text{ and hence } \quad \frac{1}{R} = \frac{\sin\theta}{2T} \cdot \frac{\mathrm{d}T}{\mathrm{d}z} .$$

The numerical values give $R = 60\,\mathrm{km}$. For a distance l we have $\Delta z = (l\tan\theta)/2$. If $l \ll R$, then $\tan\theta \approx l/R$ and $\Delta z \approx l^2/2R$. For $l = 1000\,\mathrm{m}$ then $\Delta z \approx 8\,\mathrm{m}$. An observer at an elevated position will hear the sound better than one at ground level. Also, sounds emitted by a raised source (e.g. a bell or a loudspeaker) can be heard at large distances.

In optics, the index n is defined as the ratio of velocities, so that $n = c_0/c$. The curvature can be expressed in terms of the gradient of the index, giving

$$\frac{1}{c} \cdot \frac{\mathrm{d}c}{\mathrm{d}z} = -\frac{1}{n} \cdot \frac{\mathrm{d}n}{\mathrm{d}z} .$$

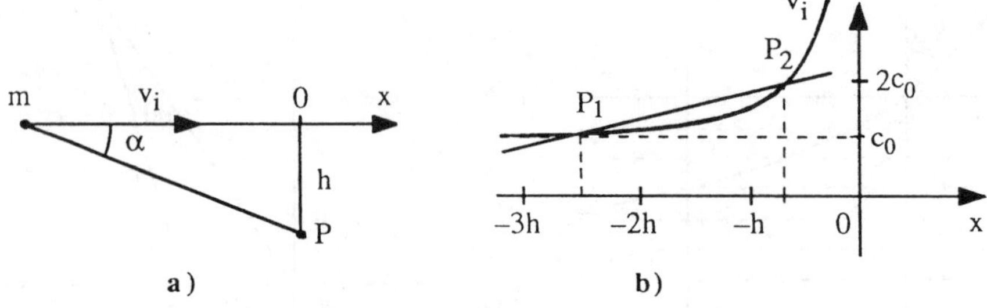

Fig. 1.34. (**a**) An object m moves to the right, in the direction Ox. (**b**) Velocity curve such that shock waves arrive simultaneously at the point P

For $\theta(z) = \pi/2$, we have $\dfrac{1}{R} = -\dfrac{1}{n} \cdot \dfrac{\mathrm{d}n}{\mathrm{d}z}$.

1.9 Shock waves. An object m in the air at height h moves to the right, with a variable velocity v_i greater than the sound velocity c_0. Hence its movement is accompanied by a rapid change of pressure which propagates at velocity c_0, this being a shock wave. How must the velocity v_i vary such that all the shock waves arrive simultaneously at a point P on the ground? Explain the double bang produced by an aircraft accelerating from an initial velocity $v < c_0$.

Solution. The required simultancity condition, Fig. 1.34a, is physically satisfied only if the object m is to the left of the origin O, where $x < 0$. From Fig. 1.34a,

$$\frac{\mathrm{d}x}{v_i} = \frac{\mathrm{d}x \cos\alpha}{c_0} = \frac{\mathrm{d}x}{c_0} \cdot \frac{-x}{\sqrt{x^2+h^2}} \quad \text{giving} \quad v_i = -c_0 \frac{\sqrt{x^2+h^2}}{x} .$$

The double bang is produced at points P_1 and P_2 determined by the intersections of the velocity curve v of the aircraft with the curve for v_i, as in Fig. 1.34b.

1.10 Doppler effect. Suppose the frequency of acoustic waves emitted by a fixed source S, measured by an observer stationary with respect to the fluid, is f_0. If c is the phase velocity of the wave, what is the frequency measured by an observer O if

(a) the observer moves with respect to the fluid with a constant velocity V_O, and the source moves with constant velocity V_S, Fig. 1.35a?
(b) the observer and source are fixed, but the waves are reflected by a mirror moving with constant velocity V_M, Fig. 1.35b?

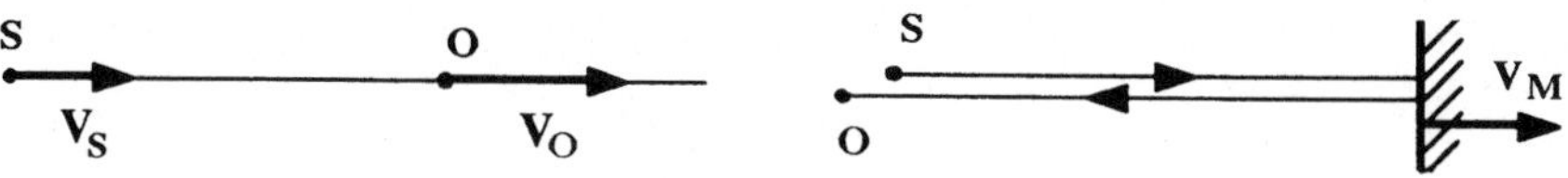

Fig. 1.35. Doppler effect

Solution.

(a) For immobile source and observer, $\lambda_0 = cT_0 = c/f_0$. When mobile, during the period T_0 (e.g. the period between two maxima), the source moves a distance $V_S T_0$, giving $\lambda = \lambda_0 - V_S T_0 = (c - V_S)/f_0$. The velocity of the maxima seen by the observer is $c - V_O$; hence the number he sees in unit time is

$$f = \frac{c - V_O}{\lambda} = \frac{c - V_O}{c - V_S} f_0 \,.$$

This formula is also valid for a light source, provided the relative velocity of the observer and source is small compared with the velocity of light. As $V_S \to c$ we have $f \to \infty$; a larger and larger number of maxima reach the observer in a short time. For $V_S = c$, they are superimposed. The energy accumulates in front of the source, giving a shock wave.

(b) Mobile mirror. The distance between two maxima of the reflected wave is $\lambda = \lambda_0 + 2V_M T_0$. They travel with velocity c toward the observer, so that the number he sees in unit time is

$$f = \frac{c}{\lambda} = \frac{c}{c + 2V_M} f_0 = \frac{f_0}{1 + 2V_M/c} \,.$$

2. Crystal Properties and Their Representation by Tensors

Elastic waves can propagate in any material medium – in a fluid (gas or liquid), as seen in the previous chapter, or in a solid, which may be homogeneous or inhomogeneous, isotropic or anisotropic. However, the amplitude decreases during propagation (Sect. 1.2.3) because the interactions between the atoms or molecules are not purely elastic. This attenuation becomes less as the medium becomes more ordered. Thus a liquid gives more loss than a solid, and an amorphous or polycrystalline solid gives more loss than a single crystal medium. Moreover, these losses increase rapidly with frequency so that liquids are hardly usable above 50 MHz, and only single crystals can be used at frequencies in the GHz region, hence their interest here. Because of the anisotropy of crystals, some particular lattice directions may be preferable for a particular wave – for example the energy propagation vector is in general parallel to the wave vector only for specific directions. Moreover, the generation of high-frequency elastic waves requires piezoelectric crystals.

This chapter therefore starts by considering the crystal structure, which is determined by the lattice and the cell contents. Crystals are classified according to their orientational symmetries, which play a role as important as the symmetries of the macroscopic physical properties. The idea of a quasicrystal is introduced. The structures of several crystals used in current applications are described.

One expects the physical and chemical properties of a crystal to depend not only on the nature of the atoms in each cell, but also on the geometrical arrangement of the cells, that is the lattice symmetry. Thus, independently of the cell contents, crystals with the same point symmetry give related behaviour for physical quantities, in corresponding orientations. Tensor analysis expresses this behaviour well, because it classifies physical quantities according to the laws for transforming their components when the axes are rotated. When a rotation corresponds to a crystal symmetry operation, the macroscopic properties are identical in the two orientations. This gives relations between the tensor components representing these properties, leading to a reduction in the number of independent components.

Accordingly, this chapter also deals with the representation of physical properties of crystals by tensors. The absence of properties represented by tensors with specific parity (evenness of rank) in various crystal classes is demonstrated. The dielectric tensor, of rank two, is chosen to illustrate the effect of the symmetry in reducing the number of independent components. The piezoelectric and elastic tensors, respectively of rank three and four, are defined and described in the following chapter.

2.1 Crystalline Structure

At the extremes, solid materials can be *amorphous or crystalline.* There are also intermediate cases, in particular quasicrystals with structures ordered over substantial, but finite, distances; and polycrystalline materials consisting of an agglomeration of randomly-oriented crystallites, so small that the material behaves isotropically on a macroscopic scale. Ordered structures are also exhibited, in one or two dimensions, by liquid crystals.

An amorphous material, such as a resin or glass, has no characteristic geometrical form. It is essentially an extremely viscous liquid which, when heated, becomes progressively more fluid. The distinction between amorphous and crystalline materials is well illustrated by the behaviour when a liquid is allowed to cool. For an amorphous material, the only visible change is a gradual increase of viscosity, and the temperature decreases smoothly with time.

In contrast, for a liquid obtained from a crystalline material, the temperature stabilizes for a time at a plateau, as in Fig. 2.1a. This corresponds to the formation of solid grains which have polyhedral forms – and these are miniature crystals. All these *polyhedra* are convex and similar; the dihedral angles between the corresponding natural faces of crystals of the same species are always the same, even though the crystals may have different appearances. For example, quartz forms polyhedra which are hexagonal prisms, terminated by a pyramid at each end, as in Fig. 2.1b. For two adjacent faces of the prism, the angle is exactly 120°; between a prism face and an adjacent pyramid face the angle is 141° 47'; and between two adjacent faces of the pyramid the angle is 133° 44'.

The crystal faces can be developed in very different forms, illustrated in Fig. 2.1c by showing a section perpendicular to the prism axis. However, the relative orientation of the faces is constant. Thus we have Romé de l'Isle's law of constant angles: *the normals to the crystal faces, drawn from a fixed point, form a geometrically invariant figure.*

The study of crystals from the same species shows that a crystalline material is *anisotropic and homogeneous.*

The well-defined orientation of the faces is one expression of anisotropy, as are other visual phenomena – the appearance of shock figures (e.g. the star shaped cracks in mica produced by a sharp point), the etch figures produced by a liquid chemically attacking particular planes preferentially (as in quartz or cadmium sulphide), and cleavage along preferred planes (mica, calcite). These effects lead R.J. Haüy to postulate, in the 18th century, that the material had an ordered structure, periodic on the atomic scale, and this was

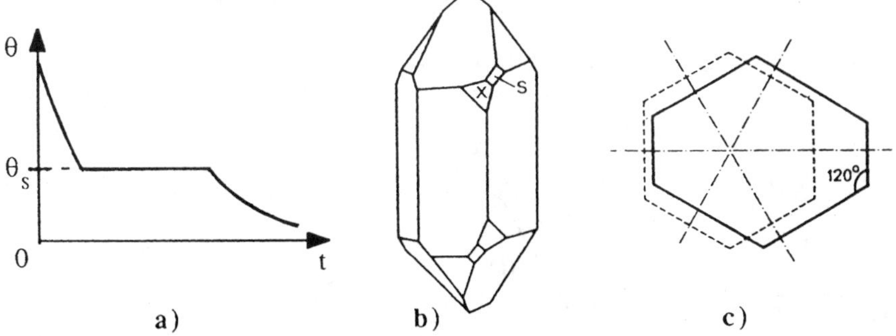

Fig. 2.1. Crystal features. (**a**) Cooling curve, $\theta(t)$. The temperature θ stabilizes while the polyhedral grains are forming. (**b**) Ideal form for a quartz crystal, showing development of all faces. Three pairs of facets s and x show that the symmetry about the prism axis is only of order 3. (**c**) Section through the prism for a practical quartz crystal. Although the shape is not a regular hexagon, the angles between its faces are 120°

verified in 1912 by X-ray diffraction (Von Laue, Friedrich). This was later confirmed by direct observation using instruments such as the electron microscope or the simple point microscope. The latter, which appeared in the 1980s [2.1], has an extremely fine point maintained very close to the surface (within 50 Å). There are several types using different interactions, such as the tunnel effect, atomic forces or capacitance. It is also possible to use evanescent waves, which might for example be optical, giving rise to 'near field' microscopy based on the principle demonstrated in 1972 using microwaves [2.2]. Some of these instruments are even able to manipulate the atoms, as well as observe them [2.3].

X-ray diffraction, cleavage and orientation are properties discontinuous with direction, and are characteristic of the crystalline state. While most crystal properties are in fact continuous, they are still subject to the anisotropy of the crystal. This applies for mechanical properties, which are our primary concern here. Note in particular that *the velocity of acoustic disturbances varies with direction.*

Crystalline materials are homogeneous, as shown by the fact that different macroscopic samples, with the same dimensions and crystalline orientation, behave identically. At the atomic scale, where the medium is discontinuous, homogeneity remains in the sense described by Bravais – there are in the crystal three distinct directions each having an infinity of discrete points which are equivalent to any one point, that is, they possess the same environment.

The external appearance and the *macroscopic* properties of crystals suggest their classification according to the symmetry shown by the normals to the natural faces, known as the *point group.* To enumerate these symmetry classes we need to know the atomic arrangement, which governs the macroscopic behaviour.

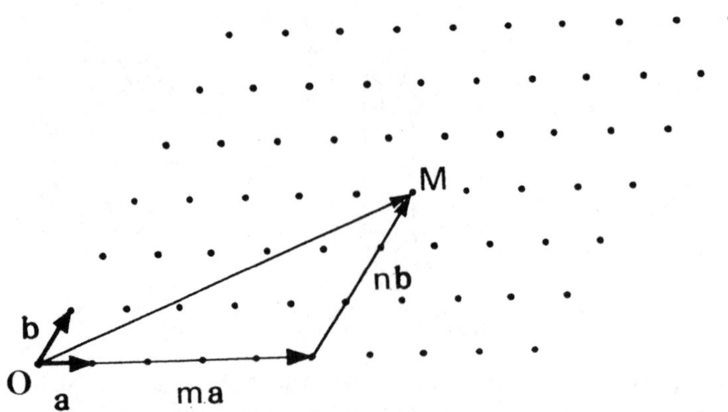

Fig. 2.2. A crystalline plane lattice. All nodes are equivalent – they have the same environment as the origin node O. A typical node position is given by $\boldsymbol{OM} = m\boldsymbol{a} + n\boldsymbol{b}$

2.1.1 Lattices, Rows, Lattice Planes and Cells

The crystalline medium is characterized by an infinity of geometrical points, each equivalent to any point O in the crystal. All of these equivalent points have the same atomic environment, and they can be deduced from one another by means of a succession of elementary translations along three vectors $\boldsymbol{a}, \boldsymbol{b}, \boldsymbol{c}$. Any point equivalent to the origin O has position given by

$$\boldsymbol{OM} = m\boldsymbol{a} + n\boldsymbol{b} + p\boldsymbol{c}\,,$$

where m, n and p are integers.

The set of all these points, called *nodes*, forms a three-dimensional *lattice* which expresses the periodicity of the crystal in all directions. Figure 2.2 represents a plane lattice, showing the node M given by $\boldsymbol{OM} = m\boldsymbol{a}+n\boldsymbol{b}$. The lattice can be decomposed into one-dimensional units (rows), two-dimensional units (lattice planes) or three-dimensional units (cells).

Lattice rows. The lattice nodes are located at the intersections of three families of parallel lines. Figure 2.3 shows, for a plane lattice, the decomposition into two families of lines. Along each of these lines, called rows, the nodes are equidistant. Three relatively prime integers u, v, w define the row $[u, v, w]$ by

$$R_{u,v,w} = u\boldsymbol{a} + v\boldsymbol{b} + w\boldsymbol{c}$$

that is, by *the vector joining one node to the one immediately following.* Thus, in Fig. 2.3 ($w = 0$), the rows of family D_3 are denoted $[2, \bar{1}, 0]$, where $\bar{1}$ signifies -1 (and is pronounced 'bar one').

Examination of this figure shows that the linear density of nodes (the number per unit length) varies with direction – it is greatest for rows with small indices such as $D_1[1, 0, 0]$ or $D_2[0, 1, 0]$, and smaller for rows with larger

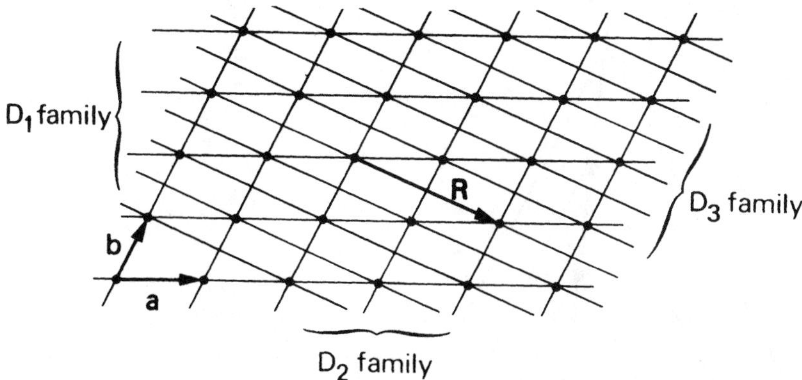

Fig. 2.3. Lattice rows. The nodes in a plane lattice are situated at the intersections of any two of the families of lines D_1, D_2, D_3, ... For family D_3 the node spacing is $\boldsymbol{R} = 2\boldsymbol{a} - \boldsymbol{b}$

indices such as $D_3[2, \bar{1}, 0]$. For a given crystal, with a fixed number of nodes per unit volume, the direction of greatest density corresponds to the largest gap between neighbouring rows. The cohesive forces are larger along the rows with small indices, since the atoms are closer. In some crystals, the difference of cohesion along one row compared with that between neighbouring rows is so large that the material becomes fibrous, as in the case of asbestos.

Lattice planes. It is possible to allocate all the nodes of a lattice to a family of *parallel equidistant planes*, called lattice planes, which can be deduced from one another by elementary translation along any row not parallel to the planes, as shown in Fig. 2.4.

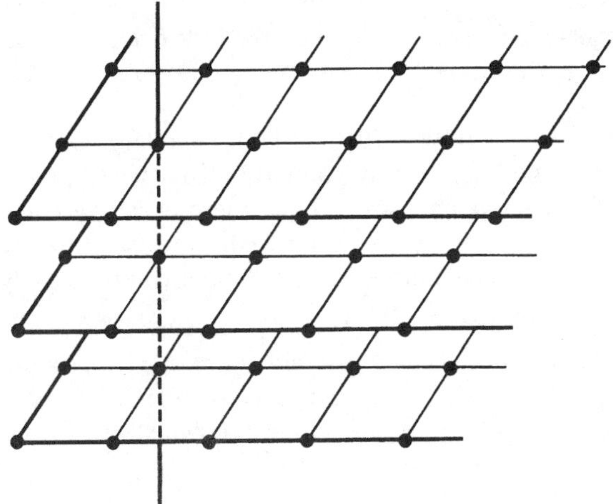

Fig. 2.4. Family of lattice planes

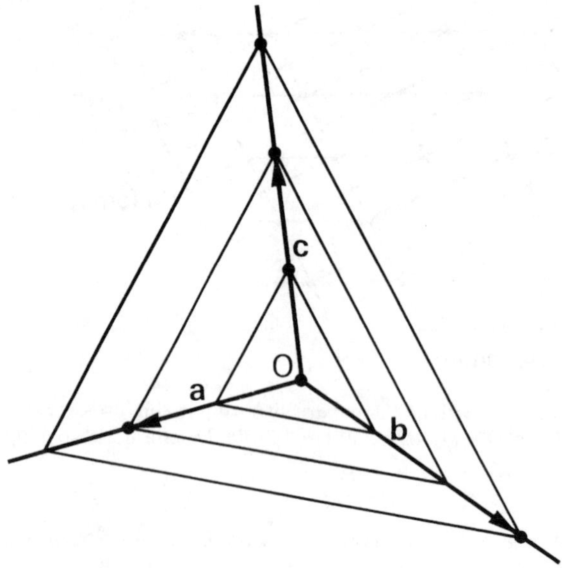

Fig. 2.5. Miller indices. For the case shown, the spacing of two neighbouring planes has vector components $a/2$ along $\boldsymbol{a}$, $b/3$ along $\boldsymbol{b}$, and $c/2$ along $\boldsymbol{c}$. The Miller indices for this family of planes are 2, 3, 2

Since there is a node at the origin and at the end of each of the vectors $\boldsymbol{a}$, $\boldsymbol{b}$, $\boldsymbol{c}$, a plane must pass through each of these four points. In general, other planes will be interlaced between these, such that *each basis vector is divided into equal segments.* If the spacing between adjacent planes is a/h along $\boldsymbol{a}$, b/k along $\boldsymbol{b}$, and c/l along $\boldsymbol{c}$, the integers h, k, l are the *Miller indices* for this family of lattice planes, as in Fig. 2.5. The indices are written (h, k, l), using when necessary an over-bar to indicate a negative index, for example $\bar{k}$. A zero index indicates that the planes are parallel to the corresponding axis.

Planes with small indices, such as (100), (010), (001), have large separations, and since the number of nodes per unit volume is fixed these planes have the largest number of nodes per unit area. Thus there is strong cohesion within these planes and weaker forces between them, and so they are potentially cleavage planes. Also, such planes reflect X-rays comparatively strongly.

Cells. The lattice can also be considered as an assembly of *parallelepipeds*, all identical to the one constructed from the basis vectors $\boldsymbol{a}$, $\boldsymbol{b}$, $\boldsymbol{c}$, as in Fig. 2.6. These parallelepipeds, whose vertices are at the nodes, are called cells.

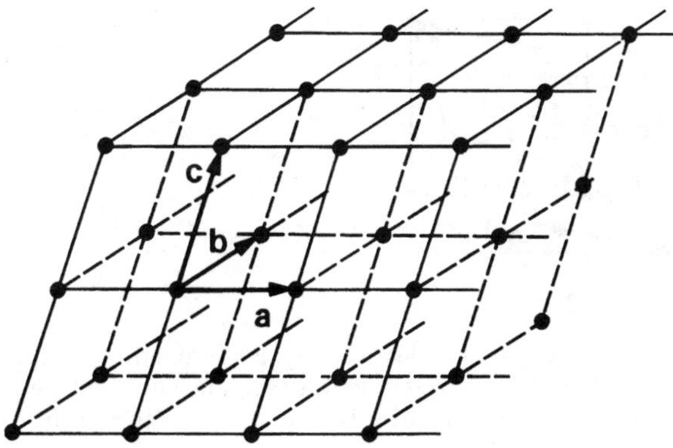

Fig. 2.6. Cells. The lattice can be regarded as an assembly of identical parallelepipeds, called cells

While the lattice of a given crystal is unique, it is possible to construct it from alternative basis vectors $\boldsymbol{a}_1$, $\boldsymbol{b}_1$, $\boldsymbol{c}_1$, and to define alternative cells, as shown for a plane lattice in Fig. 2.7.

A cell built from the three basis vectors is a *unit cell*, since it effectively contains only one node. Each of the nodes at the eight vertices is shared with seven neighbouring cells and so counts as 1/8. Consider a cell constructed with new basis vectors

$$\begin{aligned} \boldsymbol{a}' &= u_1\boldsymbol{a} + v_1\boldsymbol{b} + w_1\boldsymbol{c} \\ \boldsymbol{b}' &= u_2\boldsymbol{a} + v_2\boldsymbol{b} + w_2\boldsymbol{c} \\ \boldsymbol{c}' &= u_3\boldsymbol{a} + v_3\boldsymbol{b} + w_3\boldsymbol{c}\,. \end{aligned}$$

Its volume V' is given by the magnitude of the scalar triple product

$$V' = (\boldsymbol{a}',\ \boldsymbol{b}',\ \boldsymbol{c}')\,,$$

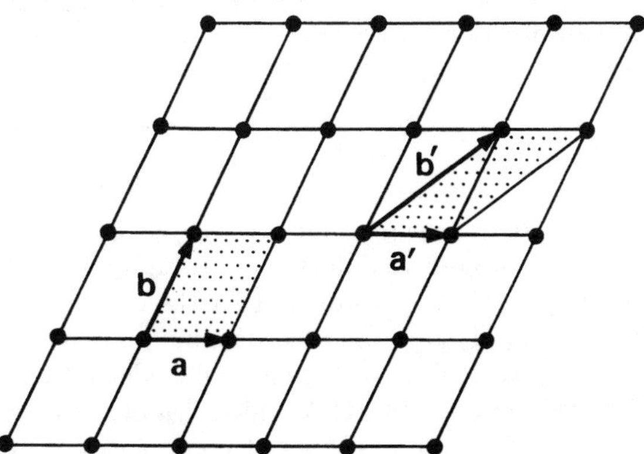

Fig. 2.7. Different cells, i.e. cells constructed using different basis vectors, can define the same plane lattice

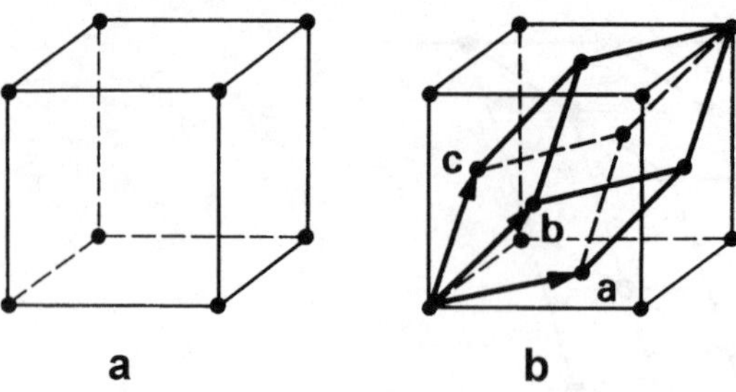

Fig. 2.8. Face-centred cubic (f.c.c.) lattice. The lattice, (**b**), is obtained by adding a node at the centre of each face of the primitive cubic cell (**a**). The cubic cell thus becomes quadruple. The unit cell is a rhombohedron based on the vectors $\boldsymbol{a}$, $\boldsymbol{b}$, $\boldsymbol{c}$

so that

$$(\boldsymbol{a}', \boldsymbol{b}', \boldsymbol{c}') = \begin{vmatrix} u_1 & v_1 & w_1 \\ u_2 & v_2 & w_2 \\ u_3 & v_3 & w_3 \end{vmatrix} (\boldsymbol{a}, \boldsymbol{b}, \boldsymbol{c}) \quad \text{giving} \quad V' = pV ,$$

where $(\boldsymbol{a}, \boldsymbol{b}, \boldsymbol{c}) \equiv \boldsymbol{a} \cdot (\boldsymbol{b} \times \boldsymbol{c})$, $V = |(\boldsymbol{a}, \boldsymbol{b}, \boldsymbol{c})|$ is the volume of a unit cell (the volume per node), and p is the magnitude of the determinant. The cell considered here is a *multiple* cell, of order p, since it contains p nodes.

A lattice is often defined in terms of multiple cells, with higher symmetry than the simplest cells. This applies for a face-centred cubic (f.c.c.) lattice, obtained by placing a node at the centre of each of the six faces of a primitive cubic lattice, as in Fig. 2.8a. We leave it to the reader to verify that this does indeed produce a lattice, that is, the environment of each node, whether at a vertex of the cube or a face centre, is the same. The cubic cell is quadruple because the six nodes at face centres count as 1/2 (they belong to two cells), so that the total number of nodes per cell is $6/2+8/8 = 4$. Considering three faces intersecting at a vertex, the unit cell, a rhombohedron, is constructed from the vectors joining the vertex to the centres of these faces, as in Fig. 2.8b.

We shall encounter other examples of multiple cells in Sect. 2.2.3, devoted to enumeration of the lattices.

2.1.2 Atomic Structure

The atomic structure of a crystal is determined unambiguously by the lattice and the *atomic group* assigned to each node, as illustrated in Fig. 2.9.

The simplest atomic group is just a single atom, so that the crystal consists of an atom at each lattice node. This is the case for numerous metals with face-centred cubic lattices, for example copper (with cube edge 3.61 Å), silver (4.071 Å), gold (4.070 Å), aluminium (4.041 Å), nickel, platinum etc. The cubic cell contains four atoms, for example those at coordinates 0 0 0, 1/2 1/2 0, 1/2 0 1/2, 0 1/2 1/2, taking the cube edge as the unit of length; other atoms belong to adjacent cells. Figure 2.10a shows this structure as seen

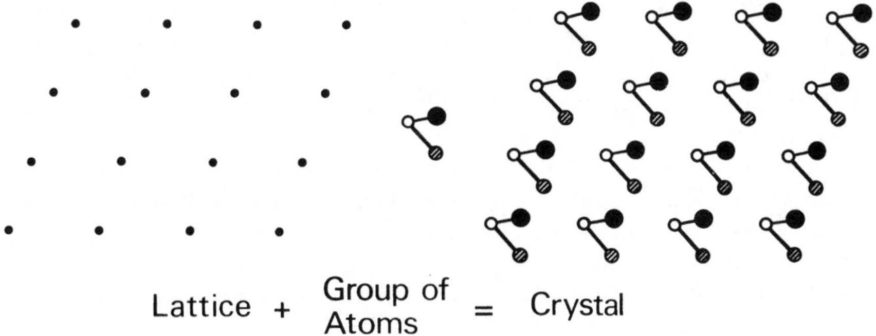

Fig. 2.9. Atomic structure of a crystal. The crystal structure is formed by assigning a group of atoms to each lattice node

from above. The number of lines in each circle specifies the vertical position of the atom, each line corresponding to 1/4 of the cube edge. Other metals, such as lithium, sodium, potassium or chromium, have an atom at each node of a body-centred cubic lattice, where a node is added at the centre of the cell of the primitive lattice. Here each cell has two atoms (nodes) with reduced coordinates 0 0 0 and 1/2 1/2 1/2, as in Fig. 2.10b.

Some monatomic crystals, such as germanium and silicon, have a more complicated structure known as the diamond type. The lattice for these three cases is face-centred cubic, but the quadruple cell contains eight atoms at positions 0 0 0, 1/2 1/2 0, 1/2 0 1/2, 0 1/2 1/2, 1/4 1/4 1/4, 3/4 3/4 1/4,

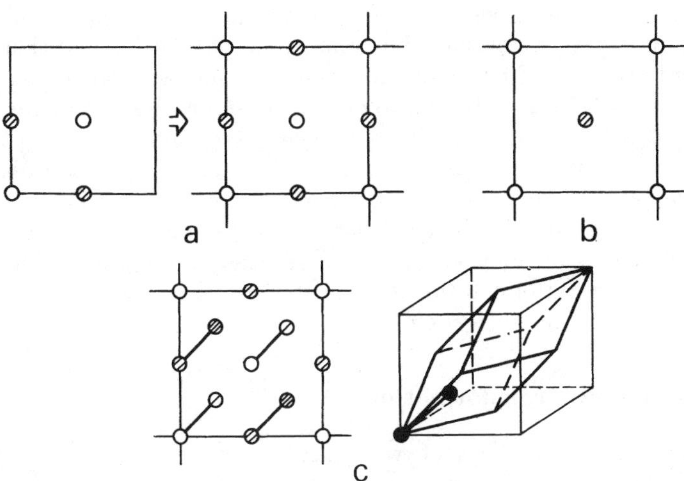

Fig. 2.10. Cubic structures. (**a**) face centred, (**b**) body centred, (**c**) face centred using a di-atomic group (as in diamond), defined either by a cubic quadruple cell or by a unit rhombohedral cell

3/4 1/4 3/4, 1/4 3/4 3/4, as in Fig. 2.10c. The four last positions can be obtained from the first four by a translation along a quarter of the diagonal. The two atoms of the atomic group can be taken as those at 0 0 0 and 1/4 1/4 1/4. This result can also be found by considering the rhombohedral unit cell containing two atoms, shown in Fig. 2.10c.

If the two atoms in the atomic group are different, the structure is that of zinc blende (ZnS) or gallium arsenide (GaAs). Another diatomic example with a simple cubic lattice is cesium chloride (CsCl), where the Cl^- ions are at the vertices of the cubes and the Cs^+ ions are at cube centres; the atomic group consists of the chlorine ion at 0 0 0 and the cesium ion at 0 1/2 1/2.

2.2 Point Groups of Crystals

The periodicity of a crystalline material is demonstrated, for example, by observations using a tunnel microscope or by X-ray diffraction measurements related to the scale of the cell size (a few Ångströms). From a *macroscopic* point of view the basic translations of the lattice are infinitesimal, so that the periodicity has no direct influence; the crystal appears to be continuous, and *only the anisotropy is relevant.* The anisotropy is limited in the sense that for one arbitrary direction there can be other equivalent directions, along which the physical properties are identical. Thus we can have a symmetry operation which leaves the crystal in a state macroscopically indistinguishable from the original. The set of all these symmetry operations for a crystal is called its *point group*, since translations are excluded here and the symmetry operations leave at least one point fixed.

In this section, we define the elements of a point group and establish some useful relations, and we then consider the point group of a lattice and that of a crystal. The lattice periodicity limits its symmetry, imposing some relations such that the lattice has to belong to one of the *seven crystal systems.* However, different lattices can have the same symmetry, and the possible ways of distributing points in space are enumerated as the fourteen Bravais lattices. *The symmetry of the crystal (lattice plus atomic group) is at most equal to that of its lattice.* It may therefore have a lower symmetry, obtained by removing particular elements from the symmetry class of the lattice. The thirty-two classes of crystal orientational symmetry are thus naturally distributed among the seven crystal systems.

2.2.1 Point Symmetry Transformations

The elements of point symmetry are of two types.

(a) A *direct* symmetry element is a direct axis of rotation. A crystal has an n-fold direct rotation axis, denoted A_n, if rotation by an angle $2\pi/n$ about this axis leaves it unchanged, n being an integer. Axes of order 2, 3, 4 and 6 are also called dyad, triad, tetrad and hexad axes respectively.

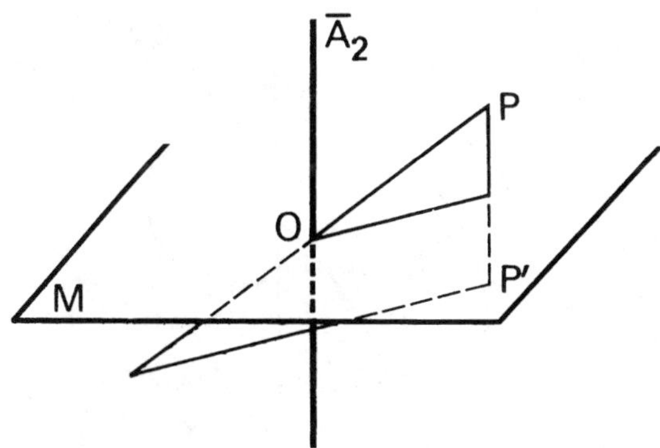

Fig. 2.11. An inverse dyad axis ($\bar{A}_2$) is equivalent to a perpendicular mirror plane

(b) An *inverse* symmetry element is either a centre of symmetry or an inverse rotation axis. The centre of symmetry, C, corresponds to the operation of symmetry about a point, also called an inversion. A crystal has an n-fold inverse rotation axis, denoted $\bar{A}_n$, if rotation by $2\pi/n$ about this axis followed by inversion through a point on the axis leaves the crystal unchanged. The sequence of these two actions can be symbolized by a dot product, and is commutative so that

$$\bar{A}_n = A_n \cdot C = C \cdot A_n .$$

Symmetry with respect to a plane is a particular case of an inverse rotation axis. Figure 2.11 shows that a *2-fold inverse axis* is equivalent to a plane of symmetry, or *mirror* plane M, perpendicular to the axis and passing through the inversion point. Thus,

$$\bar{A}_2 \equiv M .$$

Also, a centre of symmetry is the same as a 1-fold inverse axis, so that $\bar{A}_1 \equiv C$.

2.2.1.1 Stereographic Projection. To establish more complex relations between the elements of point groups it is convenient to use a planar representation, and this is commonly done using the stereographic projection which maps points on a sphere onto points on its equatorial plane (Fig. 2.12a). A point P on the northern hemisphere is represented by the point p where the line from P to the south pole S intercepts the equatorial plane E. Similarly, a point Q on the southern hemisphere is represented by q on the equatorial plane, using the north pole N. Thus, using both poles, all points on the sphere can be mapped on to points on or inside the equatorial circle. The projection conserves angles, so that two curves on the sphere intersecting at an angle α will have projections that intersect at the same angle α.

The projection of a point in the northern hemisphere or on the equator is represented by a cross, and that for a point in the southern hemisphere by a dot inside a circle. These points can represent crystal directions, defined by drawing lines from the origin O to the surface of the sphere, in the di-

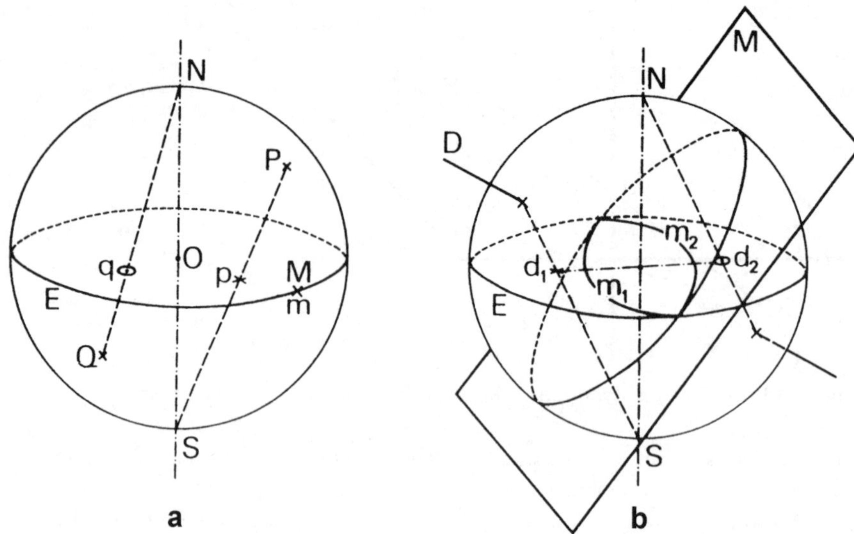

Fig. 2.12. Stereographic projection. (**a**) Triangles SOp and SPN are similar, so that $Sp/2R = R/SP$, where R is the radius of the sphere. The projection is an inversion with the property $\boldsymbol{Sp} \cdot \boldsymbol{SP} = 2R^2$. (**b**) A line D is represented by the projections d_1 and d_2 of its intersections with the sphere. A plane M is represented by the two circular arcs m_1 and m_2

rection required. A line D intersecting the sphere can be represented by the two points where it intersects the surface, as in Fig. 2.12b. A plane passing through the centre gives two circular arcs symmetric about its intersection with the equatorial plane.

In a stereogram showing the properties of a point group, the north-south axis is normally the axis with highest symmetry, and a centre of symmetry, if present, is at the centre of the sphere. Broken lines indicate geometrical constructions, while the actual projections use solid lines.

2.2.1.2 Equivalence Relations. We begin by considering relations concerning a single inverse element.

(a) *An inverse axis of odd order n is equivalent to a direct axis of order n and a centre of symmetry.* Using $(A_n)^n = I$ (the identity operation), repetition of n operations of an inverse axis gives

$$\left(\bar{A}_n\right)^n = (A_n \cdot C) \cdot (A_n \cdot C) \cdot \ldots \cdot (A_n \cdot C) = (A_n)^n \cdot C^n = C^n$$

and since n is odd ($n = 2p + 1$) this gives $C^n = C^{2p} \cdot C = C$, a simple inversion. On the other hand we have $\bar{A}_n \cdot C = A_n \cdot C^2 = A_n$. Thus an inverse axis with odd order is equivalent to a direct axis plus a centre of symmetry, so that $\bar{A}_{2p+1} \equiv A_{2p+1}C$. This is illustrated for $n = 3$ in the stereogram of Fig. 2.13a.

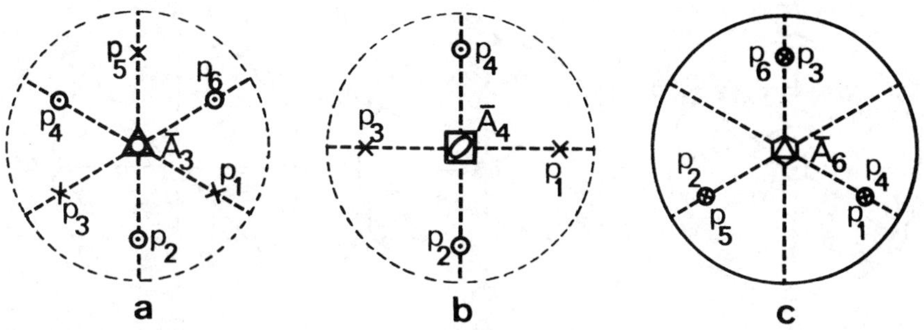

Fig. 2.13. Properties of inverse axes. The group of points p_2, p_3, ... obtained by transformation of an initial point p_1 shows that (**a**) an inverse triad axis is equivalent to a direct triad axis plus a centre of symmetry; (**b**) an inverse tetrad axis contains a direct dyad axis; (**c**) an inverse hexad axis is equivalent to a direct triad axis and a perpendicular mirror plane

Note that the notation $A_{2p+1}C$ indicates two independent elements, in contrast to the notation $\bar{A}_n \cdot C$ where the dot indicates a successive application of two operations.

(b) *An inverse axis of even order n implies a collinear direct axis of order* $n/2$. Here,

$$\left(\bar{A}_n\right)^2 = (A_n \cdot C) \cdot (A_n \cdot C) = (A_n)^2 .$$

If $n = 2p$, $(A_n)^2$ corresponds to a rotation angle $2(2\pi/2p) = 2\pi/p$, so $\left(\bar{A}_{2p}\right)^2 = A_p$. The case with $n = 4$ is shown in Fig. 2.13b.

(c) *If in addition* $p = n/2$ *is odd, there is a mirror plane perpendicular to the axis.* Since p is odd we have $C^p = C$. The operation $(\bar{A}_{2p})^p = (A_{2p})^p \cdot C^p = A_2 \cdot C = \bar{A}_2$ is identical to a perpendicular mirror plane M. Property (b) is always true, with the result that

$$\bar{A}_{2p} \equiv \frac{A_p}{M} \quad \text{if} \quad p \quad \text{is odd} .$$

The bar of the fraction indicates that the mirror is perpendicular to the axis. This result is illustrated for $n = 6$ on the stereogram of Fig. 2.13c. In the figures, the symmetry axes are represented by the following symbols:

(i) Direct axes, of order 2: (ellipse), 3: (triangle), 4: (square), 6: (hexagon).

(ii) Inverse axes, of order 1: (circle), 3: (circle within triangle), 4: (ellipse within square), 6: (triangle within hexagon). These symbols take account of the above results. Inverse axes are indicated by an over-bar.

If we add the inverse dyad axis, represented by the perpendicular mirror plane equivalent to it, these symbols are those used for the symmetry elements of crystals, see Sect. 2.2.2.

The simultaneous presence of several symmetry elements often implies other elements, as shown by the following examples.

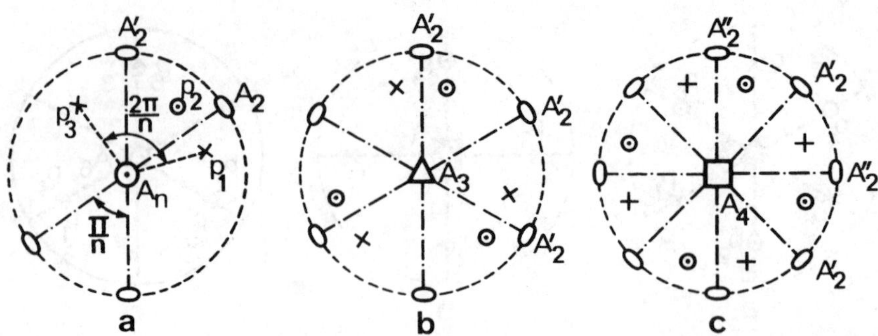

Fig. 2.14. Combination of a dyad axis with a perpendicular n-fold axis. (**a**) This combination implies that there are n dyad axes. (**b**) For odd n, all the dyad axes are equivalent. (**c**) For even n, the dyad axes are equivalent alternately

(d) *A direct axis of even order plus a centre of symmetry implies the existence of a mirror plane perpendicular to the axis.* This results from the operation $(A_{2p})^p \cdot C = A_2 \cdot C = \bar{A}_2 \equiv M$, and is written

$$A_{2p}C \quad \rightarrow \quad \frac{A_{2p}}{M}C\,.$$

(e) *A direct axis of order n plus a perpendicular binary axis implies n perpendicular binary axes with angles π/n between them.* Suppose that p_2 and p_3 are the transforms of the original point p_1, rotating respectively about the A_2 axis and by an angle $2\pi/n$ about the n-fold axis (Fig. 2.14a). Here p_3 could also be obtained by rotation about another binary axis A_2' making an angle π/n with A_2. Applying the rotation of A_n repeatedly shows the presence of n binary axes.
If n is *odd* ($n = 2p + 1$), the operation of the n-fold axis is sufficient to generate all the dyad axes starting from one of them. This follows by considering $(A_n)^p$, which has angle $p(2\pi/n) = \pi - \pi/n$. This generates a new dyad at an angle π/n, since the angle π simply gives the opposite direction. All these *equivalent* dyad axes form a set which can be represented by the symbol A_2':

$$A_{2p+1}A_2' \quad \rightarrow \quad A_{2p+1}(2p+1)A_2'\,.$$

Conversely, if n is *even* ($n = 2p$) we can consider $(A_n)^{p-1}$, giving an angle $(p-1)(2\pi/n) = \pi - 2\pi/n$, which is a multiple of $2\pi/n$. Consequently, this only generates half of the dyad axes. Hence, the dyads are only equivalent alternately and can be regarded as forming two groups, each with p elements:

$$A_{2p}A_2 \quad \rightarrow \quad A_{2p}\,pA_2'\,pA_2''\,.$$

Figures 2.14b and c illustrate the distinction for the cases $n = 3$ and 4.

(f) *A direct n-fold axis plus a mirror plane passing through this axis implies the presence of n mirrors, with angles π/n between them.*

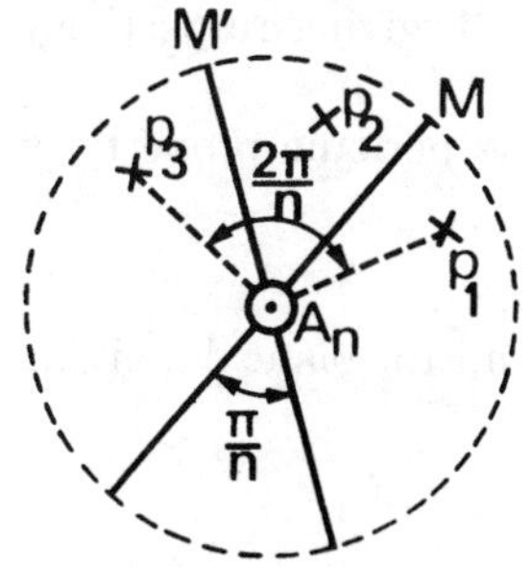

Fig. 2.15. A direct n-fold axis contained in a mirror plane implies the presence of n mirrors

This is analogous to the previous theorem, and is shown in Fig. 2.15. If n is odd, all the mirror planes are equivalent, so that

$$A_{2p+1}M' \quad \rightarrow \quad A_{2p+1}(2p+1)M' .$$

If n is even, the mirror planes are of two groups M' and M'', so that

$$A_{2p}M' \quad \rightarrow \quad A_{2p}pM'pM'' .$$

(g) *An inverse n-fold axis with n odd, plus a perpendicular dyad axis, leads to n equivalent dyad axes with n mirror planes perpendicular to them.* From properties (a), (e), and (d) above, we have

$$\bar{A}_{2p+1}A_2 \equiv A_{2p+1}CA_2' \quad \rightarrow \quad A_{2p+1}\frac{(2p+1)A_2'}{(2p+1)M'}C .$$

2.2.2 Lattice Point Groups and the Seven Crystal Systems

The lattice periodicity implies the following symmetry properties.

(a) *Any straight line parallel to an n-fold axis of the lattice, and containing a node, is itself an n-fold axis.*
(b) *Any symmetry axis passing through a node is a row of the lattice.*
(c) *Any node of the lattice is a centre of symmetry.*
(d) *A symmetry axis with order n greater than two implies the presence of n perpendicular dyad axes.*
(e) *The direct or inverse axes of a lattice can only be of order 1, 2, 3, 4 or 6.*

We derive the last property. Any translation $\boldsymbol{T}$ of the lattice is transformed through any rotation α to another translation $\boldsymbol{T}'$. Using the lattice basis vectors as units of length, the components of the vectors $\boldsymbol{T}$ and $\boldsymbol{T}'$ must be integers, and so must the coefficients of the rotation matrix. The trace of a matrix remains constant on rotation of axes (Sect. 2.5.1 below), so the trace of the matrix representing α must be an integer, giving

$$2\cos\left(\frac{2\pi}{n}\right) + 1 = p .$$

Here the integer p can only take values -1, 0, 1, 2, 3, giving respectively $n = 2$, 3, 4, 6, 1.

The classes of point symmetry consistent with the periodic property of the lattice are listed as follows.

(a) The simplest class has only a centre of symmetry, C.
(b) Adding a dyad axis to C implies a perpendicular mirror plane M, giving the symmetry

$$\frac{A_2}{M}C\,.$$

(c) If there is another dyad axis, this implies that there are three dyad axes at right angles, A_2, A_2', A_2'', and the centre of symmetry implies that there are three mirror planes M, M',M'' perpendicular to these axes, giving

$$\frac{A_2}{M}\frac{A_2'}{M'}\frac{A_2''}{M''}C\,.$$

(d) A lattice with a 3-fold axis A_3 must have three dyad axes A_2' and three mirror planes M' because of its centre of symmetry, giving

$$A_3\frac{3A_2'}{3M'}C\,.$$

(e) If the lattice has a 4-fold axis A_4, the dyad axes are equivalent only on an alternate basis, so the symmetry is

$$\frac{A_4}{M}\frac{2A_2'}{2M'}\frac{2A_2''}{2M''}C\,.$$

(f) With a 6-fold axis A_6, the dyad axes are again grouped on an alternate basis, giving

$$\frac{A_6}{M}\frac{3A_2'}{3M'}\frac{3A_2''}{3M''}C\,.$$

The six cases above include all the symmetry classes possible for a lattice that has only one axis of order greater than two.

(g) It can be shown that the only class containing more than one axis with order greater than 2 gives the complete cubic symmetry, Figure 2.16, which has
   (i) four triad axes A_3 directed along the diagonals of the cube;
   (ii) three tetrad axes A_4 perpendicular to the faces;
   (iii) six dyad axes A_2' joining the centres of opposite edges; and
   (iv) a centre of symmetry, which implies mirror planes perpendicular to the A_4 and A_2' axes.

Hence the symmetry is

$$\frac{3A_4}{3M}4A_3\frac{6A_2'}{6M'}C\,.$$

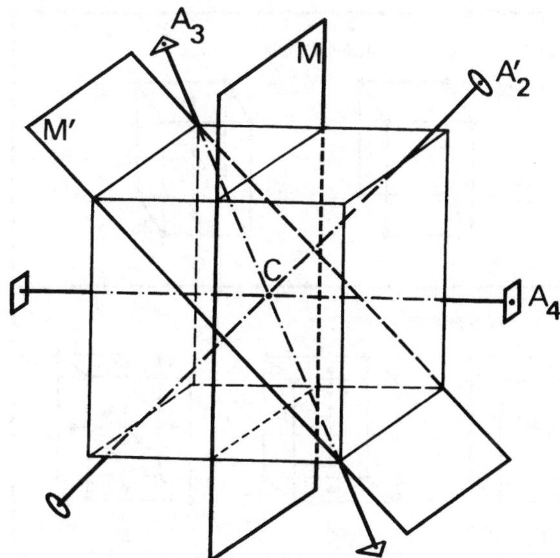

Fig. 2.16. The symmetry elements of a cube are $\frac{3A_4}{3M}4A_3\frac{6A_2'}{6M'}C$

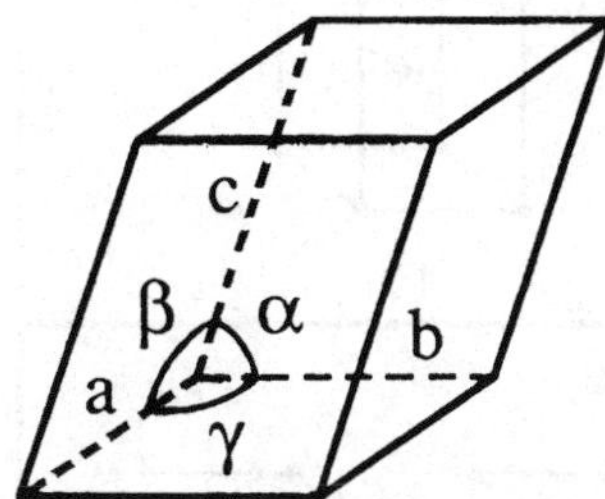

Fig. 2.17. The general parallelepiped defined by α, β, γ, a, b, c

In classifying crystals according to the point symmetry of the lattice, we define the *seven crystal systems. The lattices of all crystals belonging to the same crystal system have the same point symmetry*, that of one of the classes above. A crystal system is thus characterized by the *geometrical form of the cell*, which by repetition generates all lattices with the same symmetry. These forms vary from the most general parallelepiped (triclinic system, Fig. 2.17) to the cube (cubic system). They are as follows:

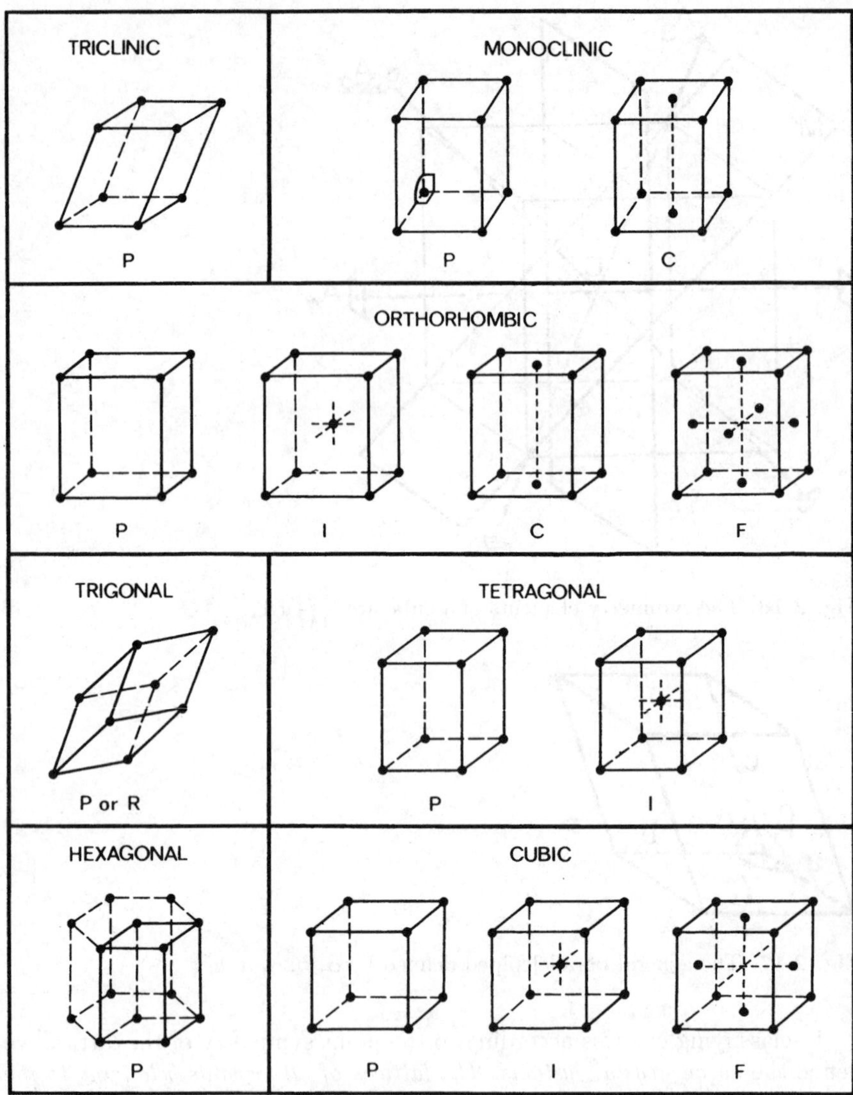

Fig. 2.18. The fourteen Bravais lattices

(1) Triclinic system	$\alpha \neq \beta \neq \gamma \neq 90°$,	$a \neq b \neq c$
(2) Monoclinic system	$\alpha = \gamma = 90°, \beta > 90°$,	$a \neq b \neq c$
(3) Orthorhombic system	$\alpha = \beta = \gamma = 90°$,	$a \neq b \neq c$
(4) Trigonal (or rhombohedral) system	$\alpha = \beta = \gamma \neq 90°$,	$a = b = c$
(5) Tetragonal (or quadratic) system	$\alpha = \beta = \gamma = 90°$,	$a = b \neq c$
(6) Hexagonal system	$\alpha = \beta = 90°, \gamma = 120°$,	$a = b \neq c$
(7) Cubic system	$\alpha = \beta = \gamma = 90°$,	$a = b = c$.

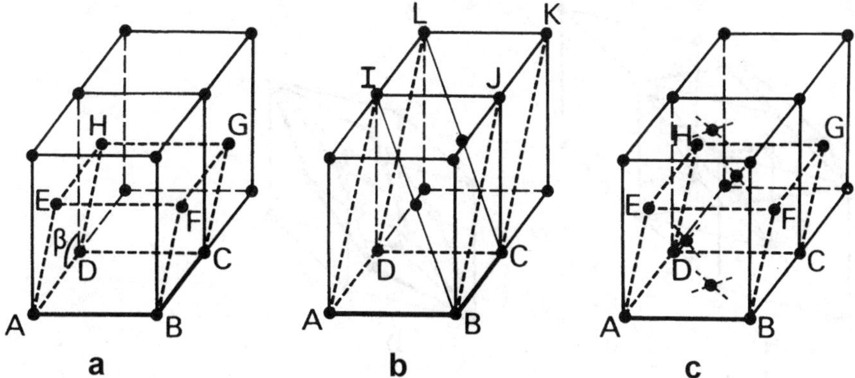

Fig. 2.19. Monoclinic system. (**a**) The lattice with oblique faces centred is a P lattice with cell ABCDEFGH. (**b**) The I lattice is a C lattice with cell ABCDIJKL. (**c**) The F lattice is a C lattice with cell ABCDEFGH

The fourteen Bravais lattices. In general, there can be several different lattices with the same point symmetry, so that crystals in the same system may have different lattices. In addition to the above *primitive* (P) lattices, which have as unit cell one of the seven parallelepipeds described, other lattices can be obtained by adding extra nodes to the cell. By definition, the environment for each node has to be the same, and this restricts the possible places for extra nodes to be at the centre of the cell (giving a body-centred lattice, denoted I), or at the centres of all faces (face-centred lattice, F), or at the centres of two opposite faces (another face-centred lattice, C). However, not all these additions can be made in each system, because some cases lead to the same type of lattice but with a modified cell.

Figure 2.18 shows the *fourteen Bravais lattices*, that is, *all the possible ways of distributing an infinity of points in space such that they all have the same environment*. The three lattices I, C, F derived from the primitive lattice are not always distinct; for example, applying any of these to the triclinic system simply gives another primitive lattice but with a smaller cell. For the monoclinic system, the only distinct cases are the primitive lattice (P), and the C lattice with rectangular faces centred. Centring the oblique faces gives another primitive lattice P with a smaller angle β (Fig. 2.19a), while the I and F lattices are equivalent to face centring of type C (Fig. 2.19b and c). For the orthorhombic system, all the lattices P, I, C, F are distinct.

The trigonal (or rhombohedral) system has only one Bravais lattice, the P lattice, sometimes called R. The I lattice is an R lattice with a smaller cell (Fig. 2.20a). A C lattice is not possible because all six faces are equivalent, and on centring all of them the result is another R lattice with a smaller cell (Fig. 2.20b).

The tetragonal lattice is of type P or I. Since the four rectangular faces are equivalent, the only possible C lattice has nodes centring the two square bases,

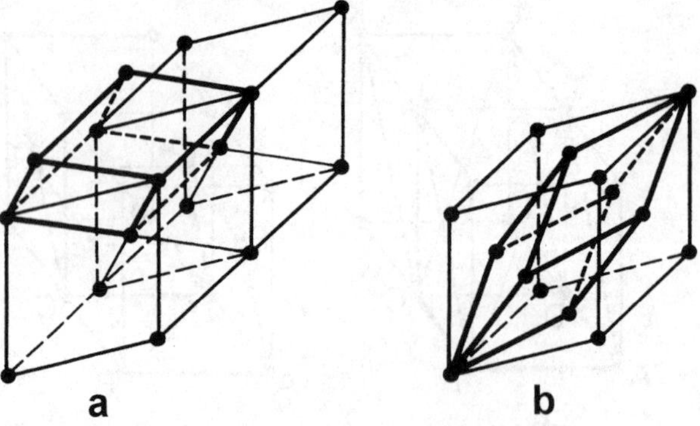

Fig. 2.20a,b. Trigonal system. There is only one lattice

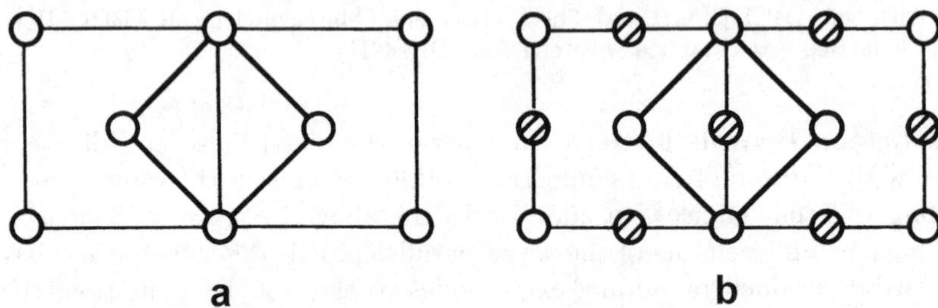

Fig. 2.21. Tetragonal system. (**a**) The C lattice is a P lattice with a smaller cell. (**b**) The F lattice is an I lattice

and this is equivalent to a P-type lattice with a smaller cell, as in Fig. 2.21a. Similarly, the F lattice is an I lattice with a smaller cell, Fig. 2.21b.

The only hexagonal lattice is P-type, because the I, C, and F lattices do not have a 6-fold axis. The cubic system has P, I and F lattices, C being excluded because the equivalence of the six faces implies that if two are centred then all must be, giving F.

2.2.3 The Thirty-Two Point Symmetry Classes of Crystals

The symmetry classes of crystals are given by their point groups. The crystal structure is obtained by assigning a group of atoms to each node of the lattice, and this can affect the symmetry, so that *the crystal symmetry is at most equal to that of the lattice*. Thus, in contrast to a lattice, *a crystal does not necessarily possess a centre of symmetry*, and an axis of order n greater than 2 does not necessarily imply n dyad axes. The various possible symmetry classes of crystals can be obtained from the symmetry of the corresponding lattices by successively deleting non-essential symmetry elements. These classes are shown in the Table of Fig. 2.22, arranged according to the seven crystal systems to which they belong. Each entry gives the symmetry elements (at the top), then the stereogram, and then the point group in Hermann–Mauguin notation at the bottom. The headings in the first column give the seven lattice systems, derived in Sect. 2.2.2. A crystal with point symmetry the same as the lattice is called holosymmetric, while one with lesser symmetry is called merohedral.

The holosymmetric *triclinic* class is $\bar{1}$, and on deleting the centre of symmetry this becomes the class with no symmetry element, denoted 1.

For the *monoclinic* lattice, the symmetry $\frac{A_2}{M}C$ is denoted as class $2/m$; retaining only the dyad axis gives class 2, and the mirror plane alone gives class m.

Similarly, the *orthorhombic* holosymmetric case $\frac{A_2}{M}\frac{A_2'}{M'}\frac{A_2''}{M''}C$ is denoted mmm. Retaining the three dyad axes gives 222, while the dyad axis A_2 and the two mirror planes M', M'' which include it give $2mm$.

The *trigonal* lattice symmetry $A_3\frac{3A_2'}{3M'}C$ gives, on deleting the centre of symmetry, the classes $A_3 3A_2'$ (denoted 32) and $A_3 3M'$ (denoted $3m$). Alternatively, removing the three dyads A_2', and consequently the three mirror planes M', gives $A_3C = \bar{A}_3$ (denoted $\bar{3}$), and removing C from this gives class A_3 (denoted 3).

The *tetragonal* system has lattice symmetry $\frac{A_4}{M}\frac{2A_2'}{2M'}\frac{2A_2''}{2M''}C$, or $\frac{\bar{A}_4}{M}\frac{2A_2'}{2M'}\frac{2A_2''}{2M''}C$, and on removing the centre we have

(a) retaining the dyad axes: $A_4 2A_2' 2A_2''$ (class 422)
(b) retaining the mirror planes: $A_4 2M' 2M''$ (class $4mm$)
(c) alternately retaining a dyad and a mirror: $\bar{A}_4 2A_2' 2M''$ (class $\bar{4}2m$)
(d) on removing all these elements, the cases A_4 (class 4) and $\bar{A}_4$ (class $\bar{4}$).

The case $\frac{A_4}{M}C$, which is class $4/m$, is the only one having a centre of symmetry.

The seven classes of the *hexagonal* system are obtained in the same manner.

The *cubic* symmetry $\frac{3A_4}{3M}4A_3\frac{6A_2'}{6M'}C$ or $\frac{3\bar{A}_4}{3M}4A_3\frac{6A_2'}{6M'}C$ gives, on removing the centre, the two cases $3A_4 4A_3 6A_2'$ (class 432) and $3\bar{A}_4 4A_3 6M'$ (class $\bar{4}3m$). The lowest symmetry compatible with the four triad axes characteristic of the cubic system is $3A_2 4A_3$ (class 23). Adding the centre gives rise to mirror planes perpendicular to the dyad axes, giving $\frac{3A_2}{3M}4A_3C$, which is class $m3$.

Each crystal system is characterized by the presence of particular symmetry elements, as follows:

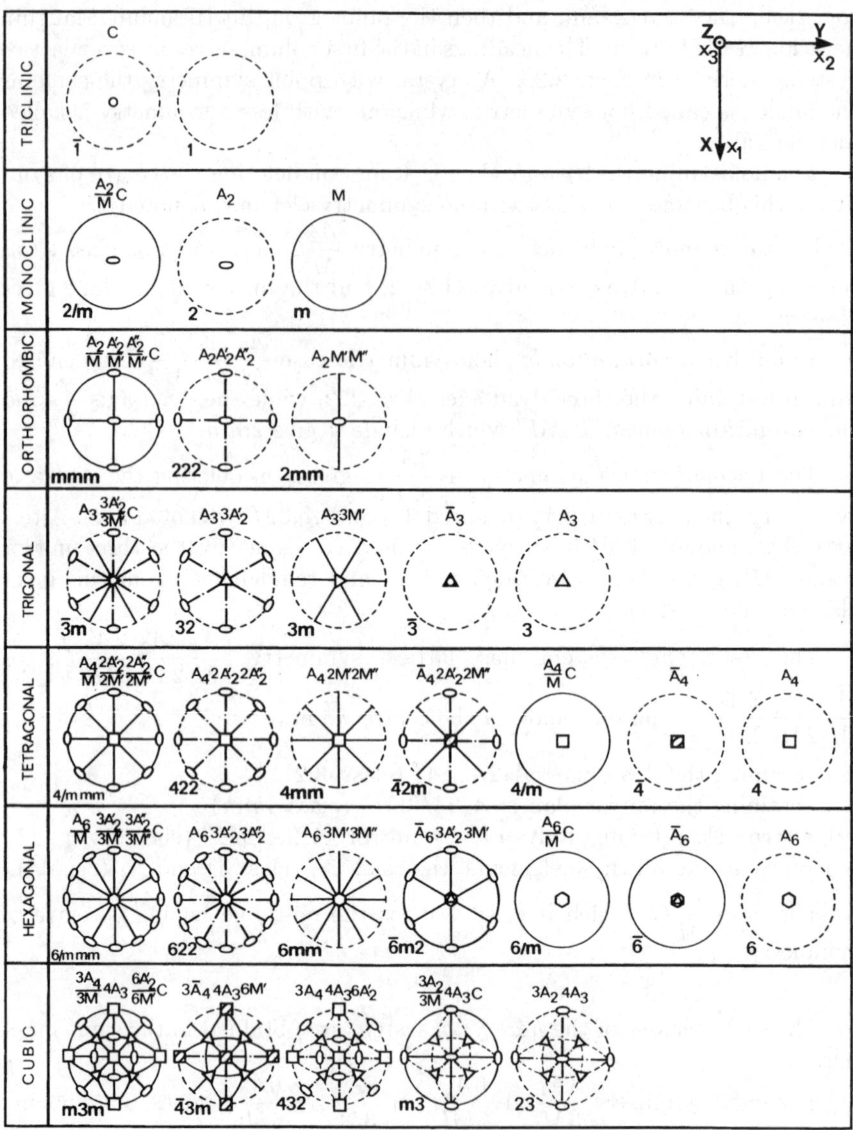

Fig. 2.22. The thirty-two point symmetry classes of crystals

triclinic:	one 1-fold axis
monoclinic:	one dyad axis
orthorhombic:	three dyad axes
trigonal:	one triad axis
tetragonal:	one tetrad axis
hexagonal:	one hexad axis
cubic:	three direct dyad axes and four direct triad axes.

Where the type of axis is not specified here, it can be direct or inverse.

The 32 possible varieties of crystal symmetry, all shown in the Table of Fig. 2.22, exist practically in either natural or artificially grown crystals.

The physical properties of crystals mostly show anisotropy, and are then *quantitatively meaningful only when related to reference axes (directions) specified in relation to the crystal symmetry.* The orthonormal reference frame is shown at the top right in Fig. 2.22. The crystallographer's notation X, Y, Z is often replaced by the notation x_1, x_2, x_3, which is more convenient for tensor calculations.

Translational symmetry. The symmetry discussed above is sufficient to describe the macroscopic properties of crystals. However, on the microscopic scale there are in general additional elements required to bring the lattice into coincidence with a previous position. These elements, not associated with points, include:

(a) a helicoidal axis $A_{n,\boldsymbol{t}}$, which gives a rotation through $2\pi/n$ followed by a translation $\boldsymbol{t}$ in the axis direction ($n = 2$, 3, 4 or 6);
(b) a glide plane $M_{\boldsymbol{t}}$, which gives symmetry about a plane followed by a translation $\boldsymbol{t}$ in the plane.

The inverse axes of order 1, 3, 4 and 6, which leave a fixed point unchanged, are not associated with a translation.

Schönflies and Fedorov have shown that crystal structures can be classified into 230 distinct groups, known as space groups.

2.3 Quasicrystals

The translational invariance of a lattice of points prohibits the existence of symmetry axes with order $n = 5$ or $n > 6$, and leads to the description of a crystal in terms of juxtaposed identical cells. These are polyhedrons which can have symmetry axes of order 2, 3, 4 or 6. These units, each containing a group of atoms (the group having any symmetry), are assembled so as to fill space completely, without any voids. This picture is confirmed by the diffraction patterns found for innumerable crystals using electrons, neutrons or X-rays – the atomic periodicity leads to preferential scattering at specific sharply-defined directions (the Bragg angles) associated with atomic planes in the crystal. The distribution of these directions is associated with the cell symmetry (the angular spacing between the 'Bragg peaks' varies inversely with cell dimensions), and shows that the axes only have the orders expected to be valid.

It was therefore surprising when, in 1984, D. Schechtman et al. [2.4] found evidence of a 'forbidden' 5-fold axis in the diffraction pattern of an aluminium-manganese alloy, produced by rapid cooling from a uniform melt. This anomaly was explained by A. Katz and M. Duneau [2.5]. At the time,

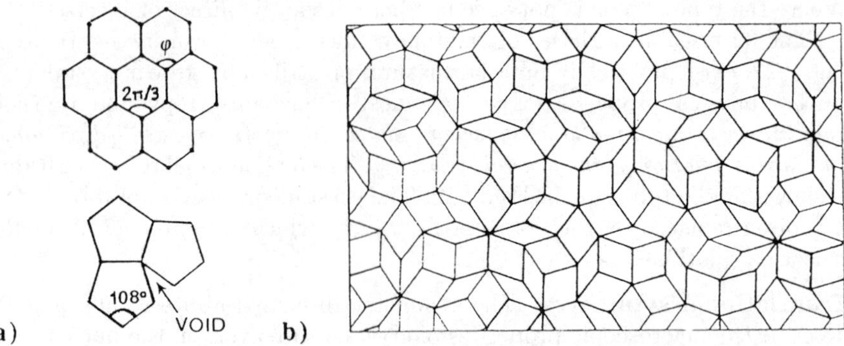

Fig. 2.23. Tiling. (**a**) A classical periodic array using juxtaposed identical polygons. The requirement for each vertex to be shared between the n polygons that surround it implies that $\varphi = 2\pi/n$ with the integer $n = 2, 3, 4$ or 6, and only allows rotation symmetries of order 2, 3, 4 or 6. A pentagon with vertex angle $\varphi = 108° = 360°/3.333\ldots$ is not acceptable. (**b**) The Penrose quasiperiodic 'lattice', regular but not periodic, is constructed from two diamonds with acute angles $\pi/5$ and $2\pi/5$. Note one of the 5-fold symmetry axes

they were concerned with a quasiperiodic 'tiling' of space (such that all features are repeated at different places, but not periodically), and in particular they sought a way of constructing the figures of Penrose [2.6]. This mathematician had tiled the infinite plane in a manner regular but not periodic, by using two diamond shapes with equal sides, with acute angles 36° ($\pi/5$) and 72° ($2\pi/5$). This geometry has a quasiperiodic appearance, with a 5-fold axis, as shown in Fig. 2.23.

Katz and Duneau showed that it was possible to tile an m-dimensional space in a quasiperiodic fashion by projecting on to it a periodic structure in an n-dimensional space (with $n > m$), and this could cause the appearance of symmetries excluded from perfect crystals. They gave the illustration of a square lattice (two-dimensional) projected onto a straight line (one-dimensional) with slope α, as in Fig. 2.24. The projection of the nodes between the main line and the broken line (which is given by sweeping the square C) gives a *quasiperiodic* array if α is an *irrational* number. The two elements a and b of the array are the projections of the sides of the square. The array is periodic if $\tan\alpha$ is a rational number.

To tile a three-dimensional space quasiperiodically, it is necessary to use a 'cubic' lattice of six dimensions. The projection of the basis hypercube is a complex polyhedron (a triacontrahedron) with icosahedral symmetry (a regular icosahedron has 20 identical equilateral triangular faces, and it has axes $6A_5$, $10A_3$, $15A_2$, giving point group 532 or, with inversion, $m\bar{3}\bar{5}$). This polyhedron is formed by a particular assembly of two rhombohedrons. The three-dimensional space is thus completely tiled by appropriate juxtaposition of these two rhombohedrons. Penrose's drawing, Fig. 2.23b, corresponds to a section of this structure in a plane normal to the 5-fold axis.

After 1984, other alloys such as AlCuFe, AlPdMn, and MnNiSi were found to give symmetries of order 5, 8, 10, 12, and were thus anomalous in comparison with perfect crystals, as in the case of AlMn studied by Schechtman et al. All such solids with quasiperiodic lattices are called quasicrystals. Their discovery and the elaboration of their lattices have lead to a clarification of the notion of crystallographic order, and have renewed interest in the concepts of *short-range order and long-range order*. The distinction can be understood by considering the example of an infinite chain of atoms (a one-dimensional crystal, Sect. 4.1). The autocorrelation function corresponding to moving the chain along itself is periodic, since any translation by a particular step a moves it into an state identical to the initial one – the atoms are exactly superimposed. This is not so for a set of atoms with random positions (an amorphous solid); these give an autocorrelation function which diminishes with displacement and tends to zero at infinity, showing that any order exists only over a short range. Now consider the question – what happens for a chain obtained from a periodic one by moving the nth atom by an amount $\sin(2\pi\beta na)$? If β is an irrational number the atoms are shifted by different amounts, so they cannot be superimposed after a finite translation. But they still have some order – their positions are given by a mathematical formula. In this case the autocorrelation function does not tend to zero, because the chain has a long-range order. This definition can be applied to three-dimensional quasiperiodic lattices, and here we note that the diffraction pattern is given by the Fourier transform of the autocorrelation function. We can conclude that a coherent diffraction pattern, consisting of a discrete set of reflections (Bragg peaks), is evidence only of *long-range* order, that is, of a lattice *ei-*

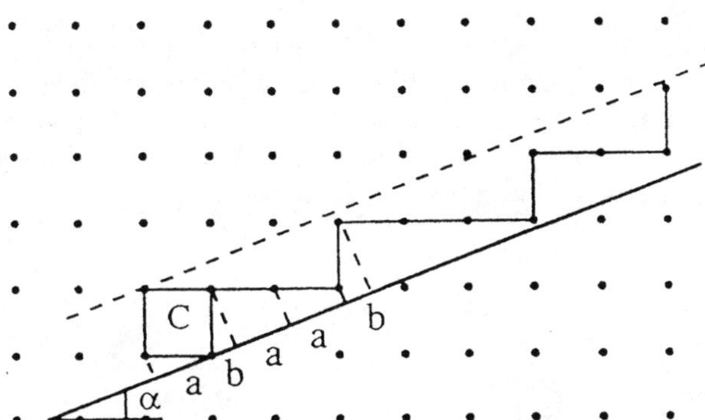

Fig. 2.24. The projection onto a one-dimensional space (*the line with slope* α) of a periodic two-dimensional structure (*the square lattice*). This gives a structure which is quasiperiodic if $\tan\alpha$ is irrational and periodic if $\tan\alpha$ is rational [2.5]

ther periodic *or* quasiperiodic. A quasicrystal thus gives a discrete diffraction pattern, of the same type as a genuine crystal.

The irrationality, with respect to cell parameters, of the slope of the 'plane' onto which the periodic lattice is projected in order to obtain a quasiperiodic lattice is related to the golden number and the Fibonacci series. For this series, each number (or segment) is equal to the sum of the previous two, and the ratio of two consecutive numbers (or segment lengths) tends towards the magic number $\tau = (1 + \sqrt{5})/2 = 1.618\ldots$. The Fibonacci series has a quasiperiodic structure, from which Penrose's drawing can be constructed. For example, Fig. 2.23 can be obtained by tracing five groups of parallel lines from the sides of a regular pentagon ($\tau = 2\cos\pi/5$), with the intervals between the lines of the same group forming a Fibonacci series, and joining the centres of gravity (in dual space) of the cells defined by these lines.

The physical study of quasicrystals [2.7, 2.8] started after Schechtman's work in 1984, though the idea of quasiperiodicity was studied by mathematicians such as Esclandon from the beginning of the century. Physical properties of quasicrystals can differ from those of perfect crystals in the same family, giving for example a higher mechanical rigidity or a lower electrical conductivity.

2.4 Examples of Structures

Close-packed structures often occur in practice because they are natural forms for monatomic crystals. Here we consider the two common cases and we describe the atomic structure of the most common crystals.

2.4.1 Close-Packed Structures

A plane of identical atoms can be represented by a set of identical spheres, assumed to be arranged in a hexagonal fashion, Fig. 2.25, since this gives the closest spacing. There are two ways of stacking such layers of spheres. Taking the first layer to have spheres centred at points A, the second layer has centres at points B, above the interstices of the first layer.

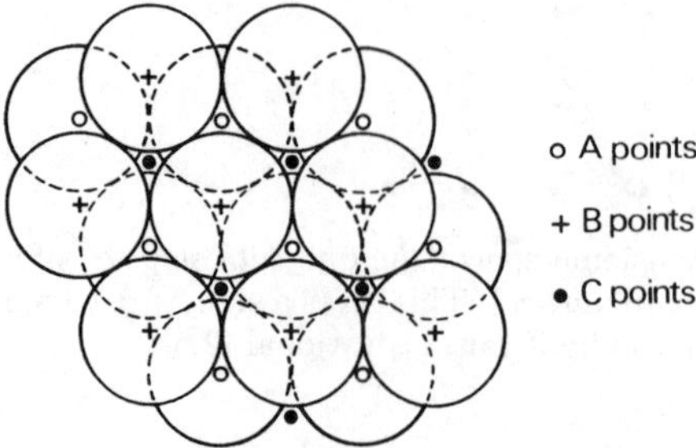

Fig. 2.25. Close-packed assembly of identical spheres

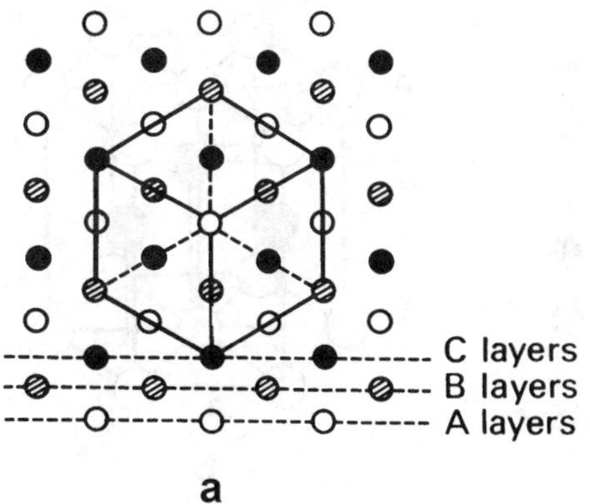

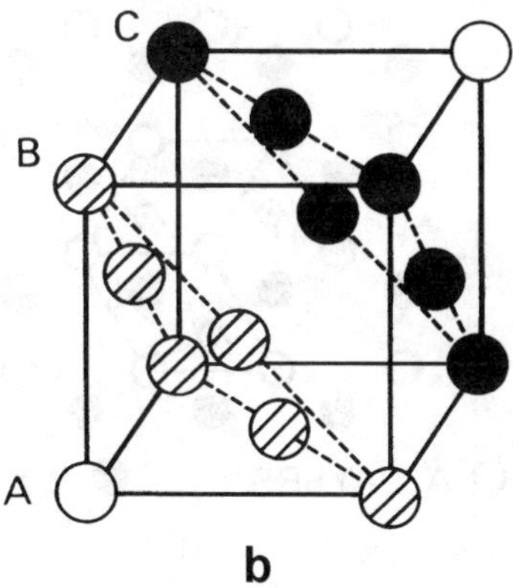

Fig. 2.26. Close-packed cubic structure. (**a**) Stacking of layers centred at A, B, C, and the cubic cell seen along a diagonal. (**b**) The f.c.c lattice cell given by this assembly. There are octahedral cavities at the cube centre and at the middle of each edge, and tetrahedral cavities at points one-quarter of the distance along each diagonal

For the third layer there are two choices – the centres can be above points C or A. In the first case the centres have a sequence ABC in each period, giving a structure called close-packed cubic. This is generated by a face-centred cubic cell, Fig. 2.26b, with the triad axis perpendicular to the plane of Fig. 2.26a. The second case has layers with centres ABAB..., giving the close-packed hexagonal structure, with the double cell (with two nodes) shown in Fig. 2.27. The ratio of the edge lengths is $c/a = 2\sqrt{2/3} \approx 1.633$ (Problem 2.11).

The reader should try, at least once, to assemble say marbles or tennis balls in these two ways, preferably with some mechanism to stop them rolling away.

These closed-packed structures have between their layers two types of cavities – *tetrahedral* cavities formed by one sphere resting on three others touching each other, and *octahedral* cavities formed by the octahedron of three touching spheres in one layer and the three nearest spheres in the next layer. In some diatomic crystals the atoms of one species form a close-packed structure, and atoms of the other species occupy the cavities.

The arrangement of the atoms is, of course, of great significance. Carbon is the classical example – one distribution of atoms gives graphite, a hexagonal structure with three coplanar bonds and one weak perpendicular bond, while another gives diamond, a cubic structure with four equivalent bonds. Recently a third structure has been found [2.9], the molecule discovered in 1985 and called fullerene because it resembles the dome built by Buckminster Fuller at the world fair in Montreal in 1967. The archetype is the C_{60} molecule consisting of 60 atoms. This has the form of a hollow sphere with diameter about 7 Å and appearance like a football, with 20 hexagonal faces

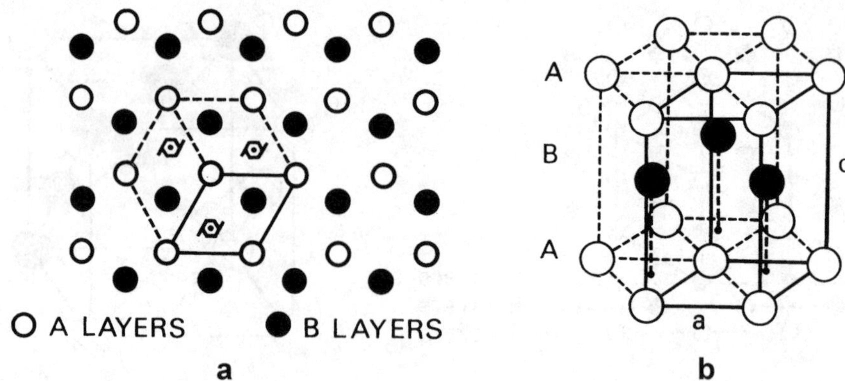

Fig. 2.27. Close-packed hexagonal structure. (**a**) Stacking of spheres with centres ABAB... and the hexagonal cell. (**b**) The duple hexagonal cell. The ratio c/a is $2\sqrt{2/3} = 1.633$

and 12 pentagonal faces. At each polygon vertex is a carbon atom. Each atom is bound to its neighbours by two single bonds and one double bond, and belongs simultaneously to one pentagon and two hexagons. Applications of fullerenes are expected in the areas of lubrication, electrochemistry and composite materials. Regarding diamond, a remarkable development is the fabrication, by a gaseous chemical reaction, of membranes. In view of the large forbidden energy gap (> 5 eV, compared with 1.12 eV for silicon), the hardness, the high thermal conductivity (five times that of copper), the optical transparency from the far infrared to the ultraviolet, and the large elastic wave velocities (the velocity of the longitudinal bulk waves in diamond is close to 18 000 m/s), these membranes must have a place in new components for electronics, optics and acoustics, with capability for high-temperature operation.

2.4.2 Useful Crystals

Before giving relevant details of the materials used for elastic waves, we note first that, in general, the crystalline structure of a solid is not unique. It can depend on the present, and perhaps past, temperature of the crystal. There are generally several varieties of the material, called *allotropes*, which give different physical properties even though the chemical formula is the same.

In particular, depending on whether the temperature is above or below a critical value, a crystal may or may not show *piezoelectricity* – the property of giving electrical polarization in response to mechanical deformation, and vice versa. Anticipating the following chapter we state here that, of the 32 crystal classes, only 20 can include piezoelectric crystals, and these are the ones with no centre of symmetry. Of these 20 classes, 10 can give *pyroelectric* crystals (Problem 2.17), and these crystals are polar, that is, they have a permanent natural electrical polarization. This polarization is normally compensated by a distribution of free charges either inside the crystal or on the surface. However, it can be observed as a current in an external circuit when the temperature is changed, since this changes the dipole moment. If the sense of the polarization can be influenced, for example reversed, by an external electric field, the pyroelectric crystal is also *ferroelectric*. A hysteresis cycle, giving polarization as a function of electric field, can be drawn. For a temperature above a critical value, called the *Curie temperature*, the polarization disappears. In this state, called paraelectric, the relation between polarization and electric field is linear. The dielectric permittivity of these materials, often high, is very sensitive to external influences (thermal, mechanical), especially near the Curie point. The difference in behaviour at different temperatures is explained by a change of crystal structure at the Curie point, the structure of the paraelectric phase being more symmetric than that of the ferroelectric phase. For example, barium titanate ($BaTiO_3$), one of the most studied ferroelectric materials, crystallizes with a cubic structure ($m3m$) for a temperature $\theta > 120\,°C$. For lower temperatures the cell changes form to give successively a tetragonal form ($10\,°C < \theta < 120\,°C$), an orthorhombic form ($-70\,°C < \theta < 10\,°C$) and then a trigonal form ($\theta < -70\,°C$).

Ferroelectric materials are divided into two groups according to their chemical constitution [2.10]. In the first group, ferroelectricity is attributed to the hydrogen bond, as in Seignette's salt (Rochelle salt, the double tartrate of sodium and potassium), and monoalkaline phosphates such as KH_2PO_4. In the second group, ferrolectricity comes from the deformation of a structure based on oxygen octahedrons. This group includes $BaTiO_3$, lithium niobate ($LiNbO_3$) and lithium tantalate ($LiTaO_3$).

In addition to lithium niobate, frequently used in the field of elastic waves for its piezoelectric qualities, we cite here a number of other important materials, mostly described in more detail below.

(1) Cadmium sulphide (CdS), a piezoelectric semiconductor formerly used to study the interaction between electrons and elastic waves;
(2) zinc oxide (ZnO), the constituent of thin-film transducers;
(3) sapphire (Al_2O_3), often chosen as the propagation medium for high-frequency elastic waves ($f > 1\,GHz$) because it gives low attenuation at little extra expense;
(4) quartz (SiO_2), an ever popular material, particularly because it has good temperature stability;
(5) gallium orthophosphate ($GaPO_4$), a fairly new piezoelectric material, like
(6) berlinite, or aluminium orthophosphate ($AlPO_4$);
(7) langasite, the silicate of gallium and lanthanum ($La_3Ga_5SiO_{14}$); and
(8) lithium tetraborate ($Li_2B_4O_7$), which have the potential to replace quartz in some applications.

(a) *Cadmium sulphide* (CdS) and *zinc oxide* (ZnO). These two materials crystallize in the hexagonal system with symmetry $6mm$. The dimensions

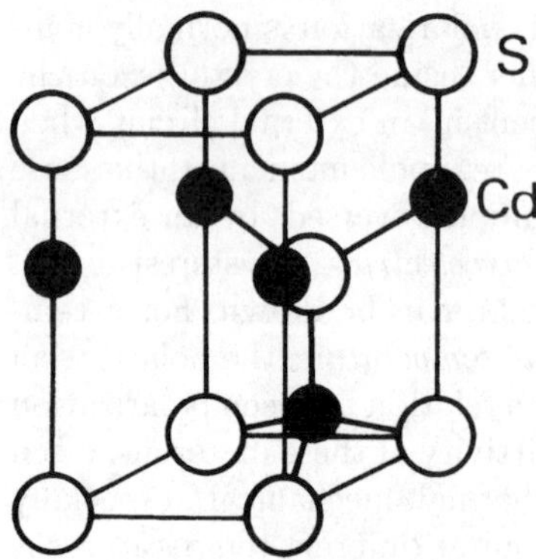

Fig. 2.28. Wurtzite-type structure (CdS, ZnO)

of the hexagonal cell are

for CdS : $a = 4.13$ Å, $c = 6.69$ Å, giving $c/a = 1.62$
for ZnO : $a = 3.24$ Å, $c = 5.19$ Å, giving $c/a = 1.60$.

In these two cases, the value of the ratio c/a is very close to that of the hexagonal close-packed structure (1.63). The anions (sulphur or oxygen) form a closed-packed structure, of which half the tetrahedral cavities are occupied by the cations (cadmium or zinc) whose diameter is about half that of the anions. The structure, known as the wurtzite type of structure, is shown in Fig. 2.28. The 6-fold axis is helicoidal, i.e. it gives a rotation followed by a translation along the axis, and it is polar (i.e. there is no mirror plane perpendicular to it).

Zinc oxide is often used in the form of *thin films*, with thickness in the micron region, deposited in a more or less crystalline form on substrates such as sapphire, silicon or gallium arsenide (GaAs), in order to generate bulk or surface waves. Aluminium nitride (AlN), which belongs to the same class $6mm$, is also used in this way.

(b) *Sapphire* (Al_2O_3). Sapphire, or alumina, crystallizes in the trigonal system with point group $\bar{3}m$. The oxygen atoms form a close-packed struc-

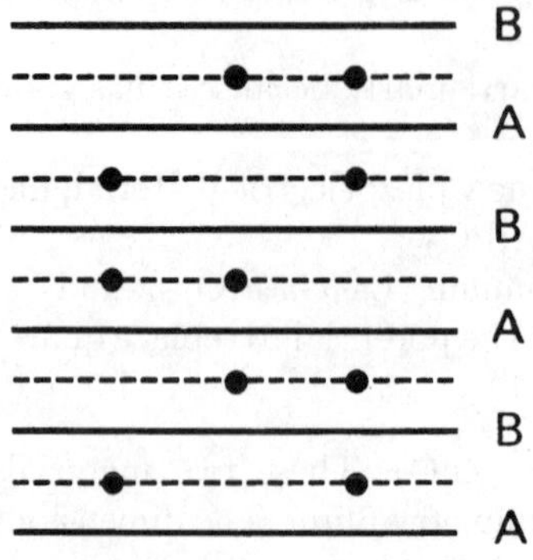

Fig. 2.29. Structure of sapphire. The aluminium atoms (*full circles*) are inserted between the oxygen layers (ABAB...)

ture of type ABAB... The aluminium atoms occupy two out of every three octahedral cavities situated between the layers A and B, giving a numerical ratio consistent with the formula Al_2O_3 (Fig. 2.29). The symmetry about the axis perpendicular to the oxygen planes is only of order 3. The trigonal cell has parameters $a = 3.51$ Å, $\alpha = 85°46'$.

(c) *Quartz* (SiO_2). Silica exists in numerous modifications – crystalline (quartz, tridymite, cristobalite) or amorphous (glass); moreover each crystalline form has two allotropes α and β. The α-quartz form (trigonal) is stable below 573 °C, while β-quartz (hexagonal) is stable from 573 °C to 870 °C. Because of its piezoelectricity, α-quartz, with trigonal class 32, is of particular interest. Because it has no mirror plane, there are two enantiomorphous forms (i.e. mirror images of each other), called right-handed (dextrorotatory) and left-handed (levorotatory) quartz. The triad axis (Z) is called the optic axis, and the three dyad axes (X) perpendicular to the edges of the hexagonal prism, called electric axes, are polar. Figure 2.30 shows the projection of the atoms on to a plane perpendicular to the helicoidal Z-axis (which gives translation $\boldsymbol{t} = \boldsymbol{c}/3$), giving an idea of the structure. The dimensions of the unit cell of the primitive hexagonal lattice are $a = 4.91$ Å and $c = 5.40$ Å.

(d) *Lithium niobate* ($LiNbO_3$), *lithium tantalate* ($LiTaO_3$) and *barium titanate* ($BaTiO_3$). Lithium niobate belongs to class $3m$. The trigonal cell ($a = 5.492$ Å, $\alpha = 55°\,53'$) contains two $LiNbO_3$ groups. As already mentioned, it is ferroelectric and its structure is based on oxygen octahedrons.

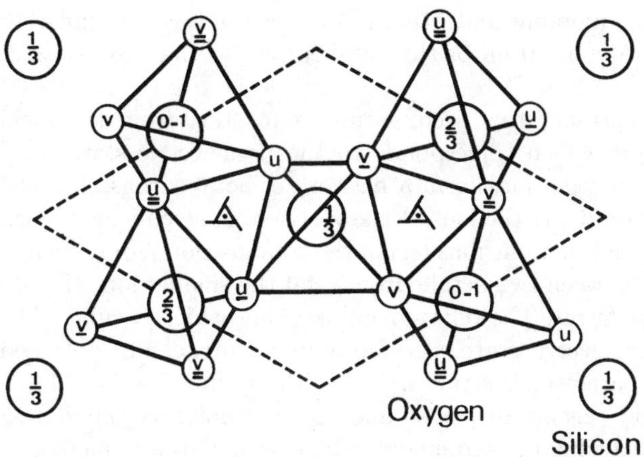

Fig. 2.30. Structure of α-quartz. The assembly of SiO_4 tetrahedrons linked at their vertices (giving the formula SiO_2) gives a helicoidal triad axis (3_1) perpendicular to the plane of the figure. The symbols shown in each circle specify its height in units of the cell height c. The silicon atoms are marked 0, 1/3, 2/3, and the oxygen atoms are marked u, $\underline{u} = u + 1/3$, $\underline{\underline{u}} = u + 2/3$ and v, $\underline{v}$, $\underline{\underline{v}}$. X-ray diffraction gives $u = 0.11$ and $v = 0.22$ [2.11]

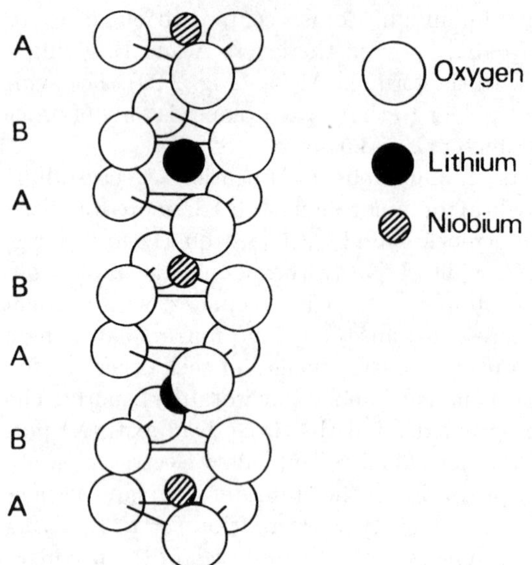

Fig. 2.31. Structure of lithium niobate (Fig. 2 of [2.13])

It can be described starting from a hexagonal close-packed structure of these ions [2.12]. As for sapphire, two out of three octahedral cavities are occupied by metallic ions (Li or Nb). Denoting an empty site by E, the filling sequence along the triad axis is Nb, Li, E, Nb, Li, E, Nb, ..., as in Fig. 2.31. The niobium and lithium ions have noticeably different diameters, causing deformation of the octahedrons so that no cation is exactly centred.

Another way to represent this architecture is to start from the ideal perovskite structure, which corresponds to the paraelectric state ($\theta > 1200\,°\mathrm{C}$). Here the oxygen ions form a network of octahedrons obtained from Fig. 2.26b by omitting the ions at the vertices of the cube, as shown in Fig. 2.32. The larger metallic ions (A) of the composition ABO_3 occupy the cube vertices, between two octahedrons, and the smaller ions (B) are at the octahedron centres. The point group is $m3m$, which is cubic. The structure of the ferroelectric state, at room temperature, can be obtained by deforming this paraelectric structure. For lithium niobate (A = Li, B = Nb), the metallic ions are displaced and create a polarization along a special direction, reducing the symmetry to class $3m$. This argument also applies to $BaTiO_3$ and to $KNbO_3$. However, for $BaTiO_3$ the displacement of the perovskite structure is along the tetrad axis (A_4), giving symmetry $4mm$ at room temperature, while for $KNbO_3$ it is along the dyad axis (A_2), giving symmetry $2mm$.

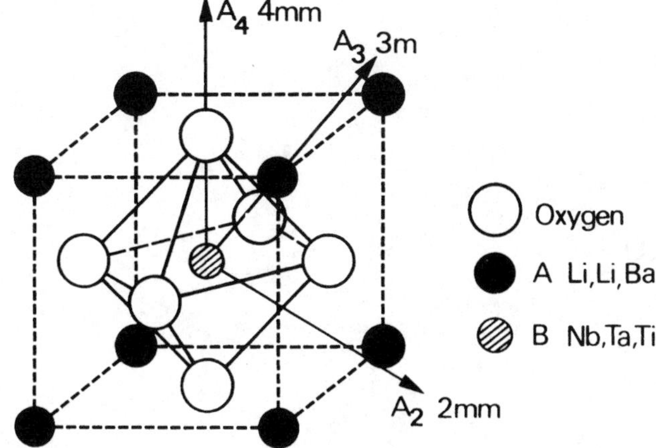

Fig. 2.32. Perovskite structure ABO_3. The octahedron formed by six oxygen anions contains one B ion, and each octahedron is linked to another at each of its vertices. The A ions are at the vertices of the cube containing an octahedron

For lithium tantalate ($LiTaO_3$), all the description above for lithium niobate applies, with the Ta ion replacing the Nb ion. The parameters of the trigonal cell are $a = 5.47$ Å and $\alpha = 56° \; 10'$.

(e) *Gallium orthophosphate* ($GaPO_4$) and *aluminium orthophosphate (berlinite* – $AlPO_4$). These two materials, whose crystal growth was developed only recently [2.14, 2.15], belong to a group of compounds of the type $M^{3+}X^{5+}O_4^{2-}$, where M^{3+} is Al, Ga, Fe or Mn, and X^{5+} is As or P. The structure, with point group 32, is of the α-quartz type. The Si atoms in quartz are alternately replaced by atoms of Ga (or Al) and P.
Figure 2.33 shows gallium phosphate. As in the case of quartz, the unit cell of the primitive hexagonal lattice ($a = 4.90$ Å, $c = 11.05$ Å) contains three molecules but, since the Ga and P atoms alternate, the cell height is doubled. The GaO_4 and PO_4 tetrahedrons are linked at their vertices, which are the sites of the oxygen atoms, and contain a gallium or phosphorous atom at the centre. The three-fold axis is helicoidal. The crystal has two enantiomorphous forms (right-handed and left-handed). Compared with quartz, gallium phosphate is of interest for its stronger piezoelectricity and the stability of the α phase right up to 933 °C. Beyond this temperature there is an irreversible transition to a structure of the crystobalite type.
The properties of berlinite ($a = 4.94$ Å, $c = 10.97$ Å) are similar to those of gallium phosphate, though it is less strongly piezoelectric [2.16].

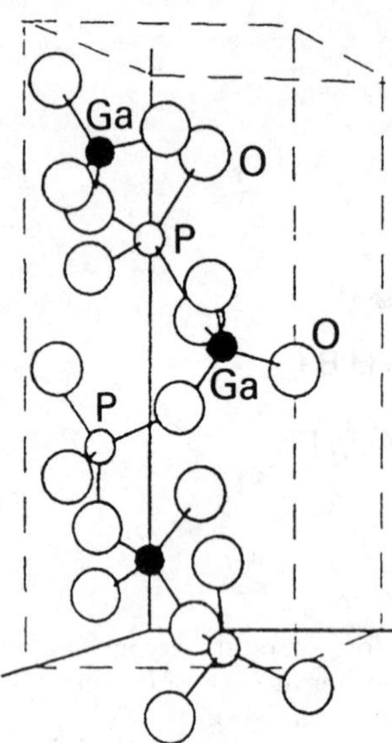

Fig. 2.33. Structure of $GaPO_4$, showing the arrangement of GaO_4 and PO_4 tetrahedrons. (From [2.14])

2.5 Representation of Physical Properties of Crystals by Tensors

The notion of a tensor appears when one sets up linear relations between causes and effects in an anisotropic medium. In a crystal, a cause applied along one direction gives, in general, an effect in another direction, as illustrated in Fig. 2.34. For example, an electric field $\boldsymbol{E}$ can give rise to a polarization $\boldsymbol{P}$ in a direction other than that of the field.

If the cause c and the effect e are vectorial quantities, the most general relation between the components e_1, e_2, e_3 of the effect and those of the cause, c_1, c_2, c_3, referred to the same system of axes, requires nine coefficients A_{ij} so that:

$$\begin{aligned} \mathsf{e}_1 &= A_{11}\mathsf{c}_1 + A_{12}\mathsf{c}_2 + A_{13}\mathsf{c}_3 \\ \mathsf{e}_2 &= A_{21}\mathsf{c}_1 + A_{22}\mathsf{c}_2 + A_{23}\mathsf{c}_3 \\ \mathsf{e}_3 &= A_{31}\mathsf{c}_1 + A_{32}\mathsf{c}_2 + A_{33}\mathsf{c}_3 \end{aligned}$$

or, more succinctly,

$$\mathsf{e}_i = A_{ij}\mathsf{c}_j \quad i,j = 1,2,3\,.$$

Here the repeated appearance of the index j on the right side is taken to imply summation over this index, by Einstein's convention; the repeated index j is called a dummy index. The nine components A_{ij} form a tensor of rank two. Generalizing, a vector such as e_i or c_i is called a tensor of rank one, and a scalar such as temperature is a tensor of rank zero.

The tensors c_i or e_i have a different nature to the tensor A_{ij}. The former, c_i or e_i, represent physical quantities. The latter, A_{ij}, is a characteristic of the

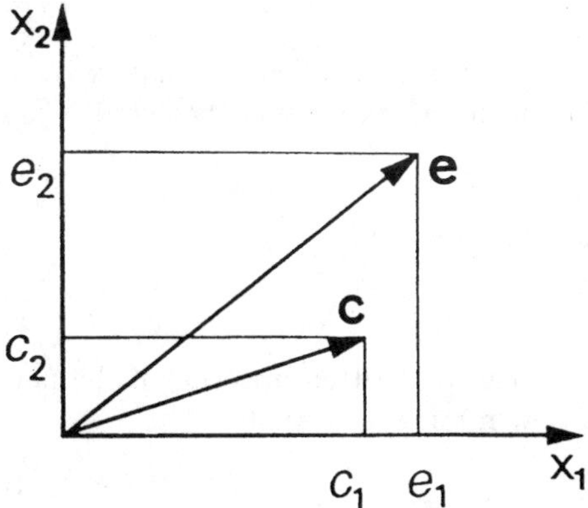

Fig. 2.34. Relation between cause and effect in a crystal. An anisotropic medium presents directions of 'lower resistance' – a cause c gives rise to an effect e(c), generally not parallel to the cause. To first order (in the linear region), the relation e(c) becomes, for vector quantities, $e_1 = A_{11}c_1 + A_{12}c_2, \quad e_2 = A_{21}c_1 + A_{22}c_2$

material – it describes the vectorial response e_i of the crystal to the vectorial cause c_i. A scalar quantity such as temperature or energy is thus represented by a tensor of rank zero; a vectorial quantity such as electric field or force by a tensor or rank one; and other more complex quantities such as stress and strain will be seen to be tensors of rank two (Sect. 3.1.1). Physical properties of crystals are expressed by tensors of rank zero (specific heat), of rank one (pyroelectricity), of rank two (permittivity, electrical conductivity, thermal expansion), of rank three (piezoelectricity), of rank four (elasticity), and so on. For example, the relation between electric displacement and electric field,

$$D_i = \varepsilon_{ij} E_j \,, \tag{2.1}$$

defines the permittivity tensor ε_{ij}.

In practice, the nine numbers A_{ij} are not sufficient to characterize the physical property considered – it is also necessary to specify the coordinate axes chosen since, for another system of axes, the tensor components will take different values A'_{ij}. However, the two sets of component values represent the same physical property, independent of the axes chosen; consequently, the two sets must have a relationship accounting for the change of axes. More specifically, the laws for transforming the components serve to define the rank of the tensors.

2.5.1 Change of Coordinate System

Vectors in the old and new coordinate systems are specified in terms of basis vectors, $\boldsymbol{e}_1$, $\boldsymbol{e}_2$, $\boldsymbol{e}_3$ for the old system, and $\boldsymbol{e}'_1$, $\boldsymbol{e}'_2$, $\boldsymbol{e}'_3$ for the new one. The new basis vectors are linearly related to the old ones so that, using the convention that a repeated dummy index is to be summed over, we have

$$\boldsymbol{e}'_i = \alpha_i^k \boldsymbol{e}_k \,. \tag{2.2}$$

The coefficients α_i^k can be written out as an array, called the matrix α, in which the ith row consists of the components α_i^k expressing the vector $\boldsymbol{e}'_i$ in terms of the basis vectors $\boldsymbol{e}_k$. Thus,

$$\alpha = \begin{pmatrix} \alpha_1^1 & \alpha_1^2 & \alpha_1^3 \\ \alpha_2^1 & \alpha_2^2 & \alpha_2^3 \\ \alpha_3^1 & \alpha_3^2 & \alpha_3^3 \end{pmatrix} \,.$$

Conversely, the original basis vectors $\boldsymbol{e}_k$ can be written in terms of the new vectors $\boldsymbol{e}'_j$ using the elements β_k^j of a new matrix β, so that

$$\boldsymbol{e}_k = \beta_k^j \boldsymbol{e}'_j \,. \tag{2.3}$$

The relation between the coefficients β_k^j and α_i^k can be obtained by substituting this expression for $\boldsymbol{e}_k$ into (2.2), giving

$$\boldsymbol{e}'_i = \alpha_i^k \beta_k^j \boldsymbol{e}'_j \,.$$

Using the Kronecker delta symbol, defined by $\delta_{ij} = 0$ for $i \neq j$ and $\delta_{ij} = 1$ for $i = j$, we have $\boldsymbol{e}'_i = \delta_{ij} \boldsymbol{e}'_j$, which gives

$$\left(\alpha_i^k \beta_k^j - \delta_{ij}\right) \boldsymbol{e}'_j = 0 \,. \tag{2.4}$$

The vectors $\boldsymbol{e}'_j$ must be independent because they form a basis. Hence the coefficients in (2.4) must be zero, so that

$$\alpha_i^k \beta_k^j = \delta_{ij} \,. \tag{2.5}$$

The coordinate systems used below are all *orthonormal*, so that

$$\boldsymbol{e}'_i \cdot \boldsymbol{e}'_j = \delta_{ij} \quad \text{and} \quad \boldsymbol{e}_k \cdot \boldsymbol{e}_l = \delta_{kl} \,.$$

In this case, the inverse matrix β can be deduced simply from α. With the aid of (2.2), the first scalar product above becomes

$$\alpha_i^k \alpha_j^l \left(\boldsymbol{e}_k \cdot \boldsymbol{e}_l\right) = \delta_{ij}$$

and hence, taking account of the second scalar product,

$$\alpha_i^k \alpha_j^k = \delta_{ij} \,.$$

Comparing this equation with (2.5) shows that

$$\beta_k^j = \alpha_j^k \,.$$

Thus the matrix β, which by definition is the reciprocal of the matrix α, can for *orthonormal* axes be obtained by interchanging the rows and columns of α. This gives the transpose of α, denoted α^{t}. The matrix α, whose reciprocal is the same as its transpose, is called *orthogonal.*

Let x_k be the coordinates of a point M, i.e. the components of the vector $\boldsymbol{x} = \boldsymbol{OM}$ in the basis $\boldsymbol{e}_k$, and x'_i the coordinates of the same point in the basis $\boldsymbol{e}'_i$. We then have

$$\boldsymbol{x} = x_k \boldsymbol{e}_k = x'_i \boldsymbol{e}'_i \,.$$

Replacing $\boldsymbol{e}_k$ by expression (2.3) gives

$$\boldsymbol{x} = x_k \beta^j_k \boldsymbol{e}'_j = x'_i \boldsymbol{e}'_i \,.$$

Since the expression of a vector in a specified basis is unique, the new coordinates x'_i can be expressed in terms of the old ones by the relation

$$x'_i = \beta^i_k x_k \,.$$

If the coordinate systems are orthonormal, so that $\beta^i_k = \alpha^k_i$, this has the form of (2.2), giving

$$x'_i = \alpha^k_i x_k \,. \tag{2.6}$$

Thus the relations for transforming the coordinates of a point, given by (2.6), are the same as those for transforming basis vectors, as in (2.2).

2.5.2 Definition of a Tensor

Physical quantities can be classified in terms of their different behaviour on changing the coordinate axes.

(a) A *scalar* is a quantity independent of the axes chosen. It is also called an *invariant*, or *tensor of rank zero*. The invariance of a scalar function $f(x_1, x_2, x_3)$ with respect to the coordinates is expressed by the equation

$$f\left(x_1, x_2, x_3\right) = f\left(x'_1, x'_2, x'_3\right) \,.$$

Examples of scalars are temperature, energy and electrostatic potential.

(b) Any group of three quantities A_i which transforms, under a change of axes, like the basis vectors $\boldsymbol{e}_i$ is a *tensor of rank one*, or *vector*. Thus,

$$A'_i = \alpha^k_i A_k \,.$$

Examples. The components of any vector in the space referenced to the three basis vectors $\boldsymbol{e}_i$ form a tensor of rank one, as shown by (2.6). The three derivatives $\partial f / \partial x_i$ of the scalar function $f(x_i)$ constitute a vector called the *gradient* of the function. In the new system the components of the gradient are

$$\frac{\partial f}{\partial x'_i} = \frac{\partial f}{\partial x_k} \cdot \frac{\partial x_k}{\partial x'_i} \,,$$

so these can be obtained from the components in the old system, $\partial f / \partial x_k$, by means of the coefficients $\partial x_k / \partial x'_i$. Inverting (2.6) gives

$$x_k = \beta^i_k x'_i = \alpha^k_i x'_i$$

showing that the components of the gradient transform according to the law

$$\frac{\partial f}{\partial x'_i} = \alpha^k_i \frac{\partial f}{\partial x_k} \,.$$

Hence the gradient is a tensor of rank one.

(c) Any set of nine quantities A_{ij} which transforms as the product of the components of two vectors is a *tensor of rank two*, so that

$$A'_{ij} = \alpha_i^k \alpha_j^l A_{kl} \,. \tag{2.7}$$

These nine components can be written out as an array, using square brackets to distinguish them from the matrix α for a change of axes:

$$A_{ij} = \begin{bmatrix} A_{11} & A_{12} & A_{13} \\ A_{21} & A_{22} & A_{23} \\ A_{31} & A_{32} & A_{33} \end{bmatrix} \,.$$

Despite their resemblance, matrices and tensors have significant differences:

(1) The elements α_i^j establish the correspondence between two sets of axes.

(2) The tensor A_{ij} is a physical (or mathematical) quantity represented in one system of axes by nine numbers.

However, for orthonormal reference frames, matrices and tensors are mathematically indistinguishable – it is possible, for example, to attribute eigenvalues and eigenvectors to a tensor.

Example. A tensor of rank two is formed by taking the derivatives $\partial A_i / \partial x_k$ of the components of a vector. The proof of this is the same as that used for the gradient above.

From a tensor A_{ij}, it is possible to form a *scalar* by summing the diagonal terms. Using the convention of repeated indices, this quantity $A_{11}+A_{22}+A_{33}$, called the *trace* of the tensor, is written A_{ii}. For a change of orthonormal axes we have

$$A'_{ii} = \alpha_i^k \alpha_i^l A_{kl} = A_{ll} \,,$$

since $\alpha_i^k \alpha_i^l = \delta_{kl}$. In particular, for the tensor $A_{ij} = A_i B_j$, the trace is the scalar $A_i B_i$, which is the scalar product of the vectors A_i and B_j. The contracted product

$$A_i A_i = (A_1)^2 + (A_2)^2 + (A_3)^2$$

denoted $(A_i)^2$ or A_i^2, represents the square of the length of the vector A_i. Equation (2.7) is written, in matrix notation,

$$[A'] = (\alpha)[A](\alpha)^{\mathrm{t}} \tag{2.8}$$

or, since there is no possibility of confusion, as $[A'] = \alpha[A]\alpha^{\mathrm{t}}$. Thus, starting for example from (2.1),

$$D_i = \varepsilon_{ij} E_j \quad \text{becomes} \quad [D] = [\varepsilon][E], \quad \text{and} \quad [D'] = [\varepsilon'][E'] \,,$$

where

$$[E'] = \alpha[E], \quad [E] = \alpha^{-1}[E'] = \alpha^{\mathrm{t}}[E'], \quad \text{and} \quad [D'] = \alpha[D] \,.$$

Thus,

$$[D'] = \alpha[\varepsilon]\alpha^{\mathrm{t}}[E']$$

and hence

$$[\varepsilon'] = (\alpha)[\varepsilon](\alpha)^{\mathrm{t}} .$$

(d) The definition is easily generalized to tensors of higher rank. A tensor of rank r is a set of 3^r components specified by r indices which, under a change of axes, transforms in the following manner:

$$A'_{\dots ijk\dots} = \dots \alpha_i^l \alpha_j^m \alpha_k^n \dots A_{\dots lmn\dots} . \tag{2.9}$$

The formation of the trace of a tensor of rank two is a particular case of a rule for contracting indices. The quantity $A_{\dots ijjl\dots}$ (with summation over the index j) is a tensor of rank $r-2$, derived from the tensor $A_{\dots ijkl\dots}$ of rank r. Under transformation of axes, the new components

$$A'_{\dots ijjl\dots} = \dots \alpha_i^m \alpha_j^n \alpha_j^p \alpha_l^q \dots A_{\dots mnpq\dots}$$

are written, using $\alpha_j^n \alpha_j^p = \delta_{np}$, as

$$A'_{\dots ijjl\dots} = \dots \alpha_i^m \alpha_l^q \dots A_{\dots mnnq\dots} .$$

To assess the tensor character of a quantity, it is often helpful to use the rule that a linear relation between a tensor $A_{\dots ij\dots}$ of rank m and a tensor $B_{\dots kl\dots}$ of rank n defines a tensor $C_{\dots ijkl\dots}$ of rank $m+n$. Thus

$$A_{\dots ij\dots} = C_{\dots ijkl\dots} B_{\dots kl\dots} . \tag{2.10}$$

To show this, consider a change of axes, which gives

$$A'_{\dots pq\dots} = \dots \alpha_p^i \alpha_q^j \dots A_{\dots ij\dots} \tag{2.11}$$

and

$$B_{\dots kl\dots} = \dots \beta_k^r \beta_l^s \dots B'_{\dots rs\dots} . \tag{2.12}$$

The matrix β expresses the old components in terms of the new ones. Since $\beta_k^r = \alpha_r^k$ we have, substituting (2.12) into (2.10) and (2.10) into (2.11),

$$A'_{\dots pq\dots} = \left(\dots \alpha_p^i \alpha_q^j \alpha_r^k \alpha_s^l \dots C_{\dots ijkl\dots}\right) B'_{\dots rs\dots} .$$

The factor in brackets is none other than $C'_{\dots pqrs\dots}$, and consequently the quantities $C_{\dots ijkl\dots}$ transform as the components of a tensor of rank $m+n$, so that

$$C'_{\dots pqrs\dots} = \dots \alpha_p^i \alpha_q^j \alpha_r^k \alpha_s^l \dots C_{\dots ijkl\dots} .$$

An example is the Kronecker delta function δ_{ij}, a tensor of rank two, since

$$A_i = \delta_{ij} A_j .$$

In *elasticity*, the stiffness constants c_{ijkl} relate the stress tensor T_{ij} to the strain tensor S_{kl}, so that

$$T_{ij} = c_{ijkl} S_{kl}$$

and thus form a tensor of rank four (Sect. 3.2.1).

2.6 Effect of Crystal Symmetry on Tensor Components

Let $A_{...ijk...}$ be the components in the coordinate system $Ox_1x_2x_3$ of a tensor expressing a physical property of a crystal, and let $A'_{...ijk...}$ be the components of this tensor in the same coordinates but for a new orientation of the crystal, obtained by applying an operation $\boldsymbol{S}$. To express the components $A'_{...ijk...}$ in terms of the $A_{...pqr...}$, we can use the equivalent process of applying the inverse operation $\boldsymbol{S}^{-1}$ to the system of axes while maintaining the crystal orientation (Fig. 2.35).

Consequently, if α is the matrix for the axis change represented by $\boldsymbol{S}^{-1}$, (2.9) gives

$$A'_{...ijk...} = \dots \alpha_i^p \alpha_j^q \alpha_k^r \dots A_{...pqr...} .$$

If $\boldsymbol{S}$ is one of the symmetry operations of the crystal point group, the new orientation of the crystal in the coordinate system $Ox_1x_2x_3$ is indistinguishable from the original one, in terms of the macroscopic physical properties. This is expressed by the equation

$$A'_{...ijk...} = A_{...ijk...} .$$

Thus, the invariance of properties under particular symmetry operations implies relations of the type

$$A_{...ijk...} = \dots \alpha_i^p \alpha_j^q \alpha_k^r \dots A_{...pqr...} \tag{2.13}$$

These reduce the number of independent components in the tensors. Since the inverse of a symmetry operation ($\boldsymbol{S} = A_n$ or $\bar{A}_n$) is another symmetry operation ($\boldsymbol{S}^{-1} = \boldsymbol{S}^{n-1}$), it follows that α is the matrix for the change of basis associated with one of the symmetry operations of the crystal point group.

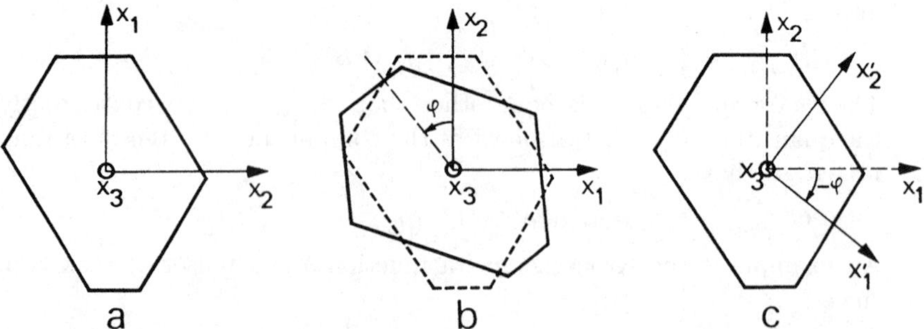

Fig. 2.35. Equivalence between a rotation of the crystal and a change of axes. (**a**) Original crystal and coordinate system. (**b**) Rotation of crystal through an angle φ about Ox_3. (**c**) Rotation of the coordinate axes through $-\varphi$

2.6.1 Matrices for Point Symmetry Transformations

The point symmetry elements of crystals are the direct or inverse axis, the mirror plane, and the centre of symmetry (Sect. 2.2.1). For a *rotation* φ about the axis Ox_3, shown in Fig. 2.36a, the new basis vectors $\boldsymbol{e}'_i$ are expressed in terms of the old ones $\boldsymbol{e}_k$ by (2.2), using the matrix

$$\alpha_{x_3,\varphi} = \begin{pmatrix} \cos\varphi & \sin\varphi & 0 \\ -\sin\varphi & \cos\varphi & 0 \\ 0 & 0 & 1 \end{pmatrix} . \tag{2.14}$$

The matrix representing a *centre of symmetry* about a point, taking the point to be at the origin, is

$$\alpha_{\mathrm{C}} = \begin{pmatrix} -1 & 0 & 0 \\ 0 & -1 & 0 \\ 0 & 0 & -1 \end{pmatrix} .$$

The matrix for an *inverse rotation axis* of order n is obtained from the relation $\bar{A}_n = A_n \cdot C$, so that

$$\bar{\alpha}_{x_3,2\pi/n} = \alpha_{x_3,2\pi/n} \cdot \alpha_{\mathrm{C}} .$$

For a *plane of symmetry* (mirror) perpendicular to Ox_3 we have

$$\alpha_{\mathrm{M}\perp x_3} = \begin{pmatrix} 1 & 0 & 0 \\ 0 & 1 & 0 \\ 0 & 0 & -1 \end{pmatrix} .$$

For a cube constructed from three orthonormal basis vectors (Fig. 2.36b), a rotation of $2\pi/3$ (triad axis) about the diagonal (the [111] direction) gives a cyclic permutation of the axes, so that

$$\boldsymbol{e}'_1 = \boldsymbol{e}_2 , \quad \boldsymbol{e}'_2 = \boldsymbol{e}_3 , \quad \boldsymbol{e}'_3 = \boldsymbol{e}_1$$

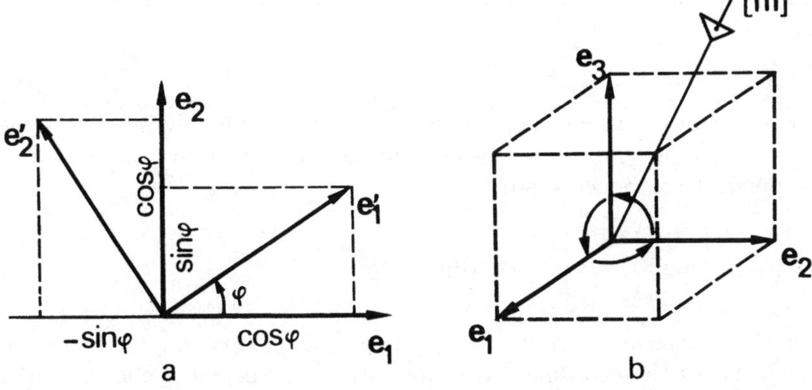

Fig. 2.36. Rotation of axes about (**a**) Ox_3, (**b**) the [111] direction of a cube

and this can be expressed by the matrix

$$\alpha_{[111],2\pi/3} = \begin{pmatrix} 0 & 1 & 0 \\ 0 & 0 & 1 \\ 1 & 0 & 0 \end{pmatrix} .$$

2.6.2 Effect of a Centre of Symmetry

For a centre of symmetry, the matrix α is diagonal (so that $\alpha_i^j = -\delta_{ij}$) and the general invariance condition of (2.13) reduces to

$$A_{...ijk...} = \ldots \alpha_i^i \alpha_j^j \alpha_k^k \ldots A_{...ijk...} .$$

Thus, for a tensor of rank n we have

$$A_{...ijk...} = (-1)^n A_{...ijk...} . \tag{2.15}$$

When n is odd we have $(-1)^n = -1$ and the above relation implies that all components are zero. Thus, *physical properties represented by tensors of odd rank must be absent in crystals belonging to the centro-symmetric classes*; in particular, this applies for *piezoelectricity* ($n = 3$).

If n is even, (2.15) is uninformative – a centre of symmetry has no bearing on physical properties represented by tensors of even rank. Moreover, since types of axes (direct or inverse) do not need to be distinguished when there is a centre, classes of the same system are restricted to a minimum number of independent components because they share one or more direct or inverse axes (Sect. 2.2.3). These comments are illustrated by the example of the dielectric tensor below.

2.6.3 Reduction of the Number of Dielectric Constants

From thermodynamic considerations analogous to those discussed later for the elastic tensor (Sect. 3.2.2), it can be shown that the dielectric tensor of (2.1) is *symmetric*, so that $\varepsilon_{ij} = \varepsilon_{ji}$. The condition of invariance (2.13) becomes

$$\varepsilon_{ij} = \alpha_i^k \alpha_j^l \varepsilon_{kl} . \tag{2.16}$$

The tensor is of *even* rank. For the *triclinic* system, the only possible symmetry, a centre of symmetry, does not impose any reduction in the number of independent components, so

$$[\varepsilon_{ij}] = \begin{bmatrix} \varepsilon_{11} & \varepsilon_{12} & \varepsilon_{13} \\ \varepsilon_{12} & \varepsilon_{22} & \varepsilon_{23} \\ \varepsilon_{13} & \varepsilon_{23} & \varepsilon_{33} \end{bmatrix} \qquad \textit{triclinic} .$$

For the *monoclinic* system, the dyad axis along Ox_3, direct or inverse (mirror), makes the coordinate transformation matrix α become diagonal, giving

$$\alpha_{A_2||x_3} = \begin{pmatrix} -1 & 0 & 0 \\ 0 & -1 & 0 \\ 0 & 0 & 1 \end{pmatrix} \quad \text{or} \quad \alpha_{M\perp x_3} = \begin{pmatrix} 1 & 0 & 0 \\ 0 & 1 & 0 \\ 0 & 0 & -1 \end{pmatrix}.$$

Thus (2.16) becomes

$$\varepsilon_{ij} = \alpha_i^i \alpha_j^j \varepsilon_{ij}.$$

This nullifies components that have one index equal to 3 (since, for example, $\alpha_i^i \alpha_3^3 = -1$ if $i \neq 3$), reducing the number of independent components to four and giving

$$[\varepsilon_{ij}] = \begin{bmatrix} \varepsilon_{11} & \varepsilon_{12} & 0 \\ \varepsilon_{12} & \varepsilon_{22} & 0 \\ 0 & 0 & \varepsilon_{33} \end{bmatrix} \quad \textit{monoclinic}.$$

Orthorhombic crystals have three perpendicular direct or inverse dyad axes. Taking these axes as the coordinate axes, and applying the above reasoning to each of them, shows that all the non-diagonal components are zero, so that

$$[\varepsilon_{ij}] = \begin{bmatrix} \varepsilon_{11} & 0 & 0 \\ 0 & \varepsilon_{22} & 0 \\ 0 & 0 & \varepsilon_{33} \end{bmatrix} \quad \textit{orthorhombic}.$$

For the *trigonal, tetragonal* or *hexagonal* systems the axis A_n or $\bar{A}_n$ (with $n > 2$) is taken to be along Ox_3, and the matrix α for rotating the axes is given by (2.14) with $\varphi = 2\pi/n < \pi$. Applying (2.16) to ε_{13} and ε_{23} gives

$$\begin{cases} \varepsilon_{13} = \alpha_1^k \alpha_3^3 \varepsilon_{k3} = \alpha_1^1 \varepsilon_{13} + \alpha_1^2 \varepsilon_{23} & \text{since} \quad \alpha_1^3 = 0 \\ \varepsilon_{23} = \alpha_2^k \alpha_3^3 \varepsilon_{k3} = \alpha_2^1 \varepsilon_{13} + \alpha_2^2 \varepsilon_{23} & \text{since} \quad \alpha_2^3 = 0 \end{cases}$$

and hence

$$\begin{cases} \varepsilon_{13}(1 - \cos\varphi) - \varepsilon_{23}\sin\varphi = 0 \\ \varepsilon_{13}\sin\varphi + \varepsilon_{23}(1 - \cos\varphi) = 0 \end{cases}.$$

Since φ is non-zero, these imply that $\varepsilon_{13} = \varepsilon_{23} = 0$. For ε_{11} we find

$$\varepsilon_{11} = \alpha_1^k \alpha_1^l \varepsilon_{kl} = \alpha_1^1 \alpha_1^2 (\varepsilon_{12} + \varepsilon_{21}) + \alpha_1^1 \alpha_1^1 \varepsilon_{11} + \alpha_1^2 \alpha_1^2 \varepsilon_{22},$$

which gives

$$\varepsilon_{11}\sin^2\varphi = (\varepsilon_{12} + \varepsilon_{21})\cos\varphi\sin\varphi + \varepsilon_{22}\sin^2\varphi.$$

On dividing by $\sin^2\varphi$, this gives

$$\varepsilon_{11} - \varepsilon_{22} = (\varepsilon_{12} + \varepsilon_{21})\cot\varphi. \tag{2.17}$$

The same method, starting from ε_{12}, leads to

$$\varepsilon_{11} - \varepsilon_{22} = -(\varepsilon_{12} + \varepsilon_{21})\tan\varphi. \tag{2.18}$$

Since φ differs from π, (2.17) and (2.18) imply that

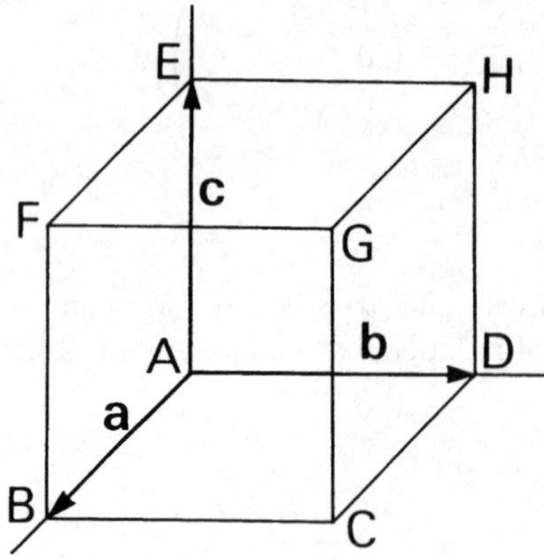

Fig. 2.37. Cubic cell

$$\varepsilon_{12} + \varepsilon_{21} = 0 \quad \text{and} \quad \varepsilon_{22} = \varepsilon_{11}$$

and since the tensor must be symmetric we must have $\varepsilon_{12} = -\varepsilon_{21} = 0$. Thus crystals in the *trigonal, tetragonal* or *hexagonal* systems have only two independent dielectric constants ε_{11} and ε_{33}, so that

$$[\varepsilon_{ij}] = \begin{bmatrix} \varepsilon_{11} & 0 & 0 \\ 0 & \varepsilon_{11} & 0 \\ 0 & 0 & \varepsilon_{33} \end{bmatrix} \quad \textit{trigonal, tetragonal, hexagonal}.$$

For *cubic* crystals with the minimal symmetry $4A_3$ and $3A_2$ (taking the latter as the coordinate axes), the non-diagonal terms are zero as in the orthorhombic case. Rotation about the triad axis [111] gives a cyclic permutation of indices, so that (1, 2, 3) → (2, 3, 1), giving $\varepsilon_{11} = \varepsilon_{22} = \varepsilon_{33} = \varepsilon$. Thus

$$[\varepsilon_{ij}] = \begin{bmatrix} \varepsilon & 0 & 0 \\ 0 & \varepsilon & 0 \\ 0 & 0 & \varepsilon \end{bmatrix} \quad \textit{cubic}.$$

A cubic crystal is therefore dielectrically *isotropic*; the electric displacement is always parallel to the electric field, so that $D_i = \varepsilon E_i$.

Problems

2.1. What are the Miller indices of the planes ACGE, BDHF and EBD and of the rows AC and AG of the cubic lattice in Fig. 2.37?

Solution. $(1, \bar{1}, 0)$, (1, 1, 0), (1, 1, 1), [110], [111].

2.2. By comparing its volume with that of an elementary rhombohedral cell, show that the cell of a f.c.c. cubic lattice is quadruple.

Solution. Using the side of the cube as the unit of length, and the notation in Fig. 2.8b, the basis vectors of the rhombohedron have components

$$\boldsymbol{a}\left(\frac{1}{2},\frac{1}{2},0\right),\quad \boldsymbol{b}\left(\frac{1}{2},0,\frac{1}{2}\right),\quad \boldsymbol{c}\left(0,\frac{1}{2},\frac{1}{2}\right)\quad \text{giving}\quad V=|(\boldsymbol{a},\boldsymbol{b},\boldsymbol{c})|=\frac{1}{4}\,.$$

2.3. Calculate the density of aluminium, which crystallizes in the f.c.c. lattice with sides $a = 4.04\,\text{Å}$, given that the atomic weight is 27 and Avogadro's number $N = 6.023 \times 10^{23}$.

Solution. Since there are 4 atoms per cell with volume a^3, we have $\rho = 4M/Na^3 = 2.72 \times 10^3\,\text{kg/m}^3$.

2.4. In calcium fluoride (CaF_2), the Ca^{2+} ions form a f.c.c. lattice with side $a = 5.45\,\text{Å}$. Calculate the density, given that the atomic weights are Ca = 40, F = 19.

Solution. The cubic cell contains four CaF_2 molecules, giving $\rho = 3.20 \times 10^3\,\text{kg/m}^3$.

2.5. Calculate the distances between the families of (100), (110), (111) planes, for the three cubic lattices. In each case, which plane has the highest density of nodes?

Solution.

$$\begin{array}{llll}
\text{Simple lattice:} & \underline{d_{100}} = a, & d_{110} = \frac{a\sqrt{2}}{2}, & d_{111} = \frac{a\sqrt{3}}{3} \\
\text{Body centred:} & d_{100} = a/2, & \underline{d_{110}} = \frac{a\sqrt{2}}{2}, & d_{111} = \frac{a\sqrt{3}}{6} \\
\text{Face centred:} & d_{100} = a/2, & d_{110} = \frac{a\sqrt{2}}{4}, & \underline{d_{111}} = \frac{a\sqrt{3}}{3}\,.
\end{array}$$

The planes with maximum density, that is, the ones with largest spacing, are underlined.

2.6. Calculate the angle α in the rhombohedral cell of the f.c.c. lattice.

Solution. From the scalar product $\boldsymbol{a} \cdot \boldsymbol{b}$ we find $\alpha = 60°$.

2.7. What is the equation of the family of lattice planes (hkl) in the coordinate system $Oxyz$ with basis vectors $\boldsymbol{a}, \boldsymbol{b}, \boldsymbol{c}$?

Solution. The nth plane in the family, counting from the origin, cuts the axes at points $A(x_n = na/h, 0, 0)$, $B(0, y_n = nb/k, 0)$, $C(0, 0, z_n = nc/l)$. Its equation is thus

$$\frac{x}{x_n} + \frac{y}{y_n} + \frac{z}{z_n} = 1 \quad \text{or} \quad h\frac{x}{a} + k\frac{y}{b} + l\frac{z}{c} = n\,.$$

2.8. In a cubic lattice, what is the distance between two consecutive planes in the family (hkl)?

Solution. From the result of the previous problem, the equation of the first lattice plane is $hx + ky + lz = a$, and its distance from the origin is thus

$$\mathrm{d}_{hkl} = \frac{a}{\sqrt{h^2 + k^2 + l^2}}\,.$$

2.9. Calculate the volume V of a cell as a function of the lengths a, b, c and the angles $\alpha = (\boldsymbol{b}, \boldsymbol{c}), \beta = (\boldsymbol{a}, \boldsymbol{c}), \gamma = (\boldsymbol{a}, \boldsymbol{b})$. Use the expression

$$V^2 = \begin{vmatrix} \boldsymbol{a}\cdot\boldsymbol{a} & \boldsymbol{a}\cdot\boldsymbol{b} & \boldsymbol{a}\cdot\boldsymbol{c} \\ \boldsymbol{b}\cdot\boldsymbol{a} & \boldsymbol{b}\cdot\boldsymbol{b} & \boldsymbol{b}\cdot\boldsymbol{c} \\ \boldsymbol{c}\cdot\boldsymbol{a} & \boldsymbol{c}\cdot\boldsymbol{b} & \boldsymbol{c}\cdot\boldsymbol{c} \end{vmatrix}$$

to put V in the form $V = abc\ F(\alpha, \beta, \gamma)$. Find $F(\alpha, \beta, \gamma)$ for each crystal system.

Solution.

$$V^2 = \begin{vmatrix} a^2 & ab\cos\gamma & ac\cos\beta \\ ab\cos\gamma & b^2 & bc\cos\alpha \\ ac\cos\beta & bc\cos\alpha & c^2 \end{vmatrix} = a^2b^2c^2 \begin{vmatrix} 1 & \cos\gamma & \cos\beta \\ \cos\gamma & 1 & \cos\alpha \\ \cos\beta & \cos\alpha & 1 \end{vmatrix}$$

giving

$$V = abc\sqrt{1 + 2\cos\alpha\cos\beta\cos\gamma - \cos^2\alpha - \cos^2\beta - \cos^2\gamma} = abc\ F(\alpha\,,\beta\,,\gamma)$$

Triclinic: $F = \sqrt{1 + 2\cos\alpha\cos\beta\cos\gamma - \cos^2\alpha - \cos^2\beta - \cos^2\gamma}$
Monoclinic: $F = \sin\beta$
Orthorhombic, tetragonal, cubic: $F = 1$
Trigonal: $F = \sqrt{1 + 2\cos^3\alpha - 3\cos^2\alpha}$
Hexagonal: $F = \sqrt{3}/2\,.$

2.10. What is the point symmetry class of a regular tetrahedron?

Solution. As shown in Fig. 2.38, there are four triad axes perpendicular to the faces, three inverse tetrad axes each perpendicular to a pair of opposite edges, and six mirror planes. The point symmetry is therefore of class $\bar{4}3m$, in the cubic system.

2.11. Calculate the ratio c/a for the hexagonal cell in the assembly ABAB (Fig. 2.27b).

Solution. Three spheres A and one sphere B form a tetrahedron with edge $a = 2r$ and height $2r\sqrt{2/3} = c/2$, giving $c/a = 2\sqrt{2/3} = 1.633$.

2.12. What is the filling factor f_r (ratio of volume occupied by the spheres to total volume) of the two close-packed structures?

Solution. For the assembly ABC (Fig. 2.26), the side of the cubic cell is such that $a\sqrt{2} = 4r$, where r is the sphere radius. Since the cell contains four spheres we have

$$f_r = \frac{4\cdot\frac{4}{3}\pi r^3}{\left(2\sqrt{2}r\right)^3} = \frac{\pi}{3\sqrt{2}} = 0.74\,.$$

For the assembly ABAB (Fig. 2.27), the hexagonal cell has volume

$$V = ca^2\sqrt{3}/2 = 8\sqrt{2}r^3$$

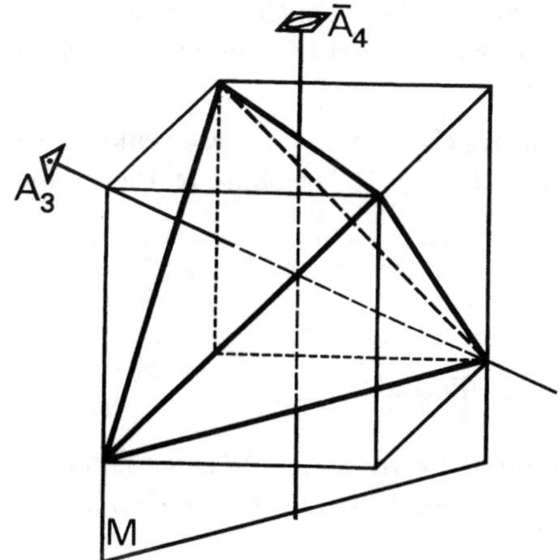

Fig. 2.38. Symmetry elements of a regular tetrahedron

and contains two spheres, so

$$f_r = \frac{2 \cdot \frac{4}{3}\pi r^3}{8\sqrt{2}r^3} = \frac{\pi}{3\sqrt{2}} = 0.74\,.$$

It is not surprising to find the same filling factor for these cases, because both structures are made of contacting spheres.

2.13. What is the number of tetrahedral and octahedral cavities per cubic cell for the close-packed structure ABC?

Solution. In Fig. 2.26b it can be seen that there are 8 tetrahedral cavities (one for each vertex) and 4 octahedral cavities (one at the cube centre and 12 at the centres of edges common to four cubes).

2.14. Show that the determinant of an orthogonal matrix α is equal to ± 1.

Solution. The determinant of the inverse matrix $\beta = \alpha^{\mathrm{t}}$ is $1/|\alpha|$, hence $|\alpha|^2 = 1$.

2.15. Write the matrices for rotation by an angle φ about Ox_1 and Ox_2.

Solution.

$$\alpha_{x_1,\varphi} = \begin{pmatrix} 1 & 0 & 0 \\ 0 & \cos\varphi & \sin\varphi \\ 0 & -\sin\varphi & \cos\varphi \end{pmatrix} \quad \alpha_{x_2,\varphi} = \begin{pmatrix} \cos\varphi & 0 & \sin\varphi \\ 0 & 1 & 0 \\ -\sin\varphi & 0 & \cos\varphi \end{pmatrix}.$$

2.16. Consider two vectors $\boldsymbol{v}$ and $\boldsymbol{u}$, where $\boldsymbol{u}$ has length unity, such that $v_i = A_{ij}u_j$ and $A_{ij} = A_{ji}$. What surface is described by the end of $\boldsymbol{v}$ as $\boldsymbol{u}$ rotates?

Solution. Since A_{ij} is a symmetric tensor, its eigenvalues $\lambda^{(i)}$ are real and its principle directions are orthogonal (Appendix B). Referencing to these, and using $u_i u_i = 1$,

$$\frac{v_1^2}{\left[\lambda^{(1)}\right]^2} + \frac{v_2^2}{\left[\lambda^{(2)}\right]^2} + \frac{v_3^2}{\left[\lambda^{(3)}\right]^2} = 1\,.$$

Thus the end of the vector $\boldsymbol{v}$ describes an ellipsoid.

2.17. A material is pyroelectric if a temperature variation $\Delta\theta$ causes an electric polarization $P_i = p_i\Delta\theta$. What classes of point symmetry are compatible with pyroelectricity?

Solution. The pyroelectric tensor p_i is zero for the 11 centrosymmetric classes

$$\bar{1}\,, 2/m\,, mmm\,, \bar{3}\,, \bar{3}m\,, 4/m\,, 4/mmm\,, 6/m\,, 6/mmm\,, m3\,, m3m\,,$$

and, for no symmetry,

$$p_i = [p_1, p_2, p_3] \quad \text{for class } 1\,.$$

Any direct symmetry axis along Ox_3 leads to $p_1 = p_2 = 0$, and a mirror plane perpendicular to Ox_3 imposes $p_3 = 0$. Hence

$$\begin{aligned} p_i &= [p_1, p_2, 0] && \text{for class } m \\ p_i &= [0, 0, p_3] && \text{for classes } 2, 3, 4, 6, 2mm, 3m, 4mm, 6mm \\ p_i &= [0, 0, 0] && \text{for all other classes}\,. \end{aligned}$$

In summary, pyroelectricity can exist only in the ten classes possessing, at most, one direct axis with no mirror plane perpendicular to it.

2.18. Show that the three independent components $A_1 = A_{23}$, $A_2 = A_{31}$, $A_3 = A_{12}$ of an antisymmetric tensor A_{ij} transform as those of a vector if the sense (handedness) of the orthonormal coordinate system is preserved.

Solution. From (2.7),

$$A'_{jk} = \alpha_j^q \alpha_k^r A_{qr}\,.$$

On collecting the terms A_{qr} and $A_{rq} = -A_{qr}$ this gives

$$A'_{jk} = \sum_{q<r} \left(\alpha_j^q \alpha_k^r - \alpha_j^r \alpha_k^q\right) A_{qr}\,.$$

Assuming $j < k$ and $q < r$, then $M_i^p = \alpha_j^q \alpha_k^r - \alpha_j^r \alpha_k^q$ is the minor of the determinant $\|\alpha\|$ of the matrix $\alpha(i \neq j$ or $k, p \neq q$ or $r)$. By definition $A'_i = A'_{jk}$ when the permutation ijk is even, so that

$$A'_i = (-1)^{i+1} A'_{jk} \quad \text{for} \quad j < k$$

and

$$A_p = (-1)^{p+1} A_{qr} \quad \text{for} \quad q < r\,.$$

Substituting into the above gives

$$A'_i = \sum_p (-1)^{i+p} M_i^p A_p\,.$$

For an orthogonal transformation, the inverse matrix is given by

$$\beta_p^i = \frac{(-1)^{i+p} M_i^p}{\|\alpha\|} = \alpha_i^p\,,$$

which gives

$$A'_i = \|\alpha\| \left(\alpha_i^p A_p\right)\,,$$

where $\|\alpha\| = \pm 1$, with sign depending on whether the transformation changes the sense of the coordinate system.

2.19. Show that the contracted product $T_{ij}U_{ij}$ of a symmetric tensor T_{ij} with any other tensor U_{ij} involves only the symmetric part S_{ij} of U_{ij}.

Solution. Writing $U_{ij} = S_{ij} + A_{ij}$ with $S_{ji} = S_{ij}$ and $A_{ji} = -A_{ij}$, the product becomes

$$T_{ij}U_{ij} = T_{ij}S_{ij} + T_{ij}A_{ij} = T_{ij}S_{ji}\,,$$

since permutation of the dummy indices i and j gives

$$T_{ij}A_{ij} = T_{ji}A_{ji} = -T_{ij}A_{ij} = 0\,.$$

3. Elasticity and Piezoelectricity

A solid body becomes deformed when subjected to external forces. If it returns to its original form when the forces are removed, it is said to be elastic. The forces which cause the deformation are not necessarily of mechanical origin, since there can exist in the solid, depending on its nature, interactions between various types of quantities – mechanical, thermal, electrical and magnetic. Also, all bodies change form on heating. Although thermoelastic effects act comparatively slowly, they can be exploited to generate high frequency elastic waves, using suitable thermal sources such as pulsed lasers. This method, which has the advantage of not requiring mechanical contact, will be taken up in Chap. 3 of Vol. II. Alternatively, in most solids an electric field can produce mechanical stresses proportional to its square (electrostrictive coupling), and in a magnetic material a magnetic field can produce stresses proportional to its square (magnetostrictive coupling). However, these quadratic effects are insignificant for small fields. Thus, for elastic wave generation it is generally more suitable to choose materials in which the stress varies linearly with the applied field, that is, materials showing piezoelectric or piezomagnetic coupling. Although the piezomagnetic effect can be used in devices at low frequencies, below 1 MHz, it is not considered here; the applications of piezoelectricity are far more numerous. The piezoelectric generation and detection of travelling or stationary elastic waves, in the bulk or along a surface, is at the heart of acousto-electric and acousto-optic components and equipment. It is thus, owing to the piezoelectric effect, that electronics benefits from the remarkable mechanical properties of certain crystals, particularly their very high quality factor. Electromechanical resonators, typically of quartz, can be inserted directly into electrical circuits, the mechanical vibration being maintained electrically. These resonators are essential elements in compact and stable filters and oscillators. At the fringe of the range of applications covered in this book are piezoelectric motors and actuators, whose main role is in instruments such as interferometers and near-field microscopes.

This chapter therefore gives an account of elasticity and piezoelectricity. We define the characteristic variables for these two areas – stresses and strains defining the mechanical state of an ordinary solid, and the familiar electric field and displacement also needed in the case of a piezoelectric solid – and then summarize the equations of mechanics and electromagnetism (in the quasistatic approximation) which these variables must satisfy both inside a solid and at the boundary between two solids.

From these equations it is possible, without prejudging whether the solid behaves linearly or non-linearly, to deduce the various forms of energy arising from the external forces acting on it. The notions of energy flow and Poynting vector, introduced in Chap. 1, are generalized to the case of a piezoelectric solid, the description being valid also for a purely elastic solid. Restricting the study to linear solids, we have proportionality between the stress and the strain (Hooke's law) and, for a piezoelectric material, the electric field. These are related by the stiffness and piezoelectric tensors.

The number of independent components in these tensors depends on the symmetry of the solid. The stiffness tensor is of rank four, but the number of independent components cannot exceed 21 because of the symmetry of the stress and strain tensors and limitations imposed by thermodynamic considerations. These 21 constants are necessary to characterize a crystal in the triclinic system. Fortunately, the symmetry of other crystals can reduce this number considerably. For a cubic crystal three constants are sufficient, while two are enough for an isotropic solid.

Piezoelectric solids are necessarily devoid of a centre of symmetry, because piezoelectricity is represented by a tensor of rank three (Sect. 2.6.2). Again, for this tensor the number of independent components is reduced by the symmetry of the crystal and that of the quantities that it relates to. The largest number, 18, occurs in the triclinic class, while the smallest, 1, occurs in the cubic class.

3.1 The Elasto-electric Field

As long as the wavelength of elastic waves is large compared with the interatomic distances, any homogeneous solid can be regarded as a material continuum. Its mechanical state is described by two tensors of rank two, the stress and the strain. These time- and space-variant quantities constitute the elastic field, whose distribution is governed by the equations of mechanics.

In piezoelectric materials the elastic field is coupled to the electromagnetic field. In practice the velocity of elastic waves is very much less than that of electromagnetic waves, and consequently only the electric field accompanying the mechanical vibrations need be considered.

3.1.1 Mechanical Variables

Under the action of external forces, a solid undergoes a deformation, or strain, if its particles are displaced relative to each other. Stresses then arise, due to intermolecular forces, tending to return the solid to the state of rest that it had before the deformation.

3.1.1.1 Strain. Before defining strain in a general solid, we consider a one-dimensional case, the elongation of a straight elastic string of length L. One end is taken to be anchored at a rigid fixed support, as in Fig. 3.1.

Under the action of a tensile force F, applied at the free end, the string elongates and its diameter decreases. The deformation persists as long as the tension is maintained. The fractional elongation is $(L' - L)/L$, where L' is the length of the deformed string. Initially, the deformation cannot be assumed to be uniform; to obtain a valid solution it is necessary to consider the elongation of a small element and then, taking the limit, the elongation in the region of a point. Consider a portion of string situated initially between points M and N, with positions x and $x + \Delta x$. When the tension is applied these points move to M' and N' with positions respectively $x' = x + u(x)$ and $x' + \Delta x'$. The displacement of M is

$$u(x) = x' - x\,,$$

and that of N is

$$u(x + \Delta x) = x' + \Delta x' - (x + \Delta x) = x' - x + \Delta(x' - x) = u + \Delta u\,.$$

The string is deformed (elongated) if $\Delta u \neq 0$, but simply translated if $\Delta u = 0$.

In the region of the point M, with initial position x, the strain S is defined as the limit of the relative elongation of the element MN, of initial length Δx, in the limit $\Delta x \to 0$. Thus

$$S = \lim_{\Delta x \to 0} \left[\frac{u(x + \Delta x) - u(x)}{\Delta x} \right] = \frac{\mathrm{d}u}{\mathrm{d}x}\,. \tag{3.1}$$

It is worth noting that the strain, an elongation here, is a dimensionless ratio, and that the origin of the abscissa need not be specified except to locate the point being studied.

Similar considerations apply to the change of diameter, another aspect of the deformation. The relative change of diameter of the string is proportional to the relative elongation (Problem 3.3).

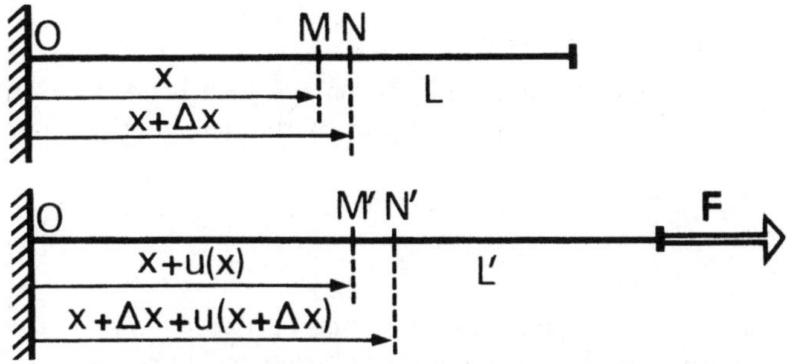

Fig. 3.1. Elongation of a string under the action of a force $\boldsymbol{F}$. Points M and N, which define an element Δx, are displaced by different amounts $u(x)$ and $u(x + \Delta x)$, becoming M' and N'. The strain S at the point M is defined as the ratio $[u(x + \Delta x) - u(x)]/\Delta x$ in the limit $\Delta x \to 0$, giving $S = \mathrm{d}u/\mathrm{d}x$

A three-dimensional solid of arbitrary shape, subject to external forces, differs from the string above in that neighbouring points M and N are not necessarily displaced in the same direction. The small segment MN can be subject not only to a length change, but also to a rotation, as in Fig. 3.2. The material point M, with rest coordinates x_i, moves under the action of the forces to the point M' with coordinates $x'_i = x_i + u_i$. The vector $\boldsymbol{u}(\boldsymbol{x}) = \boldsymbol{x}' - \boldsymbol{x}$ defines the displacement, a continuous function of the coordinates x_k. The displacement of the neighbouring material point N, with initial coordinates $x_i + \mathrm{d}x_i$, is therefore written

$$u_i\left(x_j + \mathrm{d}x_j\right) = u_i\left(x_j\right) + \frac{\partial u_i}{\partial x_j}\mathrm{d}x_j = u_i\left(x_j\right) + \mathrm{d}u_i\,, \tag{3.2}$$

where the summation from 1 to 3 over the dummy index j is understood. The medium is deformed only if different material points are displaced relative to each other, that is, if the *gradient of displacements* $\partial u_i/\partial x_j$ is non-zero. However, this tensor, of rank two, is not appropriate to express the deformation because it is non-zero in the case of a simple overall rotation, which does not affect the distances between material points and therefore leaves the internal state of the solid unchanged. For example, a rotation by an angle φ about Ox_3 is given by the matrix α_i^j (Sect. 2.6.1), so that

$$\mathrm{d}x'_i = \alpha_i^j \mathrm{d}x_j \quad \Rightarrow \quad \mathrm{d}u_i = \mathrm{d}x'_i - \mathrm{d}x_i = \left(\alpha_i^j - \delta_{ij}\right)\mathrm{d}x_j\,.$$

For a small angle φ the matrix is

$$\alpha_i^j \cong \begin{bmatrix} 1 & \varphi & 0 \\ -\varphi & 1 & 0 \\ 0 & 0 & 1 \end{bmatrix} \quad \text{giving} \quad \frac{\partial u_i}{\partial x_j} = \alpha_i^j - \delta_{ij} = \begin{bmatrix} 0 & \varphi & 0 \\ -\varphi & 0 & 0 \\ 0 & 0 & 0 \end{bmatrix}.$$

Thus the gradient of displacements is non-zero and antisymmetric. This result can be generalized to any infinitesimal rotation, because the latter can always be decomposed into three rotations about the coordinate axes. The antisymmetric part of the gradient, that is,

$$\Omega_{ij} = \frac{1}{2}\left(\frac{\partial u_i}{\partial x_j} - \frac{\partial u_j}{\partial x_i}\right) \tag{3.3}$$

must therefore be excluded. Only the symmetric part, which is zero for an overall movement, a translation or rotation, expresses the deformation. The strain is therefore defined as

$$\boxed{S_{ij} = \frac{1}{2}\left(\frac{\partial u_i}{\partial x_j} + \frac{\partial u_j}{\partial x_i}\right)}\,. \tag{3.4}$$

S_{ij} occurs naturally, to the exclusion of Ω_{ij}, in the expression for the change of length (or its square) of an element $\mathrm{d}x$. The length change is given by

$$\begin{aligned} \mathrm{d}\boldsymbol{x}' = \mathrm{d}\boldsymbol{x} + \mathrm{d}\boldsymbol{u}, \quad \text{giving} \quad \left(\mathrm{d}x'_i\right)^2 &= \left(\mathrm{d}x_i\right)^2 + 2\mathrm{d}x_i\mathrm{d}u_i + \left(\mathrm{d}u_i\right)^2 \\ &\cong \left(\mathrm{d}x_i\right)^2 + 2\mathrm{d}x_i\mathrm{d}u_i\,. \end{aligned}$$

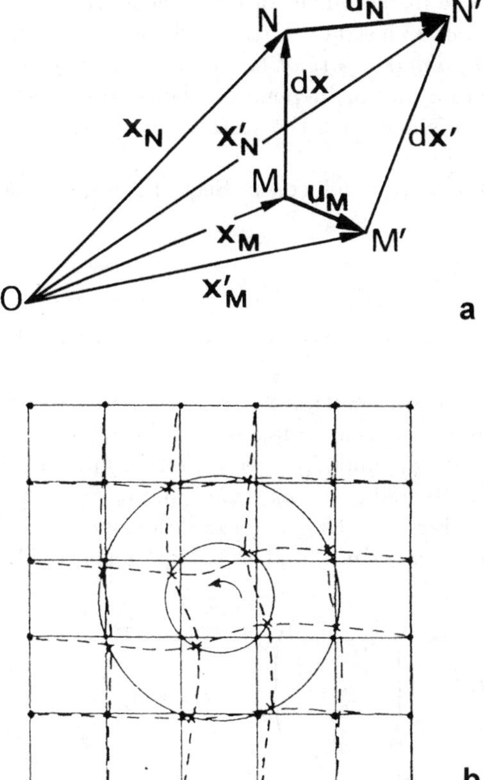

Fig. 3.2. Deformation of a solid. (**a**) Two neighbouring points M and N are displaced by different amounts in different directions. (**b**) A local rotation

Hence

$$(\mathrm{d}x_i')^2 - (\mathrm{d}x_i)^2 \cong 2\mathrm{d}x_i\mathrm{d}u_i = 2\frac{\partial u_i}{\partial x_j}\mathrm{d}x_i\mathrm{d}x_j \,.$$

Interchanging the dummy indices i and j does not affect the sum, so

$$(\mathrm{d}\boldsymbol{x}')^2 - (\mathrm{d}\boldsymbol{x})^2 = \left(\frac{\partial u_i}{\partial x_j} + \frac{\partial u_j}{\partial x_i}\right)\mathrm{d}x_i\mathrm{d}x_j = 2S_{ij}\mathrm{d}x_i\mathrm{d}x_j \,. \tag{3.5}$$

Equation (3.2) can therefore be written

$$u_i\,(x_j + \mathrm{d}x_j) = u_i\,(x_j) + S_{ij}\mathrm{d}x_j + \Omega_{ij}\mathrm{d}x_j \,. \tag{3.6}$$

The term Ω_{ij}, the antisymmetric part of the gradient of displacements, describes a local rotation which may vary with position in the solid. It plays no role in the propagation of elastic waves because it only causes a small inertial term, negligible in comparison with other terms in the dynamical equations (Sect. 3.1.2.3 below).

In summary, (3.6) shows that the displacement of an element $\mathrm{d}\boldsymbol{x}$ can be resolved, in general, into a translation u_i, a strain $S_{ij}\mathrm{d}x_j$, and a local rotation $\Omega_{ij}\mathrm{d}x_j$. *The strain tensor is the symmetric part of the gradient of displacements.* Since the values of the components S_{ij} depend on the coordinates of the point M where the deformation is analysed, the state of a strained solid is defined by a tensor field.

The *dilatation* $S = (\mathrm{d}V' - \mathrm{d}V)/\mathrm{d}V$, introduced in Sect. 1.2.1.3, can be expressed as

$$S = \operatorname{div} \boldsymbol{u} = \frac{\partial u_i}{\partial x_i} = S_{ii} \,. \tag{3.7}$$

Thus S is equal to the trace $S_{ii} = S_{11} + S_{22} + S_{33}$ of the strain tensor at the point being considered.

The terms S_{ij}, with $i \neq j$, cause displacements $\mathrm{d}u_j$ perpendicular to the element $\mathrm{d}x_i$, and correspond to *shear* motion. Suppose, for example, that the only non-diagonal components are S_{12} and $S_{21} (= S_{12})$. Two infinitesimal elements MN_1 and MN_2, initially directed along axes Ox_1 and Ox_2, do not remain perpendicular, as shown in Fig. 3.3. Under deformation, points M, N_1 and N_2 undergo displacements $\boldsymbol{u}, \boldsymbol{u} + \mathrm{d}\boldsymbol{u}^{(1)}$ and $\boldsymbol{u} + \mathrm{d}\boldsymbol{u}^{(2)}$ respectively, with

$$\mathrm{d}\boldsymbol{u}^{(1)} = \begin{pmatrix} \dfrac{\partial u_1}{\partial x_1}\mathrm{d}x_1 \\ \dfrac{\partial u_2}{\partial x_1}\mathrm{d}x_1 \\ 0 \end{pmatrix} \qquad \mathrm{d}\boldsymbol{u}^{(2)} = \begin{pmatrix} \dfrac{\partial u_1}{\partial x_2}\mathrm{d}x_2 \\ \dfrac{\partial u_2}{\partial x_2}\mathrm{d}x_2 \\ 0 \end{pmatrix} .$$

since $\partial u_3/\partial x_1 = \partial u_3/\partial x_2 = 0$ (by hypothesis, $S_{13} = S_{23} = 0$).

The angle γ_{12} between these two segments, under deformation, is

$$\gamma_{12} = \frac{\pi}{2} - \gamma_1 - \gamma_2$$

with

$$\tan\gamma_1 = \frac{\overline{P_1 N_1'}}{\overline{M' P_1}} = \frac{\dfrac{\partial u_2}{\partial x_1}\mathrm{d}x_1}{\mathrm{d}x_1 + \dfrac{\partial u_1}{\partial x_1}\mathrm{d}x_1} = \frac{\dfrac{\partial u_2}{\partial x_1}}{1 + \dfrac{\partial u_1}{\partial x_1}} .$$

Since the strains are very small, we have $\dfrac{\partial u_i}{\partial x_j} \ll 1$, giving

$$\tan\gamma_1 \cong \gamma_1 \cong \frac{\partial u_2}{\partial x_1} \quad \text{and} \quad \gamma_2 \cong \frac{\partial u_1}{\partial x_2} .$$

Hence

$$\gamma_{12} = \frac{\pi}{2} - \left(\frac{\partial u_2}{\partial x_1} + \frac{\partial u_1}{\partial x_2} \right) = \frac{\pi}{2} - 2S_{12} \,. \tag{3.8}$$

Thus, under a shear displacement S_{ij}, the angle between two elements $\mathrm{d}x_i$ and $\mathrm{d}x_j$, initially perpendicular, decreases by $2S_{ij}$.

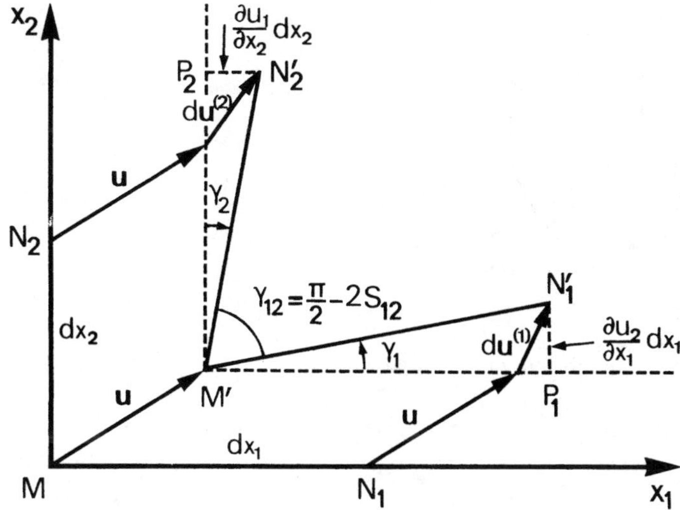

Fig. 3.3. Physical interpretation of the shear strain component S_{12}. Two elements MN_1 and MN_2, initially perpendicular, rotate under deformation, reducing the included angle by $2S_{12}$

3.1.1.2 Stress. External forces, which must be present if a solid is deformed, can arise from mechanical contact at the surface or, in the volume, from an applied field. A field in the interior of the material may cause a density of force per unit volume (for example the gravitational field), or a density of moment per unit volume (an electric field in a polar crystal). Mechanical traction forces appear in the deformed solid, tending to return it to its rest state and ensuring mechanical equilibrium. These forces arise from interactions between neighbouring particles. They are transmitted progressively through the solid by interatomic forces whose radius of action, a few interatomic distances, is negligible on the macroscopic scale. Consequently, the forces exerted between neighbouring particles (macroscopically small but large compared with the interatomic distance) can be considered to act across *surfaces* within the solid.

For an arbitrary surface element within the solid, with area $\mathrm{d}s$, the *mechanical traction* vector $\boldsymbol{T}$ is defined as the force per unit area, $\mathrm{d}\boldsymbol{F}/\mathrm{d}s$, exerted by the material on one side of the surface on the material on the other side. To define the sense of the force, we introduce the unit vector $\boldsymbol{l} = (l_1, l_2, l_3)$ normal to the surface, as in Fig. 3.4. Then $\boldsymbol{T}$ is defined such that the material which $\boldsymbol{l}$ points into exerts a force $\boldsymbol{T}$ per unit area on the material on the other side of the surface.

For a solid, unlike a liquid, shear forces can be transmitted across a surface, so the traction $\boldsymbol{T}$ is not necessarily parallel to $\boldsymbol{l}$. Since its magnitude and direction are both dependent on the orientation of the element $\mathrm{d}s$, the traction is written as

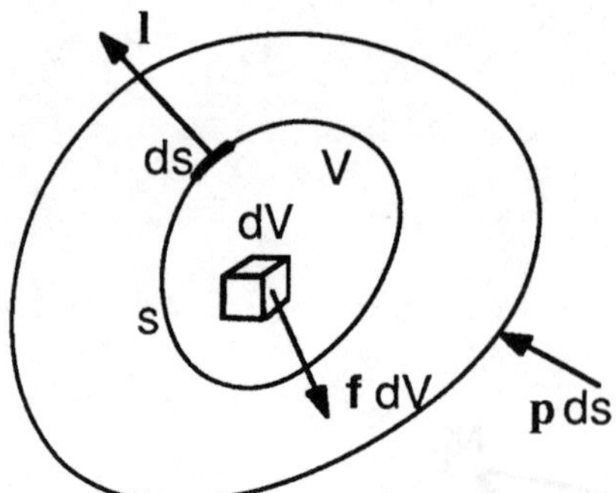

Fig. 3.4. Mechanical equilibrium of a volume within a solid

$$\boldsymbol{T}(\boldsymbol{l}) = \lim_{\mathrm{d}s \to 0} \left(\frac{\mathrm{d}\boldsymbol{F}}{\mathrm{d}s} \right) . \tag{3.9}$$

To define the *mechanical stress*, one method consists of writing the equilibrium conditions for a volume V inside the solid, subject to applied surface forces with density p_i and to internal forces with density f_i per unit volume, as in Fig. 3.4. The material surrounding the volume acts on it only by means of the forces applied to its surface. The resultant $\boldsymbol{F}$ of the forces applied to the volume V is therefore the sum of the mechanical tractions $\boldsymbol{T}(\boldsymbol{l})$ on its surface s and the volume forces acting in the interior, so that

$$F_i = \int_s T_i(\boldsymbol{l})\mathrm{d}s + \int_V f_i \mathrm{d}V = 0 . \tag{3.10}$$

From Green's theorem, f_i is the negative of the divergence of a tensor of rank two, so that

$$f_i = -\frac{\partial T_{ik}}{\partial x_k} \tag{3.11}$$

and (C.1) in Appendix C then gives

$$\int_s T_i(\boldsymbol{l})\mathrm{d}s = \int_s T_{ik} l_k \mathrm{d}s .$$

This gives

$$\boxed{T_i(\boldsymbol{l}) = T_{ik} l_k} . \tag{3.12}$$

This relation, with implied summation over the dummy index k, expresses the mechanical traction $\boldsymbol{T}(\boldsymbol{l})$ on a surface with any orientation $\boldsymbol{l}$, in terms of the quantities T_{ik}, called stresses. These nine quantities constitute the *stress tensor*, which has rank two since $T_i(\boldsymbol{l})$ and l_k are the components of two vectors. The force per unit area exerted on a surface by the material on the side which the normal $\boldsymbol{l}$ points into is thus $T_i = T_{ij} l_j$. If the surface element is perpendicular to an axis (so that $l_j = \delta_{jk}$), then $T_i = T_{ik}$; hence T_{ik} *is the*

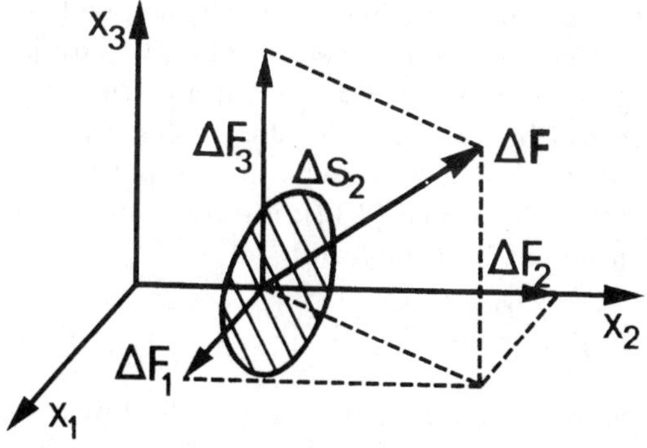

Fig. 3.5. Surface element perpendicular to a coordinate axis. On a surface element Δs_2, perpendicular to the Ox_2 axis, the three stresses acting are T_{12}, T_{22} and T_{32}

i-th component of the force acting per unit area on a surface perpendicular to the $\mathrm{O}x_k$ axis, that is,

$$T_{ik} = \lim_{\Delta s_k \to 0} \left(\frac{\Delta F_i}{\Delta s_k} \right) .$$

For example see Fig. 3.5,

$$T_{12} = \lim_{\Delta x_2 \to 0} \left(\frac{\Delta F_1}{\Delta s_2} \right) .$$

3.1.2 Field Equations

From the above relations, it is possible to express the conditions for static and dynamic equilibrium of a given volume within the solid, in terms of the stress tensor. Applying basic dynamics for a local rotation, we show here that this tensor must be symmetric, in the absence of an externally applied torque. When the solid is piezoelectric the field includes electrical components, and the equations which these must satisfy are specified, in the quasi-static approximation.

3.1.2.1 Static Case. If the forces acting in the volume arise, for example, from the gravitational field g_i, then $f_i = \rho g_i$ in (3.11), giving

$$\frac{\partial T_{ik}}{\partial x_k} + \rho g_i = 0 , \tag{3.13}$$

where ρ is the mass density of the solid. In the absence of any volume forces, this *equation of equilibrium* reduces to

$$\frac{\partial T_{ik}}{\partial x_k} = 0 .$$

3.1.2.2 Dynamic Case. So far, it has been assumed that all the points of a deformed solid remain at rest, so that the displacement u_i of each material point is a function only of its initial coordinates. However, if a disturbance travels through the medium there will be local movements, so that the displacement is also a function of time. In this case, the equation of motion can be obtained using Newton's law. From (3.10) and (3.12), the resultant of the forces acting on the volume V can be put into the form

$$F_i = \int_s T_{ik} l_k \mathrm{d}s + \int_V f_i \mathrm{d}V = \int_V \left(\frac{\partial T_{ik}}{\partial x_k} + f_i \right) \mathrm{d}V \,. \tag{3.14}$$

We can now write that, for a solid with internal stresses T_{ij}, the force per unit volume must equal the mass density ρ multiplied by the acceleration, which has components $\partial^2 u_i / \partial t^2$, so that

$$\boxed{\frac{\partial T_{ik}}{\partial x_k} + f_i = \rho \frac{\partial^2 u_i}{\partial t^2}} \,. \tag{3.15}$$

This is the fundamental elastodynamic equation of motion for a continuous solid medium. It is analogous to (1.43), which was derived for a fluid medium.

3.1.2.3 Local Rotations. We apply basic dynamics to the rotation of a small volume δV through an angle θ_k about the coordinate axis Ox_k, taking the volume to have moment of inertia I_k about this axis (Fig. 3.6). With respect to this axis, the torque due to the traction $\boldsymbol{T}$ is $\boldsymbol{r} \wedge \boldsymbol{T}$, and can be expressed with the aid of an antisymmetric tensor ε_{ijk} defined by

$$\varepsilon_{ijk} = \begin{cases} 1 & \text{if the permutation}(i,\ j,\ k) \text{ is even}\,, \\ 0 & \text{if two or three indices are equal}\,, \\ -1 & \text{if the permutation } (i,\ j,\ k) \text{ is odd}\,. \end{cases}$$

This gives

$$(\boldsymbol{r} \wedge \boldsymbol{T})_k = \varepsilon_{ijk} x_i T_j \,. \tag{3.16}$$

The volume δV is subject to tractions applied by the material outside it, giving a torque whose components M_k are obtained by integration over the surface δs, so that

$$M_k = \int_{\delta s} \varepsilon_{ijk} x_i T_j \mathrm{d}s$$

and, taking account of (3.12),

$$M_k = \int_{\delta s} \varepsilon_{ijk} x_i T_{jm} l_m \mathrm{d}s \,.$$

This surface integral can be transformed into a volume integral by Green's theorem, giving

$$M_k = \int_{\delta \mathrm{V}} \frac{\partial \left(\varepsilon_{ijk} x_i T_{jm} \right)}{\partial x_m} \mathrm{d}V = \int_{\delta \mathrm{V}} \left(\varepsilon_{ijk} x_i \frac{\partial T_{jm}}{\partial x_m} + \varepsilon_{ijk} \frac{\partial x_i}{\partial x_m} T_{jm} \right) \mathrm{d}V \,.$$

Since the coordinates x_i and x_m are independent, this leads to

$$\begin{aligned} M_k &= \int_{\delta V} \left(\varepsilon_{ijk} x_i \frac{\partial T_{jm}}{\partial x_m} + \varepsilon_{ijk} \delta_{im} T_{jm} \right) \mathrm{d}V \\ &= \int_{\delta V} \left(\varepsilon_{ijk} x_i \frac{\partial T_{jm}}{\partial x_m} + \varepsilon_{ijk} T_{ji} \right) \mathrm{d}V \,. \end{aligned} \tag{3.17}$$

If the solid is subject to an applied torque, with density G_k per unit volume, the equation for rotational motion of the volume δV is

$$I_k \frac{\partial^2 \theta}{\partial t^2} = \int_{\delta V} \left(\varepsilon_{ijk} x_i \frac{\partial T_{jm}}{\partial x_m} + \varepsilon_{ijk} T_{ji} + G_k \right) \mathrm{d}V$$

with

$$I_k = \int_{\delta V} r^2 \rho \mathrm{d}V \,.$$

As the volume δV tends towards zero, the integrands containing the terms r^2 and x become negligible, so that the equation reduces to

$$\varepsilon_{ijk} T_{ji} + G_k = 0 \quad \text{giving} \quad T_{ji} - T_{ij} + G_k = 0 \,, \tag{3.18}$$

where the index k is such that the permutation (ijk) is even. It follows that local rotations play no part in the equations of motion since they do not involve any inertial effects. They do, however, have a role in phenomena such as acousto-optic interactions where the crystal orientation, with respect to the light polarization, is relevant (Sect. 3.1.3 of Vol. II).

If the solid is not subject to an applied torque, so that $G_k = 0$, the fact that the three torque components are zero in (3.18) implies that

$$\boxed{T_{ji} = T_{ij}} \,. \tag{3.19}$$

Thus *the stress tensor is symmetric*, so that the number of its independent components is reduced to six. These are

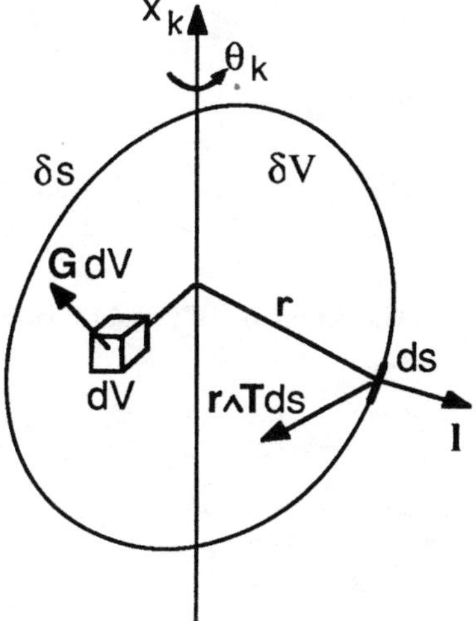

Fig. 3.6. A volume element δV subject to a torque

(a) T_{11}, T_{22}, T_{33}: stresses normal to faces, corresponding to tractions or compressions;
(b) T_{12}, T_{13}, T_{23}: tangential stresses, corresponding to shear forces (Fig. 3.5).

The case where the stress tensor is not symmetric is found, in practice, when a polar crystal is subjected to an electric field, for example, when it is placed between electrodes in order to form a transducer. However, since the torque is weak, the tensor can still be taken to be symmetric for practical purposes.

3.1.2.4 Electrical Variables. 'Quasi-static' Approximation. In a piezoelectric medium, the coupling between the elastic field and the electromagnetic field introduces electrical terms into the development of the mechanical equations, and mechanical terms into the development of Maxwell's equations. The distributions of these fields can in principle be determined only by simultaneously solving these coupled equations. In practice, since elastic fields involve displacements of the material, the velocity with which a stress or strain propagates is much less than that of an electric field. As already seen in Sect. 1.2.1.1, the velocity of elastic waves is 10^4 to 10^5 times smaller than the velocity of electromagnetic waves. Consequently, the magnetic field associated with mechanical vibrations plays little part; for example, the magnetic energy produced by a strain is negligible in comparison with the electrical energy. This implies that the electromagnetic field associated with an elastic field is quasi-static, so that Maxwell's equations reduce to

$$\operatorname{curl} \boldsymbol{E} = -\frac{\partial \boldsymbol{B}}{\partial t} \cong 0\,.$$

The only electrical variables needed for the description of a piezoelectric solid are the electric field E_i and the electric displacement D_i. As in electrostatics, the *electric field* is derived from a scalar potential Φ, so that

$$E_i = -\frac{\partial \Phi}{\partial x_i} \tag{3.20}$$

and the *electric displacement* obeys Poisson's equation

$$\boxed{\frac{\partial D_j}{\partial x_j} = \rho_{\mathrm{e}}}\,, \tag{3.21}$$

where ρ_{e} is the density of free charges per unit volume.

If the material is a conductor, one also needs the equation of conservation of charge, that is

$$\frac{\partial J_i}{\partial x_i} + \frac{\partial \rho_{\mathrm{e}}}{\partial t} = 0\,, \tag{3.22}$$

where J_i is the density of conduction current, per unit surface area.

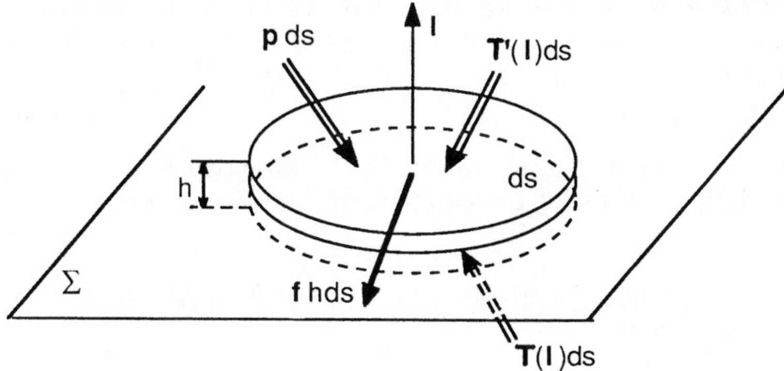

Fig. 3.7. The mechanical traction T_i is continuous across a boundary Σ

3.1.3 Boundary Conditions

In practice, any solid must have finite dimensions. The solutions of the above equations must satisfy boundary conditions which relate the mechanical and electrical variables at the boundary between two materials M and M'.

Mechanical Conditions. If the two materials are solids rigidly bound to each other, there is no possibility of one sliding over the other; hence the *material displacement* is continuous at all points on the boundary Σ separating them. Thus

$$u_i = u_i' \,. \tag{3.23}$$

It should be noted that, for the particular case of a boundary between a solid and a nonviscous liquid, such as water, this condition applies only to the normal component of displacement, since the liquid is free to move in any direction parallel to the surface.

For the *mechanical traction* $T_i(\boldsymbol{l})$, (3.15) is applied to a cylinder with small height h crossing the boundary, as in Fig. 3.7. Taking account of the surface force p_i per unit area, this gives

$$\left(T'_{ij}l_j - T_{ij}l_j\right) \mathrm{d}s + f_i h \,\mathrm{d}s + p_i \mathrm{d}s = \frac{\rho + \rho'}{2} h \,\mathrm{d}s \frac{\partial^2 u_i}{\partial t^2} \,.$$

As h tends to zero, the terms due to applied volume forces (f_i) and inertial forces disappear. The condition for mechanical equilibrium, satisfied at all points on the surface, is

$$T_{ij}l_j - T'_{ij}l_j = p_j \quad \text{giving} \quad T_i - T_i' = p_i \,. \tag{3.24}$$

Thus, in the absence of any external force acting on the boundary, the mechanical traction $T_i = T_{ij}l_j$ is continuous. If the second medium, using primed variables, is a vacuum, then $T_i' = 0$, so that

$$T_i = T_{ij}l_j = 0 \quad \text{on a } \textit{free surface} \,. \tag{3.25}$$

Electrical Conditions. The electrical conditions are as in electrostatics. The potential Φ and the *tangential component of the electric field* E_t are continuous, so that

$$E'_t = E_t, \quad \Phi' = \Phi. \tag{3.26}$$

The *normal component of displacement*, D_n, is discontinuous if there are charges on the surface. Taking the surface charge density as σ we have

$$(D'_i - D_i)\, l_i = \sigma, \quad \text{i.e.,} \quad D'_n - D_n = \sigma. \tag{3.27}$$

If there is a current density, with normal component $J_n = J_i l_i$, conservation of charge requires, from (3.22),

$$J_n - J'_n = \frac{\mathrm{d}\sigma}{\mathrm{d}t}. \tag{3.28}$$

In all these relations the surface normal is directed from the solid under consideration into the external medium, the latter having primed variables, as in Fig. 3.7.

3.1.4 Poynting's Theorem and Energy Conservation

The work done by mechanical forces is $\mathrm{d}w_m = f_i \mathrm{d}u_i$, and that done by electric fields is $\mathrm{d}w_e = \Phi\, \mathrm{d}\rho_e$. Hence the instantaneous power, per unit volume, supplied by the volume densities of force (f_i) and charge (ρ_e) is

$$\frac{\mathrm{d}w}{\mathrm{d}t} = f_i \dot{u}_i + \Phi \dot{\rho}_e. \tag{3.29}$$

Using the elasto-electric field relations (3.15) and (3.21), this becomes

$$\frac{\mathrm{d}w}{\mathrm{d}t} = \rho \ddot{u}_i \dot{u}_i - \frac{\partial T_{ij}}{\partial x_j} \dot{u}_i + \Phi \frac{\partial \dot{D}_j}{\partial x_j}$$

and since $E_j = -\partial\Phi/\partial x_j$ this gives

$$\frac{\mathrm{d}w}{\mathrm{d}t} = \rho \ddot{u}_i \dot{u}_i + T_{ij} \frac{\partial \dot{u}_i}{\partial x_j} + E_j \dot{D}_j + \frac{\partial}{\partial x_j}\left(-T_{ij}\dot{u}_i + \Phi \dot{D}_j\right).$$

Since the stress tensor T_{ij} is symmetric, the second term on the right contains only the symmetric part of the gradient of displacements (Problem 2.19), that is, the strain $S_{ij} = \frac{1}{2}\left(\partial u_i/\partial x_j + \partial u_j/\partial x_i\right)$. Hence,

$$\frac{\mathrm{d}w}{\mathrm{d}t} = \rho \ddot{u}_i \dot{u}_i + T_{ij} \dot{S}_{ij} + E_j \dot{D}_j + \frac{\partial}{\partial x_j}\left(-T_{ij}\dot{u}_i + \Phi \dot{D}_j\right). \tag{3.30}$$

We can identify the physical significance of each term on the right of this equation. The first term is the time derivative of the *kinetic energy* e_k per unit volume, given by

$$e_k(x_i, t) = \frac{1}{2}\rho \dot{u}_i^2 \quad \text{so that} \quad \rho \ddot{u}_i \dot{u}_i = \frac{\mathrm{d}e_k}{\mathrm{d}t}. \tag{3.31}$$

The second and third terms are related to the stored *mechanical* and *electrical potential energies*, per unit volume. The total potential energy is denoted e_p, defined in terms of its differential $\mathrm{d}e_\mathrm{p}$ which is

$$\boxed{\mathrm{d}e_\mathrm{p} = T_{ij}\mathrm{d}S_{ij} + E_j\mathrm{d}D_j} \quad \text{giving} \quad \frac{\mathrm{d}e_\mathrm{p}}{\mathrm{d}t} = T_{ij}\frac{\mathrm{d}S_{ij}}{\mathrm{d}t} + E_j\frac{\mathrm{d}D_j}{\mathrm{d}t}\,. \tag{3.32}$$

The last term of (3.30) is the divergence of the instantaneous *Poynting vector*, defined by

$$\boxed{P_j(x_i,t) = -T_{ij}\frac{\partial u_i}{\partial t} + \Phi\frac{\partial D_j}{\partial t}}\,. \tag{3.33}$$

Here the first term is the acoustic part of the Poynting vector, corresponding to the product $v\delta p$ in a fluid with particle velocity v (Sect. 1.2.2). The second term represents the electrical contribution.

With these definitions, (3.30) can be put into the form

$$\frac{\mathrm{d}w}{\mathrm{d}t} = \frac{\mathrm{d}}{\mathrm{d}t}(e_\mathrm{k} + e_\mathrm{p}) + \frac{\partial P_j}{\partial x_j}\,. \tag{3.34}$$

This equation expresses locally, at all points, the law of energy conservation.

As in Sect. 1.2.2, we designate W as the work done by the sources, and E_k and E_p as the kinetic and potential energies, for a volume V of the material. With these, integration of (3.34) leads to the relation

$$\boxed{\frac{\mathrm{d}W}{\mathrm{d}t} = \frac{\mathrm{d}}{\mathrm{d}t}(E_\mathrm{k} + E_\mathrm{p}) + \int_s P_j l_j \mathrm{d}s}\,, \tag{3.35}$$

which is a generalization of (1.50). The last term represents the power P_R radiated across the surface s of the volume V. From (3.33), P_R can be expressed in terms of the mechanical traction $T_i(\boldsymbol{l})$ and the normal component of electric displacement, D_n, giving

$$P_\mathrm{R} = \int_s \left[-T_i(\boldsymbol{l})\dot{u}_i + \Phi\dot{D}_\mathrm{n}\right] \mathrm{d}s\,. \tag{3.36}$$

These results are valid for both linear and non-linear materials.

3.2 Linear Behaviour of an Elastic Solid

A solid is elastic if it returns to its initial state when the external forces responsible for deformations are removed. This return to the initial state is the work of internal stresses. In the rest state, the stresses and strains are all zero.

3.2.1 Hooke's Law

In a non-piezoelectric elastic solid, there is a one-to-one relation between the stresses and the strains. Experimental evidence shows that the elastic behaviour of most substances is well described, assuming small deformations, by the first-order term in the following Taylor expansion for the stress:

$$T_{ij}(S_{kl}) = T_{ij}(0) + \left(\frac{\partial T_{ij}}{\partial S_{kl}}\right)_{S_{kl}=0} S_{kl} + \frac{1}{2}\left(\frac{\partial^2 T_{ij}}{\partial S_{kl}\partial S_{mn}}\right)_{\substack{S_{kl}=0\\ S_{mn}=0}} S_{kl}S_{mn} + \ldots .$$

This gives, since $T_{ij}(0) = 0$,

$$\boxed{T_{ij} = c_{ijkl}S_{kl}} \tag{3.37}$$

where we define

$$c_{ijkl} = \left(\frac{\partial T_{ij}}{\partial S_{kl}}\right)_{S_{kl}=0} . \tag{3.38}$$

The coefficients c_{ijkl} are components of the *stiffness tensor*, of rank four, which expresses the most general possible linear relation between the second-rank tensors T_{ij} and S_{kl}. The law of proportionality between stress and strain was first stated in the 17th century by Hooke, for the case of a stretched elastic string.

A tensor of rank four has $3^4 = 81$ components. Since the tensors T_{ij} and S_{kl} are symmetric, the elastic constants defined by (3.37) are unaffected when either the first two or the last two indices are interchanged, so that

$$c_{ijkl} = c_{jikl} \quad \text{and} \quad c_{ijkl} = c_{ijlk} . \tag{3.39}$$

In terms of displacements, Hooke's law (3.37) becomes

$$T_{ij} = \frac{1}{2}c_{ijkl}\frac{\partial u_k}{\partial x_l} + \frac{1}{2}c_{ijkl}\frac{\partial u_l}{\partial x_k}$$

and since $c_{ijkl} = c_{ijlk}$ the two summations on the right are equal, so that

$$\boxed{T_{ij} = c_{ijkl}\frac{\partial u_l}{\partial x_k}} . \tag{3.40}$$

The above symmetry relations reduce the number of independent elastic constants from 81 to 36. Indeed, a pair of unordered indices (i, j) can give only six independent values. These are numbered 1 to 6 as follows:

$$\begin{array}{ccc} (11) \leftrightarrow 1 & (22) \leftrightarrow 2 & (33) \leftrightarrow 3 \\ (23) = (32) \leftrightarrow 4 & (31) = (13) \leftrightarrow 5 & (12) = (21) \leftrightarrow 6 \end{array} . \tag{3.41}$$

The independent elastic moduli can thus be represented in terms of only two indices α and β, with values 1 to 6, corresponding to a 6×6 square array with 36 entries (Sect. 3.2.3), such that

$$c_{\alpha\beta} = c_{ijkl} ,$$

where α is related to (ij), and β to (kl), in accordance with (3.41). For example,

$$c_{14} = c_{1123} = c_{1132}$$
$$c_{56} = c_{1312} = c_{1321} = c_{3121} = c_{3112} \, .$$

This notation, called the matrix notation to distinguish it from the tensor notation c_{ijkl}, can be extended to the stresses and strains. Thus Hooke's law can be written

$$T_\alpha = c_{\alpha\beta} S_\beta \quad \alpha, \beta = 1, 2, \ldots, 6 \, . \tag{3.42}$$

It is usual to adopt the convention

$$T_\alpha = T_{ij}, \quad \text{with} \quad \alpha \leftrightarrow (ij) \quad \text{as in (3.41)} \, . \tag{3.43}$$

However, it must be noted that this requires S_β to be defined as

$$S_1 = S_{11}, S_2 = S_{22}, S_3 = S_{33}, S_4 = 2S_{23}, S_5 = 2S_{13}, S_6 = 2S_{12} \tag{3.44}$$

in order that (3.42) gives Hooke's law correctly. For example, the expansion for T_{11} is, from (3.37),

$$T_{11} = c_{11kl} S_{kl} \, ,$$

that is,

$$T_{11} = c_{1111} S_{11} + c_{1122} S_{22} + c_{1133} S_{33} + 2c_{1123} S_{23} \\ + 2c_{1113} S_{13} + 2c_{1112} S_{12} \, .$$

This can be identified with T_1 from (3.42), given by

$$T_1 = c_{11} S_1 + c_{12} S_2 + c_{13} S_3 + c_{14} S_4 + c_{15} S_5 + c_{16} S_6$$

provided the relations (3.44) are satisfied.

Hooke's law can also be inverted in order to express the strains in terms of the stresses, giving

$$S_{ij} = s_{ijkl} T_{kl} \, . \tag{3.45}$$

The *compliance coefficients* s_{ijkl} form a tensor of rank four with the same symmetry properties as the tensor c_{ijkl}, so that

$$s_{ijkl} = s_{jikl} \quad \text{and} \quad s_{ijkl} = s_{ijlk} \, .$$

Similarly, (3.42) can be expressed in terms of S_α, leading to the solution

$$S_\alpha = s_{\alpha\beta} T_\beta \, . \tag{3.46}$$

Here the matrix $s_{\alpha\beta}$ is the reciprocal of the matrix $c_{\alpha\beta}$, so that

$$s_{\alpha\beta} = (c_{\alpha\beta})^{-1}$$

that is,

$$s_{\alpha\beta} \, c_{\beta\gamma} = \delta_{\alpha\gamma} \, ,$$

where $\delta_{\alpha\gamma}$ is the six-dimensional Kronecker delta.

To relate the $s_{\alpha\beta}$ to the s_{ijkl} we use (3.45) to expand, for example, S_{13}, giving

$$S_{13} = s_{1311}T_{11} + s_{1322}T_{22} + s_{1333}T_{33} + 2s_{1323}T_{23} + 2s_{1313}T_{13} + 2s_{1312}T_{12} \,.$$

Following the correspondence rule of (3.43), and with $S_5 = 2S_{13}$, comparison with (3.46) gives

$$\begin{array}{lll} s_{51} = 2s_{1311} & s_{52} = 2s_{1322} & s_{53} = 2s_{1333} \\ s_{54} = 4s_{1323} & s_{55} = 4s_{1313} & s_{56} = 4s_{1312} \end{array} .$$

In general, if p is the number of indices greater than 3 in the pair (α, β), then

$$s_{\alpha\beta} = 2^p s_{ijkl} \,. \tag{3.47}$$

Thermodynamic considerations reduce further the number of independent elastic constants, as will be seen below.

3.2.2 Elastic Energy of a Strained Solid. Maxwell's Relations

The work done by external forces during a deformation must be stored in the form of elastic potential energy, since, when the forces are removed, this energy is returned by the internal forces which restore the solid to its original state. For a strain change $\mathrm{d}S_{ij}$ (Sect. 3.1.4), the change of elastic potential energy per unit volume is, from (3.32),

$$\mathrm{d}e_\mathrm{p} = T_{ij}\mathrm{d}S_{ij} = \delta w \,.$$

Thus the change of *internal energy* U per unit volume is

$$\mathrm{d}U = \delta w + \delta Q \,, \tag{3.48}$$

where δQ is the heat gained per unit volume. From the first law of thermodynamics, U is a function of state for the solid and $\mathrm{d}U$ is an exact differential; this is not the case individually for δw or δQ. The second law of thermodynamics shows that, for a reversible change,

$$\delta Q = \theta \, \mathrm{d}\sigma \,,$$

where θ is the absolute temperature and σ the entropy per unit volume.

Equation (3.48) thus becomes

$$\mathrm{d}U = \theta \, \mathrm{d}\sigma + T_{ij}\mathrm{d}S_{ij} \,. \tag{3.49}$$

Hence the internal energy of a strained body depends on the entropy and the strains, so that $U = U(\sigma, S_{ij})$. It follows that the stress can be expressed as

$$T_{ij} = \left(\frac{\partial U}{\partial S_{ij}}\right)_\sigma \,,$$

where the subscript σ indicates that the partial derivative is to be taken with the entropy held constant. Expressions for the elastic constants c_{ijkl} can be obtained from the definition in (3.38), which gives

$$c_{ijkl} = \left(\frac{\partial T_{ij}}{\partial S_{kl}}\right) \quad \text{and} \quad c_{klij} = \left(\frac{\partial T_{kl}}{\partial S_{ij}}\right) ,$$

and on substituting for T_{ij} and T_{kl} we find

$$c_{ijkl}^{(\sigma)} = \left(\frac{\partial^2 U}{\partial S_{ij} \partial S_{kl}}\right)_\sigma = c_{klij}^{(\sigma)} . \tag{3.50}$$

Thus the interchange of the first two indices with the last two does not affect the value of the elastic constants, in the *adiabatic* case ($\sigma =$ constant) considered here. These are the moduli involved in elastic wave propagation, because the wave velocities are high enough to ensure that little heat can flow over a distance of one wavelength during one cycle. We therefore have $\delta Q = 0$, and hence $\mathrm{d}\sigma = 0$ if the process is reversible.

The symmetry relation (3.50), which is Maxwell's relation, is also valid in other thermodynamic conditions. For example, for isothermal changes ($\theta =$ constant), we consider another function of state, the *free energy* $F = U - \theta\sigma$. Taking account of (3.49), the change of F is

$$\mathrm{d}F = -\sigma \mathrm{d}\theta + T_{ij} \mathrm{d}S_{ij} . \tag{3.51}$$

The free energy depends on the temperature θ and the strains S_{ij}, so that

$$F = F(\theta, S_{ij}) .$$

Consequently,

$$T_{ij} = \left(\frac{\partial F}{\partial S_{ij}}\right)_\theta$$

and the isothermal elastic constants are

$$c_{ijkl}^{(\theta)} = \left(\frac{\partial^2 F}{\partial S_{ij} \partial S_{kl}}\right)_\theta = c_{klij}^{(\theta)} . \tag{3.52}$$

When Hooke's law is valid, the change of internal energy is, from (3.49),

$$\mathrm{d}U = \theta\, \mathrm{d}\sigma + c_{ijkl}^{(\sigma)} S_{kl} \mathrm{d}S_{ij}$$

and this can be written, permuting pairs of dummy indices (ij) and (kl),

$$\mathrm{d}U = \theta\, \mathrm{d}\sigma + \frac{1}{2}\left(c_{ijkl}^{(\sigma)} S_{kl} \mathrm{d}S_{ij} + c_{klij}^{(\sigma)} S_{ij} \mathrm{d}S_{kl}\right) .$$

Using Maxwell's relation (3.50), this gives

$$\mathrm{d}U = \theta\, \mathrm{d}\sigma + \frac{1}{2}\, c_{ijkl}^{(\sigma)}\, \mathrm{d}(S_{ij} S_{kl}) .$$

Integrating this, the energy per unit volume of a strained solid is expressed as

$$U(\sigma, S_{ik}) = U_0(\sigma) + \frac{1}{2}\, c_{ijkl}^{(\sigma)}\, S_{ij} S_{kl} , \tag{3.53}$$

where $U_0(\sigma) \equiv U(\sigma, S_{ij} = 0)$ is the internal energy before deformation.

The free energy per unit volume is obtained in the same way from (3.51), giving

$$F(\theta, S_{ik}) = F_0(\theta) + \frac{1}{2}\, c^{(\theta)}_{ijkl}\, S_{ij} S_{kl} \,.$$

Thus the mechanical contribution to the internal energy or the free energy, for either thermodynamic condition, arises from an elastic potential energy e_p (per unit volume) which is a quadratic function of the strain, given by

$$\boxed{e_\mathrm{p} = \frac{1}{2}\, c_{ijkl}\, S_{ij} S_{kl}} \,. \tag{3.54}$$

Depending on whether the process of deformation is adiabatic or isothermal, this quantity gives the increase of internal energy or free energy per unit volume in an elastic solid.

3.2.3 Effect of Crystal Symmetry on Elastic Constants

Maxwell's relation, valid for all solids, can be written in matrix notation as

$$c_{\alpha\beta} = c_{\beta\alpha} \,.$$

The 6×6 matrix of the coefficients $c_{\alpha\beta}$ is therefore symmetric about its main diagonal, so that

$$(c_{\alpha\beta}) = \begin{vmatrix} c_{11} & c_{12} & c_{13} & c_{14} & c_{15} & c_{16} \\ c_{12} & c_{22} & c_{23} & c_{24} & c_{25} & c_{26} \\ c_{13} & c_{23} & c_{33} & c_{34} & c_{35} & c_{36} \\ c_{14} & c_{24} & c_{34} & c_{44} & c_{45} & c_{46} \\ c_{15} & c_{25} & c_{35} & c_{45} & c_{55} & c_{56} \\ c_{16} & c_{26} & c_{36} & c_{46} & c_{56} & c_{66} \end{vmatrix} \qquad \textit{triclinic} \,. \tag{3.55}$$

This property reduces the number of independent components to 21 (= 6 + 5 + 4 + 3 + 2 + 1). Crystals of the *triclinic system* are richest from this point of view; a centre of symmetry imposes no restriction, so all these crystals have 21 independent elastic constants. Fortunately, in other crystal systems the symmetry intervenes to reduce the number of independent constants. We first consider the other extreme, showing that the mechanical properties of an isotropic solid, which has the highest symmetry possible, are completely described in terms of only two coefficients.

3.2.3.1 Isotropic Solid. The physical constants of an isotropic solid are, by definition, independent of the choice of the coordinate axes, assumed orthonormal here. In particular, the elastic tensor c_{ijkl} must be invariant for any change of axes – rotation, or symmetry about a point or a plane. Now, the only quantities unaffected by these orthogonal transformations are a scalar or the unit tensor δ_{ij}. Consequently, each component c_{ijkl} can be expressed in terms of components of the unit tensor. Because of the symmetry $\delta_{ij} = \delta_{ji}$, there are only three distinct combinations containing the four indices $ijkl$, and these are

$$\delta_{ij}\delta_{kl}, \quad \delta_{ik}\delta_{jl}, \quad \delta_{il}\delta_{jk} \,.$$

The tensor therefore takes the form

$$c_{ijkl} = \lambda\delta_{ij}\delta_{kl} + \mu_1\delta_{ik}\delta_{jl} + \mu_2\delta_{il}\delta_{jk}\,,$$

where λ, μ_1 and μ_2 are constants. Moreover, the condition $c_{ijkl} = c_{jikl}$ demands that $\mu_1 = \mu_2 = \mu$. Other required symmetry relations are satisfied by this, and so we have

$$c_{ijkl} = \lambda\delta_{ij}\delta_{kl} + \mu(\delta_{ik}\delta_{jl} + \delta_{il}\delta_{jk})\,. \tag{3.56}$$

Thus, the properties of an *isotropic* solid are specified by *two independent constants*, such as the Lamé constants λ and μ used here.

Assigning values 1 to 6 to the pairs (ij) and (kl) gives

$$\begin{aligned} c_{11} &= c_{22} = c_{33} = \lambda + 2\mu \\ c_{12} &= c_{23} = c_{13} = \lambda \\ c_{44} &= c_{55} = c_{66} = \mu = (c_{11} - c_{12})/2\,. \end{aligned} \tag{3.57}$$

The other twelve of the moduli c_{ijkl} are zero since they have an odd number of distinct indices, for example $c_{25} = c_{2213}$. Expressing all the components in terms of c_{11} and c_{12}, the matrix $c_{\alpha\beta}$ takes the following form:

$$(c_{\alpha\beta}) = \left|\begin{array}{cccccc} c_{11} & c_{12} & c_{12} & 0 & 0 & 0 \\ c_{12} & c_{11} & c_{12} & 0 & 0 & 0 \\ c_{12} & c_{12} & c_{11} & 0 & 0 & 0 \\ 0 & 0 & 0 & c_{66} & 0 & 0 \\ 0 & 0 & 0 & 0 & c_{66} & 0 \\ 0 & 0 & 0 & 0 & 0 & c_{66} \end{array}\right| \quad \textit{isotropic},\ c_{66} = (c_{11} - c_{12})/2\,. \tag{3.58}$$

In an isotropic solid, Hooke's law is much simplified. In terms of the Lamé constants, the *normal* stresses (T_{11}, T_{22}, T_{33}) are given by

$$T_{ii} = c_{iikl}S_{kl} = (\lambda\delta_{kl} + 2\mu\delta_{ik}\delta_{il})S_{kl}\,,$$

so that

$$T_{ii} = \lambda(S_{11} + S_{22} + S_{33}) + 2\mu S_{ii} \tag{3.59}$$

and the *tangential* stresses (T_{ij}, with $i \neq j$) are

$$T_{ij} = c_{ijkl}S_{kl} = \mu(\delta_{ik}\delta_{jl} + \delta_{il}\delta_{jk})S_{kl}$$

giving

$$T_{ij} = 2\mu S_{ij}, \quad i \neq j\,. \tag{3.60}$$

Equations (3.59) and (3.60) can be combined to give

$$\boxed{T_{ij} = \lambda S\,\delta_{ij} + 2\mu S_{ij}}\,, \tag{3.61}$$

where $S = S_{11} + S_{22} + S_{33}$ is the dilatation $\Delta V/V$.

Equation (3.60) shows that μ is the proportionality coefficient between a shear stress T_{ij} and the change $2S_{ij}$ of the angle between two directions i and j which are initially perpendicular (Fig. 3.3).

	E, ν	E, μ	λ, μ	c_{11}, c_{12}
λ	$\frac{E\nu}{(1+\nu)(1-2\nu)}$	$\frac{\mu(E-2\mu)}{3\mu-E}$	λ	c_{12}
μ	$\frac{E}{2(1+\nu)}$	μ	μ	$\frac{c_{11}-c_{12}}{2}$
E	E	E	$\frac{\mu(3\lambda+2\mu)}{\lambda+\mu}$	$c_{11}-2\frac{c_{12}^2}{c_{11}+c_{12}}$
B	$\frac{E}{3(1-2\nu)}$	$\frac{\mu E}{3(3\mu-E)}$	$\lambda+\frac{2}{3}\mu$	$\frac{c_{11}+2c_{12}}{3}$
ν	ν	$\frac{E-2\mu}{2\mu}$	$\frac{\lambda}{2(\lambda+\mu)}$	$\frac{c_{12}}{c_{11}+c_{12}}$

Fig. 3.8. Relations between the various elastic constants and Poisson's ratio ν for an isotropic solid

Of course, an isotropic solid is equally well characterized by an alternative pair of parameters, the Young's modulus E and the Poisson's ratio ν (see Problem 3.3), instead of the elastic constants c_{11} and c_{12} or the Lamé constants λ, μ. The relationships between these coefficients are shown in the table of Fig. 3.8. The volume elastic modulus B arises when the solid is subject to a hydrostatic pressure, so that $T_{ij} = -p\delta_{ij}$, and from (3.61) this leads to

$$S_{ij} = \frac{S}{3}\delta_{ij} \quad \text{and} \quad p = -\left(\lambda + \frac{2}{3}\mu\right) S = -BS\,.$$

3.2.3.2 Crystalline Media. Since the axes are taken to be orthonormal, the general invariance condition (2.13) can be written, for the stiffness tensor, as

$$c_{ijkl} = \alpha_i^p \alpha_j^q \alpha_k^r \alpha_l^s c_{pqrs}\,. \tag{3.62}$$

As already stated, crystals in the *triclinic* system have 21 *independent elastic constants*, arranged as in the matrix of (3.55).

For the *monoclinic* system the dyad axis, direct or inverse (mirror), is taken to be parallel to the Ox_3 axis, and the axis rotation matrix α has the diagonal form

$$\alpha = \pm \begin{pmatrix} -1 & 0 & 0 \\ 0 & -1 & 0 \\ 0 & 0 & 1 \end{pmatrix}.$$

Equation (3.62), which becomes

$$c_{ijkl} = \alpha_i^i \alpha_j^j \alpha_k^k \alpha_l^l c_{ijkl}\,,$$

implies cancellation of the constants in which the index 3 occurs an odd number of times; these have $\alpha_i^i \alpha_j^j \alpha_k^k \alpha_l^l = -1$. There are therefore 13 *independent constants*, those in which the index 3 occurs 0, 2 or 4 times in the subscripts, and the matrix becomes

$$(c_{\alpha\beta}) = \begin{vmatrix} c_{11} & c_{12} & c_{13} & 0 & 0 & c_{16} \\ c_{12} & c_{22} & c_{23} & 0 & 0 & c_{26} \\ c_{13} & c_{23} & c_{33} & 0 & 0 & c_{36} \\ 0 & 0 & 0 & c_{44} & c_{45} & 0 \\ 0 & 0 & 0 & c_{45} & c_{55} & 0 \\ c_{16} & c_{26} & c_{36} & 0 & 0 & c_{66} \end{vmatrix} \quad \textit{monoclinic}. \tag{3.63}$$

Crystals in the *orthorhombic* system are characterized by having three orthogonal dyad axes, direct or inverse, which are of course taken as the coordinate axes. Applying the above reasoning to each of these axes, that is, to each index, leads to the matrix of (3.64) below, where the only remaining components are those having indices which repeat an even number of times. Thus there are only *nine* independent constants.

$$(c_{\alpha\beta}) = \begin{vmatrix} c_{11} & c_{12} & c_{13} & 0 & 0 & 0 \\ c_{12} & c_{22} & c_{23} & 0 & 0 & 0 \\ c_{13} & c_{23} & c_{33} & 0 & 0 & 0 \\ 0 & 0 & 0 & c_{44} & 0 & 0 \\ 0 & 0 & 0 & 0 & c_{55} & 0 \\ 0 & 0 & 0 & 0 & 0 & c_{66} \end{vmatrix} \quad \textit{orthorhombic}. \tag{3.64}$$

For *cubic* crystals, which have at least four triad axes and three direct dyad axes, the latter are taken as the coordinate axes. The non-zero components are the same as those given above for the orthorhombic system. A rotation by $2\pi/3$ about the triad axis directed along [111] transforms Ox_1 into Ox_2, Ox_2 into Ox_3, and Ox_3 into Ox_1. The c_{ijkl} are therefore invariant for the cyclic permutation of the indices

$$(123) \quad \to \quad (231) \quad \to \quad (312).$$

This implies the relations

$$\begin{aligned} & c_{1111} = c_{2222} & & c_{2222} = c_{3333} \\ \text{and} \quad & c_{1122} = c_{2233} = c_{3311} & & c_{1212} = c_{2323} = c_{3131} \end{aligned}$$

thus reducing the number of independent constants to *three* for the cubic system. The matrix (3.64) becomes

$$(c_{\alpha\beta}) = \begin{vmatrix} c_{11} & c_{12} & c_{12} & 0 & 0 & 0 \\ c_{12} & c_{11} & c_{12} & 0 & 0 & 0 \\ c_{12} & c_{12} & c_{11} & 0 & 0 & 0 \\ 0 & 0 & 0 & c_{44} & 0 & 0 \\ 0 & 0 & 0 & 0 & c_{44} & 0 \\ 0 & 0 & 0 & 0 & 0 & c_{44} \end{vmatrix} \quad \textit{cubic}. \tag{3.65}$$

Principal Axis with Order Greater than Two. Crystals in the trigonal, tetragonal or hexagonal systems have a single axis, direct or inverse, with order n greater than 2. The rotation matrix α about this principal axis, taken as Ox_3, is no longer diagonal:

$$\alpha = \begin{pmatrix} \cos\varphi & \sin\varphi & 0 \\ -\sin\varphi & \cos\varphi & 0 \\ 0 & 0 & 1 \end{pmatrix} \quad \text{with} \quad \varphi = \frac{2\pi}{n} \neq \pi\,. \tag{3.66}$$

The invariance condition (3.62) is more complex in this case because it involves many of the components. So we consider instead a diagonal rotation matrix obtained by using the eigenvectors $\xi^{(1)}$, $\xi^{(2)}$, $\xi^{(3)}$ of the rotation as basis vectors. The matrix is diagonalized by solving the following set of equations, analogous to (B.3) (Appendix B):

$$\left(\alpha_i^k - \lambda\delta_{ik}\right)\xi_k = 0\,. \tag{3.67}$$

The eigenvalues λ are obtained from the compatibility condition as in (B.4), so that

$$\begin{vmatrix} \cos\varphi - \lambda & \sin\varphi & 0 \\ -\sin\varphi & \cos\varphi - \lambda & 0 \\ 0 & 0 & 1-\lambda \end{vmatrix} = 0\,.$$

Expanding the determinant gives

$$[(\lambda - \cos\varphi)^2 + \sin^2\varphi](1-\lambda) = 0$$

with the result

$$\lambda^{(1)} = \mathrm{e}^{\mathrm{i}\varphi}, \quad \lambda^{(2)} = \mathrm{e}^{-\mathrm{i}\varphi}, \quad \lambda^{(3)} = 1\,. \tag{3.68}$$

For each of these solutions there is a corresponding eigenvector $\xi^{(i)}$ with components obtained by solving (3.67). Thus, for $\lambda^{(1)}$ we have

$$\begin{cases} \xi_1(-\mathrm{i}\sin\varphi) + \xi_2\sin\varphi = 0 \\ (1-\mathrm{e}^{\mathrm{i}\varphi})\xi_3 = 0 \end{cases}$$

and hence $\xi_2 = \mathrm{i}\xi_1, \xi_3 = 0$.

For $\lambda^{(2)}$ we find

$$\xi_1 = \mathrm{i}\xi_2, \quad \xi_3 = 0$$

and for $\lambda^{(3)}$,

$$\xi_1 = \xi_2 = 0, \quad \xi_3 \text{ arbitrary}\,.$$

Since these components are complex, they are normalized using the Hermitian product $\xi_i\xi_i^*$, giving

$$\xi^{(1)}\left(\frac{1}{\sqrt{2}}, \frac{\mathrm{i}}{\sqrt{2}}, 0\right), \quad \xi^{(2)}\left(\frac{\mathrm{i}}{\sqrt{2}}, \frac{1}{\sqrt{2}}, 0\right), \quad \xi^{(3)}\,(0,0,1)\,. \tag{3.69}$$

In the orthonormal basis $\xi^{(1)}$, $\xi^{(2)}$, $\xi^{(3)}$, the elastic moduli are denoted by γ_{ijkl}. The invariance relation (3.62) becomes

$$\gamma_{ijkl} = \lambda^{(i)}\lambda^{(j)}\lambda^{(k)}\lambda^{(l)}\gamma_{ijkl}\,,$$

since the axis rotation matrix is diagonal in this basis. If ν_1 and ν_2 are the numbers of indices equal to 1 and 2 in the combination $ijkl$, this gives, taking account of (3.68),

$$\lambda^{(i)}\lambda^{(j)}\lambda^{(k)}\lambda^{(l)} = \exp\left[\mathrm{i}(\nu_1 - \nu_2)\frac{2\pi}{n}\right]\,.$$

Consequently, γ_{ijkl} is non-zero only when $(\nu_1 - \nu_2)$ is a multiple of the order n of the axis, since the product $\lambda^{(i)}\lambda^{(j)}\lambda^{(k)}\lambda^{(l)}$ is then equal to unity. This is always the case for the five moduli

$$\gamma_{1122},\quad \gamma_{1212},\quad \gamma_{3312},\quad \gamma_{2313},\quad \gamma_{3333}\,,$$

which have $\nu_1 = \nu_2$. For crystals in the *hexagonal* system ($n = 6$) there are no other cases, so there are only *five* independent moduli.

Trigonal crystals have seven independent constants, since γ_{1113} and γ_{2223} are also non-zero ($\nu_1 - \nu_2 = \pm 3$). Similarly, in *tetragonal* crystals ($n = 4$) we find that γ_{1111} and γ_{2222} are non-zero ($\nu_1 - \nu_2 = \pm 4$).

Conversion back to the constants c_{ijkl} is made with the aid of a relation changing the axes:

$$c_{ijkl} = a_i^p a_j^q a_k^r a_l^s\, \gamma_{pqrs}\,. \tag{3.70}$$

Here the matrix a converts from the system with basis vectors $\xi^{(1)}$, $\xi^{(2)}$, $\xi^{(3)}$, with constants γ_{ijkl}, to the system $Ox_1x_2x_3$, with constants c_{ijkl}. Its ith row consists of the components (3.69) of the eigenvector $\xi^{(i)}$, so that

$$a = \begin{pmatrix} 1/\sqrt{2} & \mathrm{i}/\sqrt{2} & 0 \\ \mathrm{i}/\sqrt{2} & 1/\sqrt{2} & 0 \\ 0 & 0 & 1 \end{pmatrix}\,. \tag{3.71}$$

The reduction in the number of components γ_{ijkl} simplifies the expansion of (3.70). Moreover, $a_3^3 = 1$ is the only non-zero component of a with an index equal to 3, and it follows that each constant c_{ijkl} depends only on constants γ_{ijkl} in which the number 3 has the same distribution among the indices.

We begin with the *trigonal* system, in which only γ_{3333}, γ_{3312}, γ_{2313}, γ_{1113}, γ_{2223}, γ_{1122} and γ_{1212} are non-zero, and consider the constants in order of decreasing number of 3's.

(a) For four 3's, we have $c_{3333} = \gamma_{3333}$, so $c_{33} \neq 0$.
(b) For three 3's, c_{3313} and c_{3323} are zero since $\gamma_{3313} = \gamma_{3323} = 0$, giving $c_{35} = c_{34} = 0$.
(c) For two 3's, the moduli c_{ij33} and c_{i3k3} can be expressed in terms of $\gamma_{1233} (= \gamma_{2133})$ and $\gamma_{1323} (= \gamma_{2313})$, respectively. For the former,

$$c_{ij33} = \left(a_i^1 a_j^2 + a_i^2 a_j^1\right)\gamma_{1233}\,.$$

We find $a_1^1 a_1^2 + a_1^2 a_1^1 = a_2^1 a_2^2 + a_2^2 a_2^1$, giving $c_{1133} = c_{2233}$, or $c_{13} = c_{23}$.

Also, we have $a_1^1 a_2^2 + a_1^2 a_2^1 = 0$, giving $c_{1233} = 0$, or $c_{36} = 0$. Similarly,

$$c_{i3k3} = \left(a_i^1 a_k^2 + a_i^2 a_k^1\right) \gamma_{1323}$$

leads to

$$c_{1313} = c_{2323}, \quad \text{or} \quad c_{55} = c_{44}\,,$$

and

$$c_{2313} = 0, \qquad \text{or} \quad c_{45} = 0\,.$$

(d) For a single 3, (3.70) gives

$$c_{ijk3} = a_i^1 a_j^1 a_k^1 \gamma_{1113} + a_i^2 a_j^2 a_k^2 \gamma_{2223}\,,$$

from which

$$c_{22k3} = -\frac{1}{2} a_k^1 \gamma_{1113} + \frac{1}{2} a_k^2 \gamma_{2223} = -c_{11k3}$$

and

$$c_{i223} = -\frac{1}{2} a_i^1 \gamma_{1113} + \frac{1}{2} a_i^2 \gamma_{2223} = -c_{i113}\,.$$

These give

$$c_{14} = -c_{24} = c_{56}, \quad \text{and} \quad c_{25} = -c_{15} = c_{46}\,,$$

where the first two relations are for $k = 2$ and $k = 1$ respectively, and the last two are for $i = 1$ and $i = 2$.

(e) For no 3's, the moduli $c_{1111}, c_{2222}, c_{1112}, c_{2221}, c_{1122}$ and c_{1212} can be expressed in terms of γ_{1122} and γ_{1212}, giving

$$\begin{aligned} c_{ijkl} &= (a_i^1 a_j^1 a_k^2 a_l^2 + a_i^2 a_j^2 a_k^1 a_l^1)\gamma_{1122} \\ &+ (a_i^1 a_j^2 + a_i^2 a_j^1)(a_k^1 a_l^2 + a_k^2 a_l^1)\gamma_{1212} \end{aligned} \tag{3.72}$$

This gives

$$c_{iiii} = 2(a_i^1 a_i^2)^2(\gamma_{1122} + 2\gamma_{1212}) = -\frac{1}{2}(\gamma_{1122} + 2\gamma_{1212})$$

and hence

$$c_{11} = c_{22} = -\frac{1}{2}(\gamma_{1122} + 2\gamma_{1212})\,. \tag{3.73}$$

The modulus c_{1112} can be written

$$c_{1122} = a_1^1 a_1^2 (a_1^1 a_2^2 + a_1^2 a_2^1)(\gamma_{1122} + 2\gamma_{1212}) = 0$$

since $a_1^1 a_2^2 + a_1^2 a_2^1 = 0$. Similarly, it can be shown that $c_{2221} = 0$, and so

$$c_{16} = c_{26} = 0\,.$$

Since there are only two independent constants with no index equal to 3, there must be a relation between c_{1111}, c_{1122} and c_{1212}. This follows from (3.72), which gives

$$c_{1122} = \frac{1}{2}(\gamma_{1122} - 2\gamma_{1212}) = c_{12} \quad \text{and} \quad c_{1212} = -\frac{1}{2}\gamma_{1122} = c_{66}\,.$$

Using (3.73) we find

$$c_{66} = \frac{c_{11} - c_{12}}{2}\,.$$

Summarizing, the matrix takes the following form:

$$(c_{\alpha\beta}) = \begin{vmatrix} c_{11} & c_{12} & c_{13} & c_{14} & -c_{25} & 0 \\ c_{12} & c_{11} & c_{13} & -c_{14} & c_{25} & 0 \\ c_{13} & c_{13} & c_{33} & 0 & 0 & 0 \\ c_{14} & -c_{14} & 0 & c_{44} & 0 & c_{25} \\ -c_{25} & c_{25} & 0 & 0 & c_{44} & c_{14} \\ 0 & 0 & 0 & c_{25} & c_{14} & \frac{c_{11}-c_{12}}{2} \end{vmatrix} \quad \textit{trigonal}. \tag{3.74}$$

For classes 32, $3m$ and $\bar{3}m$ the dyad axes, perpendicular to the principal axis, impose extra conditions. Taking Ox_1 to be along one of these dyad axes, constants c_{ijkl} with an odd number of indices equal to unity become zero, in a manner similar to the monoclinic case. This reduces the number of independent elastic constants to six, since $c_{15} = c_{1113} = 0$.

For the *tetragonal* system, the relations between moduli where four, three or two indices take the value 3 are the same as those established above, because the same non-zero components γ_{ijkl} occur in these two cases. Constants with a single index equal to 3 are zero, since $\gamma_{1113} = \gamma_{2223} = 0$, so

$$c_{14} = c_{24} = c_{15} = c_{25} = c_{46} = c_{56} = 0\,.$$

The remaining moduli, with no index 3, are determined by $\gamma_{1111}, \gamma_{2222}, \gamma_{1122}$ and γ_{1212} as follows:

$$c_{ijkl} = a_i^1 a_j^1 a_k^1 a_l^1 \gamma_{1111} + a_i^2 a_j^2 a_k^2 a_l^2 \gamma_{2222} + c_{ijkl}^{(3)}\,,$$

where $c_{ijkl}^{(3)}$ is the constant as given by (3.72) when the order of the axis, n, is equal to 3. With (3.73), this gives

$$c_{iiii} = \frac{1}{4}(\gamma_{1111} + \gamma_{2222}) - \frac{1}{2}(\gamma_{1122} + 2\gamma_{1212}) \quad \text{or} \quad c_{11} = c_{22}\,.$$

Also, $c_{1112}^{(3)}$ and $c_{2221}^{(3)}$ are zero, so

$$c_{1112} = \frac{\mathrm{i}}{4}(\gamma_{1111} - \gamma_{2222}) = -c_{2221} \quad \text{or} \quad c_{26} = -c_{16}\,.$$

The four independent constants with no index equal to 3 are c_{11}, c_{12}, c_{16} and c_{66}, and with c_{13}, c_{33} and c_{44} these give the form shown in (3.75) below.

For classes 422, $4mm$, $\bar{4}2m$ and $4/mmm$, with the Ox_1 axis along one of the dyad axes perpendicular to the principal axis, the constant $c_{1112} = c_{16}$ is zero because it has an odd number of indices equal to unity.

$$(c_{\alpha\beta}) = \begin{vmatrix} c_{11} & c_{12} & c_{13} & 0 & 0 & c_{16} \\ c_{12} & c_{11} & c_{13} & 0 & 0 & -c_{16} \\ c_{13} & c_{13} & c_{33} & 0 & 0 & 0 \\ 0 & 0 & 0 & c_{44} & 0 & 0 \\ 0 & 0 & 0 & 0 & c_{44} & 0 \\ c_{16} & -c_{16} & 0 & 0 & 0 & c_{66} \end{vmatrix} \quad \textit{tetragonal}. \tag{3.75}$$

For the *hexagonal* system the principle axis has order six, behaving as a dyad axis combined with a triad. The matrix $c_{\alpha\beta}$ thus has a form combining the features of the monoclinic and trigonal systems, from (3.63) and (3.74), giving

$$(c_{\alpha\beta}) = \begin{vmatrix} c_{11} & c_{12} & c_{13} & 0 & 0 & 0 \\ c_{12} & c_{11} & c_{13} & 0 & 0 & 0 \\ c_{13} & c_{13} & c_{33} & 0 & 0 & 0 \\ 0 & 0 & 0 & c_{44} & 0 & 0 \\ 0 & 0 & 0 & 0 & c_{44} & 0 \\ 0 & 0 & 0 & 0 & 0 & \frac{c_{11}-c_{12}}{2} \end{vmatrix} \quad \textit{hexagonal}. \tag{3.76}$$

This tensor is invariant for all rotations about the 6-fold axis. This property, demonstrated in Problem 3.4, can also be expressed in other ways – for example, planes perpendicular to the principle axis are elastically isotropic. Also, all planes which include the principle axis are equivalent, as are all directions equally inclined to this axis.

The above results for the reduction of the number of independent components in the various systems are summarized in the table of Fig. 3.9. The relationships, expressed in the symbolism introduced by K.S. Van Dyke, apply directly to the stiffness constants c_{ijkl} or $c_{\alpha\beta}$. For orthonormal coordinate axes, the compliance tensor s_{ijkl} reduces in the same way as c_{ijkl}. However, relations between the matrix components $s_{\alpha\beta}$ need to be found using (3.47), that is,

$$s_{\alpha\beta} = 2^p s_{ijkl} \,,$$

where p is the number of indices greater than 3 in the pair $(\alpha,\ \beta)$. For example, in the trigonal or hexagonal systems, or in isotropic materials, we have

$$c_{66} = \frac{c_{11} - c_{12}}{2} \quad \text{since} \quad c_{1212} = \frac{c_{1111} - c_{1122}}{2} \,.$$

The same relation occurs between the corresponding compliance components s_{ijkl}, so that

$$s_{1212} = \frac{s_{1111} - s_{1122}}{2} \,.$$

However, since $s_{66} = 4s_{1212}$, $s_{11} = s_{1111}$ and $s_{12} = s_{1122}$, this gives

$$s_{66} = 2(s_{11} - s_{12}) \,. \tag{3.77}$$

Figure 3.10 gives the stiffness constants and mass densities for a variety of materials, organized by crystal system. For the piezoelectric cases, marked by asterisks, the entries refer to $c^{\mathrm{E}}_{\alpha\beta}$, defined in Sect. 3.3.2 below. Numerically, *stiffness constants are of the order* of $10^{11}\,\mathrm{N/m^2}$ (100 GPa).

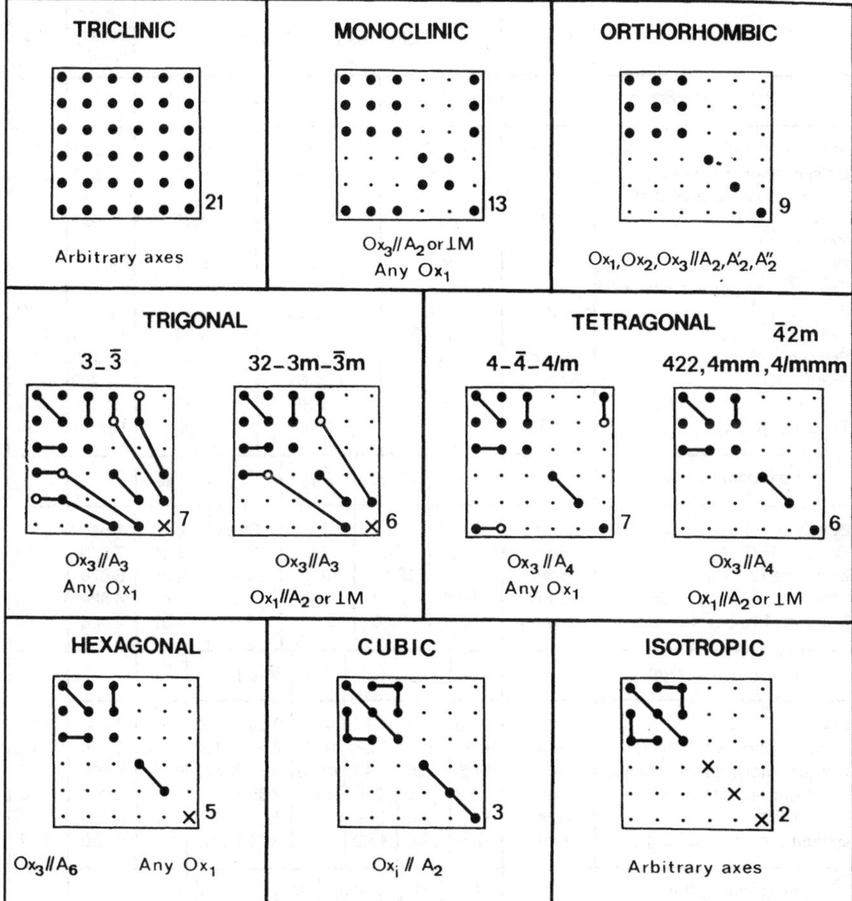

Fig. 3.9. Relations between components $c_{\alpha\beta}$ of the elastic stiffness tensor for different crystal classes, with coordinate axes as in Fig. 2.22.
• ○ : non-zero component, . : zero component,
•—• : equal components, •—○ : equal and opposite components,
× : component equal to $(c_{11} - c_{12})/2$.
Symmetry about the main diagonal is not shown. The number of independent components is shown at the bottom right for each case

3.3 Piezoelectric Solids

A solid is piezoelectric if it becomes electrically polarized under the action of a mechanical force (the direct effect), and becomes mechanically deformed when an electric field is applied (the inverse effect). The discovery of the direct effect, by Pierre and Jacques Curie in 1880, followed directly from the application of principles relating to the symmetry of causes and effects. Clarifying the conditions imposed by symmetry, which had previously been

Material	Class	Stiffness (10^{10} N/m^2)			Mass density (kg/m^3)	Ref.
Isotropic, cubic system		c_{11}	c_{12}	c_{44}	ρ	
Aluminium (Al)	$m3m$	10.73	6.08	2.83	2 702	[3.1]
Gallium arsenide (GaAs)	$\bar{4}3m$	11.88	5.38	5.94	5 307	[3.2]
Yttrium Aluminium Garnet ($Y_3Al_5O_{12}$), YAG	$m3m$	33.2	11.07	11.50	4 550	[3.3]
Yttrium Iron Garnet ($Y_3Fe_5O_{12}$), YIG	$m3m$	26.9	10.77	7.64	5 170	[3.3]
Bismuth and germanium oxide ($Bi_{12}GeO_{20}$)	23	12.8	3.05	2.55	9 230	[3.4]
Gold (Au)	$m3m$	19.25	16.30	4.24	19 300	[3.5]
Platinum (Pt)	$m3m$	34.7	25.1	7.65	21 400	[3.6]
Silica (SiO_2)	*isotropic*	7.85	1.61	3.12	2 203	
Silicon (Si)	$m3m$	16.56	6.39	7.95	2 329	[3.7]
Tungsten (W)	$m3m$	52.24	20.44	16.06	19 260	[3.8]

Hexagonal system		c_{11}	c_{12}	c_{13}	c_{33}	c_{44}		
Beryllium (Be)	$6/mmm$	29.23	2.67	1.4	33.64	16.25	1 848	[3.9]
Ceramic PZT–4	$6mm$	13.9	7.8	7.4	11.5	2.56	7 500	[3.25]
Zinc oxide (ZnO)	$6mm$	20.97	12.11	10.51	21.09	4.25	5 676	[3.10]
Cadmium sulphide (CdS)	$6mm$	8.56	5.32	4.62	9.36	1.49	4 824	[3.11]
Titanium (Ti)	$6/mmm$	16.24	9.20	6.90	18.07	4.67	4 506	[3.12]

Tetragonal system		c_{11}	c_{12}	c_{13}	c_{33}	c_{44}	c_{66}	c_{16}		
Indium (In)	$4/mmm$	4.53	4.00	4.15	4.51	0.65	1.21	0	7 280	[3.13]
Lead molybdate ($PbMoO_4$)	$4/m$	10.92	6.83	5.28	9.17	2.67	3.37	1.36	6 950	[3.14]
Calcium molybdate ($CaMoO_4$)	$4/m$	14.5	6.6	4.46	12.65	3.69	4.5	1.3	4 255	[3.15]
Paratellurite (TeO_2)	422	5.6	5.1	2.2	10.6	2.65	6.6	0	6 000	[3.16]
Rutile (TiO_2)	$4/mmm$	27.3	17.6	14.9	48.4	12.5	19.4	0	4 250	[3.17]
Barium titanate ($BaTiO_3$)	$4mm$	27.5	17.9	15.2	16.5	5.43	11.3	0	6 020	[3.18]

Trigonal system		c_{11}	c_{12}	c_{13}	c_{33}	c_{44}	c_{14}		
Alumina (Al_2O_3)	$\bar{3}m$	49.7	16.3	11.1	49.8	14.7	−2.35	3 986	[3.19]
Lithium niobate ($LiNbO_3$)	$3m$	20.3	5.3	7.5	24.5	6.0	0.9	4 700	[3.20]
Lithium tantalate ($LiTaO_3$)	$3m$	23.3	4.7	8.0	27.5	9.4	−1.1	7 450	[3.20]
α–Quartz (SiO_2)	32	8.67	0.70	1.19	10.72	5.79	−1.79	2 648	[3.21]
Tellurium (Te)	32	3.27	0.86	2.49	7.22	3.12	1.24	6 250	[3.22]

Orthorhombic system		c_{11}	c_{12}	c_{13}	c_{22}	c_{23}	c_{33}	c_{44}	c_{55}	c_{66}	ρ	
α–iodic acid (HIO_3)	222	3.01	1.61	1.11	5.80	0.80	4.29	1.69	2.06	1.58	4 640	[3.23]
Barium sodium niobate ($Ba_2NaNb_5O_{15}$)	$2mm$	23.9	10.4	5.0	24.7	5.2	13.5	6.5	6.6	7.6	5 300	[3.24]

Fig. 3.10. Elastic stiffness constants and mass densities for various crystals, organized according to crystal class

implemented in a more or less intuitive way, Pierre Curie put forward the two following principles:

(1) A phenomenon must possess all the symmetry elements of the causes which give rise to it. It may have higher symmetry than the causes.
(2) Any asymmetry of a phenomenon must already exist in its causes.

Each phenomenon has an associated characteristic symmetry. For example, a scalar such as temperature has an infinite number of isotropic axes, each denoted A_∞, an infinite number of mirror planes M, and a centre C; thus the point group is $\frac{\infty A_\infty}{\infty M} C$, as for a sphere. A polar vector, such as electric field, has an isotropic axis A_∞ with an infinite number of mirrors passing through it, giving $A_\infty \infty M$. For a compression or tension, resulting from two equal and opposite forces as in Fig. 3.11, the symmetry is $\frac{A_\infty}{M} \frac{\infty A_2}{\infty M'} C$.

Curie's two principles can be summarized as follows: *a phenomenon can exist only in a system having a symmetry which is a sub-group of the characteristic symmetry of the phenomenon.*

For the direct piezoelectric effect, in which an applied compression gives rise to an electric polarization, the system comprising the crystal plus the compression must have a symmetry lower than or equal to the point group $A_\infty \infty M$ of the polarization. In particular, the system cannot have a centre of symmetry, so the centre present in the compression must be eliminated by the absence of a centre in the crystal. Thus, *centro-symmetric crystals cannot be piezoelectric.* This conclusion, first stated by Curie, was confirmed experimentally. He found piezoelectric crystals in 20 classes, these being the 21 non-centrosymmetric classes with the special case of class 432 excluded (Sect. 3.3.3.5 below).

Symmetry principles can also be exploited to predict directions for which a mechanical force will give an electric polarization. Thus, for α *quartz* (class

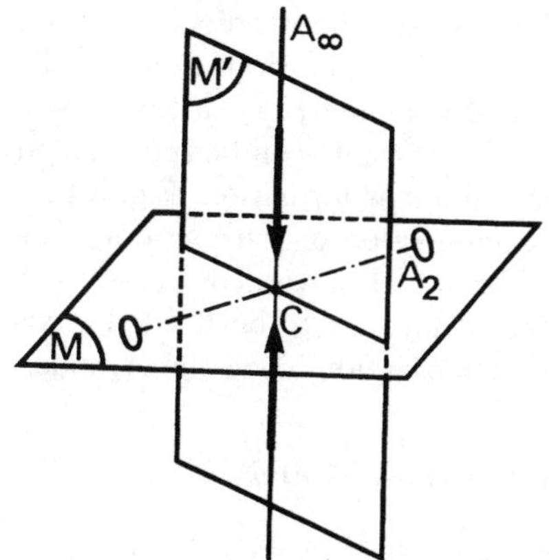

Fig. 3.11. The characteristic symmetry of a compression or tension, resulting from the action of two opposing forces, is $\frac{A_\infty}{M} \frac{\infty A_2}{\infty M'} C$

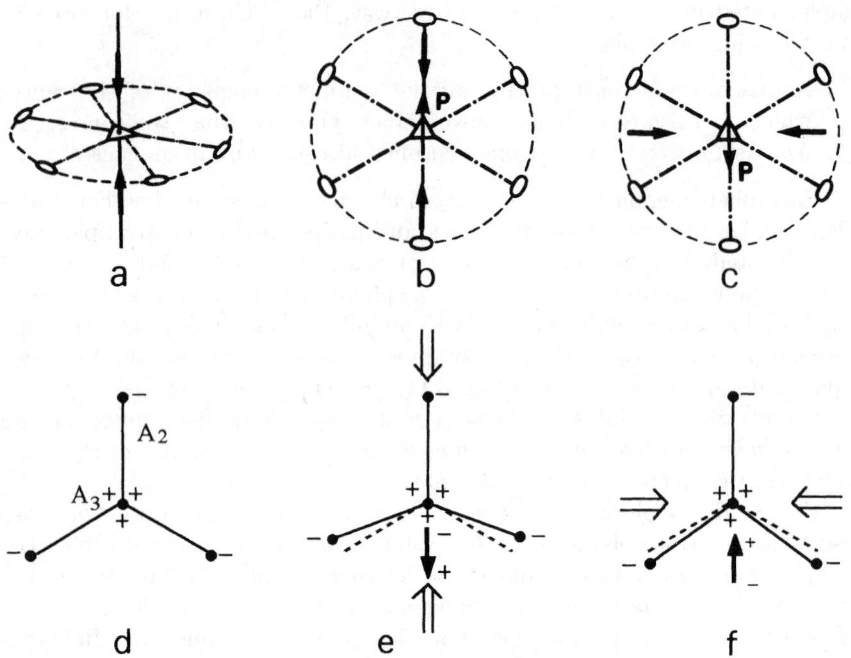

Fig. 3.12a–f. Application of Curie's symmetry principles to quartz

32), a compression (or tension) along the triad axis gives no polarization, because the combined symmetry of the crystal and the compression ($A_3 3A_2$) is not a sub-group of that of the polarization ($A_\infty \infty M$), as shown in Fig. 3.12a. However, for compression along a dyad axis A_2, the combined symmetry of the crystal and the compression is compatible with the appearance of a polarization $\boldsymbol{P}$ along this axis, as in Fig. 3.12b. Compression perpendicular to both a dyad axis and the triad axis, as in Fig. 3.12c, gives polarization along the dyad axis, which is the only symmetry element common to both the crystal and the compression. As a simple reminder, these results can be represented in terms of a group of three coplanar dipoles in a star formation, Fig. 3.12d, having the symmetry of quartz ($A_3 3A_2$). Compression or tension along the triad axis does not affect the electrical equilibrium. However, compression or tension along an A_2 axis (Fig. 3.12e), or perpendicular to both an A_2 axis and the A_3 axis (Fig. 3.12f), can generate a polarization along the A_2 axis.

3.3.1 Physical Mechanism – One-Dimensional Model

As an example, we consider a cadmium sulphide crystal which, as explained in Sect. 2.4.2a above, consists of successive layers of sulphur and cadmium ions. The direction perpendicular to the layers is the hexad axis, that is, the helicoidal 6_3 axis of Fig. 2.28, and in this direction the ions repeat along identical rows. The composition of one of these rows is represented in Fig. 3.13a.

For an external agency, such as stress or electric field, acting along the hexad axis, it is sufficient to consider only one of these rows, since all the ions in one layer are displaced by the same amount. Let $-q$ and $+q$ be the effective charges of the sulphur and cadmium ions, which are imagined to be linked to each other by springs. Since the neighbours nearest to one ion are not symmetric with respect to it, the springs on either side have different force constants K_1 and K_2. The chain is divided into cells of length a, each containing two dipoles with moments $q(a-b)/2$ and $-qb/2$, where b is one of the interatomic distances as in Fig. 3.13a. The dipole moment of one molecule is therefore

$$p_0 = \frac{q(a-2b)}{2} .$$

At rest, the polarization P_0 per unit volume is non-zero because b differs from $a/2$. The crystal is therefore polar. Taking n as the number of CdS molecules per unit volume gives

$$P_0 = \frac{nq(a-2b)}{2} .$$

This one-dimensional model illustrates, in a simple manner, the relations between mechanical and electrical properties in a piezoelectric material. Under the action of a stress, the chain is stretched or compressed; the changes in the lengths a and b lead to a change of polarization given by

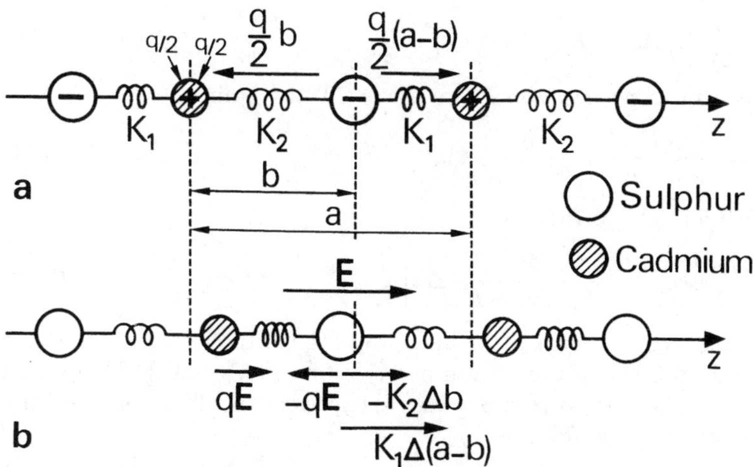

Fig. 3.13. Piezoelectricity in a crystal of cadmium sulphide. (**a**) Distribution of sulphur and cadmium ions along a row parallel to the principal axis. (**b**) Deformation of the chain under the action of an electric field E – the Cd and S ions are displaced in opposite directions. The cell length changes by Δa

$$P = \Delta P_0 = \frac{nq(\Delta a - 2\Delta b)}{2} . \quad (3.78)$$

This is the *direct* piezoelectric effect. On the other hand, an applied electric field causes the positive and negative ions to be displaced in opposite directions, so that if $(a - b)$ is increased then b is reduced. Since the springs have different stiffnesses, the result is a deformation. This is the *inverse* piezoelectric effect, also called the Lippman effect.

We consider the relations between the mechanical variables, the stress T and strain S, and the electrical variables – the field E and the induced polarization P or electric displacement $D = \varepsilon_0 E + P$. The equilibrium of each ion requires, from Fig. 3.13b,

$$-qE + K_1\Delta(a - b) - K_2\Delta b = 0 . \quad (3.79)$$

To express the traction T, we consider a section perpendicular to the A_6 axis. The axis is helicoidal (6_3), so that two neighbouring rows are shifted by a half-period $a/2$, as in Fig. 2.28. Hence a plane perpendicular to the A_6 axis alternately cuts the springs with stiffness K_1 and K_2. The force exerted by the right part of the chain on the left part is, for the first case, $F_1 = K_1\Delta(a - b)$ and for the second case, $F_2 = K_2\Delta b$.

Taking N as the number of rows per unit area perpendicular to the A_6 axis, the mechanical traction, or force per unit area, is

$$T = \frac{N}{2}K_1\Delta(a - b) + \frac{N}{2}K_2\Delta b ,$$

since there is an equal number of springs of each type. Since $N = na$ this becomes

$$T = \frac{na}{2}\left[K_1\Delta a + (K_2 - K_1)\,\Delta b\right] . \quad (3.80)$$

For the *induced polarization* P, we use (3.79) to obtain

$$\Delta b = -\frac{qE}{K_1 + K_2} + \frac{K_1}{K_1 + K_2}\Delta a \quad (3.81)$$

and substitution into (3.78) gives

$$P = \frac{nq}{2}\left[\frac{2qE}{K_1 + K_2} + \frac{K_2 - K_1}{K_1 + K_2}\Delta a\right] . \quad (3.82)$$

Thus P consists of two terms, the first, proportional to electric field, giving the ionic susceptibility of the crystal as $\chi_{\mathrm{ion}} = nq^2/(K_1 + K_2)$. The second term, proportional to the strain $S = \Delta a/a$, expresses the direct piezoelectric effect. Thus P can be written as

$$P = \chi_{\mathrm{ion}}E + eS ,$$

where we define e, the piezoelectric constant, as

$$e = \frac{nq}{2}\frac{K_2 - K_1}{K_2 + K_1}a . \quad (3.83)$$

The total polarization, including the electronic susceptibility χ_e, is

$$P = (\chi_{\text{ion}} + \chi_e)E + eS\,.$$

Introducing the dielectric constant $\varepsilon = \varepsilon_0 + \chi_{\text{ion}} + \chi_e$, the electric displacement D can be written

$$D = \varepsilon E + eS\,. \tag{3.84}$$

The *stress* is obtained by substituting Δb from (3.81) into (3.80), giving

$$T = \frac{na}{2}\left(\frac{2K_2K_1}{K_2+K_1}\Delta a - \frac{K_2-K_1}{K_2+K_1}qE\right)\,.$$

This can be expressed as

$$T = cS - eE\,, \tag{3.85}$$

where e is the piezoelectric constant derived above, and c is the stiffness constant given by $c = na^2K_1K_2/(K_1+K_2)$. Equation (3.83) shows that, in this one-dimensional model, the existence of the piezoelectric effect is associated with asymmetry in the spring constants, since $e = 0$ if $K_1 = K_2$. Equation (3.85) shows that there can be different forms of behaviour. If the ends of the chain are fixed there is no strain, and an electric field gives rise to a stress $T = -eE$. Conversely, if there is no stress an electric field causes a strain $S = eE/c$.

In practice, the constant e is one component of a tensor of rank three, since the displacement D_i is a tensor of rank one (a vector), as is the electric field, while the strain S is a tensor of rank two, as is the stress. Taking the Ox_3 coordinate axis to be along the hexad axis, (3.84) becomes

$$D_3 = \varepsilon_{33}E_3 + e_{333}S_{33}\,.$$

This model explains the origin of piezoelectricity in asymmetric ionic crystals. In practice, the mechanism is more complex, since the effect also occurs in monatomic crystals such as tellurium (Te) and selenium (Se). In such cases the electric polarization produced by a strain is attributed to changes in the electronic distribution. Regarding polycrystalline materials, such as ceramics, these can become piezoelectric only after particular processing. These ceramics, containing lead, zirconium and titanium in the case of PZT, consist of ferroelectric domains (Sect. 2.4.2), oriented at random during fabrication. Application of a large electric field at high temperature, typically 20 kV/cm at 100 °C, causes an alignment of the domains and an elongation, and these features persist partially at zero field and at ambient temperature. The permanent polarization, and hence the piezoelectricity, disappears above the Curie temperature, which is generally between 200 and 400 °C. Concerning the symmetry, the polarization direction is an isotropic axis. All planes which contain this axis are equivalent; the material is transversally isotropic. *Polymers*, such as polyvinylidine fluoride (PVF_2), also need special processing to become piezoelectric – these materials are made in the form of thin sheets, and are simultaneously stretched and subjected to an electric field.

3.3.2 Tensor Expressions

Generalising (3.84) to three dimensions defines a tensor e_{ijk} of rank three, so that

$$D_i = \varepsilon_{ij}E_j + e_{ijk}S_{jk} \,. \tag{3.86}$$

The piezoelectric constants e_{ijk} relate changes of displacement D_i to strain S_{jk} in the solid, with the electric field held constant, so that

$$e_{ijk} = \left(\frac{\partial D_i}{\partial S_{jk}}\right)_E \,. \tag{3.87}$$

They are expressed in units of C/m^2. Since $S_{jk} = S_{kj}$, the tensor e_{ijk} is symmetric with respect to the last two indices j and k, so that

$$e_{ijk} = e_{ikj} \,.$$

Thus *the number of independent constants is reduced from 27 (for an arbitrary tensor of rank three) to 18.* Indeed, the last two indices, j and k, form a pair which can only take six distinct values represented by the number α, which conforms to the convention (3.41) adopted in Sect. 3.2.1. Thus

$$e_{i\alpha} = e_{ijk} \quad i = 1,2,3 \quad \alpha = (j,k) = 1,2,\ldots,6 \,.$$

The piezoelectric constants can thus be set out in an array with three rows $(i = 1,2,3)$ and six columns $(\alpha = 1,2,\ldots,6)$:

$$c_{i\alpha} = \begin{vmatrix} e_{11} & e_{12} & e_{13} & e_{14} & e_{15} & e_{16} \\ e_{21} & e_{22} & e_{23} & e_{24} & e_{25} & e_{26} \\ e_{31} & e_{32} & e_{33} & e_{34} & e_{35} & e_{36} \end{vmatrix} \,. \tag{3.88}$$

Extending the matrix notation to the strain, and complying with the convention of (3.44), equation (3.86) becomes

$$D_i = \varepsilon_{ij}E_j + e_{i\alpha}S_\alpha \,. \tag{3.89}$$

The inverse piezoelectric coefficient, generalising (3.85) to relate the stress T_{jk} to the electric field E_i, is obtained from the constants e_{ijk} using thermodynamic considerations. The variation of internal energy per unit volume is given by (3.49), established in Sect. 3.2.2 for a reversible transformation in an elastic solid. For a piezoelectric material, the change of electric potential energy $E_i \mathrm{d}D_i$ must be added, giving

$$\mathrm{d}U = \theta\,\mathrm{d}\sigma + T_{jk}\mathrm{d}S_{jk} + E_i\mathrm{d}D_i \,, \tag{3.90}$$

where θ is the absolute temperature and σ the entropy. U is a function of the entropy, the strains and the electric displacement. To obtain functions of the variables σ, S_{jk} and E_i we introduce the thermodynamic potential G, defined by

$$G = U - E_iD_i \,. \tag{3.91}$$

The variation of this function of state,

$$\mathrm{d}G = \theta\,\mathrm{d}\sigma + T_{jk}\mathrm{d}S_{jk} - D_i\mathrm{d}E_i \tag{3.92}$$

is an exact differential, so that

$$T_{jk} = \left(\frac{\partial G}{\partial S_{jk}}\right)_{\sigma,E_i} \quad \text{and} \quad D_i = -\left(\frac{\partial G}{\partial E_i}\right)_{\sigma,S_{jk}} .$$

Since the second derivatives are independent of the order of differentiation, we have

$$\left(\frac{\partial T_{jk}}{\partial E_i}\right)_{\sigma,S} = -\left(\frac{\partial D_i}{\partial S_{jk}}\right)_{\sigma,E} = \left(\frac{\partial^2 G}{\partial E_i \partial S_{jk}}\right)_{\sigma} .$$

This relation shows that, if a strain produces an electric displacement for a constant electric field, given by $(\partial D_i/\partial S_{jk})_{\sigma,E} = e_{ijk}$, then an applied electric field will produce a stress at constant strain (i.e. when the solid is rigidly fixed), given by

$$\left(\frac{\partial T_{jk}}{\partial E_i}\right)_{\sigma,S} = -e_{ijk} . \tag{3.93}$$

Thus *the proportionality coefficients for the two effects are opposite of each other.* The inverse piezoelectric effect is a thermodynamic consequence of the direct effect. In the linear region, where the coefficients e_{ijk} are constants by definition, integration of (3.93) with σ and S held constant leads to

$$T_{jk} = -e_{ijk}E_i .$$

If the solid is also subject to a strain S_{lm}, then

$$T_{jk} = -e_{ijk}E_i + c^E_{jklm}S_{lm} . \tag{3.94}$$

Here the superscript E indicates that the stiffness constant in the generalized Hooke's law

$$c^E_{jklm} = \left(\frac{\partial T_{jk}}{\partial S_{lm}}\right)_{\sigma,E} \tag{3.95}$$

relates stresses and strains when the electric field is held constant. Indeed, since the material is piezoelectric, the values of all mechanical constants are affected by the electrical conditions, so the latter must be specified. Also, the coefficient ε_{ij} which occurs in (3.89) must be taken as the permittivity at constant strain, ε^S_{ij}. Equations (3.89) and (3.94) constitute a primary system of equations of state. In matrix notation they become

$$\boxed{\begin{aligned} T_\alpha &= c^E_{\alpha\beta}S_\beta - e_{i\alpha}E_i \\ D_i &= \varepsilon^S_{ij}E_j + e_{i\alpha}S_\alpha \end{aligned}} \qquad i,j = 1,2,3 \quad \text{and} \quad \alpha,\beta = 1,\ldots,6 . \tag{3.96a,b}$$

These give the mechanical stress and electric displacement, in terms of the independent variables electric field and strain.

The equations of state for a piezoelectric material can be expressed in various forms by taking other pairs of variables to be independent – D and S, D and T, or E and T. For example, we can use the thermodynamic potential

$$I = G - T_{jk}S_{jk} = U - E_i D_i - T_{jk}S_{jk} ,$$

which has differential

$$\mathrm{d}I = \theta\,\mathrm{d}\sigma - S_{jk}\mathrm{d}T_{jk} - D_i\mathrm{d}E_i .$$

This shows that I is a function of the variables T_{jk} and E_i, so that

$$I = I(\sigma, T_{jk}, E_i) ,$$

and

$$S_{jk} = -\left(\frac{\partial I}{\partial T_{jk}}\right)_{\sigma,E} \quad \text{and} \quad D_i = -\left(\frac{\partial I}{\partial E_i}\right)_{\sigma,T} .$$

We can equate the second derivatives, so that

$$\left(\frac{\partial S_{jk}}{\partial E_i}\right)_{\sigma,T} = \left(\frac{\partial D_i}{\partial T_{jk}}\right)_{\sigma,E} = d_{ijk} . \tag{3.97}$$

These are associated with the definitions of compliance at constant electric field, s^E_{jklm}, and permittivity at constant stress, ε^T_{ij}, which are

$$s^E_{jklm} = \left(\frac{\partial S_{jk}}{\partial T_{lm}}\right)_E \quad \text{and} \quad \varepsilon^T_{ij} = \left(\frac{\partial D_i}{\partial E_j}\right)_T . \tag{3.98}$$

After integration, these lead to a second system of equations of state, using as independent variables the electric field and the stresses:

$$\begin{aligned} S_{jk} &= d_{ijk}E_i + s^E_{jklm}T_{lm} \\ D_i &= \varepsilon^T_{ij}E_j + d_{ijk}T_{jk} . \end{aligned}$$

These can also be expressed in the matrix notation, giving

$$\boxed{\begin{aligned} S_\alpha &= s^E_{\alpha\beta}T_\beta + d_{i\alpha}E_i \\ D_i &= \varepsilon^T_{ij}E_j + d_{i\alpha}T_\alpha \end{aligned}} , \tag{3.99a,b}$$

where T_α and S_α satisfy the conditions (3.43) and (3.44). This requires $d_{i\alpha}$ to be defined as

$$\begin{aligned} d_{i\alpha} &= d_{ijk} \quad \text{if } \alpha \le 3 \\ d_{i\alpha} &= 2d_{ijk} \quad \text{if } \alpha > 3 . \end{aligned} \tag{3.100}$$

The moduli $d_{i\alpha}$ express the proportionality relation between the strains and the electric field components; they also give the mechanical displacement when a voltage is applied between two electrodes (Problems 3.9 and 3.10). They are of the order of 10^{-12} m/V.

The piezoelectric and mechanical constants, $e_{i\alpha}$, $d_{i\alpha}$, $c^E_{\alpha\beta}$ and $s^E_{\alpha\beta}$, are not independent. Substituting the expression

$$S_\beta = d_{i\beta}E_i + s^E_{\beta\gamma}T_\gamma$$

into (3.96a) gives

$$T_\alpha = \left(-e_{i\alpha} + c^E_{\alpha\beta}d_{i\beta}\right) E_i + c^E_{\alpha\beta}s^E_{\beta\gamma}T_\gamma = \delta_{\alpha\gamma}T_\gamma$$

and here the independent variables E_i and T_γ have arbitrary values, so that

$$\boxed{c^E_{\alpha\beta}s^E_{\beta\gamma} = \delta_{\alpha\gamma}} \tag{3.101}$$

and

$$\boxed{e_{i\alpha} = d_{i\beta}c^E_{\beta\alpha}}\,. \tag{3.102}$$

The first relation shows that $s^E_{\alpha\beta}$ is the inverse of $c^E_{\alpha\beta}$, as in a non-piezoelectric material. The second shows that the constants $e_{i\alpha}$ can be calculated from the moduli $d_{i\alpha}$, or the reverse using the formula

$$\boxed{d_{i\beta} = e_{i\alpha}s^E_{\alpha\beta}}\,. \tag{3.103}$$

To relate ε^T_{ij} to ε^S_{ij}, we use (3.96a) for the stress, so that

$$T_\alpha = -e_{j\alpha}E_j + c^E_{\alpha\beta}S_\beta$$

and upon substituting the result into (3.99b), we find

$$D_i = \left(\varepsilon^T_{ij} - d_{i\alpha}e_{j\alpha}\right) E_j + d_{i\alpha}c^E_{\alpha\beta}S_\beta\,.$$

Comparison with (3.96b) shows that

$$\boxed{\varepsilon^T_{ij} - \varepsilon^S_{ij} = d_{i\alpha}e_{j\alpha} = d_{i\alpha}c^E_{\alpha\beta}d_{j\beta}}\,. \tag{3.104}$$

The permittivity ε^S_{ij} for a solid rigidly fixed is very difficult to measure, but it can be calculated, using the above equation, from the piezoelectric constants and the more commonly used permittivity ε^T_{ij} for a solid with no stress. The difference between ε^T_{ij} and ε^S_{ij} can be significant for strongly piezoelectric materials. For example, lithium niobate ($LiNbO_3$) has

$$\varepsilon^T_{11} = 74.3 \times 10^{-11}\mathrm{F/m} \quad \text{and} \quad \varepsilon^S_{11} = 38.9 \times 10^{-11}\mathrm{F/m}\,.$$

The systems (3.96a,b) and (3.99a,b) are those most used in practice. The two other groups of state equations, in which the electric independent variable is D_i, can be obtained by considering the internal energy $U(\sigma, S_{jk}, D_i)$ and the enthalpy $H(\sigma, T_{jk}, D_i) = U - T_{jk}S_{jk}$. Thus, with the variables D_i and S_{jk}, the result is

$$\boxed{\begin{aligned} E_i &= \beta^S_{ij}D_j - h_{ijk}S_{jk} \\ T_{jk} &= c^D_{jklm}S_{lm} - h_{ijk}D_i \end{aligned}}\,. \tag{3.105}$$

Here c^D_{jklm} is the stiffness tensor for constant electric displacement and β^S_{ij} is the reciprocal of ε^S_{ij}. The piezoelectric constant h_{ijk} is defined by

$$h_{ijk} = -\left(\frac{\partial E_i}{\partial S_{jk}}\right)_D = -\left(\frac{\partial T_{jk}}{\partial D_i}\right)_S = -\left(\frac{\partial^2 U}{\partial D_i \partial S_{jk}}\right)_\sigma$$

and is written in matrix notation as $h_{ijk} = h_{i\alpha}$. The last system of equations is

$$\boxed{\begin{aligned} E_i &= \beta^T_{ij} D_j - g_{ijk} T_{jk} \\ S_{jk} &= s^D_{jklm} T_{lm} + g_{ijk} D_i \end{aligned}} \,. \tag{3.106}$$

Here the piezoelectric modulus g_{ijk} is the negative of the second derivative of the enthalpy H, with respect to the independent variables D_i and T_{jk}, so that

$$g_{ijk} = -\left(\frac{\partial^2 H}{\partial D_i \partial T_{jk}}\right)_\sigma \,. \tag{3.107}$$

In matrix notation, $g_{i\alpha} = g_{ijk}$ if $\alpha = 1, 2, 3$, and $g_{i\alpha} = 2g_{ijk}$ if $\alpha > 3$.

3.3.3 Effect of Crystal Symmetry on Piezoelectric Constants

From (2.13), the invariance condition for the components of the piezoelectric tensor under a symmetry operation, using the axis rotation matrix α, is

$$e_{ijk} = \alpha^l_i \alpha^m_j \alpha^n_k \, e_{lmn} \,. \tag{3.108}$$

We examine the effects of various point symmetry operations of crystals.

3.3.3.1 Centre of Symmetry. Since $\alpha^k_i = -\delta_{ik}$, (3.108) gives

$$e_{ijk} = (-1)^3 e_{ijk} \quad \text{and hence} \quad e_{ijk} = 0 \,.$$

Hence, *the piezoelectric tensor is zero in the eleven centrosymmetric classes:*

$$\bar{1}, 2/m, mmm, \bar{3}, \bar{3}m, 4/m, 4/mmm, 6/m, 6/mmm, m3 \text{ and } m3m \,.$$

This result has already been obtained from Curie's symmetry principles, Sect. 3.3.

3.3.3.2 Plane of Symmetry. The axis rotation matrix is still diagonal. Taking Ox_3 to be perpendicular to the mirror plane, we have

$$\alpha = \begin{pmatrix} 1 & 0 & 0 \\ 0 & 1 & 0 \\ 0 & 0 & -1 \end{pmatrix} .$$

The condition (3.108) gives

$$e_{ijk} = \alpha^i_i \alpha^j_j \alpha^k_k e_{ijk} \tag{3.109}$$

showing that all constants with one or three indices equal to 3 are zero. Thus for *class m* the number of non-zero moduli $e_{i\alpha}$ is reduced to ten, giving

$$e_{i\alpha} = \begin{vmatrix} e_{11} & e_{12} & e_{13} & 0 & 0 & e_{16} \\ e_{21} & e_{22} & e_{23} & 0 & 0 & e_{26} \\ 0 & 0 & 0 & e_{34} & e_{35} & 0 \end{vmatrix} \quad \text{class}\, m\, (M \perp x_3)\,. \qquad (3.110)$$

3.3.3.3 Direct Dyad Axis. The matrix for rotation by an angle π about Ox_3 is

$$\alpha = \begin{pmatrix} -1 & 0 & 0 \\ 0 & -1 & 0 \\ 0 & 0 & 1 \end{pmatrix}.$$

Equation (3.109) leads to cancellation of components with an even number of indices equal to 3, so that for *class 2* the array $e_{i\alpha}$ is complementary to that for class m. Thus,

$$e_{i\alpha} = \begin{vmatrix} 0 & 0 & 0 & e_{14} & e_{15} & 0 \\ 0 & 0 & 0 & e_{24} & e_{25} & 0 \\ e_{31} & e_{32} & e_{33} & 0 & 0 & e_{36} \end{vmatrix} \quad \text{class}\, 2\, (A_2 \| x_3)\,. \qquad (3.111)$$

3.3.3.4 Several Dyad Axes. For crystals in *class 222*, the coefficients e_{ijk} are non-zero only if each of the values 1, 2, 3 occurs an odd number of times in the indices. This applies for e_{123}, e_{213} and e_{312}, and hence

$$e_{i\alpha} = \begin{vmatrix} 0 & 0 & 0 & e_{14} & 0 & 0 \\ 0 & 0 & 0 & 0 & e_{25} & 0 \\ 0 & 0 & 0 & 0 & 0 & e_{36} \end{vmatrix} \quad \text{class}\, 222\,. \qquad (3.112)$$

For *class 2mm*, the mirror planes perpendicular to Ox_1 and Ox_2 cancel the coefficients with an odd number of indices equal to 1 or 2. This applies to $e_{14} = e_{123}, e_{25} = e_{213}$ and $e_{36} = e_{312}$ in the array of (3.111), which becomes

$$e_{i\alpha} = \begin{vmatrix} 0 & 0 & 0 & 0 & e_{15} & 0 \\ 0 & 0 & 0 & e_{24} & 0 & 0 \\ e_{31} & e_{32} & e_{33} & 0 & 0 & 0 \end{vmatrix} \quad \text{class}\, 2mm\,. \qquad (3.113)$$

In *classes 23* and *$\bar{4}3m$* of the cubic system, the three dyad axes define the coordinate axes $Ox_1x_2x_3$. Since a rotation by an angle $2\pi/3$ about the triad axis along [111] permutes x_1, x_2 and x_3, the three constants e_{123}, e_{312} and e_{231} of class 222 must be equal, so that

$$e_{i\alpha} = \begin{vmatrix} 0 & 0 & 0 & e_{14} & 0 & 0 \\ 0 & 0 & 0 & 0 & e_{14} & 0 \\ 0 & 0 & 0 & 0 & 0 & e_{14} \end{vmatrix} \quad \text{classes 23 and}\, \bar{4}3m\,. \qquad (3.114)$$

3.3.3.5 Direct Axis with Order n Greater than Two. To find the form of the piezoelectric tensor in crystals having a direct principal axis of order 3, 4 or 6, we apply the method used in Sect. 3.2.3.2 for the stiffness tensor. With the same notations, the invariance condition for the new moduli η_{ijk} is written in the basis of the eigenvectors $\xi^{(1)}$, $\xi^{(2)}$, $\xi^{(3)}$ of the rotation matrix (3.66), so that

$$\eta_{ijk} = \lambda^{(i)}\lambda^{(j)}\lambda^{(k)}\eta_{ijk} ,$$

where the eigenvalues are

$$\lambda^{(1)} = \mathrm{e}^{\mathrm{i}2\pi/n} \quad \lambda^{(2)} = \mathrm{e}^{-\mathrm{i}2\pi/n} \quad \lambda^{(3)} = 1 .$$

Taking ν_1 as the number of indices equal to 1, and ν_2 as the number equal to 2, the component η_{ijk} is non-zero when the difference $(\nu_1 - \nu_2)$ is a multiple of the order n of the axis, since the product $\lambda^{(i)}\lambda^{(j)}\lambda^{(k)} = \exp[\mathrm{i}(\nu_1 - \nu_2)2\pi/n]$ is then equal to unity. This is always the case for the moduli η_{123}, η_{213}, η_{312} and η_{333}, for which $\nu_1 = \nu_2$. There are no more cases when $n = 4$ or 6. However, for $n = 3$ the components η_{111} and η_{222} are also non-zero, since they give $\nu_1 - \nu_2 = \pm 3$.
To revert to the constants e_{ijk} we use the relation

$$e_{ijk} = a_i^l a_j^m a_k^n \, \eta_{lmn} , \tag{3.115}$$

where a is the matrix of (3.71), converting from the system $Ox_1x_2x_3$ (to which the e_{ijk} are referred) to the system of the basis vectors $\xi^{(1)}$, $\xi^{(2)}$, $\xi^{(3)}$. This matrix is given by

$$a = \begin{pmatrix} 1/\sqrt{2} & \mathrm{i}/\sqrt{2} & 0 \\ \mathrm{i}/\sqrt{2} & 1/\sqrt{2} & 0 \\ 0 & 0 & 1 \end{pmatrix} .$$

We begin with *classes 4 and 6*, in which the tensors e_{ijk} have the same form because the same components η_{ijk} are non-zero. Since an A_4 or A_6 axis contains a dyad axis, only the non-zero components in the array of (3.111), for class 2, need to be considered. Consider constants with one index equal to 3. Constants e_{3jk}, with j and k not equal to 3, can be expressed in terms of $\eta_{312} = \eta_{321}$, giving

$$e_{3jk} = \left(a_j^1 a_k^2 + a_j^2 a_k^1\right) \eta_{312} . \tag{3.116}$$

In particular,

$$e_{3jj} = 2a_j^1 a_j^2 \eta_{312} = \mathrm{i}\eta_{312} \quad \text{for} \quad j \neq 3, \quad \text{giving} \quad e_{31} = e_{32}$$

and

$$e_{312} = (a_1^1 a_2^2 + a_1^2 a_2^1)\eta_{312} = 0 \quad \text{giving} \quad e_{36} = 0 .$$

Still considering constants having one index equal to 3, the constants e_{ij3}, with i and j unequal to 3, are given by

$$e_{ij3} = a_i^1 a_j^2 \eta_{123} + a_i^2 a_j^1 \eta_{213} .$$

This leads to

$$e_{jj3} = a_j^1 a_j^2 (\eta_{123} + \eta_{213}) = \mathrm{i}(\eta_{123} + \eta_{213})/2 \quad \text{for} \quad j \neq 3$$

and hence $e_{15} = e_{24}$.
For e_{123} and e_{213}, we have, from (3.115),

$$e_{123} = (\eta_{123} - \eta_{213})/2 \quad \text{and} \quad e_{213} = (-\eta_{123} + \eta_{213})/2$$

showing that these moduli are opposite, giving $e_{14} = -e_{25}$. Finally, $e_{333} = \eta_{333}$ is non-zero, so the array for $e_{i\alpha}$ has the form

$$e_{i\alpha} = \begin{vmatrix} 0 & 0 & 0 & e_{14} & e_{15} & 0 \\ 0 & 0 & 0 & e_{15} & -e_{14} & 0 \\ e_{31} & e_{31} & e_{33} & 0 & 0 & 0 \end{vmatrix} \quad \text{classes 4 and 6}. \tag{3.117}$$

For *classes 422 and 622*, which possess three perpendicular dyad axes (Fig. 2.22), it is sufficient to combine the arrays of (3.112) and (3.117), giving

$$e_{i\alpha} = \begin{vmatrix} 0 & 0 & 0 & e_{14} & 0 & 0 \\ 0 & 0 & 0 & 0 & -e_{14} & 0 \\ 0 & 0 & 0 & 0 & 0 & 0 \end{vmatrix} \quad \text{classes 422 and 622}. \tag{3.118}$$

For *classes 4mm and 6mm* the method is similar, starting from the array (3.113) for class $2mm$, and hence

$$e_{i\alpha} = \begin{vmatrix} 0 & 0 & 0 & 0 & e_{15} & 0 \\ 0 & 0 & 0 & e_{15} & 0 & 0 \\ e_{31} & e_{31} & e_{33} & 0 & 0 & 0 \end{vmatrix} \quad \text{classes } 4mm \text{ and } 6mm. \tag{3.119}$$

Problem 3.8 shows that the piezoelectric tensor for classes $4mm$ and $6mm$ is invariant for any rotation about the Ox_3 axis. A similar remark was made earlier regarding the stiffness tensor of a hexagonal crystal (Sect. 3.2.3.2 and Problem 3.4); thus we can conclude that the elasto-electric properties of a crystal of class $6mm$ are invariant for any rotation about the principal axis. In *piezoelectric ceramics*, this *transverse isotropy* is found perpendicular to the direction of polarization, so it is appropriate to represent the elastic, electric and piezoelectric properties by tensors with the same form as those for class $6mm$.

For *class 432* in the cubic system we have all the symmetry elements of class 422, in which the only non-zero piezoelectric coefficients are $e_{14} = -e_{25}$. In addition, however, the triad axis along [111] gives rise to permutation of the indices 1, 2, 3, so that $e_{123} = e_{231}$, and hence

$$e_{14} = e_{25} = -e_{25} = 0\,.$$

All components of $e_{i\alpha}$ are therefore zero. Thus, *piezoelectricity cannot exist in crystals of class 432, despite the absence of a centre of symmetry*, i.e.

$$e_{i\alpha} = 0 \quad \text{class 432}.$$

For crystals having a *triad axis*, the relations established above for classes 4 or 6, regarding moduli with either one or three indices equal to 3, are valid. However, moduli with no index equal to 3 are no longer zero; they can be expressed as the following function of η_{111} and η_{222}:

$$e_{ijk} = a_i^1 a_j^1 a_k^1 \eta_{111} + a_i^2 a_j^2 a_k^2 \eta_{222}\,. \tag{3.120}$$

In particular,

$$e_{i11} = (a_i^1 \eta_{111} - a_i^2 \eta_{222})/2 \quad \text{and} \quad e_{i22} = (-a_i^1 \eta_{111} + a_i^2 \eta_{222})/2 = -e_{i11}\,.$$

Taking i to be 1 or 2 gives $e_{12} = -e_{11}$ and $e_{22} = -e_{21}$. Similarly, (3.120) also leads to

$$e_{122} = \frac{1}{2\sqrt{2}}(i\eta_{111} - \eta_{222}) = -e_{222}$$

and

$$e_{212} = \frac{1}{2\sqrt{2}}(-\eta_{111} + i\eta_{222}) = -e_{111}\,.$$

These results give the following array:

$$e_{i\alpha} = \begin{vmatrix} e_{11} & -e_{11} & 0 & e_{14} & e_{24} & -e_{22} \\ -e_{22} & e_{22} & 0 & e_{15} & -e_{14} & -e_{11} \\ e_{31} & e_{31} & e_{33} & 0 & 0 & 0 \end{vmatrix} \quad \text{class } 3\,. \tag{3.121}$$

An additional dyad axis along Ox_1 cancels all the moduli in the above array having an even number of indices equal to 1 (Sect. 3.3.3.3), that is

$$e_{15} = e_{24} \quad e_{16} = -e_{22} \quad e_{21} = -e_{22} \quad e_{31} = e_{32}\ e_{33}\,.$$

Hence, crystals of *class 32* have only two independent piezoelectric constants e_{11} and e_{14}, giving

$$e_{i\alpha} = \begin{vmatrix} e_{11} & -e_{11} & 0 & e_{14} & 0 & 0 \\ 0 & 0 & 0 & 0 & -e_{14} & -e_{11} \\ 0 & 0 & 0 & 0 & 0 & 0 \end{vmatrix} \quad \text{class } 32\,. \tag{3.122}$$

For *class 3m*, the mirror plane perpendicular to Ox_1 cancels the constants in (3.121) with one or three indices equal to 1, that is, e_{14} and e_{11}. The tensor becomes

$$e_{i\alpha} = \begin{vmatrix} 0 & 0 & 0 & 0 & e_{15} & -e_{22} \\ -e_{22} & e_{22} & 0 & e_{15} & 0 & 0 \\ e_{31} & e_{31} & e_{33} & 0 & 0 & 0 \end{vmatrix} \quad \text{class } 3m\,. \tag{3.123}$$

3.3.3.6 Inverse Principal Axis. An inverse hexad axis is equivalent to $3/m$ (Sect. 2.2.1.2c). The tensor for crystals in *class* $\bar{6}$ therefore follows directly from (3.110) and (3.121), giving

$$e_{i\alpha} = \begin{vmatrix} e_{11} & -e_{11} & 0 & 0 & 0 & -e_{22} \\ -e_{22} & e_{22} & 0 & 0 & 0 & -e_{11} \\ 0 & 0 & 0 & 0 & 0 & 0 \end{vmatrix} \quad \text{class } \bar{6}\,. \tag{3.124}$$

For *class* $\bar{6}m2$ the dyad axis along Ox_1 eliminates from the above array the constant e_{22} which has no index equal to 1 (Sect. 3.3.3.3), giving

$$e_{i\alpha} = \begin{vmatrix} e_{11} & -e_{11} & 0 & 0 & 0 & 0 \\ 0 & 0 & 0 & 0 & 0 & -e_{11} \\ 0 & 0 & 0 & 0 & 0 & 0 \end{vmatrix} \quad \text{class } \bar{6}m2\,. \tag{3.125}$$

To treat the case of an inverse tetrad axis, we use the general method. The matrix for rotation and inversion, $\bar{\alpha}$, is equal to $-\alpha$, and the eigenvectors $\xi^{(i)}$

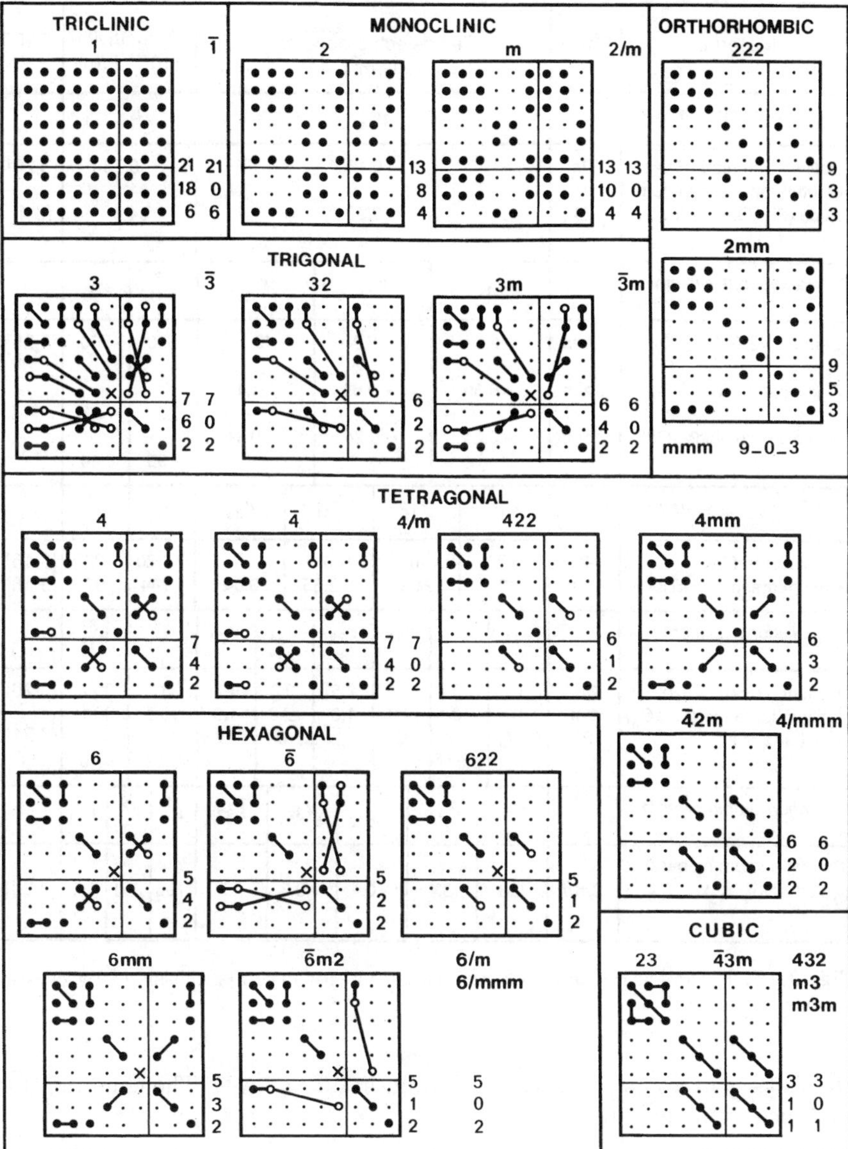

Fig. 3.14. Components of the elastic, piezoelectric and dielectric tensors, organised according to symmetry class, using the coordinate axes as in Fig. 2.22.
• ○ : non-zero component, . : zero component,
•—• : equal components, •—○ : equal and opposite components,
× : component equal to $(c_{11} - c_{12})/2$.
Symmetry about the main diagonal is not shown. The number of independent elastic, piezoelectric and dielectric components is shown at the right for each case

Material	Class	Piezoelectric constant (C/m^2)						Permittivity (10^{-11} F/m)			Ref.
Cubic system		e_{14}						ε^S			
Gallium arsenide (GaAs)	$\bar{4}3m$	−0.16						9.73			[3.26]
Bismuth and germanium oxide ($Bi_{12}GeO_{20}$)	23	0.99						34.2			[3.4]
Hexagonal system		e_{15}	e_{31}	e_{33}				ε^S_{11}	ε^S_{33}		
Ceramic PZT–4	6*mm*	12.7	−5.2	15.1				650	560		[3.25]
Zinc oxide (ZnO)	6*mm*	−0.59	−0.61	1.14				7.38	7.83		–
Cadmium sulphide (CdS)	6*mm*	−0.21	−0.24	0.44				7.99	8.44		–
Tetragonal system		e_{14}	e_{15}	e_{31}	e_{33}			ε^S_{11}	ε^S_{33}		
Paratellurite (TeO_2)	422	0.22	0	0	0			20	22		[3.16]
Barium titanate ($BaTiO_3$)	4*mm*	0	21.3	−2.65	3.64			1 744	97		[3.18]
Trigonal system		e_{11}	e_{14}	e_{15}	e_{22}	e_{31}	e_{33}	ε^S_{11}	ε^S_{33}		
Lithium niobate ($LiNbO_3$)	3*m*	0	0	3.7	2.5	0.2	1.3	38.9	25.7		[3.20]
Lithium tantalate ($LiTaO_3$)	3*m*	0	0	2.6	1.6	≅0	1.9	36.3	38.1		—
α–Quartz (SiO_2)	32	0.171	−0.041	0	0	0	0	3.92	4.10		[3.21]
Orthorhombic system		e_{15}	e_{24}	e_{31}	e_{32}	e_{33}		ε^S_{11}	ε^S_{22}	ε^S_{33}	
Barium sodium niobate ($Ba_2NaNb_5O_{15}$)	2*mm*	2.8	3.4	−0.4	−0.3	4.3		196	201	28	[3.24]

Fig. 3.15. Piezoelectric and dielectric constants for crystals, classified by crystal system

are also eigenvectors of $\bar{\alpha}$, with eigenvalues $\bar{\lambda}^{(i)} = -\lambda^{(i)}$. The condition for existence of the moduli η_{ijk},

$$\bar{\lambda}^{(i)}\bar{\lambda}^{(j)}\bar{\lambda}^{(k)} = 1$$

is equivalent to

$$\lambda^{(i)}\lambda^{(j)}\lambda^{(k)} = \exp\left[\mathrm{i}(\nu_1 - \nu_2)\frac{2\pi}{n}\right] = -1\,.$$

When $n = 4$ this is satisfied by the moduli η_{113}, η_{223}, η_{311} and η_{322}, for which $\nu_1 - \nu_2 = \pm 2$. Since an $\bar{A}_4$ axis contains a dyad axis, it is sufficient to consider the elements e_{ijk} in the array (3.111). Applying (3.115) to e_{ii3} gives

$$e_{ii3} = a_i^1 a_i^1 \eta_{113} + a_i^2 a_i^2 \eta_{223} \,,$$

which shows that $e_{24} = -e_{15}$. Similarly, the expansion

$$e_{3ii} = a_i^1 a_i^1 \eta_{311} + a_i^2 a_i^2 \eta_{322}$$

leads to $e_{32} = -e_{31}$, and equality of e_{25} and e_{14} follows from the formula

$$e_{ij3} = a_i^1 a_j^1 \eta_{113} + a_i^2 a_j^2 \eta_{223} = e_{ji3} \,.$$

Since e_{33} is zero ($\eta_{333} = 0$), the tensor $e_{i\alpha}$ becomes

$$e_{i\alpha} = \left| \begin{matrix} 0 & 0 & 0 & e_{14} & e_{15} & 0 \\ 0 & 0 & 0 & -e_{15} & e_{14} & 0 \\ e_{31} & -e_{31} & 0 & 0 & 0 & e_{36} \end{matrix} \right| \quad \text{class } \bar{4} \,. \tag{3.126}$$

For *class* $\bar{4}\,2m$, the dyad axis along Ox_1 cancels the constants e_{15} and e_{31} which have an even number of indices equal to 1, giving

$$e_{i\alpha} = \left| \begin{matrix} 0 & 0 & 0 & e_{14} & 0 & 0 \\ 0 & 0 & 0 & 0 & e_{14} & 0 \\ 0 & 0 & 0 & 0 & 0 & e_{36} \end{matrix} \right| \quad \text{class } \bar{4}2m \,. \tag{3.127}$$

These results are applicable to any tensor A_{ijk} of rank three that is symmetric in j and k, and therefore to the components $e_{ijk}, d_{ijk}, h_{ijk}$ and g_{ijk}. They remain true for the constants $e_{i\alpha} = e_{ijk}$ and $h_{i\alpha} = h_{ijk}$ with two indices. However, to find the relations for $d_{i\alpha}$ or $g_{i\alpha}$ it is necessary to take account of the formulae (3.100). For example, crystals in class 32 have $e_{26} = -e_{11}$ but $d_{26} = -2d_{11}$. The table in Fig. 3.14 summarizes the results for the reduction in the number of components of the elastic, dielectric and piezoelectric tensors for crystals belonging to the various point groups. The entries in the table are set out as in the following pattern:

c_{11}	c_{12}	·	·	·	·	e_{11}	e_{21}	e_{31}
c_{12}	c_{22}	·	·	·	·	e_{12}	·	·
·	·	·	·	·	·	·	·	·
·	·	·	$c^E_{\alpha\beta}$	·	·	·	$e_{i\alpha}$	·
·	·	·	$\left(c^D_{\alpha\beta}\right)$ ·		·	·	$(h_{i\alpha})$	·
·	·	·	·	·	·	·	·	·
e_{11}	e_{12}	·	·	·	·	ε^S_{11}	·	·
e_{21}	·	·	$e_{i\alpha}$	·	·	·	ε^S_{ij}	·
e_{31}	·	·	$(h_{i\alpha})$ ·		·	·	$\left(\beta^S_{ij}\right)$	·

The table in Fig. 3.15 gives the piezoelectric constants $e_{i\alpha}$ and dielectric constants ε^S_{ij} for several crystals of practical importance.

Problems

3.1. A shaft with symmetry of revolution is suspended vertically as shown in Fig. 3.16. How should the cross-section area $s(x)$ vary with x in order that the average tension T due to the weight is independent of x?

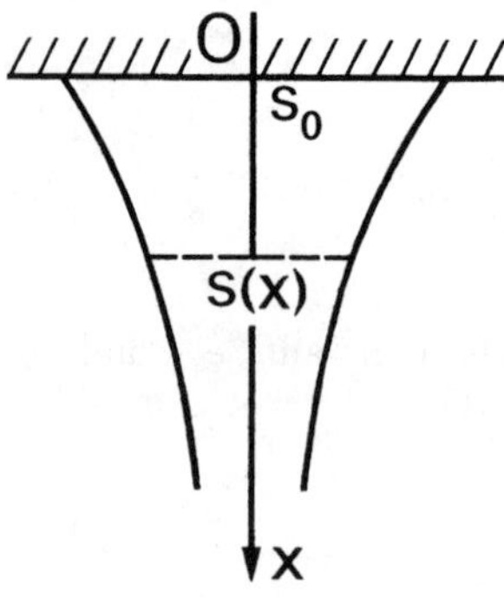

Fig. 3.16. Shaft with varying cross-section

Solution. We have

$$T = \frac{\rho g \int_x^\infty s(x)\mathrm{d}x}{s(x)} = \text{constant} = T_0 \,.$$

Differentiating gives

$$-\rho g s(x) = T_0 \frac{\mathrm{d}s}{\mathrm{d}x} \quad \text{and hence} \quad s = s_0 \exp\left(-\frac{\rho g}{T_0} x\right) \,.$$

3.2. Calculate the compressibility coefficient $\chi = -S/p$, where S is the dilatation, for a crystal subject to a hydrostatic pressure p.
Solution. With $T_{kl} = -p\delta_{kl}$, Hooke's law gives $S_{ij} = s_{ijkl}T_{kl} = -ps_{ijkk}$, and from (3.7) the dilatation $S = \Delta v/v$ is given by $S = S_{ii} = -ps_{iikk}$. Hence

$$\chi = s_{iikk} = s_{11} + s_{22} + s_{33} + 2\left(s_{12} + s_{23} + s_{31}\right) \,.$$

3.3. Express Young's modulus E and Poisson's ratio ν in terms of stiffness constants, given that for an isotropic solid cylinder of cross-section area s they are respectively defined by the following expressions for the relative elongation $\Delta l/l$ and the relative change of diameter $\Delta d/d$ when a uniform stress F/s is applied along the axis x_1:

$$\frac{\Delta l}{l} = \frac{1}{E}\frac{F}{s}, \quad \frac{\Delta d}{d} = -\nu\frac{\Delta l}{l} = -\frac{\nu}{E}\frac{F}{s} \,.$$

Solution. The strains are $S_{11} = T_{11}/E$ and $S_{22} = S_{33} = -\nu T_{11}/E$. For an isotropic solid,

$$T_{22} = c_{12}(S_{11} + S_{33}) + c_{11}S_{22} = 0 \quad \text{giving} \quad \nu = \frac{c_{12}}{c_{11} + c_{12}}$$

and

$$T_{11} = c_{11}S_{11} + c_{12}(S_{22} + S_{33}) \quad \text{giving} \quad E = c_{11} - 2c_{12}\nu = c_{11} - 2\frac{c_{12}^2}{c_{11} + c_{12}} .$$

3.4. Show that the stiffness constants of a crystal belonging to the hexagonal system are invariant for any rotation about the hexad axis.

Solution. We examine the behaviour of the five independent moduli $c_{33}, c_{13}, c_{44}, c_{12}$ and c_{11} under a change of axes according to the matrix (2.14), which gives

$$c'_{ijkl} = \alpha_i^p \alpha_j^q \alpha_k^r \alpha_l^s c_{pqrs} ;$$

(a) $c'_{3333} = \alpha_3^3 \alpha_3^3 \alpha_3^3 \alpha_3^3 c_{3333}$ giving $c'_{33} = c_{33}$.

(b) $c'_{1133} = \alpha_1^p \alpha_1^q \alpha_3^3 \alpha_3^3 c_{pq33}$, in which the only non-zero terms have $p = q = 1$ or 2, giving
$c'_{13} = c_{13}\cos^2\varphi + c_{23}\sin^2\varphi = c_{13}$ since $c_{23} = c_{13}$.

(c) $c'_{2323} = \alpha_2^p \alpha_2^r c_{p3r3}$ giving
$c'_{44} = c_{44}\cos^2\varphi + c_{55}\sin^2\varphi = c_{44}$ since $c_{55} = c_{44}$.

(d) $c'_{1122} = \alpha_1^p \alpha_1^q \alpha_2^r \alpha_2^s c_{pqrs}$, with p, q, r and s not equal to 3, leading to
$c'_{12} = (c_{11} + c_{22} - 4\,c_{66})\cos^2\varphi\sin^2\varphi + c_{21}\sin^4\varphi + c_{12}\cos^4\varphi$

and replacing c_{66} by $(c_{11} - c_{12})/2$ gives

$$c'_{12} = (2\cos^2\varphi\sin^2\varphi + \sin^4\varphi + \cos^4\varphi)c_{12} = c_{12} .$$

(e) Similarly, $c'_{11} = c_{11}$.

3.5. Express the compliances $s_{\alpha\beta}$ in terms of the stiffnesses $c_{\alpha\beta}$ for crystals in the tetragonal system, with classes $4/mmm, 422, 4mm$ and $\bar{4}2m$, and in the hexagonal and cubic systems.

Solution. Inverting the matrix $c_{\alpha\beta}$ gives directly $s_{44} = 1/c_{44}$ and $s_{66} = 1/c_{66}$. The other relations follow simply from the following transformation using the first three of equations (3.42):

$$\begin{cases} (1)\ T_1 = c_{11}S_1 + c_{12}S_2 + c_{13}S_3 \\ (2)\ T_2 = c_{12}S_1 + c_{11}S_2 + c_{13}S_3 \\ (3)\ T_3 = c_{13}S_1 + c_{13}S_2 + c_{33}S_3 \end{cases}$$

$$(1) - (2) \Rightarrow \begin{cases} T_1 - T_2 = (c_{11} - c_{12})(S_1 - S_2) \\ T_2 \qquad\quad = c_{12}(S_1 - S_2) + (c_{11} + c_{12})S_2 + c_{13}S_3 \\ T_3 \qquad\quad = c_{13}(S_1 - S_2) + 2c_{13}S_2 + c_{13}S_3 \end{cases} .$$

The same transformation applied to (3.46), using the variables

$$\tau_1 = T_1 - T_2, \quad \tau_2 = T_2, \quad \tau_3 = T_3 ,$$

$$\sigma_1 = S_1 - S_2, \quad \sigma_2 = S_2, \quad \sigma_3 = S_3 ,$$

shows that the matrix

$$\begin{pmatrix} s_{11}-s_{12} & 0 & 0 \\ s_{12} & s_{11}+s_{12} & s_{13} \\ s_{13} & 2s_{13} & s_{33} \end{pmatrix} \quad \text{is the inverse of} \quad \begin{pmatrix} c_{11}-c_{12} & 0 & 0 \\ c_{12} & c_{11}+c_{12} & c_{13} \\ c_{13} & 2c_{13} & c_{33} \end{pmatrix}$$

giving the relations

$$s_{11}-s_{12} = \frac{1}{c_{11}-c_{12}}\,, \quad s_{11}+s_{12} = \frac{c_{33}}{c^2}\,, \quad s_{13} = -\frac{c_{13}}{c^2}\,, \quad s_{33} = \frac{c_{11}+c_{12}}{c^2}$$

where

$$c^2 = (c_{11}+c_{12})c_{33} - 2c_{13}^2\,.$$

3.6. Using Curie's principles, find the crystal classes for which an electric polarization can appear under the action of hydrostatic pressure.

Solution. A hydrostatic pressure acts in all directions with equal intensity, so it has spherical symmetry. The crystal symmetry must therefore belong to a subgroup of the characteristic symmetry $A_\infty \infty M$ of the polarization. This is the case for crystals in the ten polar classes $1, 2, m, 2mm, 3, 3m, 4, 4mm, 6$ and $6mm$, which also show pyroelectricity.

3.7. Express the constants $c^D_{\alpha\beta}$, in terms of the stiffnesses $c^E_{\alpha\beta}$ and the moduli $e_{i\alpha}$.

Solution. Multiplying (3.96b) by β^S_{kl} gives

$$\beta^S_{ki} D_i = E_k + \beta^S_{ki} e_{i\beta} S_\beta$$

and hence

$$E_i = \beta^S_{ik} D_k - \beta^S_{ik} e_{k\beta} S_\beta\,.$$

Substituting into (3.96a) gives

$$T_\alpha = \left(c^E_{\alpha\beta} + e_{i\alpha}\beta^S_{ik} e_{k\beta}\right) S_\beta - e_{i\alpha}\beta^S_{ik} D_k$$

and hence

$$c^D_{\alpha\beta} = c^E_{\alpha\beta} + e_{i\alpha}\beta^S_{ik} e_{k\beta}\,.$$

3.8. Show that the piezoelectric tensor for classes $4mm$ and $6mm$ is invariant for any rotation about the Ox_3 axis.

Solution. The new components e'_{ijk} are found using the axis rotation matrix of (2.14), so that

$$e'_{ijk} = \alpha^p_i \alpha^q_j \alpha^r_k e_{pqr}$$

From the array of (3.119), the only non-zero components are

$$e_{113} = e_{223} = e_{15}, \quad e_{333} = e_{33} \quad \text{and} \quad e_{311} = e_{322} = e_{31}$$

and with $\alpha^3_k = \delta_{k3}$ we have

$$e'_{ijk} = \left(\alpha^1_i \alpha^1_j + \alpha^2_i \alpha^2_j\right)\delta_{k3} e_{15} + \delta_{i3}\delta_{j3}\delta_{k3} e_{33} + \left(\alpha^1_j \alpha^1_k + \alpha^2_j \alpha^2_k\right)\delta_{i3} e_{31}\,.$$

It is sufficient to apply this to the non-zero constants in (3.117), for classes 4 and 6, giving

$$e'_{123} = \left(\alpha_1^1\alpha_2^1 + \alpha_1^2\alpha_2^2\right) e_{15} = 0 \quad \text{and hence } e'_{14} = 0\,;$$

$$e'_{113} = \left(\alpha_1^1\alpha_1^1 + \alpha_1^2\alpha_1^2\right) e_{15} = e_{15} \text{ and hence } e'_{15} = e_{15}\,;$$

$$e'_{333} = e_{333} \quad \text{giving} \quad e'_{33} = e_{33} \quad \text{and}$$
$$e'_{311} = \left(\alpha_1^1\alpha_1^1 + \alpha_1^2\alpha_1^2\right) e_{31} = e_{31} \quad \text{giving} \quad e'_{31} = e_{31}\,.$$

These equalities show that the new tensor is identical to the old one.

3.9. One end of a quartz rod, with length L parallel to the X axis, is rigidly fixed to an immobile support, with the other end and the lateral faces being free. What piezoelectric modulus $d_{i\alpha}$ gives the mechanical displacement of the face at the free end when the rod vibrates in response to an alternating voltage of amplitude U applied between the ends, which are metallized? Can the frequency be chosen at will? Given that the measured amplitude of the displacement is 0.25 Å when the voltage is 10.55 V, what is the value of the modulus? How can mechanical displacements of the order of Ångströms be measured without mechanical contact?

Solution. We have $S_\alpha = s_{\alpha\beta}T_\beta + d_{i\alpha}E_i$, with $\alpha, \beta = 1, 2, \ldots, 6$, and $i = 1, 2, 3$. For the *free faces* $T_\beta = 0$, giving $S_{11} = d_{11}E_1$. Integration from 0 to L with $u(0) = 0$ gives

$$u(L) = -d_{11}U, \quad \text{and hence} \quad d_{11} = -u(L)/U\,.$$

The frequency must be chosen to be well below the resonant frequency of the rod, because otherwise the Q-factor needs to be taken into account. For example, if the resonant frequency is 80 kHz, a suitable measurement frequency might be 10 kHz. Numerically, $d_{11} = -0.25/10.55\,[\text{Å/V}] = -2.38\,\text{pm/V}$. The generally accepted value is $-2.31\,\text{pm/V}$.

Normal displacements of a surface vibrating at a frequency above a few kHz may be measured using a capacitive probe, or more often using an optical interferometer – see Sect. 3.2 of Vol. II.

3.10. A thin piezoelectric platelet of thickness h has one face, perpendicular to x_3, rigidly bonded to a fixed support, with the opposite face free. It is excited by an alternating voltage of amplitude U applied between these two faces, at a frequency well below resonance. Express the strains S_α in terms of the stresses T_β and the electric field E_3. Apply the result to a Z-cut plate (perpendicular to x_3) of lithium niobate (class $3m$). Deduce from the expressions for strains that are zero at the free surface, the relations between stresses and electric field. Show that the proportionality coefficient relating S_3 to E_3 is not d_{33} but rather $\mathrm{d}'_{33} = \mathrm{d}_{33} - \delta_{33}$, and derive a formula for δ_{33}. Calculate d'_{33}, given that the measured amplitude of the free-surface displacement is 1.46 Å when the applied potential has amplitude 16.40 V.

Solution. On the face $x_3 = 0$, bonded to the fixed support, we have $u_1 = u_2 = 0$, giving $S_1 = S_2 = S_6 = 0$. For $f \ll f_r$, that is for $\lambda \gg \lambda_r \approx 4h$, we further have $S_1 = S_2 = S_6 = 0$ throughout the thickness h. At the free face $x_3 = h$ we have $T_3 = T_4 = T_5 = 0$, and this also applies (for $f \ll f_r$) throughout the thickness h. Hence

$$S_\alpha = s^E_{\alpha 1} T_1 + s^E_{\alpha 2} T_2 + s^E_{\alpha 6} T_6 + d_{3\alpha} E_3 \,.$$

For $LiNbO_3$ (class $3m$), the equation $S_1 = S_2 = S_6 = 0$ leads to $T_6 = 0$ and

$$T_1 = T_2 = -\frac{d_{31}}{s^E_{11} + s^E_{12}} E_3 \,,$$

giving

$$S_3 = \left(d_{33} - \frac{d_{31} s^E_{13}}{s^E_{11} + s^E_{12}} \right) E_3 = d'_{33} E_3, \quad \text{hence} \quad \delta_{33} = \frac{d_{31} s^E_{13}}{s^E_{11} + s^E_{12}} \,.$$

Integrating as in Problem 3.9 gives $d'_{33} = u_3(h)/U$.

Numerically, $d'_{33} = 1.46/16.4\,[\text{Å/V}] = 8.9 \times 10^{-12}\,\text{m/V}$, and with $\delta_{33} = 0.62\,\text{pm/V}$ this gives $\text{d}_{33} = 9.5\,\text{pm/V}$.

4. Plane Waves in Crystals

In Chap. 1 we introduced ideas relating to waves in general, whatever the nature of the vibration or mechanism of propagation, and the specific case of elastic waves propagating in fluids, i.e. acoustic waves. Elastic waves, since they arise from particle displacements, can propagate only in material media, unlike electromagnetic waves which can also propagate in a vacuum. Because a fluid consists of freely-moving particles, its properties are scalar parameters – the mass density ρ, compressibility coefficient χ and the mean free path (average distance travelled between collisions) – and hence the wave analysis could be developed using scalars. The propagating waves can be entirely specified by one scalar quantity, such as pressure, dilatation or the displacement or velocity potential. In brief, waves in *fluids* have the following features:

(a) The wave polarization, expressed by the particle displacement, is necessarily longitudinal, that is, parallel to the wave vector; there cannot be any shear motion when there is no viscosity.
(b) The velocity of propagation is given by $c = (\rho\chi)^{-1/2}$.
(c) The Poynting vector, showing the direction of energy propagation, is parallel to the wave vector.
(d) When a wave is incident on a surface separating two media with different impedances, the transmitted and reflected waves have the same polarization as the incident wave. The relative wave amplitudes, and the propagation directions given by the Snell–Descartes law, depend only on the impedances of the media and the angle of incidence.
(e) The wave cannot propagate when the distance between a pressure maximum and an adjacent minimum ($\lambda/2$) becomes comparable to the mean free path of the particles.

This chapter is concerned with propagation in *crystals*, ordered assemblies of atoms whose properties are described by tensors having rank up to four here. Fortunately, the structure (lattice plus cell contents) has symmetry elements, belonging to one of the thirty-two crystal classes, that reduce the number of independent tensor components, and also cause some components to vanish (Sect. 2.6). Nevertheless, it is still true that a response in a crystal (assumed to behave linearly) does not generally have the same direction as the cause giving rise to it. For a crystal, taken to be infinite, we can thus anticipate the following:

(a) the possibility of propagation of waves other than longitudinal, with different velocities;
(b) a possible lack of parallelism between the wave vector and the energy propagation vector for each mode;
(c) generation by reflection or refraction of waves with polarization different to that of the incident wave;
(d) in a piezoelectric crystal, the presence of an electric field accompanying the mechanical vibration;
(e) a limit on the wave frequency, associated with the interatomic distance. When the wavelength becomes comparable to this very small distance, the crystal is no longer continuous on the scale of the spatial variation of the mechanical variables.

We can also expect that the results will be simpler in the particular cases of propagation along an axis or in a plane of symmetry, or in the extreme case of an isotropic material.

The objective of Sect. 4.1 is to determine the frequency limitation for waves in a monatomic crystal, and to show that this frequency is several orders of magnitude above the highest frequency used in the applications described later (10 GHz). Having established this, the crystal is regarded as a continuum from Sect. 4.2 onwards. The solutions of the wave equation are expressed as three surfaces called the *slowness surfaces.* These surfaces, highlighting the anisotropy of the solid, are illustrated for crystals belonging to various symmetry classes. Section 4.3 deals with propagation in a piezoelectric crystal, taking the electric field to be quasi-static (Sect. 3.1.2.4), and the slowness surfaces for several crystals are shown. The electro-mechanical coupling coefficient is emphasized. Finally, Sect. 4.4 covers reflection and refraction at a solid–solid or solid–liquid interface. This generalizes results derived in Chap. 1.

To avoid confusion with the elastic constants, the wave velocities in a solid are here designated by V, instead of the c used earlier for the case of a fluid.

4.1 Monatomic One-Dimensional Model

The atoms in a crystal are located at the nodes of a three-dimensional lattice (Sect. 2.1.1), that is, at the intersections of three families of parallel lines with three principal directions $\boldsymbol{a}$, $\boldsymbol{b}$, $\boldsymbol{c}$, assumed here to be perpendicular. This assumption does not affect the generality of the results. When a plane longitudinal elastic wave propagates along one of these directions, such as $\boldsymbol{a}$, all the atoms in a plane normal to the wave vector vibrate simultaneously. Hence the rows parallel to $\boldsymbol{a}$ all move in the same manner (Fig. 4.1) and it is sufficient to consider only one chain of identical equidistant atoms, as in Fig. 4.2.

If one atom in this chain is displaced, the two neighbours to which it is linked will also be set in motion, and so on. The disturbance propagates

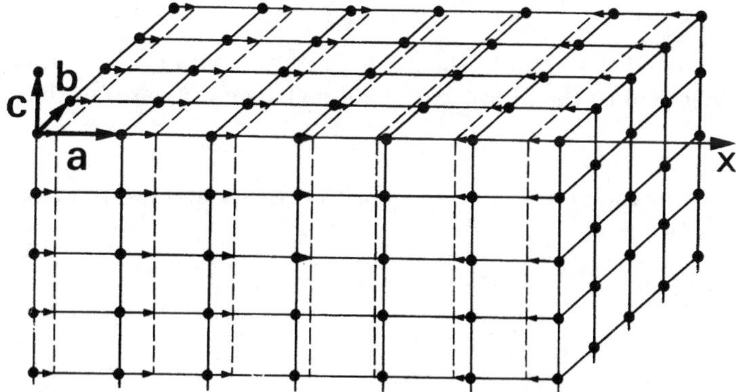

Fig. 4.1. Deformation of a crystalline lattice during propagation of a longitudinal wave parallel to the row vector $\boldsymbol{a}$. It is sufficient to consider the displacements of the atoms in one row

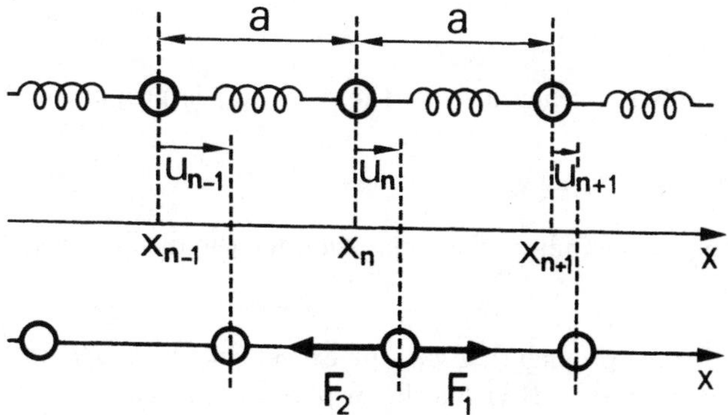

Fig. 4.2. Propagation of a disturbance along a chain of atoms. The springs represent the interactions between atoms

step by step, distorting the chain. We assume that the interactions are only between nearest neighbours, and we suppose that the deformations are very small, such that the two restoring forces F_1 and F_2 acting on each atom are proportional to the changes of the interatomic distances from their equilibrium value a.

Let u_n be the displacement of atom n, with rest position x_n. The forces exerted by atoms $n-1$ and $n+1$ on the nth atom are respectively

$$F_1 = K\,(u_{n-1} - u_n) \quad \text{and} \quad F_2 = K\,(u_{n+1} - u_n)\ .$$

The equation of motion for this nth atom, of mass M, is

$$M\frac{\partial^2 u_n}{\partial t^2} = K\,(u_{n+1} + u_{n-1} - 2u_n)\ . \tag{4.1}$$

We impose the condition that one atom, taken to be that with $n = 0$, has a sinusoidal motion, so that

$$u_0 = A\mathrm{e}^{\mathrm{i}\omega t}\,.$$

In the steady-state condition, and in the absence of attenuation, the motion of each atom is identical to the above except for a change of phase, so that

$$u_n = A\mathrm{e}^{\mathrm{i}(\omega t+\varphi_n)}\,.$$

Substituting this expression for u_n into (4.1), with corresponding expressions for u_{n-1} and u_{n+1}, we find

$$\left(-M\omega^2 + 2K\right) A\mathrm{e}^{\mathrm{i}(\omega t+\varphi_n)} = KA\mathrm{e}^{\mathrm{i}\omega t}\left(\mathrm{e}^{\mathrm{i}\varphi_{n+1}} + \mathrm{e}^{\mathrm{i}\varphi_{n-1}}\right)\,,$$

which gives

$$-M\omega^2 + 2K = K\left[\mathrm{e}^{\mathrm{i}(\varphi_{n+1}-\varphi_n)} + \mathrm{e}^{\mathrm{i}(\varphi_{n-1}-\varphi_n)}\right]\,. \tag{4.2}$$

In order that the right side of this equation should be real, in agreement with the left, the phase-shift $\Delta\varphi$ between neighbouring atoms has to be a constant along the chain, so that

$$\varphi_n - \varphi_{n-1} = \varphi_{n+1} - \varphi_n = \Delta\varphi\,.$$

Since the atoms are equidistant when at rest, the phase φ_n is proportional to the distance x_n, so

$$\varphi_n = -kx_n \quad \text{and} \quad \Delta\varphi = -ka\,,$$

where k is defined as the wavenumber. The expression for the displacement,

$$u_n = A\mathrm{e}^{\mathrm{i}(\omega t-kx_n)}$$

is that of a longitudinal wave propagating with phase velocity $V = \omega/k$.

Substituting $\Delta\varphi = -ka$ into (4.2) yields the dispersion relation

$$M\omega^2 = 2K(1-\cos ka) = 4K\sin^2(ka/2)$$

or, alternatively,

$$\boxed{\omega = 2\sqrt{\frac{K}{M}}\left|\sin\frac{ka}{2}\right|}\,. \tag{4.3}$$

The dispersion curve, which is periodic, is sketched in Fig. 4.3 for the interval $-\pi/a < k < \pi/a$, called the first Brillouin zone. For small values of the wavenumber, such that $ka \ll 1$, this becomes the straight line

$$\omega \cong \sqrt{\frac{K}{M}}ak\,,$$

where the slope $V_0 = a\sqrt{K/M}$ is the phase velocity of the elastic waves at low frequencies. When the wavelength is comparable to the interatomic distance a, so that ka is in the region of π, the medium is dispersive. Since $|\sin ka/2| < 1$, elastic waves can propagate in a crystal only when the frequency is less than a *cut-off frequency* f_c, given by

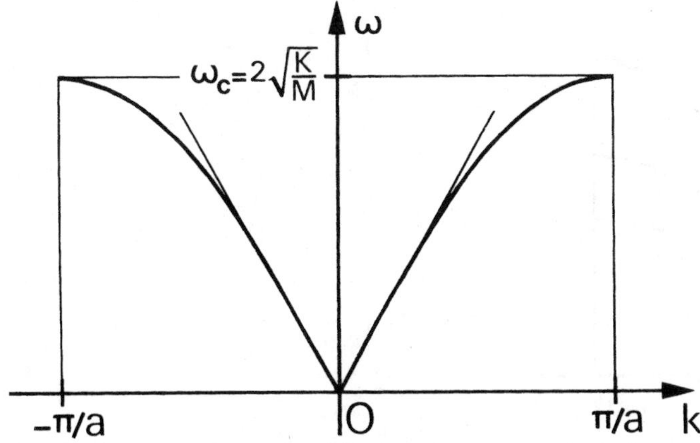

Fig. 4.3. Dispersion curve $\omega(k)$ for the longitudinal mode in a monatomic chain. For any frequency below the cut-off frequency ω_c, two modes can propagate, with opposite directions and opposite wavenumbers

$$f_c = \frac{1}{\pi}\sqrt{\frac{K}{M}} = \frac{V_0}{\pi a}.$$

At this frequency, the group velocity $V_g = \mathrm{d}\omega/\mathrm{d}k$ is zero because the tangent to the dispersion curve is horizontal. The ratio $u_{n+1}/u_n = \exp(-ika)$ takes the value -1, showing that neighbouring atoms vibrate with opposite phase.

Typical Values. The velocity V_0 of low-frequency elastic waves in solids is typically between 1000 and 10 000 m/s. Interatomic distances, a, are a few Å. Taking $V_0 = 5000$ m/s and $a = 5$ Å $= 5 \times 10^{-10}$ m, the cut-off frequency is

$$f_c = 3.2 \times 10^{12}\,\mathrm{Hz} = 3\,200\,\mathrm{GHz}.$$

This frequency is so high that the frequency range considered in this book, a few kHz to at most 10 GHz, is *very close to the origin of the dispersion curve*, where the frequency is proportional to the wavenumber. The wavelength $\lambda = V_0/f$ is in the range from a fraction of a metre to a few microns (µm) and is much greater than the interatomic distance; thus, to the wave, *the solid appears to be continuous.*

Transverse Waves. It is possible for the atomic displacement (i.e., the wave polarization) to be transverse, for example along one of the basis vectors $\boldsymbol{b}$ or $\boldsymbol{c}$ of the atomic lattice (Fig. 4.1). For these cases the restoring forces will be different, and hence for propagation along $\boldsymbol{a}$ there will be three distinct dispersion curves, as in Fig. 4.4, referring to three waves:

(a) a longitudinal wave L, polarized along $\boldsymbol{a}$;
(b) a transverse wave T_1, polarized along $\boldsymbol{b}$; and
(c) another transverse wave T_2, polarized along $\boldsymbol{c}$.

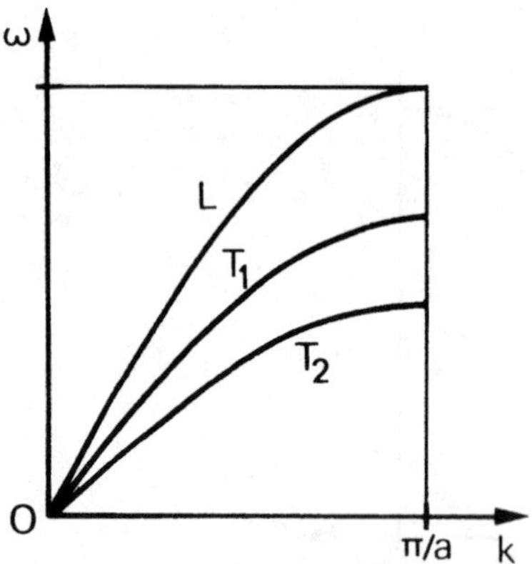

Fig. 4.4. Dispersion for a monatomic chain. For propagation along $\boldsymbol{a}$, with $k > 0$, there are three curves

The transverse waves are also known as *shear waves*. The three waves can also propagate, with different velocities, along $\boldsymbol{b}$ or $\boldsymbol{c}$. The three dispersion curves of Fig. 4.4 will generally change for these waves – the phase velocity depends on the direction of the wave vector. Thus, for frequencies well below the cut-off frequency, *a crystal is non-dispersive for elastic waves, but it is anisotropic.*

More generally, the wave vector may be inclined to the vector $\boldsymbol{a}$. In this case there are still three waves that may propagate, but the atomic displacements are generally neither parallel nor perpendicular to the propagation direction. The wave with polarization closest to the propagation vector is called the *quasi-longitudinal* wave; the other two have polarizations perpendicular to this and are called *quasi-transverse* waves. In the particular case where the 'springs' linking the atoms are identical for the directions $\boldsymbol{b}$ and $\boldsymbol{c}$, the two transverse waves with propagation along $\boldsymbol{a}$ have the same velocity. A propagation direction such as this is called an *acoustic axis.*

Optical Branches. If the chain consists of two types of atoms, the above comments remain valid for the centre of mass of the molecule, giving an *acoustic* branch such as that in Fig. 4.3. However, there is also another curve, with very high frequency whatever the wavenumber, called the *optical* branch (Problem 4.2). The two branches are particularly differentiated at long wavelengths ($ka \ll 1$). Vibration of the molecules, with an overall motion, gives a low-frequency wave, the acoustic branch; vibration of the atoms about a stationary centre of mass gives a high-frequency solution, the optical branch, corresponding to the eigenfrequency for longitudinal vibration of a molecule. Including the transverse waves, the dispersion relation now has six branches. This result can be generalized – if the cell contains p atoms, the dispersion relation has $3p$ branches, of which $3p - 3$ are optical.

Incoherent Thermoelastic Waves. The waves considered above are coherent waves, related to the overall movements of atoms and molecules. However, in all materials there are also thermoelastic waves, that is, elastic waves occurring spontaneously, associated with random motions of the atoms. These are related to the thermal behaviour of the solid, in particular the variation of specific heat with temperature.

4.2 Anisotropic Solids

From the above observations, we see that elastic waves in a crystal can be characterized by three surfaces giving the velocities V_1, V_2 and V_3 of the one (quasi-) longitudinal and the two (quasi-) transverse waves as functions of direction. For this purpose the crystal, an anisotropic solid, is regarded as a continuous medium of infinite extent, that is, its dimensions are much larger than those of the beams of the waves. In practice, surfaces giving the reciprocals of the velocities, $1/V_1$, $1/V_2$ and $1/V_3$, are more useful. These surfaces, called *slowness surfaces*, are analogous to the index surfaces of optics. They are of interest for showing the direction of energy propagation. These results are found naturally as solutions of the basic dynamical equations, applied to a continuous solid.

4.2.1 Wave Equation

By hypothesis, the solid has a disturbance propagating through it, so it is locally in motion. The displacement u_i of an arbitrary point in the solid, with coordinates x_k, varies with time, so that $u_i = u_i(x_k, t)$. The equation of motion follows from Newton's law. Neglecting gravitation, Sect. 3.1.2.2 showed this equation to be

$$\rho \frac{\partial^2 u_i}{\partial t^2} = \frac{\partial T_{ij}}{\partial x_j} . \tag{4.4}$$

From (3.40), Hooke's law can be written as

$$T_{ij} = c_{ijkl} \frac{\partial u_l}{\partial x_k} ,$$

and combining this with (4.4) leads to the wave equation

$$\boxed{\rho \frac{\partial^2 u_i}{\partial t^2} = c_{ijkl} \frac{\partial^2 u_l}{\partial x_j \partial x_k}} . \tag{4.5}$$

This system of three second-order differential equations is a generalization, allowing for three dimensions and for anisotropy, of the wave equation for a fluid. The latter was shown in (1.25) to be

$$\rho \frac{\partial^2 u}{\partial t^2} = \frac{1}{\chi} \frac{\partial^2 u}{\partial x^2}$$

with the general solution

$$u = F(t - x/V), \quad \text{with} \quad V^2 = 1/\rho\chi\,.$$

By analogy, we seek a solution for the solid in the form of a plane wave propagating in the direction of the unit vector $\boldsymbol{n} = (n_1, n_2, n_3)$, the latter being perpendicular to the wavefronts given by $\boldsymbol{n} \cdot \boldsymbol{x} =$ constant. We thus consider the solution

$$u_i = {}^\circ u_i F\left(t - \frac{\boldsymbol{n} \cdot \boldsymbol{x}}{V}\right) = {}^\circ u_i F\left(t - \frac{n_j x_j}{V}\right), \tag{4.6}$$

where the constant ${}^\circ u_i$, independent of x_i and t, is the *wave polarization*, giving the direction of the particle displacement. To determine the phase velocity V and the polarization ${}^\circ u_i$, we substitute (4.6) into the wave equation (4.5). Denoting the second derivative of the function F by F'', this gives

$$\frac{\partial^2 u_i}{\partial t^2} = {}^\circ u_i F'' \quad \text{and} \quad \frac{\partial u_l}{\partial x_j} = -\,{}^\circ u_l \frac{n_j}{V} F',$$

from which

$$\frac{\partial^2 u_l}{\partial x_j \partial x_k} = {}^\circ u_l \frac{n_j n_k}{V^2} F''\,.$$

Thus,

$$\rho\,{}^\circ u_i F'' = c_{ijkl} n_j n_k\,{}^\circ u_l \frac{F''}{V^2}\,,$$

which leads to

$$\boxed{\rho V^2\,{}^\circ u_i = c_{ijkl} n_j n_k\,{}^\circ u_l}\,. \tag{4.7}$$

This is known as *Christoffel's equation.*

Introducing the second-rank tensor (the Christoffel tensor)

$$\boxed{\Gamma_{il} = c_{ijkl} n_j n_k}\,, \tag{4.8}$$

Christoffel's equation becomes

$$\Gamma_{il}\,{}^\circ u_l = \rho V^2\,{}^\circ u_i\,. \tag{4.9}$$

Assuming familiarity with the results of Appendix B, this shows that the polarization ${}^\circ u_i$ is an eigenvector of the tensor Γ_{il}, with eigenvalue $\gamma = \rho V^2$. Thus, *the phase velocities and polarizations of plane waves propagating in the direction $\boldsymbol{n}$ in a crystal with stiffness c_{ijkl} are given by the eigenvalues and eigenvectors of the tensor $\Gamma_{il} = c_{ijkl} n_j n_k$.*

For a given direction, there are in general three phase velocities which are the solutions of the secular equation

$$\left|\Gamma_{il} - \rho V^2 \delta_{il}\right| = 0 \tag{4.10}$$

expressing the compatibility condition for the three homogeneous equations (4.9). For each velocity there is a corresponding eigenvector giving the direction of the displacement of the material, i.e. the wave polarization. By virtue of the symmetry properties of the stiffness tensor, Sect. 3.2.2, the tensor Γ_{il} is *symmetric*, since

$$\Gamma_{li} = c_{ljki} n_j n_k = c_{kilj} n_j n_k = c_{ikjl} n_j n_k = c_{ijkl} n_j n_k = \Gamma_{il} \,.$$

Its eigenvalues are therefore real, and its eigenvectors are orthogonal. Moreover, the eigenvalues $\gamma = \rho V^2$ are found to be positive, as required since V must be real.

To demonstrate the last property, we return to (4.9) and premultiply by ${}^\circ u_i$ to obtain

$$\gamma = \frac{{}^\circ u_i \Gamma_{il} \, {}^\circ u_l}{{}^\circ u_i^2} \,.$$

The sign of γ is that of the numerator

$$N = {}^\circ u_i \Gamma_{il} \, {}^\circ u_l = {}^\circ u_i n_j c_{ijkl} n_k \, {}^\circ u_l \,.$$

Now, the density of elastic potential energy is, from (3.54), $e_{\mathrm{p}} = \frac{1}{2} c_{ijkl} S_{ij} S_{kl}$, and this must be positive whatever the stress, that is, for any symmetric tensor S_{ij} of rank two. Considering an arbitrary tensor U_{ij}, which can be decomposed into a symmetric part S_{ij} and an antisymmetric part A_{ij}, the contracted product

$$P = U_{ij} c_{ijkl} U_{kl} = S_{ij} c_{ijkl} S_{kl}$$

involves only S_{ij} because the tensor c_{ijkl} is symmetric (Problem 2.19). Applying this result to the tensor $U_{ij} = {}^\circ u_i n_j$ shows that N, and hence $\gamma = \rho V^2$, is positive.

We conclude that, in general, *three plane waves with orthogonal polarizations can propagate in the same direction, with different velocities.* This situation is illustrated by Fig. 4.5. The displacement vector $\boldsymbol{u}$ is not generally parallel or perpendicular to the propagation direction $\boldsymbol{n}$. The wave with polarization closest to $\boldsymbol{n}$ is called *quasi-longitudinal*, and the others are called *quasi-transverse*. These last usually travel more slowly than the quasi-longitudinal wave. Only in particular propagation directions are the waves purely longitudinal or transverse [4.1].

Before studying propagation in directions related to the crystal symmetry, we give explicit expressions for the components of the Christoffel tensor $\Gamma_{il} = c_{ijkl} n_j n_k$, which occur in the general equation

$$[\Gamma_{il}] \, [\, {}^\circ u_l] = \rho V^2 \, [\, {}^\circ u_i] \,.$$

Expanding the sum over the indices j and k gives

$$\begin{aligned} \Gamma_{il} = {} & c_{i11l} n_1^2 + c_{i22l} n_2^2 + c_{i33l} n_3^2 + (c_{i12l} + c_{i21l}) n_1 n_2 \\ & + (c_{i13l} + c_{i31l}) n_1 n_3 + (c_{i23l} + c_{i32l}) n_2 n_3 \,, \end{aligned}$$

from which we have the following six components:

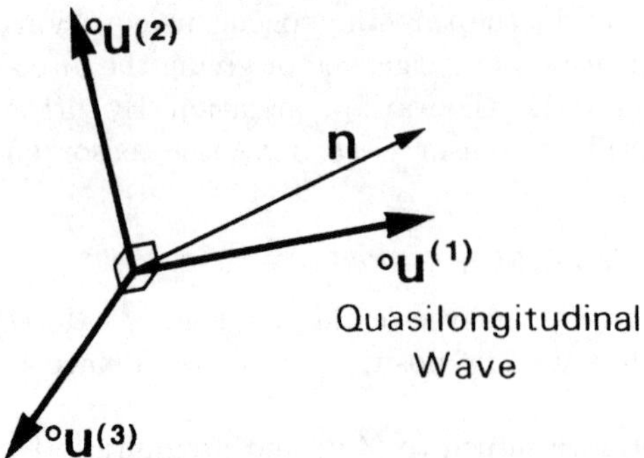

Fig. 4.5. Propagation in a crystal, an anisotropic solid. In general, three plane waves can propagate in the same direction $\boldsymbol{n}$, each with its characteristic velocity. The one whose polarization (material displacement) $°\boldsymbol{u}^{(1)}$ is closest to the wave vector $\boldsymbol{n}$ is called the quasi-longitudinal wave. Its velocity is usually higher than that of the other waves, called quasi-transverse, which have polarizations $°\boldsymbol{u}^{(2)}$ and $°\boldsymbol{u}^{(3)}$. The three vectors $°\boldsymbol{u}^{(1)}$, $°\boldsymbol{u}^{(2)}$, $°\boldsymbol{u}^{(3)}$ are mutually perpendicular

$$\begin{aligned}
\Gamma_{11} &= c_{11}n_1^2+c_{66}n_2^2+c_{55}n_3^2+2c_{16}n_1n_2+2c_{15}n_1n_3+2c_{56}n_2n_3 \\
\Gamma_{12} &= c_{16}n_1^2+c_{26}n_2^2+c_{45}n_3^2+(c_{12}+c_{66})n_1n_2+(c_{14}+c_{56})n_1n_3+(c_{46}+c_{25})n_2n_3 \\
\Gamma_{13} &= c_{15}n_1^2+c_{46}n_2^2+c_{35}n_3^2+(c_{14}+c_{56})n_1n_2+(c_{13}+c_{55})n_1n_3+(c_{36}+c_{45})n_2n_3 \\
\Gamma_{22} &= c_{66}n_1^2+c_{22}n_2^2+c_{44}n_3^2+2c_{26}n_1n_2+2c_{46}n_1n_3+2c_{24}n_2n_3 \\
\Gamma_{23} &= c_{56}n_1^2+c_{24}n_2^2+c_{34}n_3^2+(c_{46}+c_{25})n_1n_2+(c_{36}+c_{45})n_1n_3+(c_{23}+c_{44})n_2n_3 \\
\Gamma_{33} &= c_{55}n_1^2+c_{44}n_2^2+c_{33}n_3^2+2c_{45}n_1n_2+2c_{35}n_1n_3+2c_{34}n_2n_3 \\
\Gamma_{21} &= \Gamma_{12}; \quad \Gamma_{31} = \Gamma_{13}; \quad \Gamma_{32} = \Gamma_{23}\,.
\end{aligned} \tag{4.11}$$

4.2.2 Directions Linked to Symmetry Elements

Propagation along an axis. For propagation along the Ox_3 axis ($n_1 = n_2 = 0, n_3 = 1$), the tensor Γ_{il} becomes $\Gamma_{il} = c_{i33l}$ or, explicitly,

$$\Gamma_{x_3} = \begin{bmatrix} c_{55} & c_{45} & c_{35} \\ c_{45} & c_{44} & c_{34} \\ c_{35} & c_{34} & c_{33} \end{bmatrix}.$$

For *monoclinic* crystals, with the dyad axis along Ox_3, (3.63) shows that $c_{35} = c_{34} = 0$, so

$$\Gamma_{x_3\|A_2} = \begin{bmatrix} c_{55} & c_{45} & 0 \\ c_{45} & c_{44} & 0 \\ 0 & 0 & c_{33} \end{bmatrix}.$$

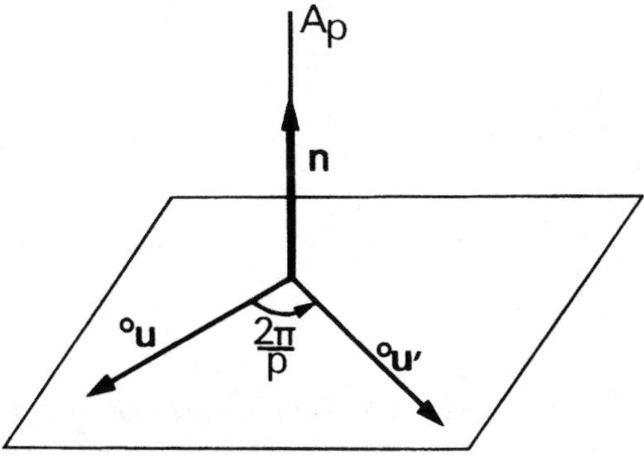

Fig. 4.6. Propagation along a symmetry axis A_p, with $p > 2$. Any vector normal to A_p can be decomposed into components along $^\circ\boldsymbol{u}$ and $^\circ\boldsymbol{u}'$ corresponding to a rotation by $2\pi/p$, and is an eigenvector of Γ with the same eigenvalue as $^\circ\boldsymbol{u}$ and $^\circ\boldsymbol{u}'$

Hence the longitudinal wave propagates with velocity $V_1 = (c_{33}/\rho)^{1/2}$ and the two transverse waves have different velocities (Appendix B).

For *orthorhombic* crystals, with Ox_3 taken as one of the dyad axes, c_{45} is also zero. The transverse wave with velocity $V_2 = (c_{44}/\rho)^{1/2}$ is polarized along Ox_2, and that with velocity $V_3 = (c_{55}/\rho)^{1/2}$ is polarized along Ox_1. Note that the elastic constants can be obtained by measuring these velocities.

If there is an *axis with order p higher than* 2, we have $c_{44} = c_{55}$, so *the two transverse waves have the same velocity* – the polarization has an arbitrary direction in the plane normal to the axis because the eigenvalues are identical, and the waves are degenerate. Such a direction, along which a transverse wave has an arbitrary polarization direction, is called an *acoustic axis.* However, there may be directions other than these symmetry axes such that the two transverse waves have equal velocities.

These conclusions can also be found from symmetry considerations. If $\boldsymbol{n}$ is along the axis A_p, the combined symmetry of the crystal and the propagation direction is that of the axis, A_p. The polarizations of the waves must necessarily be invariant for a rotation by an angle $2\pi/p$ about $\boldsymbol{n}$, and this is possible only if the waves are purely transverse or longitudinal. Moreover, when p is greater than 2 the displacement vector $^\circ\boldsymbol{u}$ of a transverse wave is not parallel to that of its transform $^\circ\boldsymbol{u}'$, as shown in Fig. 4.6. These correspond to the same velocity, and therefore have the same eigenvalue of the tensor Γ_{il}. Any vector perpendicular to A_p, which can be decomposed into components along $^\circ\boldsymbol{u}$ and $^\circ\boldsymbol{u}'$, is therefore an eigenvector of Γ_{il} with the same eigenvalue. Hence the transverse waves propagate with the same velocity and with arbitrary polarization directions.

Propagation in a plane of symmetry, or in a plane normal to an axis with even order. In these two cases there is a dyad axis. Since the elastic properties and the propagation direction are invariant for an inver-

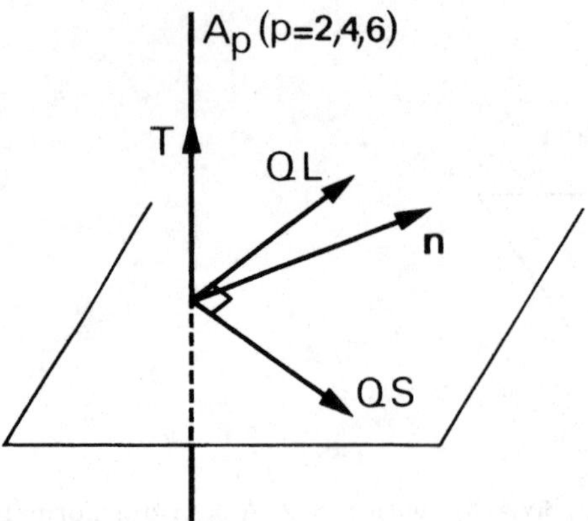

Fig. 4.7. For any direction $\boldsymbol{n}$ perpendicular to an axis A_p of even order (direct or inverse), there is a transverse wave polarized along A_p

sion (Sect. 2.2.1), and bearing in mind the equivalence $A_2C = \bar{A}_2 = M$, the combination of the crystal and the wave is symmetric about the plane perpendicular to the axis A_p. The polarization must therefore be either parallel to the axis, as for a shear wave T, or in the plane perpendicular to the axis, as for a quasi- longitudinal wave QL and a quasi-shear wave QT (Fig. 4.7).

The above arguments show that, if the propagation is in the x_1x_2 plane, perpendicular to the A_p axis and therefore to Ox_3, the elements Γ_{13} and Γ_{23} are zero, so that

$$\Gamma_{x_1x_2\perp A_p} = \begin{bmatrix} \Gamma_{11} & \Gamma_{12} & 0 \\ \Gamma_{12} & \Gamma_{22} & 0 \\ 0 & 0 & \Gamma_{33} \end{bmatrix}.$$

For a tetrad or hexad axis, we use (4.11) with $n_3 = 0$ to obtain

$$\Gamma_{33} = c_{55}n_1^2 + c_{44}n_2^2 + 2c_{45}n_1n_2$$

and from (3.75) and (3.76) the axis imposes $c_{45} = 0$ and $c_{55} = c_{44}$, so that

$$\Gamma_{33} = c_{44}\left(n_1^2 + n_2^2\right) = c_{44}\,.$$

Thus *the velocity* $V_3 = (\Gamma_{33}/\rho)^{1/2} = (c_{44}/\rho)^{1/2}$ *of a transverse wave propagating in a plane normal to a tetrad or hexad axis is independent of direction.*

4.2.3 Isotropic Solid

For an isotropic solid, the tensor Γ is independent of the propagation direction, being the same as that for propagation along an axis of order $p > 2$ in the Ox_3 direction. Using (3.58) for the stiffness constants, with $c_{33} = c_{11}$, gives

$$\Gamma = \begin{bmatrix} c_{44} & 0 & 0 \\ 0 & c_{44} & 0 \\ 0 & 0 & c_{11} \end{bmatrix} \quad \text{with} \quad c_{44} = \frac{c_{11} - c_{12}}{2}\,,$$

and hence

$$\begin{cases} c_{44}\,{}^{\circ}u_1 = \rho V^2\,{}^{\circ}u_1 \\ c_{11}\,{}^{\circ}u_3 = \rho V^2\,{}^{\circ}u_3\,. \end{cases}$$

Thus all directions are acoustic axes. The velocities V_{T} and V_{L} of the transverse and longitudinal waves, with polarizations ${}^{\circ}u_1$ and ${}^{\circ}u_3$ respectively, are independent of the propagation direction. They are given by

$$\boxed{V_{\mathrm{T}} = \sqrt{\frac{c_{44}}{\rho}} = \sqrt{\frac{(c_{11}-c_{12})}{2\rho}}\,, \qquad V_{\mathrm{L}} = \sqrt{\frac{c_{11}}{\rho}}}\,. \tag{4.12}$$

Also, since c_{12} is positive, the inequality

$$V_{\mathrm{T}} < \frac{V_{\mathrm{L}}}{\sqrt{2}}$$

holds for all isotropic solids.

This decomposition, into two waves with different velocities, is valid for more general waves, not merely for plane waves. To show this we return to (4.4), the equation of motion, that is,

$$\rho\frac{\partial^2 u_i}{\partial t^2} = \frac{\partial T_{ij}}{\partial x_j}\,.$$

From (3.57) and (3.61) the stresses are given by

$$T_{ij} = (c_{11}-2c_{44})S\delta_{ij} + 2c_{44}S_{ij} = (c_{11}-2c_{44})S\delta_{ij} + c_{44}\left(\frac{\partial u_i}{\partial x_j} + \frac{\partial u_j}{\partial x_i}\right), \tag{4.13}$$

where S, with no index, is the dilatation defined by $S = S_{ii} = \operatorname{div}\boldsymbol{u} = \partial u_i/\partial x_i$. Substituting into the equation of motion, the latter becomes

$$\rho\frac{\partial^2 u_i}{\partial t^2} = \frac{\partial}{\partial x_i}\left[(c_{11}-2c_{44})\frac{\partial u_i}{\partial x_i}\right] + c_{44}\frac{\partial^2 u_i}{\partial x_j^2} + c_{44}\frac{\partial}{\partial x_i}\left(\frac{\partial u_j}{\partial x_j}\right).$$

This elasto-dynamic equation, which governs the displacement in an isotropic solid, can be written in vector notation as

$$\rho\frac{\partial^2 \boldsymbol{u}}{\partial t^2} = (c_{11}-c_{44})\,\mathrm{grad}(\operatorname{div}\boldsymbol{u}) + c_{44}\Delta\boldsymbol{u}\,,$$

where $\Delta = \partial^2/\partial x_k^2$ is the Laplacian. Introducing the nabla operator $\nabla = (\partial/\partial x_1, \partial/\partial x_2, \partial/\partial x_3)$, the equation becomes

$$\boxed{\rho\frac{\partial^2 \boldsymbol{u}}{\partial t^2} = (c_{11}-c_{44})\nabla(\nabla\cdot\boldsymbol{u}) + c_{44}\nabla^2\boldsymbol{u}}\,. \tag{4.14}$$

In this differential equation, the three components of the displacement are coupled. With a view to obtaining decoupled equations, it is useful to define a scalar potential ϕ and a vector potential Ψ, such that

$$\boxed{\boldsymbol{u} = \nabla\phi + \nabla\wedge\Psi}\,. \tag{4.15}$$

We recall that for a non-viscous fluid, which excludes transverse motion, it is sufficient to use a scalar potential (Sect. 1.3.1). For the present case, we substitute (4.15) into (4.14) and make use of the relations

$$\nabla \wedge (\nabla \phi) = 0 \quad \text{and} \quad \nabla \cdot (\nabla \wedge \Psi) = 0$$

giving the result

$$\rho \frac{\partial^2 \nabla \phi}{\partial t^2} + \rho \frac{\partial^2 (\nabla \wedge \Psi)}{\partial t^2} = (c_{11} - c_{44}) \nabla (\nabla^2 \phi) + c_{44} \nabla^2 (\nabla \phi) + c_{44} \nabla^2 (\nabla \wedge \Psi)$$

and this becomes

$$\nabla \left(\rho \frac{\partial^2 \phi}{\partial t^2} - c_{11} \nabla^2 \phi \right) + \nabla \wedge \left(\rho \frac{\partial^2 \Psi}{\partial t^2} - c_{44} \nabla^2 \Psi \right) = 0 \, .$$

This can be split into the following two equations, one with a scalar variable and the other with a vector variable:

$$\boxed{\rho \frac{\partial^2 \phi}{\partial t^2} - c_{11} \nabla^2 \phi = 0} \quad \text{and} \quad \boxed{\rho \frac{\partial^2 \Psi}{\partial t^2} - c_{44} \nabla^2 \Psi = 0} \, . \tag{4.16a,b}$$

These equations show that the two potentials propagate independently of each other, with velocities $V_L = (c_{11}/\rho)^{1/2}$ and $V_T = (c_{44}/\rho)^{1/2}$ respectively. The first equation, with a scalar variable whose gradient is the displacement, corresponds to longitudinal waves since it gives $\nabla \wedge \boldsymbol{u} = 0$. The second, with a vector variable, corresponds to transverse waves since it gives $\nabla \cdot \boldsymbol{u} = 0$. However, since the decomposition (4.15) expresses the original vector in terms of a scalar and a new vector, the number of variables involved has been increased by one – the three components u_1, u_2, u_3 of the displacement have been replaced by the four variables ϕ, ψ_1, ψ_2 and ψ_3. It is therefore possible to impose a relation between these four, for example $\nabla \cdot \Psi = 0$.

It is also possible to decompose the displacement vector $\boldsymbol{u}$ into a divergence-free vector $\boldsymbol{u}_T$ and an irrotational vector $\boldsymbol{u}_L$, so that

$$\boldsymbol{u} = \boldsymbol{u}_T + \boldsymbol{u}_L \, .$$

Substituting into (4.14), and incorporating the velocities V_L and V_T, lead to two equations

$$\boxed{\frac{\partial^2 \boldsymbol{u}_L}{\partial t^2} - V_L^2 \nabla^2 \boldsymbol{u}_L = 0} \quad \text{and} \quad \boxed{\frac{\partial^2 \boldsymbol{u}_T}{\partial t^2} - V_T^2 \nabla^2 \boldsymbol{u}_T = 0} \, , \tag{4.16c,d}$$

which are similar to the previous two. This confirms the independent propagation of the two components $\boldsymbol{u}_L$ and $\boldsymbol{u}_T$, with velocities V_L and V_T.

These conclusions are not surprising since they are already apparent on the microscopic scale, as discussed in Sect. 4.1. Propagation of a longitudinal wave along $\boldsymbol{a}$ involves a variation of the distances between the parallel planes of atoms, so that the volume occupied by a given number of atoms varies. Hence the rotation of the displacement vector is zero, curl $\boldsymbol{u}_L = 0$, but its divergence is finite, so div $\boldsymbol{u}_L \neq 0$. For a transverse wave the atomic displacement is perpendicular to $\boldsymbol{a}$; the parallel planes slide across each other,

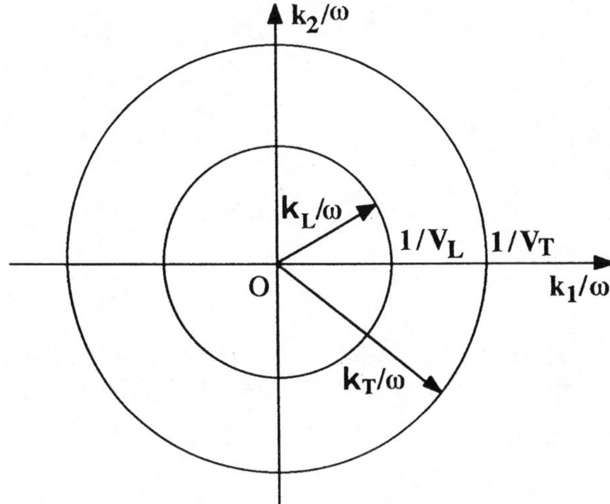

Fig. 4.8. An isotropic solid is characterized, for the purpose of elastic waves, by two spheres with radii V_T and V_L or $1/V_\mathrm{T}$ and $1/V_\mathrm{L}$, shown here in the $(k_1/\omega, k_2/\omega)$ plane

so there is no change of volume. Thus the rotation of the displacement is finite, curl $\boldsymbol{u}_\mathrm{T} \neq 0$, but its divergence is zero, so div $\boldsymbol{u}_\mathrm{T} = 0$.

In summary, elastic wave propagation in an isotropic solid is characterized by by *two spheres*, of radius either V_L and V_T or $1/V_\mathrm{L}$ and $1/V_\mathrm{T}$ (Fig. 4.8). These two representations are equivalent here, but this is not so for an anisotropic solid.

4.2.4 Energy Velocity

The energy velocity vector is, by definition, equal to the Poynting vector divided by the energy per unit volume (Sect. 3.1.4). For a plane wave, (4.6) gives the relations

$$\frac{\partial u_i}{\partial t} = {}^\circ u_i F' \quad \text{and} \quad \frac{\partial u_i}{\partial x_j} = -n_j \, {}^\circ u_i \frac{F'}{V}, \tag{4.17}$$

so the kinetic energy density can be expressed as

$$e_\mathrm{k} = \frac{1}{2}\rho \left(\frac{\partial u_i}{\partial t}\right)^2 = \frac{1}{2}\rho \, {}^\circ u_i^2 F'^2 . \tag{4.18}$$

The potential energy density is, from (3.54),

$$e_\mathrm{p} = \frac{1}{2} c_{ijkl} S_{ij} S_{kl} = \frac{1}{2} c_{ijkl} \frac{\partial u_i}{\partial x_j} \frac{\partial u_l}{\partial x_k}, \tag{4.19}$$

giving

$$e_{\mathrm{p}} = \frac{1}{2} c_{ijkl} n_j n_k \,{}^\circ u_i \,{}^\circ u_l \frac{F'^2}{V^2} \,,$$

moreover, multiplication of Christoffel's equation (4.7) by ${}^\circ u_i$ gives

$$c_{ijkl} n_j n_k \,{}^\circ u_i \,{}^\circ u_l = \rho V^2 \,{}^\circ u_i^2 \,, \tag{4.20}$$

and this simplifies e_{p} to

$$e_{\mathrm{p}} = \frac{1}{2} \rho \,{}^\circ u_i^2 F'^2 \,. \tag{4.21}$$

Comparing (4.18) and (4.21) shows that *the kinetic and potential energy densities of a plane wave are equal*, a classic result. The total energy density is thus

$$e = \rho \,{}^\circ u_i^2 F'^2 \,. \tag{4.22}$$

The Poynting vector

$$P_i = -T_{ij} \frac{\partial u_j}{\partial t} = -c_{ijkl} \frac{\partial u_l}{\partial x_k} \frac{\partial u_j}{\partial t}$$

becomes, with the aid of (4.17),

$$P_i = c_{ijkl} \,{}^\circ u_j \,{}^\circ u_l \frac{n_k}{V} F'^2 \,.$$

The energy velocity V_i^{e} is obtained by dividing P_i by the energy density e (Sect. 1.2.2), giving

$$V_i^{\mathrm{e}} = \frac{c_{ijkl} \,{}^\circ u_j \,{}^\circ u_l n_k}{\rho \,{}^\circ u_i^2 V} \tag{4.23}$$

and, for a displacement with unit magnitude, this becomes

$$\boxed{V_i^{\mathrm{e}} = \frac{c_{ijkl} \,{}^\circ u_j \,{}^\circ u_l n_k}{\rho V}} \quad \text{for} \quad u_i^2 = 1 \,. \tag{4.24}$$

The energy velocity $\boldsymbol{V}^{\mathrm{e}}$ shows the direction of energy transport, that is, the direction of the *acoustic ray. When this ray is perpendicular to the wavefronts, and therefore parallel to* $\boldsymbol{n}$*, the mode (the wave) is described as 'pure'.* However, many writers in English designate as 'pure' a wave whose polarization is rigorously longitudinal or transverse. The two definitions coincide only for a longitudinal wave – see Problem 4.3 and the example given later in this section.

Pure mode directions are of practical importance. Crystals used for applications such as delay lines take the form of cylinders, or bars with square or rectangular section, bearing a transducer at each end (Sect. 1.1 of Vol. II). The beam of waves generated by one transducer needs to be received by the other, and this requires the Poynting vector to be parallel to the axis of the cylinder or bar; normally, this vector needs to be parallel to the wave

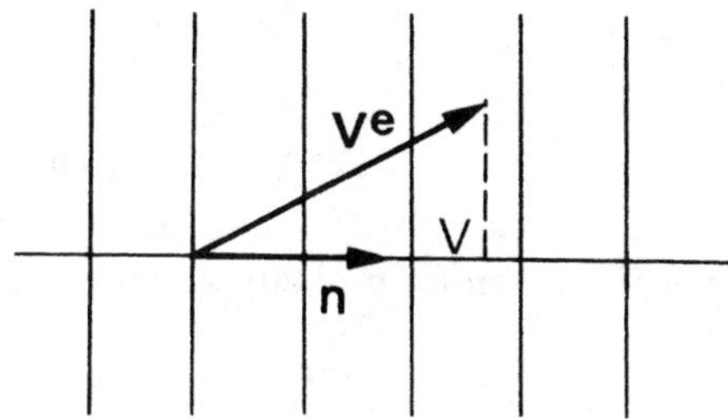

Fig. 4.9. The projection of the energy velocity $\boldsymbol{V}^{\mathrm{e}}$ onto the propagation direction $\boldsymbol{n}$ is equal to the phase velocity V of the plane wave

vector, which is perpendicular to the end faces of the crystal. The pure modes were the subject of many early systematic searches in cubic, hexagonal and tetragonal crystals.

If we form the scalar product

$$\boldsymbol{V}^{\mathrm{e}} \cdot \boldsymbol{n} = V_i^{\mathrm{e}} n_i = \frac{c_{ijkl}\ {}^{\circ}u_j\ {}^{\circ}u_l n_i n_k}{\rho\ {}^{\circ}u_i^2 V}$$

and replace the numerator using (4.20), we obtain

$$\boxed{\boldsymbol{V}^{\mathrm{e}} \cdot \boldsymbol{n} = V}\,. \tag{4.25}$$

Thus, for a plane wave, the projection of the energy velocity onto the propagation direction is equal to the phase velocity V, as in Fig. 4.9. The energy velocity is always larger than, or equal to, the phase velocity.

We can compare this energy velocity with the *group velocity* V^{g} defined by (1.12) in Sect. 1.1.3, that is $V^{\mathrm{g}} = \mathrm{d}\omega/\mathrm{d}k$. In the present case, the group velocity has the three components

$$V_j^{\mathrm{g}} = \frac{\partial \omega}{\partial k_j} \quad \text{with} \quad k_j = k n_j\,.$$

The secular equation (4.10) can be written

$$\left| c_{ijkl} n_j n_k - \rho V^2 \delta_{il} \right| = 0 \tag{4.26}$$

and its solution gives the phase velocity as a function of the propagation direction n_j, so that

$$V = f(n_j).$$

Multiplying (4.26) by k^6, that is, multiplying each line of the determinant by k^2, gives

$$\left| c_{ijkl} k n_j k n_k - \rho k^2 V^2 \delta_{il} \right| = 0\,.$$

This is the same as (4.26) except that $kV(=\omega)$ has replaced V, and $kn_j(=k_j)$ has replaced n_j. The dependence of ω on k_j is therefore the same as the above dependence of V on n_j, so that

$$\omega = f(k_j)\,.$$

Consequently,

$$V_j^{\mathrm{g}} = \frac{\partial \omega}{\partial k_j} = \frac{\partial V}{\partial n_j}\,. \tag{4.27}$$

We can obtain $\partial V/\partial n_j$ by differentiating V^2, taken from (4.20), giving

$$2V\frac{\partial V}{\partial n_j} = 2c_{ijkl}\frac{n_k\,{}^\circ u_i\,{}^\circ u_l}{\rho\,{}^\circ u_i^2}$$

and hence

$$V_j^{\mathrm{g}} = \frac{\partial V}{\partial n_j} = c_{ijkl}\frac{n_k\,{}^\circ u_i\,{}^\circ u_l}{\rho\,{}^\circ u_i^2 V}\,. \tag{4.28}$$

Thus the group velocity $\boldsymbol{V}^{\mathrm{g}}$ is equal to the energy transport velocity $\boldsymbol{V}^{\mathrm{e}}$ given by (4.23).

Note that the equation $\boldsymbol{V}^{\mathrm{e}} \cdot \boldsymbol{n} = V$ is a consequence of Euler's theorem on homogeneous functions. If f is a homogeneous function of the n_i with degree p, then

$$n_i\frac{\partial f}{\partial n_i} = pf\,.$$

The relation $\omega = Vk = f(kn_i)$ shows that $V = f(n_i)$ is a homogeneous function of the n_i with degree unity, so that

$$n_i\frac{\partial V}{\partial n_i} = V \quad \text{and} \quad \boldsymbol{n} \cdot \boldsymbol{V}^{\mathrm{e}} = V\,.$$

Applying the same theorem to the function $V_i^{\mathrm{e}} = \partial V/\partial n_i$, a homogeneous function of n_i with degree zero, gives

$$n_k\frac{\partial V_i^{\mathrm{e}}}{\partial n_k} = 0\,.$$

Example. To clarify the notion of the energy velocity, consider the direction of the acoustic ray for a wave propagating along the triad axis of a sapphire crystal (Al_2O_3, class $\bar{3}m$), a material often used for delay lines owing to its low attenuation coefficient (Sect. 4.2.6). The tensor $c_{\alpha\beta}$ is shown in (3.74). Since $n_k = \delta_{k3}$, the energy velocity is, from (4.24),

$$V_i^{\mathrm{e}} = \frac{c_{ij3l}\,{}^\circ u_j\,{}^\circ u_l}{\rho V}\,, \qquad {}^\circ u_i^2 = 1\,. \tag{4.29}$$

The *longitudinal* wave, with velocity $V_1 = (c_{33}/\rho)^{1/2}$ from Sect. 4.2.2, gives

$$u_i = \delta_{i3} \quad \text{and hence} \quad \boldsymbol{V}^{\mathrm{e}} = \left[\frac{c_{53}}{\rho V_1} = 0\,, \frac{c_{43}}{\rho V_1} = 0\,, \frac{c_{33}}{\rho V_1} = V_1\right]\,.$$

This is therefore a *pure* mode.

The *transverse* wave, with velocity $V_2 = (c_{44}/\rho)^{1/2}$, is degenerate. Its displacement has an arbitrary direction in the Ox_1x_2 plane, as in Fig. 4.10, and can be written as

$$u_1 = \cos\varphi,\ u_2 = \sin\varphi,\ u_3 = 0\,.$$

Expanding (4.29) gives

$$\rho V_2 V_i^{\mathrm{e}} = c_{i131}\cos^2\varphi + (c_{i132} + c_{i231})\cos\varphi\sin\varphi + c_{i232}\sin^2\varphi$$

and hence the components of the energy velocity are given by

$$\begin{cases} \rho V_2 V_1^{\mathrm{e}} = c_{15}\cos^2\varphi + (c_{14} + c_{56})\cos\varphi\sin\varphi + c_{46}\sin^2\varphi \\ \rho V_2 V_2^{\mathrm{e}} = c_{56}\cos^2\varphi + (c_{46} + c_{25})\cos\varphi\sin\varphi + c_{24}\sin^2\varphi \\ \rho V_2 V_3^{\mathrm{e}} = c_{55}\cos^2\varphi + (c_{54} + c_{45})\cos\varphi\sin\varphi + c_{44}\sin^2\varphi\,. \end{cases}$$

Here the constants c_{15}, c_{46}, c_{25} and c_{45} are zero, and the others are related by $c_{56} = c_{14} = -c_{24}$ and $c_{55} = c_{44}$. This leads to

$$\boldsymbol{V}^{\mathrm{e}} = \left(\frac{c_{14}}{\rho V_2}\sin 2\varphi,\ \frac{c_{14}}{\rho V_2}\cos 2\varphi,\ \frac{c_{44}}{\rho V_2}\right)\,.$$

The angle θ^{e} between $\boldsymbol{V}^{\mathrm{e}}$ and the triad axis is given by

$$\tan\theta^{\mathrm{e}} = \frac{|c_{14}|}{c_{44}} \tag{4.30}$$

and is independent of φ. Hence *the energy velocity vector describes a cone* with vertex angle $2\theta^{\mathrm{e}}$ as the polarization of the transverse wave rotates in the Ox_1x_2 plane, as shown in Fig. 4.10. For sapphire, the values $c_{14} = -2.35 \times 10^{10}\,\mathrm{N/m^2}$ and $c_{44} = 14.74 \times 10^{10}\,\mathrm{N/m^2}$ lead to $\theta^{\mathrm{e}} = 9°$.

The triad axis is the only case of a symmetry axis which is not always a pure mode propagation direction. For the other symmetry axes A_n, with $n = 2$, 4 or 6, the modulus c_{14} is zero, implying that $\theta^{\mathrm{e}} = 0$.

4.2.5 Characteristic Surfaces

Several characteristic surfaces are used to illustrate the propagation of elastic waves in crystals. They are analogous to the surfaces with the same names in crystal optics, referring to electromagnetic waves.

4.2.5.1 Definitions and Properties. The *velocity surface* is defined by the end of the vector $\boldsymbol{V} = V\boldsymbol{n}$, drawn from an origin O in varied directions, whose length equals the phase velocity V for waves with wavefronts perpendicular to it. Generally there are three such surfaces, one for the quasi-longitudinal wave (with velocity V_1), and one for each of the two quasi-transverse waves (with velocities V_2 and V_3). Normally, the longitudinal wave is the fastest of the three for all directions, and the surface for V_1 surrounds those for V_2 and V_3 without intersecting them. However, there are exceptions such as tellurium dioxide, TeO_2, as shown later in the table of Fig. 4.22. The surfaces

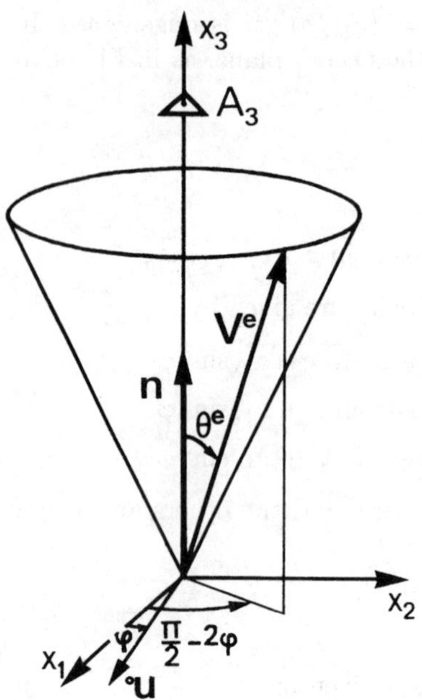

Fig. 4.10. Propagation along a triad axis. The energy velocity $\boldsymbol{V}^{\mathrm{e}}$ of a degenerate transverse wave with polarization $^\circ\boldsymbol{u}$ makes a constant angle θ^{e} with the propagation direction $\boldsymbol{n}$, which is along the triad axis. For sapphire, $\theta^{\mathrm{e}} = 9°$

for V_2 and V_3 have common points in the directions of the acoustic axes, for example along a symmetry axis of order $p > 2$. For an isotropic solid there are two spherical velocity surfaces with radii V_{L} and V_{T}, as in Fig. 4.8.

The *slowness surface* (L) is that defined by the end of the vector $\boldsymbol{m} = \boldsymbol{n}/V$ drawn from the origin. Since $\boldsymbol{m}$ and $\boldsymbol{V}$ are collinear and $mV = 1$, the velocity and slowness surfaces are related by inversion through the origin. The slowness surface, analogous to the *index* surface in optics, plays an important role in reflection and refraction problems, as seen later in Sect. 4.4.1.1. It is also helpful in providing, in addition to the phase velocity, the direction of energy flow. To see this, consider two radius vectors $\boldsymbol{m}$ and $\boldsymbol{m} + \mathrm{d}\boldsymbol{m}$, corresponding to two infinitesimally close propagation directions $\boldsymbol{n}$ and $\boldsymbol{n} + \mathrm{d}\boldsymbol{n}$. The vector

$$\mathrm{d}\boldsymbol{m} = \frac{\partial \boldsymbol{m}}{\partial n_k}\mathrm{d}n_k$$

is tangential to the slowness surface, as shown in Fig. 4.11. We write the differential of m_i as

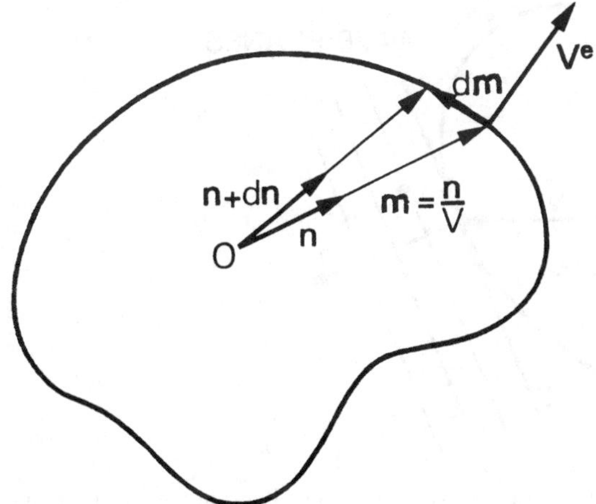

Fig. 4.11. Slowness surface and energy velocity. The energy velocity is, at all points, normal to the slowness surface, so that $\boldsymbol{V}^{\mathrm{e}} \cdot \mathrm{d}\boldsymbol{m} = 0$

$$\frac{\partial m_i}{\partial n_k} = \frac{\partial \left(\frac{n_i}{V}\right)}{\partial n_k} = \frac{\delta_{ik}}{V} - \frac{n_i}{V^2}\frac{\partial V}{\partial n_k}$$

and multiply by the components V_i^{e} of the energy velocity to obtain

$$V_i^{\mathrm{e}} \frac{\partial m_i}{\partial n_k} = \frac{1}{V}\left(V_k^{\mathrm{e}} - \frac{V_i^{\mathrm{e}} n_i}{V}\frac{\partial V}{\partial n_k}\right) .$$

Recalling the relations $V_i^{\mathrm{e}} n_i = V$ from (4.25), and $V_k^{\mathrm{e}} = V_k^{\mathrm{g}} = \partial V/\partial n_k$ from (4.28), we find

$$V_i^{\mathrm{e}} \frac{\partial m_i}{\partial n_k} = 0$$

and hence

$$\boldsymbol{V}^{\mathrm{e}} \cdot \mathrm{d}\boldsymbol{m} = V_i^{\mathrm{e}} \frac{\partial m_i}{\partial n_k} \mathrm{d}n_k = 0 \quad \forall\, \mathrm{d}\boldsymbol{m} .$$

Since this orthogonality relation applies for any vector $\mathrm{d}\boldsymbol{m}$ in the plane tangential to the slowness surface, *the energy velocity is, at all points, normal to the slowness surface.* As before, there are three such surfaces, and here the one for the quasi-longitudinal wave is normally enclosed within those for the quasi-transverse waves.

The third surface, called the *wave surface*, is the locus of points traced by the end of the *energy velocity vector* $\boldsymbol{V}^{\mathrm{e}}$, drawn from a fixed point O, as the propagation direction varies. The radius vector joining the origin O to a point on the surface represents the distance travelled by the elastic energy in

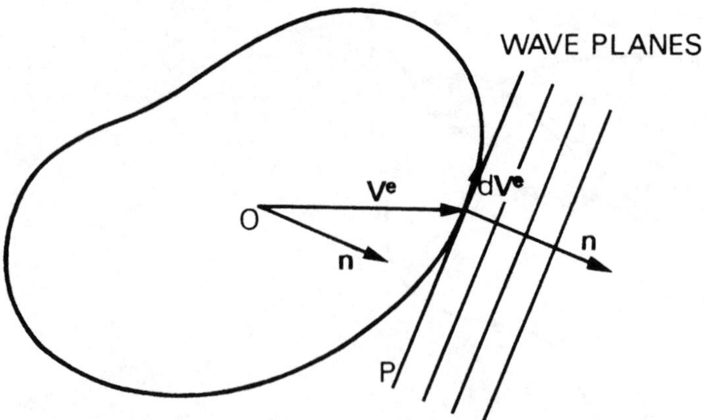

Fig. 4.12. Wave surface and the propagation direction. The wavefront P is the plane tangential to the wave surface at the end of the energy velocity vector $\boldsymbol{V}^{\mathrm{e}}$

unit time. The wave surface therefore gives the points reached, at unit time, by the vibrations emitted at time zero by a point elastic source at the origin. This is also an *equi-phase* surface, since all the points vibrate at the same time. Moreover, the propagation direction $\boldsymbol{n}$ of a plane wave with energy velocity $\boldsymbol{V}^{\mathrm{e}}$ is normal to the wave surface at the corresponding point. This follows from the relations

$$\boldsymbol{V}^{\mathrm{e}} \cdot \boldsymbol{m} = 1 \quad \text{and} \quad \boldsymbol{V}^{\mathrm{e}} \cdot \mathrm{d}\boldsymbol{m} = 0$$

which imply

$$\boldsymbol{m} \cdot \mathrm{d}\boldsymbol{V}^{\mathrm{e}} = 0 \quad \text{and hence} \quad \boldsymbol{n} \cdot \mathrm{d}\boldsymbol{V}^{\mathrm{e}} = 0$$

since $\boldsymbol{m} = \boldsymbol{n}/V$. The vector $\boldsymbol{n}$ is therefore perpendicular to any vector $\mathrm{d}\boldsymbol{V}\mathrm{e}$ in the plane tangential to the wave surface at the end of the vector $\boldsymbol{V}^{\mathrm{e}}$, as in Fig. 4.12.

There is a simple construction to obtain the velocity surface from the wave surface, and vice versa, as shown in Fig. 4.13. Let P be the plane tangential to the wave surface at a point A. The relation $\boldsymbol{V}^{\mathrm{e}} \cdot \boldsymbol{n} = V$ shows that the phase velocity V is the projection of the radius vector $\boldsymbol{OA} = \boldsymbol{V}^{\mathrm{e}}$ onto the vector $\boldsymbol{n}$, which is normal to the wave surface. The vector $\boldsymbol{V}$ is $\boldsymbol{OB}$, the line dropped perpendicularly from O to the plane P.
The velocity surface is therefore the surface obtained by projecting the point O orthogonally onto planes tangential to the wave surface. Conversely, the wave surface is the envelope of the family of planes constructed normally at the end of the vector $\boldsymbol{V}$ drawn from O.

4.2.5.2 Examples of Slowness Surfaces. The examples here refer to crystals in the cubic, tetragonal, hexagonal and trigonal systems. The drawings show plane sections of the surfaces. If the plane of the drawing is a symmetry plane for the elastic properties (a mirror plane or a plane normal to an axis

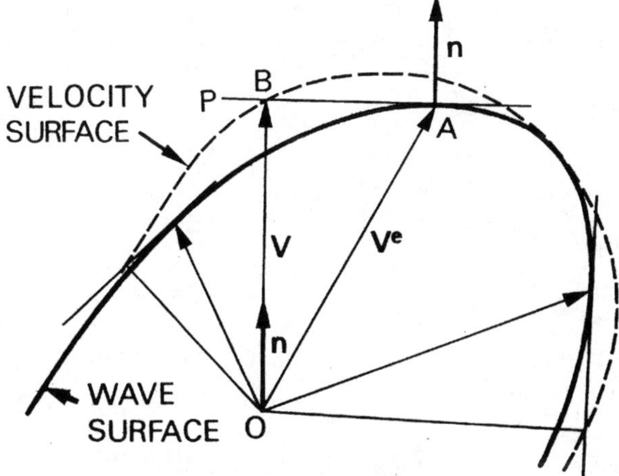

Fig. 4.13. Velocity surface and wave surface. The velocity surface is the locus of the normal projection of the point O onto planes tangential to the wave surface. Conversely, the wave surface is the envelope of the family of planes perpendicular to the vector $\boldsymbol{V}$ drawn from the point O

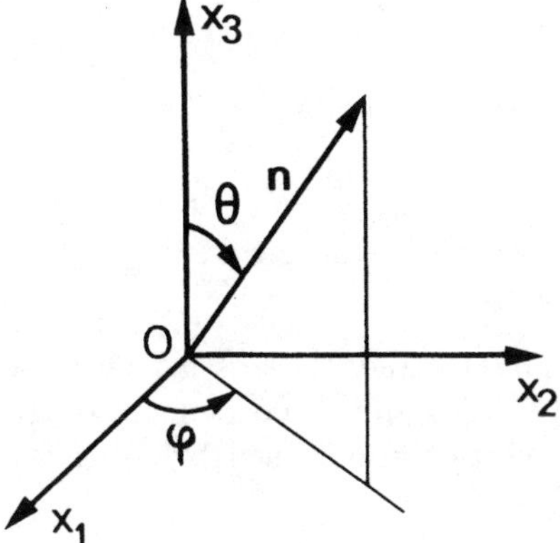

Fig. 4.14. The propagation direction $\boldsymbol{n}$ is specified by the angles θ and φ

of even order), it also contains the surface normal and therefore the energy vector. The propagation direction is specified by the two polar angles θ and φ, as shown in Fig. 4.14, so that

$$n_1 = \sin\theta\cos\varphi\,, \quad n_2 = \sin\theta\sin\varphi\,, \quad n_3 = \cos\theta\,. \tag{4.31}$$

The secular equation is analytically solvable only for particular directions or planes, such that at least two of the three non-diagonal components of the tensor Γ_{il} are zero. The determinant can then be factorized into a product of polynomials in V^2 with degree one or two, as shown in Appendix B.

(a) Cubic System. For this system, the stiffness tensor $c_{\alpha\beta}$ has the form of (3.65). Using this, the Christoffel tensor components Γ_{il} of (4.11) simplify to give

$$\begin{aligned}
\Gamma_{11} &= c_{11}n_1^2 + c_{44}(n_2^2 + n_3^2) \\
\Gamma_{12} &= \Gamma_{21} = (c_{12} + c_{44})n_1 n_2 \\
\Gamma_{13} &= \Gamma_{31} = (c_{12} + c_{44})n_1 n_3 \\
\Gamma_{22} &= c_{44}(n_1^2 + n_3^2) + c_{11}n_2^2 \\
\Gamma_{23} &= \Gamma_{32} = (c_{12} + c_{44})n_2 n_3 \\
\Gamma_{33} &= c_{44}(n_1^2 + n_2^2) + c_{11}n_3^2 \,.
\end{aligned} \tag{4.32}$$

Propagation along a cube face. For propagation in the plane of a cube face, the equations are simplified further. For example, for propagation in the (001) plane we have

$$n_1 = \cos\varphi\,, \quad n_2 = \sin\varphi\,, \quad n_3 = 0 \quad \text{giving} \quad \Gamma_{13} = \Gamma_{23} = 0\,.$$

Thus Γ_{il} has the form

$$\Gamma_{il} = \begin{vmatrix} \Gamma_{11} & \Gamma_{12} & 0 \\ \Gamma_{12} & \Gamma_{22} & 0 \\ 0 & 0 & \Gamma_{33} \end{vmatrix}\,,$$

where

$$\begin{aligned}
\Gamma_{11} &= c_{11}\cos^2\varphi + c_{44}\sin^2\varphi \qquad \Gamma_{12} = (c_{12} + c_{44})\sin\varphi\cos\varphi \\
\Gamma_{22} &= c_{11}\sin^2\varphi + c_{44}\cos^2\varphi \qquad \Gamma_{33} = c_{44}\,.
\end{aligned}$$

For any propagation direction in the (001) plane, there is a transverse wave polarized along $\mathrm{O}x_3$, with velocity $V_3 = (c_{44}/\rho)^{1/2}$ independent of the angle φ. Solutions for the other two velocities are obtained by using the eigenvalues $\gamma = \rho V^2$ which are, from (B.9),

$$\gamma_{\frac{1}{2}} = \frac{\Gamma_{11} + \Gamma_{22}}{2} \pm \frac{1}{2}\sqrt{(\Gamma_{11} - \Gamma_{22})^2 + 4\Gamma_{12}^2}\,, \tag{4.33}$$

where the upper sign refers to V_1 and the lower sign to V_2.

The velocities V_1 and V_2, as functions of the angle φ, are given by the expressions

$$2\rho V_1^2 = c_{11} + c_{44} + \sqrt{(c_{11} - c_{44})^2\cos^2 2\varphi + (c_{12} + c_{44})^2\sin^2 2\varphi} \tag{4.34}$$

$$2\rho V_2^2 = c_{11} + c_{44} - \sqrt{(c_{11} - c_{44})^2\cos^2 2\varphi + (c_{12} + c_{44})^2\sin^2 2\varphi}\,. \tag{4.35}$$

In general, the polarization is neither longitudinal nor transverse. However, for $\varphi = 0$ or $\pi/2$, the velocity $V_1 = (c_{11}/\rho)^{1/2}$ refers to a pure longitudinal wave and $V_2 = (c_{44}/\rho)^{1/2}$ refers to a pure transverse wave. In other directions, (4.34) and (4.35) give velocities V_1 and V_2 for quasi-longitudinal and quasi-transverse waves.

Figure 4.15 shows a section of the slowness surfaces of silicon (class $m3m$) in the (001) plane, obtained from the elastic constants in Fig. 3.10. These curves exhibit the maximal symmetry of the cubic system – the stiffness tensor has the same form for all five cubic classes, and so is invariant for the symmetry operations in the holosymmetric class. The velocity of the quasi-transverse wave has extrema in the [100] and [110] directions, given by

$$V_2[100] = \sqrt{\frac{c_{44}}{\rho}}, \quad V_2[110] = \sqrt{\frac{c_{11} - c_{12}}{2\rho}}.$$

The ratio of these values,

$$\frac{V_2[100]}{V_2[110]} = \sqrt{\frac{2c_{44}}{c_{11} - c_{12}}},$$

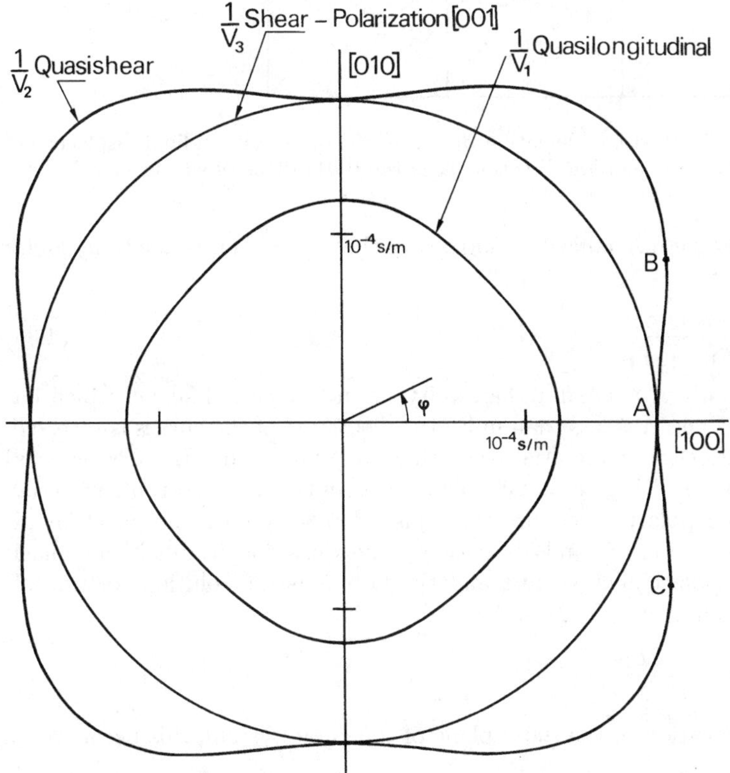

Fig. 4.15. Cubic system. Section of the slowness surfaces for silicon ($m3m$, anisotropy factor 1.565) in the (001) plane. The point A corresponds to a velocity of 5843 m/s

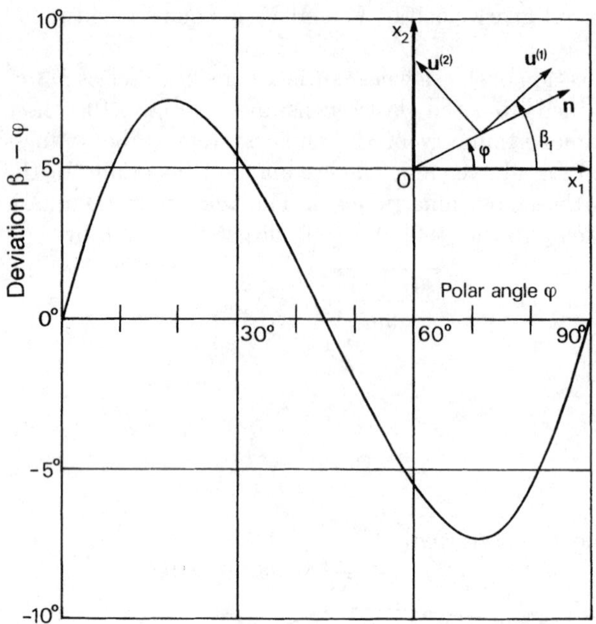

Fig. 4.16. Cubic system. Deviation $\beta_1 - \varphi$ of the quasi-longitudinal displacement vector from the propagation direction φ, in the (001) plane of silicon

characterizes the mechanical anisotropy of the crystal. The *anisotropy factor* A is defined by

$$A = \frac{2c_{44}}{c_{11} - c_{12}} . \tag{4.36}$$

Most commonly A is greater than unity, as in the case of silicon which has $A = 1.565$. When A is less than unity the distortion of the curves is reversed, as in the piezoelectric material bismuth germanium oxide ($Bi_{12}GeO_{20}$, class 23), shown later in Fig. 4.32. Of course, A is unity for an isotropic material.

The three polarization directions also depend on the propagation direction. The angles β_1 and β_2 between the quasi-longitudinal and quasi-transverse displacement vectors and the Ox_1 axis are solutions of the following relation, from (B.12) in Appendix B:

$$\tan 2\beta = \frac{2\Gamma_{12}}{\Gamma_{11} - \Gamma_{22}} .$$

For propagation in the (001) plane of the cubic system, this becomes

$$\tan 2\beta = \frac{c_{12} + c_{44}}{c_{11} - c_{44}} \tan 2\varphi . \tag{4.37}$$

Figure 4.16 shows $\beta_1 - \varphi$, the deviation of the quasi-longitudinal displacement vector from the propagation direction, for propagation in the (001)

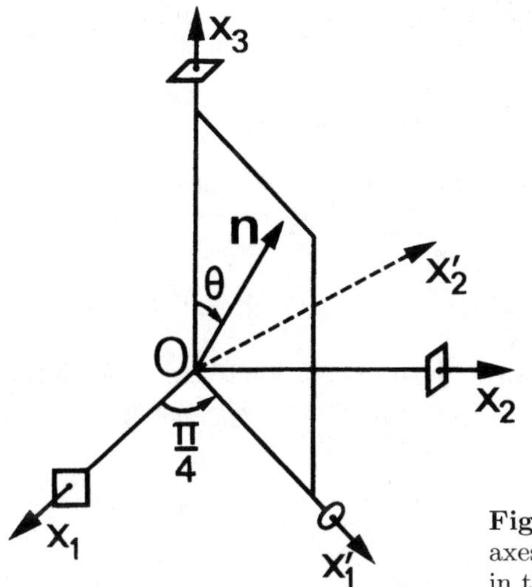

Fig. 4.17. Cubic system. New coordinate axes $Ox_1'x_2'x_3'$ for calculation of velocities in the diagonal plane $(1\bar{1}0)$

plane of silicon. The deviation is a maximum at an angle $\varphi_{\mathrm{M}} = 18°\,52'$, such that

$$\tan 2\varphi_{\mathrm{M}} = \sqrt{\frac{c_{11} - c_{44}}{c_{12} + c_{44}}},$$

and the maximum deviation, $7°\,16'$, is given by $\beta_1 - \varphi_{\mathrm{M}} = \pi/4 - 2\varphi_{\mathrm{M}}$.

Propagation in a diagonal plane. We now consider the diagonal plane $(1\bar{1}0)$. The appropriate axes $Ox_1'x_2'x_3'$ can be obtained from the crystallographic axes $\mathrm{O}x_1x_2x_3$ by rotation about $\mathrm{O}x_3$ through $\pi/4$, as in Fig. 4.17.

The new stiffness matrix, $c'_{\alpha\beta}$, is analogous to that of a tetragonal crystal with $c'_{16} = 0$, since Ox_1' is along a dyad axis. From the stiffness matrix of (3.75), and with $n_1' = \sin\theta$, $n_2' = 0$ and $n_3' = \cos\theta$, the components of the Christoffel tensor Γ'_{il} in (4.11) reduce to

$$\Gamma'_{il} = \begin{vmatrix} \Gamma'_{11} & 0 & \Gamma'_{13} \\ 0 & \Gamma'_{22} & 0 \\ \Gamma'_{13} & 0 & \Gamma'_{33} \end{vmatrix},$$

where

$$\Gamma'_{11} = c'_{11}\sin^2\theta + c'_{44}\cos^2\theta \qquad \Gamma'_{13} = \frac{c'_{13} + c'_{44}}{2}\sin 2\theta$$
$$\Gamma'_{22} = c'_{66}\sin^2\theta + c'_{44}\cos^2\theta \qquad \Gamma'_{33} = c'_{44}\sin^2\theta + c'_{33}\cos^2\theta\,.$$

Thus there is a transverse wave polarized along Ox_2', that is, along $[1\bar{1}0]$, with velocity

$$V_3 = \sqrt{\frac{\Gamma'_{22}}{\rho}} = \sqrt{\frac{c'_{66}\sin^2\theta + c'_{44}\cos^2\theta}{\rho}}\,.$$

The other waves are quasi-longitudinal and quasi-transverse and their velocities, V_1 and V_2 respectively, are given by

$$2\rho V_{\frac{1}{2}}^2 = \Gamma'_{11} + \Gamma'_{33} \pm \sqrt{(\Gamma'_{11} - \Gamma'_{33})^2 + 4\,(\Gamma'_{13})^2}\,.$$

On replacing the Γ'_{il} by the above expressions this becomes

$$\begin{aligned} 2\rho V_{\frac{1}{2}}^2 &= c'_{44} + c'_{11}\sin^2\theta + c'_{33}\cos^2\theta \\ &\quad \pm \sqrt{\left[(c'_{11} - c'_{44})\sin^2\theta + (c'_{44} - c'_{33})\cos^2\theta\right]^2 + (c'_{13} + c'_{44})^2 \sin^2 2\theta}\,. \end{aligned}$$

Since the elastic constants are always given in coordinates referred to the crystal axes, the constants $c'_{\alpha\beta}$ need to be replaced by $c_{\alpha\beta}$ using the axis rotation matrix

$$\alpha_i^j = \begin{pmatrix} 1/\sqrt{2} & 1/\sqrt{2} & 0 \\ -1/\sqrt{2} & 1/\sqrt{2} & 0 \\ 0 & 0 & 1 \end{pmatrix}.$$

The relation

$$c'_{ijkl} = \alpha_i^p \alpha_j^q \alpha_k^r \alpha_l^s c_{pqrs}$$

leads to the following:

$$c'_{3333} = c_{3333} \quad \text{giving} \quad c'_{33} = c_{33} = c_{11}$$

$$\begin{aligned} c'_{2323} &= \alpha_2^1\alpha_2^1 c_{1313} + \alpha_2^2\alpha_2^2 c_{2323} = \left(\alpha_2^1\alpha_2^1 + \alpha_2^2\alpha_2^2\right) c_{44} \quad \text{giving}\, c'_{44} = c_{44}\,, \\ c'_{1133} &= \alpha_1^1\alpha_1^1 c_{1133} + \alpha_1^2\alpha_1^2 c_{2233} = \left(\alpha_1^1\alpha_1^1 + \alpha_1^2\alpha_1^2\right) c_{13} \quad \text{giving}\, c'_{13} = c_{13} = c_{12}\,, \end{aligned}$$

$$\begin{aligned} c'_{1212} &= \tfrac{1}{4}c_{11} + \tfrac{1}{4}c_{22} - \tfrac{1}{2}c_{12} \qquad \text{giving} \quad c'_{66} = \tfrac{1}{2}(c_{11} - c_{12})\,, \\ c'_{1111} &= \tfrac{1}{4}c_{11} + \tfrac{1}{4}c_{22} + c_{66} + \tfrac{1}{2}c_{12} \;\text{giving} \quad c'_{11} = \tfrac{1}{2}(c_{11} + c_{12}) + c_{44}\,. \end{aligned}$$

These enable the velocities to be expressed in terms of c_{11}, c_{12} and c_{44}. The velocity of the transverse wave is

$$V_3 = \sqrt{\frac{c_{44}}{\rho}\cos^2\theta + \frac{c_{11} - c_{12}}{2\rho}\sin^2\theta}\,, \tag{4.38}$$

and the velocities of the quasi-longitudinal and quasi-transverse waves, respectively V_1 and V_2, are given by

$$\begin{aligned} 2\rho V_{\frac{1}{2}}^2 &= c_{44} + \left(c_{44} + \frac{c_{11} + c_{12}}{2}\right)\sin^2\theta + c_{11}\cos^2\theta \\ &\pm\sqrt{\left[\frac{c_{11} + c_{12}}{2}\sin^2\theta + (c_{44} - c_{11})\cos^2\theta\right]^2 + (c_{12} + c_{44})^2 \sin^2 2\theta}\,. \end{aligned} \tag{4.39}$$

Curves for the slownesses $1/V_1$, $1/V_2$ and $1/V_3$, still referring to silicon, are shown in Fig. 4.18.

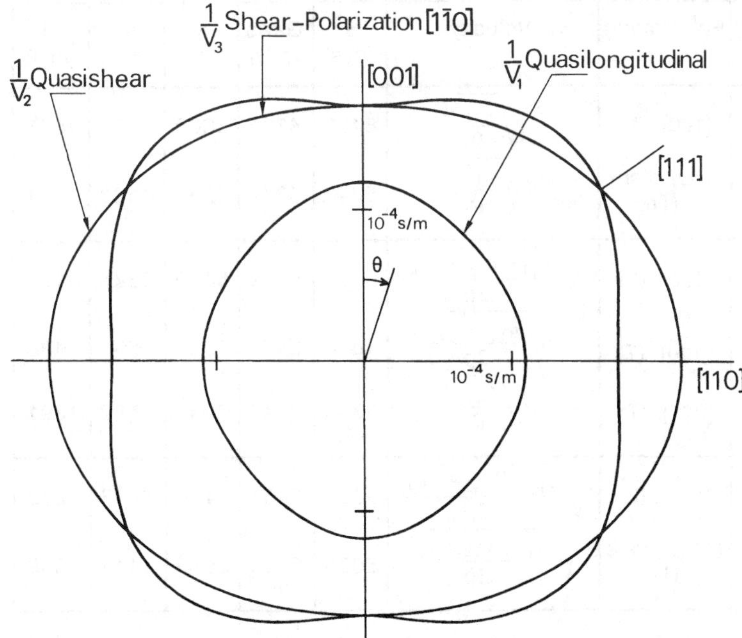

Fig. 4.18. Cubic system. Section of the slowness surfaces of silicon in the $(1\bar{1}0)$ plane

The table in Fig. 4.19 gives expressions for the velocities for selected propagation directions and polarizations, with numerical values for several cubic crystals – silicon, gallium arsenide (ignoring the weak piezoelectricity), aluminium, gold and platinum.

Wave Surface. The energy vector has the property that it can have several values for one direction. For example, Fig. 4.15 shows a section of the slowness surfaces for silicon in the symmetry plane (001), and here the energy vector for the quasi-transverse wave will be in the plane of the figure and normal to the surface for $1/V_2$. It can be seen that the energy vector is parallel to [100] when the wave vector is not only along OA, but also along OB and OC. It thus follows that, for the direction OA, the energy velocity has two values (two rather than three because of the symmetry about OA). For a direction near to OA there are three values. The energy carried by these quasi-transverse waves, with wave vector $\boldsymbol{k}$ between OB and OC, tends to be concentrated near the line OA, the [100] direction. Figure 4.20 illustrates these points, showing the construction of a wave surface from a slowness surface like that of silicon, but with the curvature exaggerated.

The 'focusing' of energy, produced when the direction of the energy velocity tends to remain the same while that of the phase velocity varies (implying a variation of curvature of the slowness curves), is of course observed in prac-

Propagation direction	Polarization	Velocity	Si (m/s)	GaAs (m/s)	Al (m/s)	Au (m/s)	Pt (m/s)
[100]	[100] (L)	$\sqrt{\frac{c_{11}}{\rho}}$	8433	4735	6300	3158	4025
	(100) plane (T)	$\sqrt{\frac{c_{44}}{\rho}}$	5843	3347	3236	1482	1891
[110]	[100] (L)	$\sqrt{\frac{c_{11}+c_{12}+2c_{44}}{2\rho}}$	9134	5242	6450	3376	4189
	[1$\bar{1}$0] (T)	$\sqrt{\frac{c_{11}-c_{12}}{2\rho}}$	4673	2478	2935	874	1498
	[001] (T)	$\sqrt{\frac{c_{44}}{\rho}}$	5843	3347	3236	1482	1891
[111]	[111] (L)	$\sqrt{\frac{c_{11}+2c_{12}+4c_{44}}{3\rho}}$	9360	5401	6496	3447	4242
	(111) plane (T)	$\sqrt{\frac{c_{11}-c_{12}+c_{44}}{3\rho}}$	5085	2798	3039	1114	1639

Fig. 4.19. Cubic crystals. Velocities and polarizations of waves propagating in the [100], [110], and [111] directions

tice only if the radiation diagram of the source is spread out, that is, if the source is localized.

The other two modes, in silicon, do not give concentration of the energy near the OA direction. Indeed, for the longitudinal wave, this direction corresponds to an energy depletion – the curvature of the $1/V_1$ curve has sign opposite to that of the $1/V_2$ curve. For the pure transverse wave, with velocity V_3, the energy density is independent of the propagation direction.

Figure 4.21b shows a section of the wave surface for quasi-transverse waves in rutile, in the (001) plane. Here the folding due to the change of curvature in the slowness surface is very marked because of the large anisotropy factor, $A = 4$.

(b) Tetragonal System. From the tetragonal stiffness matrix of (3.75), the components of the Christoffel tensor in (4.11) become

$$\begin{aligned}
\Gamma_{11} &= c_{11}n_1^2 + c_{66}n_2^2 + c_{44}n_3^2 + 2c_{16}n_1n_2 \\
\Gamma_{12} &= c_{16}(n_1^2 - n_2^2) + (c_{12} + c_{66})n_1n_2 \\
\Gamma_{13} &= (c_{13} + c_{44})n_1n_3 \\
\Gamma_{22} &= c_{66}n_1^2 + c_{11}n_2^2 + c_{44}n_3^2 - 2c_{16}n_1n_2 \\
\Gamma_{23} &= (c_{13} + c_{44})n_2n_3 \\
\Gamma_{33} &= c_{44}(n_1^2 + n_2^2) + c_{33}n_3^2 \,.
\end{aligned} \tag{4.40}$$

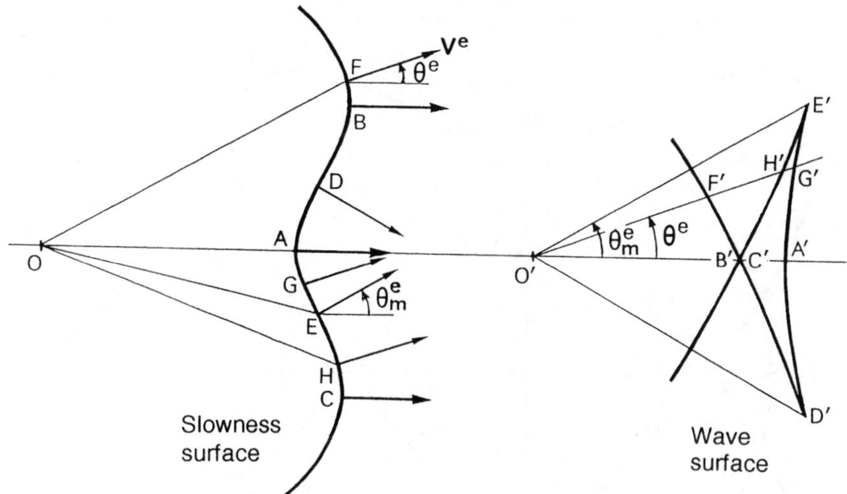

Fig. 4.20. Section of the wave surface. A variation of curvature in the slowness surface can give, for a particular direction, several values for the energy velocity. The inflection points D and E of the slowness surface correspond to the peaks D' and E' of the 'horns' on the wave surface

For propagation in a plane normal to the tetrad axis, the propagation direction is given by

$$n_1 = \cos\varphi\,, \quad n_2 = \sin\varphi\,, \quad n_3 = 0 \quad \text{giving} \quad \Gamma_{13} = \Gamma_{23} = 0\,.$$

The tensor Γ_{il}, and therefore the secular equation, has the same form as for propagation in the plane of a cube face in the cubic system (Sect. 4.2.5.2a), but the non-zero elements take the different expressions

$$\Gamma_{11} = c_{11}\cos^2\varphi + c_{66}\sin^2\varphi + c_{16}\sin 2\varphi$$

$$\Gamma_{12} = c_{16}\cos 2\varphi + (c_{12} + c_{66})\frac{\sin 2\varphi}{2}$$

$$\Gamma_{22} = c_{66}\cos^2\varphi + c_{11}\sin^2\varphi - c_{16}\sin 2\varphi$$

$$\Gamma_{33} = c_{44}\,.$$

The transverse wave, polarized along Ox_3, has velocity $V_3 = (c_{44}/\rho)^{1/2}$. The two other waves, polarized in the (001) plane, are the quasi-longitudinal and quasi-transverse waves, with velocities V_1 and V_2 respectively. These are given by (4.33), so that

$$2\rho V^2_{\substack{1\\2}} = \Gamma_{11} + \Gamma_{22} \pm \sqrt{(\Gamma_{11} - \Gamma_{22})^2 + 4\,(\Gamma_{12})^2}\,.$$

Replacing the Γ_{il} by the above expressions leads to

$$2\rho V^2_{\substack{1\\2}} = c_{11} + c_{66} \pm\sqrt{(c_{11} - c_{66})^2\cos^2 2\varphi + (c_{12} + c_{66})^2\sin^2 2\varphi + 2c_{16}(c_{11} + c_{12})\sin 4\varphi + 4c_{16}^2}\,. \tag{4.41}$$

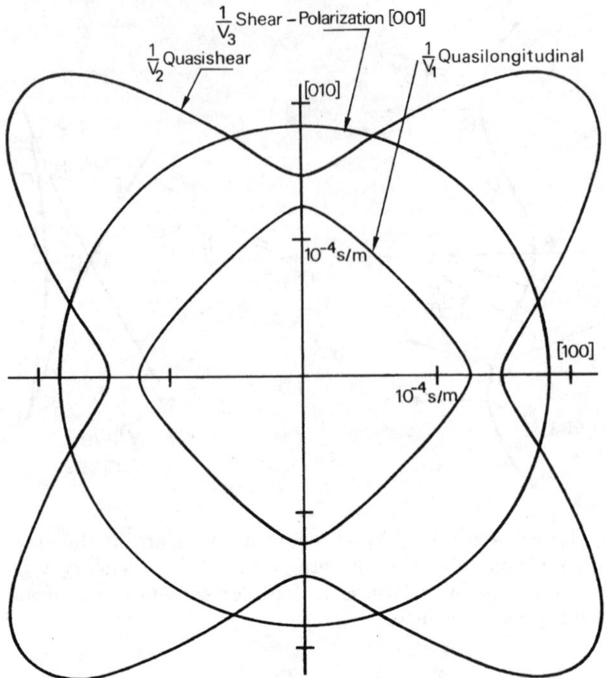

a

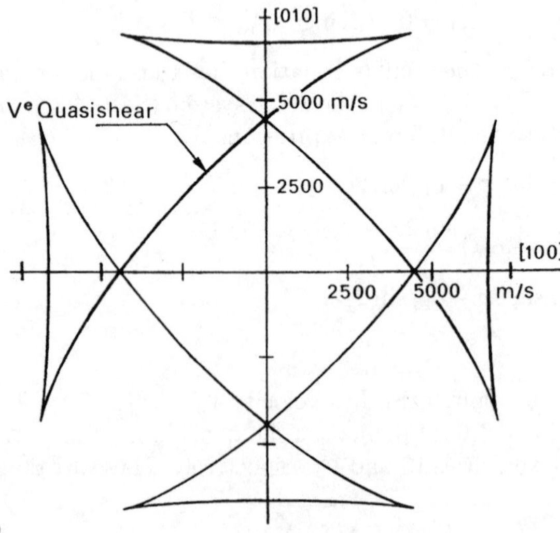

b

Fig. 4.21. Tetragonal system, classes 422, $4mm$, $\bar{4}2m$ and $4/mmm$. (**a**) Section through the slowness surfaces of rutile ($4/mmm$) in the (001) plane. (**b**) Section of the wave surface for the quasi-transverse wave

Propagation direction	Polarization	Velocity (m/s)	TiO_2 (m/s)	TeO_2 (m/s)	In (m/s)
[100]	[100] (L)	$\sqrt{\frac{c_{11}}{\rho}}$	8014	3050	2495
	[010] (T)	$\sqrt{\frac{c_{66}}{\rho}}$	6756	3317	1290
	[001] (T)	$\sqrt{\frac{c_{44}}{\rho}}$	5424	2100	946
[110]	[110] (L)	$\sqrt{\frac{c_{11}+c_{12}+2c_{66}}{2\rho}}$	9923	4663	2740
	$[1\bar{1}0]$ (T)	$\sqrt{\frac{c_{11}-c_{12}}{2\rho}}$	3378	616	603
	[001] (T)	$\sqrt{\frac{c_{44}}{\rho}}$	5424	2100	946
[001]	[001] (L)	$\sqrt{\frac{c_{33}}{\rho}}$	10671	4200	2490
	(001) plane (T)	$\sqrt{\frac{c_{44}}{\rho}}$	5424	2100	946

Fig. 4.22. Tetragonal system, classes 422, $4mm$, $\bar{4}2m$ and $4/mmm$. Velocities and polarizations for waves propagating in the [100], [110] and [001] directions. For the [100] direction in paratellurite (TeO_2), the transverse wave is faster than the longitudinal wave

The modulus c_{16} is *zero* in classes 422, $4mm$, $\bar{4}2m$ and $4/mmm$, and for these the formula simplifies to

$$2\rho V_{\frac{1}{2}}^2 = c_{11} + c_{66} \pm \sqrt{(c_{11} - c_{66})^2 \cos^2 2\varphi + (c_{12} + c_{66})^2 \sin^2 2\varphi}\,. \quad (4.42)$$

Slowness curves for rutile (TiO_2, class $4/mmm$), obtained from the constants in Fig. 3.10, are shown in Fig. 4.21a. These curves show *very strong anisotropy*, associated with the large value of the ratio

$$A = \frac{2c_{66}}{c_{11} - c_{12}} = 4\,.$$

For the quasi-transverse wave, the velocity in the [100] direction is twice as large as that in the [110] direction.

The table in Fig. 4.22 gives velocities for several crystals in classes for which $c_{16} = 0$ (with the coordinate axes along the crystal axes), namely rutile (TiO_2), paratellurite (TeO_2) and indium (In).

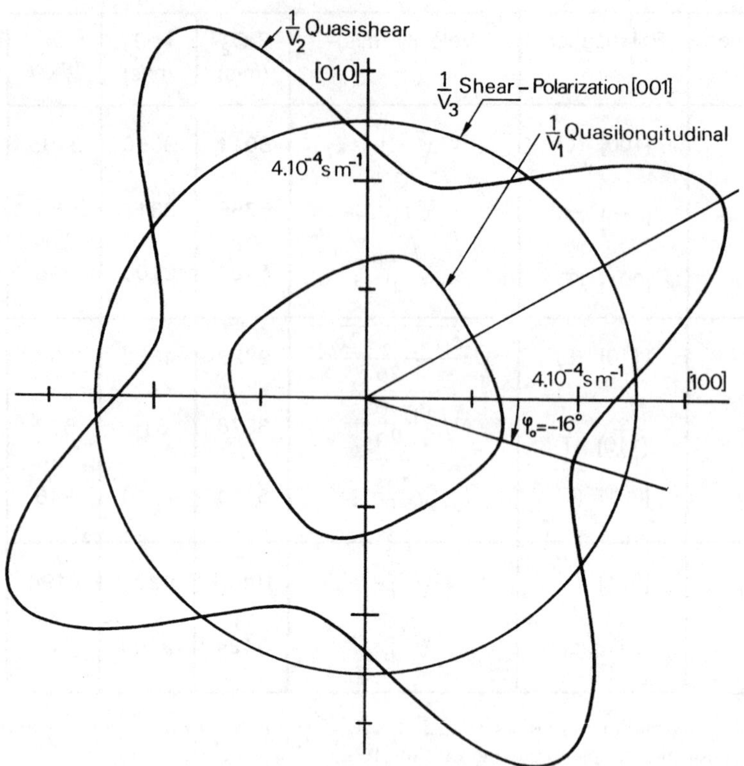

Fig. 4.23. Tetragonal system, classes 4, $\bar{4}$ and $4/m$. Section of the slowness surfaces of lead molybdate ($4/m$) in the (001) plane

The modulus c_{16} is *non-zero* in classes 4, $\bar{4}$ and $4/m$. Figure 4.23 shows slowness curves in the (001) plane for lead molybdate ($PbMoO_4$, class $4/m$), using the constants of Fig. 3.10. These curves have a similar appearance to those of rutile, but rotated in the (001) plane through an angle φ_0. The resemblance is not accidental – it arises from the fact that we can revert to the previous case by rotating the axes in such a way that c_{16} becomes zero. Problem 4.7 shows that c'_{16} is zero for coordinate axes $Ox'_1x'_2x'_3$ defined by the angle $(\mathrm{O}x_1, Ox'_1) = \varphi_0$, such that

$$\tan 4\varphi_0 = \frac{4c_{16}}{c_{11} - c_{12} - 2c_{66}} . \tag{4.43}$$

Thus, in the Ox'_1 and Ox'_2 directions, and in the directions half way between them, the waves polarized in the (001) plane are longitudinal or transverse; in these cases, the waves are pure. For lead molybdate $\varphi_0 = -16°$, and the pure wave directions make angles of $-16°$ and $+29°$ $(= -16° + 45°)$ with the [100] crystal axis.

To recap, the wave velocities in these crystal classes are obtainable by two methods – (a) from the constants $c_{\alpha\beta}$ in the crystallographic axis system, using the full formula (4.41), or (b) by rotating the axes through the polar angle φ_0 so that c'_{16} becomes zero, and then using the simpler formula (4.42) with the new constants $c'_{\alpha\beta}$. The second method has the advantage of yielding directly the pure wave directions, important in practice, and simple expressions for the velocities in these directions. Along Ox'_1, for example, $V_1 = (c'_{11}/\rho)^{1/2}$ and $V_2 = (c'_{66}/\rho)^{1/2}$, giving for lead molybdate $V_1 = 3780\,\mathrm{m/s}$ and $V_2 = 2460\,\mathrm{m/s}$.

(c) Hexagonal System. For a hexagonal crystal, all planes containing the principal axis (i.e., meridian planes) are elastically equivalent, as shown in Problem 3.4. The characteristic surfaces are therefore surfaces of revolution about this axis. For the same reason, the velocities of waves propagating in the plane perpendicular to the axis are independent of direction (Sect. 4.2.2). To calculate them, we first note that the tensor Γ_{il} becomes

$$
\begin{aligned}
\Gamma_{11} &= c_{11}n_1^2 + c_{66}n_2^2 + c_{44}n_3^2 \\
\Gamma_{12} &= (c_{12} + c_{66})n_1 n_2 \\
\Gamma_{13} &= (c_{13} + c_{44})n_1 n_3 \\
\Gamma_{22} &= c_{66}n_1^2 + c_{11}n_2^2 + c_{44}n_3^2 \quad \text{with} \quad c_{66} = \frac{c_{11} - c_{12}}{2} \\
\Gamma_{23} &= (c_{13} + c_{44})n_2 n_3 \\
\Gamma_{33} &= c_{44}(n_1^2 + n_2^2) + c_{33}n_3^2 \,.
\end{aligned}
\tag{4.44}
$$

It is convenient to choose [100] as the propagation direction, so that Γ has the diagonal form

$$
\Gamma = \begin{vmatrix} c_{11} & 0 & 0 \\ 0 & c_{66} & 0 \\ 0 & 0 & c_{44} \end{vmatrix} .
$$

Thus, for propagation in the (001) plane, the three waves are

- a longitudinal wave with velocity $V_1 = (c_{11}/\rho)^{1/2}$;
- a transverse wave polarized in the plane, with velocity $V_2 = (c_{66}/\rho)^{1/2}$;
- a transverse wave polarized parallel to the A_6 axis, with velocity $V_3 = (c_{44}/\rho)^{1/2}$.

The slowness curves are circles with radii $(\rho/c_{11})^{1/2}$, $(\rho/c_{66})^{1/2}$ and $(\rho/c_{44})^{1/2}$.

For propagation in a *meridian* plane, the velocities depend on the angle θ between the A_6 axis ($\mathrm{O}x_3$) and the propagation direction $\boldsymbol{n}$. Choosing the $\mathrm{O}x_2x_3$ plane, we have

$$
n_1 = 0\,, \quad n_2 = \sin\theta\,, \quad n_3 = \cos\theta\,, \quad \text{and hence} \quad \Gamma_{12} = \Gamma_{13} = 0\,.
$$

The Γ tensor becomes

$$\Gamma = \begin{vmatrix} \Gamma_{11} & 0 & 0 \\ 0 & \Gamma_{22} & \Gamma_{23} \\ 0 & \Gamma_{23} & \Gamma_{33} \end{vmatrix} . \tag{4.45}$$

where

$$\begin{aligned} \Gamma_{11} &= c_{66} \sin^2\theta + c_{44}\cos^2\theta \quad & \Gamma_{22} &= c_{11}\sin^2\theta + c_{44}\cos^2\theta \\ \Gamma_{23} &= (c_{13}+c_{44})\sin\theta\cos\theta \quad & \Gamma_{33} &= c_{44}\sin^2\theta + c_{33}\cos^2\theta , \end{aligned} \tag{4.46}$$

A transverse wave, with polarization normal to the meridian plane, has velocity

$$V_3 = \sqrt{\frac{\Gamma_{11}}{\rho}} = \sqrt{\frac{c_{66}\sin^2\theta + c_{44}\cos^2\theta}{\rho}} \quad \text{where} \quad c_{66} = \frac{c_{11}-c_{12}}{2} , \tag{4.47}$$

The velocities of the other waves, polarized in the meridian plane, are given by

$$2\rho V_{\substack{1\\2}}^2 = \Gamma_{22} + \Gamma_{33} \pm \sqrt{(\Gamma_{22}-\Gamma_{33})^2 + 4(\Gamma_{23})^2} . \tag{4.48}$$

Substituting for Γ_{22}, Γ_{23} and Γ_{33} from (4.46) leads to

$$\begin{aligned} 2\rho V_{\substack{1\\2}}^2 &= c_{44} + c_{11}\sin^2\theta + c_{33}\cos^2\theta \\ &\pm \sqrt{\left[(c_{11}-c_{44})\sin^2\theta + (c_{44}-c_{33})\cos^2\theta\right]^2 + (c_{13}+c_{44})^2\sin^2 2\theta} \end{aligned} \tag{4.49}$$

with the upper sign for the quasi-longitudinal wave, and the lower for the quasi-transverse wave.

The above analysis applies to any solid having hexagonal symmetry. Although we are primarily concerned with crystals here, it can also be applied to the slowness curves of a composite material, a stratified assembly of carbon fibres immersed in a thermo-setting epoxy resin, as in Figs. 4.24 and 4.25. Typically, the fibres have diameter 7 µm, the thickness of one layer (or ply) is 130 µm, and in each ply all the fibres are oriented in the same direction. Here, this direction changes by 45° from one layer to the next, with orientations 0°, 45°, 90°, −45°. Thus the composite contains an 8-fold symmetry axis normal to the plies, elastically equivalent to a 6-fold axis, and it will behave as a

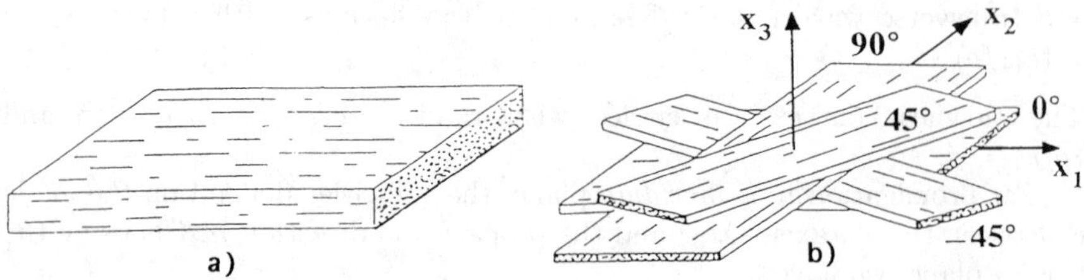

Fig. 4.24. Composite of crossed carbon fibre layers and thermo-setting epoxy resin. (**a**) One ply, a pre-impregnated layer. (**b**) Fibre orientations, with angles 0, 45°, 90°, −45° in the four layers which constitute one period of a periodic stack

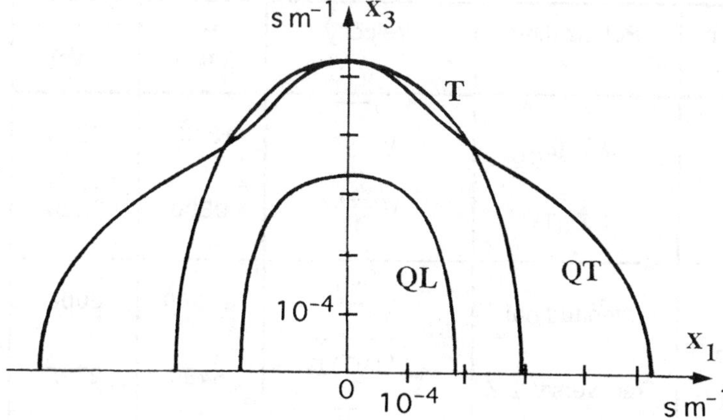

Fig. 4.25. Hexagonal system. Section of the slowness surfaces in the meridian plane, for a composite layered material. These curves, symmetric about the x_1-axis, are plotted using $c_{11} = 4.34$, $c_{33} = 1.35$, $c_{44} = 0.54$, $c_{66} = 1.68$, $c_{13} = 0.62(\times 10^{10}\,\mathrm{N/m^2})$

homogeneous solid provided the wavelength is large compared with the ply thickness.

These composites, which may have different orientation sequences $(0, \pi; 0, \pi/2)$, are of course not normally used as propagation media for elastic waves. However, they are used in this way when being ultrasonically tested for their homogeneity. This *non-destructive evaluation*, which enables one to verify, in particular, the adhesion between layers, requires a knowledge of the wave behaviour in the material. The interest in such materials, with applications in the aircraft and automobile industries, arises from the low weight ($\rho = 1550\,\mathrm{kg/m^3}$) for a given strength.

The table in Fig. 4.26 shows numerical values for beryllium (Be) and titanium (Ti). For beryllium, the velocity for longitudinal waves along the principle axis (Z) is almost 13500 m/s.

(d) Trigonal System. When the modulus c_{25} is non-zero, the calculation is complicated for this system. However, the common materials sapphire, quartz and lithium niobate belong to classes $\bar{3}m$, 32 and $3m$, for which $c_{25} = 0$. For these, the stiffness constants are given in (3.74), and the elements of Γ_{il} are

$$
\begin{aligned}
\Gamma_{11} &= c_{11}n_1^2 + c_{66}n_2^2 + c_{44}n_3^2 + 2c_{14}n_2n_3 \\
\Gamma_{12} &= (c_{12} + c_{66})n_1n_2 + 2c_{14}n_1n_3 \\
\Gamma_{13} &= 2c_{14}n_1n_2 + (c_{13} + c_{44})n_1n_3 \\
\Gamma_{22} &= c_{66}n_1^2 + c_{11}n_2^2 + c_{44}n_3^2 - 2c_{14}n_2n_3 \\
\Gamma_{23} &= c_{14}(n_1^2 - n_2^2) + (c_{13} + c_{44})n_2n_3 \\
\Gamma_{33} &= c_{44}(n_1^2 + n_2^2) + c_{33}n_3^2\,.
\end{aligned}
\qquad c_{66} = (c_{11} - c_{12})/2 \qquad (4.50)
$$

Propagation direction	Polarization	Velocity	Be (m/s)	Ti (m/s)
Z	// Z (L)	$\sqrt{\frac{c_{11}}{\rho}}$	13490	6330
	⊥ Z (T)	$\sqrt{\frac{c_{44}}{\rho}}$	9380	3220
Normal to Z-axis	Longitudinal	$\sqrt{\frac{c_{11}}{\rho}}$	12580	6000
	Transversal ⊥ Z	$\sqrt{\frac{c_{11}-c_{12}}{2\rho}}$	8480	2795
	Transversal // Z	$\sqrt{\frac{c_{44}}{\rho}}$	9380	3220

Fig. 4.26. Hexagonal crystals, beryllium (Be) and titanium (Ti). Velocities of waves propagating along the principal axis (Z) and along an arbitrary direction normal to Z

For a general propagation direction perpendicular to the triad axis (so that $n_1 = \cos\varphi, n_2 = \sin\varphi$ and $n_3 = 0$) all of the above components are non-zero, and the eigenvalue equation can only be solved numerically. However, for propagation along Ox_1 or Ox_2 there are algebraic solutions.

For propagation along Ox_1, with $n_1 = 1$ and $n_2 = n_3 = 0$, the tensor becomes

$$\Gamma = \begin{vmatrix} c_{11} & 0 & 0 \\ 0 & c_{66} & c_{14} \\ 0 & c_{14} & c_{44} \end{vmatrix}$$

showing that there is a longitudinal wave with velocity $V_1 = (c_{11}/\rho)^{1/2}$ and two transverse waves with velocities V_2 and V_3 given by

$$2\rho V_{\frac{2}{3}}^2 = c_{66} + c_{44} \pm \sqrt{(c_{66} - c_{44})^2 + 4c_{14}^2}\,. \tag{4.51}$$

For propagation along Ox_2, with $n_2 = 1$ and $n_1 = n_3 = 0$, we find

$$\Gamma = \begin{vmatrix} c_{66} & 0 & 0 \\ 0 & c_{11} & -c_{14} \\ 0 & -c_{14} & c_{44} \end{vmatrix}\,.$$

In this case, there is a transverse wave polarized along Ox_1, with velocity $V_3 = (c_{66}/\rho)^{1/2}$, a quasi-longitudinal wave with velocity V_1 given by

$$2\rho V_1^2 = c_{11} + c_{44} + \sqrt{(c_{11} - c_{44})^2 + 4c_{14}^2} \tag{4.52}$$

and a quasi-transverse wave with velocity V_2 given by

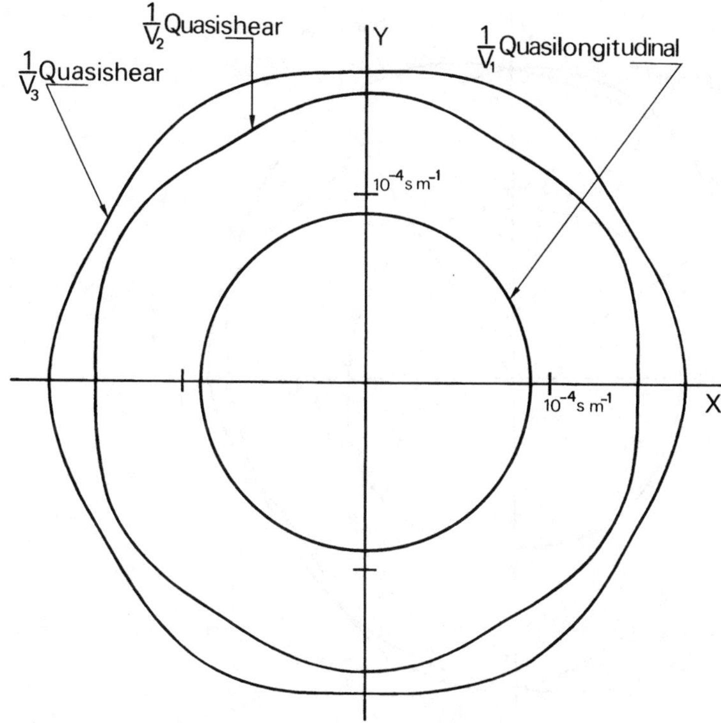

Fig. 4.27. Trigonal system, classes $\bar{3}m$, 32 and $3m$. Section of the slowness surfaces for sapphire ($\bar{3}m$) in the XY plane. The surface normal is not in this plane because the plane is not a mirror – the energy vector is inclined to the XY plane, as shown by Fig. 4.28. Thus, the Y-axis is not a pure mode direction

$$2\rho V_2^2 = c_{11} + c_{44} - \sqrt{(c_{11} - c_{44})^2 + 4c_{14}^2}\,. \tag{4.53}$$

Figure 4.27 shows slowness curves for sapphire, obtained using the constants in Fig. 3.10.

If the propagation is in the Ox_2x_3 plane, i.e. the symmetry plane YZ, then $n_1 = 0$, $n_2 = \sin\theta$, $n_3 = \cos\theta$, and Γ takes the form of (4.45), with components

$$\begin{aligned}
\Gamma_{11} &= c_{66}\sin^2\theta + c_{44}\cos^2\theta + c_{14}\sin 2\theta \\
\Gamma_{22} &= c_{11}\sin^2\theta + c_{44}\cos^2\theta - c_{14}\sin 2\theta \\
\Gamma_{23} &= -c_{14}\sin^2\theta + \frac{c_{13} + c_{44}}{2}\sin 2\theta \\
\Gamma_{33} &= c_{44}\sin^2\theta + c_{33}\cos^2\theta\,.
\end{aligned} \tag{4.54}$$

This gives a transverse wave, polarized along Ox_1, with velocity

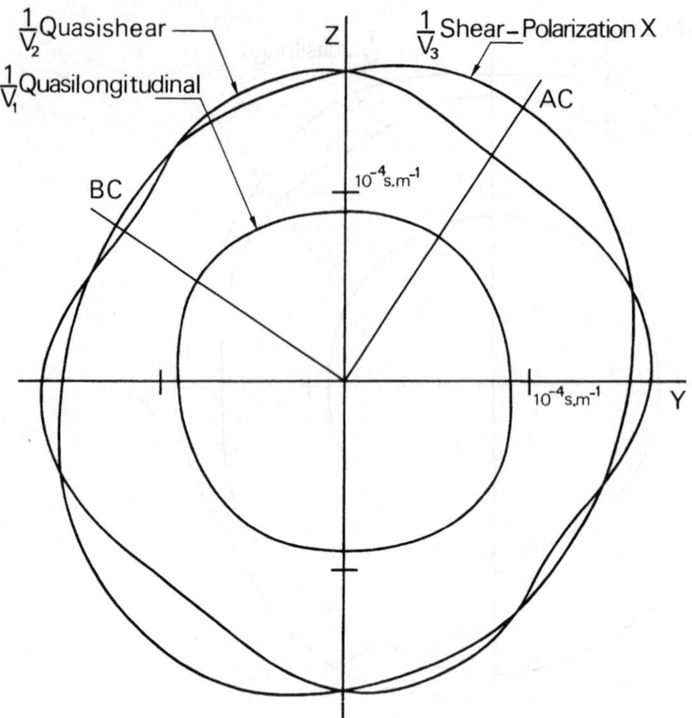

Fig. 4.28. Trigonal system, classes $\bar{3}m$, 32, $3m$. Section of the slowness surfaces for sapphire ($\bar{3}m$) in the YZ plane

$$V_3 = \sqrt{\frac{\Gamma_{11}}{\rho}} = \sqrt{\frac{c_{44}\cos^2\theta + c_{66}\sin^2\theta + c_{14}\sin 2\theta}{\rho}} . \tag{4.55}$$

The other waves are quasi-longitudinal and quasi-transverse, with velocities V_1 and V_2 respectively. These are given by (4.48), so that

$$\begin{aligned} 2\rho V_{\substack{1\\2}}^2 &= c_{44} + c_{11}\sin^2\theta + c_{33}\cos^2\theta - c_{14}\sin 2\theta \\ &\pm \left((c_{11}\sin^2\theta - c_{33}\cos^2\theta + c_{44}\cos 2\theta - c_{14}\sin 2\theta)^2\right. \\ &\quad \left. + [(c_{13}+c_{44})\sin 2\theta - 2c_{14}\sin^2\theta]^2\right)^{1/2} \end{aligned} \tag{4.56}$$

Figure 4.28, which refers to sapphire, shows that the triad axis is not a pure mode direction for transverse waves, as discussed in the example in Sect. 4.2.4. It can be seen that the normals of the two curves are inclined to the Z axis where they cross it. However, the transverse wave polarized along X is pure for some directions in the YZ plane. To determine these, we calculate the energy velocity using (4.24), giving

$$V_i^{\mathrm{e}} = c_{i1k1}\frac{n_k}{\rho V_3} \quad \text{since} \quad {}^\circ u_j = \delta_{j1} .$$

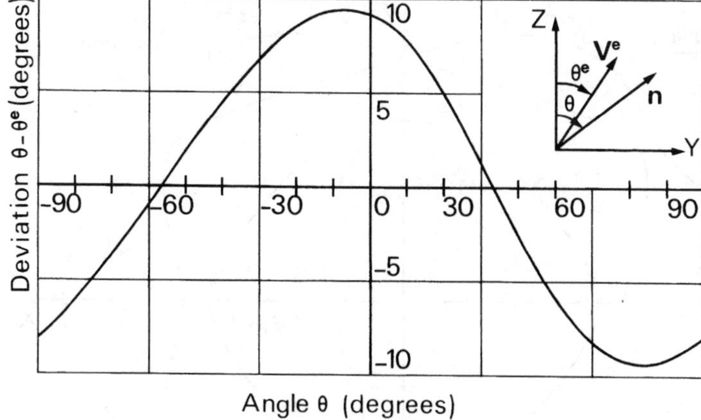

Fig. 4.29. Angular deviation of the acoustic ray velocity $\boldsymbol{V}^{\mathrm{e}}$ from the propagation direction $\boldsymbol{n}$, for the transverse wave propagating in the YZ plane of sapphire

With $n_1 = 0$, $n_2 = \sin\theta$ and $n_3 = \cos\theta$, the angular dependence is given by

$$\boldsymbol{V}^{\mathrm{e}} = \left[V_1^{\mathrm{e}} = 0\,,\; V_2^{\mathrm{e}} = \frac{c_{66}\sin\theta + c_{14}\cos\theta}{\rho V_3}\,,\; V_3^{\mathrm{e}} = \frac{c_{14}\sin\theta + c_{44}\cos\theta}{\rho V_3} \right] .$$

The energy velocity makes an angle θ^{e} with the Z axis, such that

$$\tan\theta^{\mathrm{e}} = \frac{V_2^{\mathrm{e}}}{V_3^{\mathrm{e}}} = \frac{c_{66}\sin\theta + c_{14}\cos\theta}{c_{14}\sin\theta + c_{44}\cos\theta} . \tag{4.57}$$

Figure 4.29 shows the deviation $\theta - \theta^{\mathrm{e}}$ of the acoustic ray direction from the propagation direction, the latter being in the YZ plane. This transverse wave becomes pure when $\theta^{\mathrm{e}} = \theta$, that is, for angles θ_1 and θ_2 that are solutions of

$$\tan 2\theta = \frac{2c_{14}}{c_{44} - c_{66}} . \tag{4.58}$$

For sapphire, $\theta_1 = 33^\circ\, 28'$ and $\theta_2 = -56^\circ\, 32'$. These mutually perpendicular directions, denoted AC and BC on Fig. 4.28, are of interest because waves in these directions are both pure and transverse.

Figure 4.30 shows a section of the slowness surfaces of quartz (SiO_2, class 32), in the XZ plane. The piezoelectric effect in quartz is weak, as seen in Fig. 3.15, and its effect on the curves is indistinguishable on the scale of the figure.

4.2.6 Attenuation

The relation $[T] = [c][S]$ between stress and strain (or, for the one-dimensional case, the simpler form $T = cS$) implies that any stress instantaneously

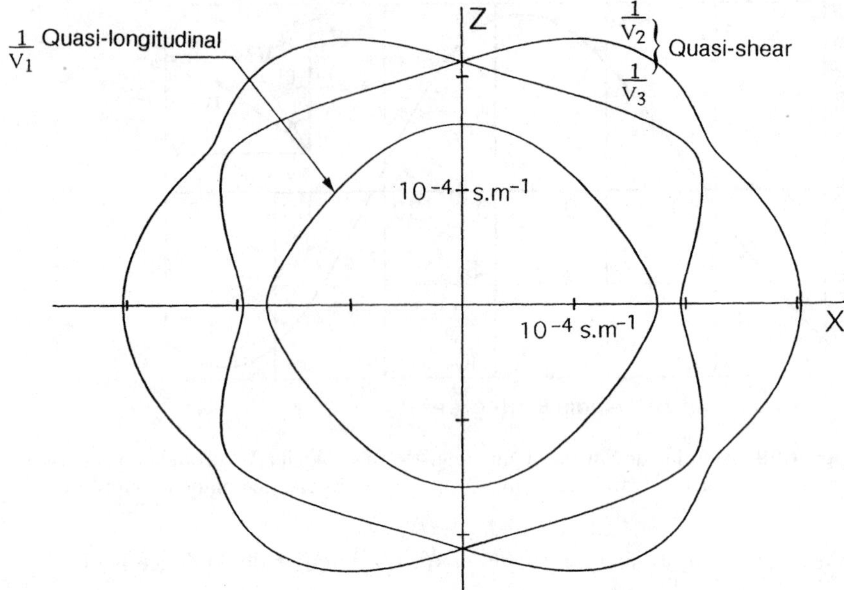

Fig. 4.30. Trigonal system. Section of the slowness surfaces for quartz (SiO_2, class 32), in the XZ plane (courtesy of J. Détaint, CNET). The symmetry of the curves shows that the X axis is a dyad axis. The three modes propagating along this axis are pure. As for sapphire, the Z axis is not a pure mode axis for the degenerate transverse waves

gives rise to a strain, and vice versa; for the sinusoidal case these variables will be in phase. This hypothesis is in conflict with the observation that the amplitude of a wave decreases with distance, by an amount depending on the nature of the propagation medium. Thus, the vibrations of an elastic-wave resonator will in practice die away, unless maintained by a supply of energy.

It is therefore reasonable, for a solid medium, to introduce a loss term into the equation for Hooke's law, which thus becomes

$$T = cS + \eta \frac{\partial S}{\partial t}, \tag{4.59}$$

where η is a viscosity coefficient, with units N·s/m^2. This formulation is analogous to that of Sect. 1.2.3, which referred to liquids (and c denoted the wave velocity). It describes a first-order process involving a characteristic time τ. We can write

$$\tau \frac{\partial S}{\partial t} + S(t) = KT \quad \text{on defining} \quad \tau = \eta/c \quad \text{and} \quad K = 1/c. \tag{4.60}$$

For the harmonic case, where $T(t) = T_{\mathrm{m}} \exp(\mathrm{i}\omega t)$, (4.60) gives

$$(1 + \mathrm{i}\omega\tau) S_{\mathrm{m}} = KT_{\mathrm{m}}.$$

The product of angular frequency and relaxation time, $\omega\tau = 2\pi\,\tau/T = \omega\eta/c$, appears naturally as a characteristic quantity. However, for solids it is usual to quote its reciprocal Q, called the elastic quality factor, so that

$$Q = \frac{1}{\omega\tau} = \frac{c}{\omega\eta}\,. \tag{4.61}$$

For a crystal, this generalization of Hooke's law takes the form

$$T_{ij} = c_{ijkl}S_{kl} + \eta_{ijkl}\frac{\partial S_{kl}}{\partial t}\,. \tag{4.62}$$

A crystal is therefore characterized by many relaxation times, or many quality factors, relating to the nature of the wave and the propagation direction. Since it relates two tensors of rank two, the viscosity tensor η_{ijkl} has rank four. The crystal symmetry limits the number of independent components η_{ijkl}, applying the same restrictions as for the elastic constants c_{ijkl}.

Substituting (4.62) into Newton's law (4.4) leads to the wave equation

$$\rho\frac{\partial^2 u_i}{\partial t^2} = c_{ijkl}\frac{\partial^2 u_l}{\partial x_j \partial x_k} + \eta_{ijkl}\frac{\partial^3 u_l}{\partial x_j \partial x_k \partial t}\,. \tag{4.63}$$

We consider the most symmetric case, an isotropic solid, whose description requires a viscosity coefficient η_{11} appropriate for compression and another coefficient η_{44} for shear. Hooke's law (4.13) becomes

$$T_{ij} = (c_{11} - 2c_{44})\,\delta_{ij}S + (\eta_{11} - 2\eta_{44})\delta_{ij}\frac{\partial S}{\partial t} + 2c_{44}S_{ij} + 2\eta_{44}\frac{\partial S_{ij}}{\partial t}\,.$$

This leads to

$$\begin{aligned}\frac{\partial T_{ij}}{\partial x_j} &= (c_{11} - 2c_{44})\frac{\partial S}{\partial x_i} + (\eta_{11} - 2\eta_{44})\frac{\partial^2 S}{\partial x_i \partial t} + c_{44}\left(\frac{\partial^2 u_i}{\partial x_j^2} + \frac{\partial^2 u_j}{\partial x_i \partial x_j}\right)\\ &\quad + \eta_{44}\frac{\partial^3 u_i}{\partial x_j^2 \partial t} + \eta_{44}\frac{\partial^3 u_j}{\partial x_i \partial x_j \partial t}\end{aligned}$$

and with $S = \dfrac{\partial u_j}{\partial x_j}$ we find

$$\rho\frac{\partial^2 u_i}{\partial t^2} = (c_{11} - c_{44})\frac{\partial S}{\partial x_i} + c_{44}\frac{\partial^2 u_i}{\partial x_j^2} + (\eta_{11} - \eta_{44})\frac{\partial^2 S}{\partial x_i \partial t} + \eta_{44}\frac{\partial^3 u_i}{\partial x_j^2 \partial t}\,. \tag{4.64}$$

Considering first a plane longitudinal wave propagating along x_1, so that $i = j = 1$, the above equation simplifies to

$$\rho\frac{\partial^2 u_1}{\partial t^2} = c_{11}\frac{\partial^2 u_1}{\partial x_1^2} + \eta_{11}\frac{\partial^3 u_1}{\partial x_1^2 \partial t}\,.$$

For a harmonic wave with the form

$$u_1 = {}^{\circ}u_1\,\mathrm{e}^{-\alpha_{\mathrm{L}} x_1}\,\mathrm{e}^{\mathrm{i}\omega(t - x_1/V_{\mathrm{L}})}$$

the formulae of (1.57), with $\omega\tau \ll 1$, lead to the following:

$$V_{\mathrm{L}} \cong \sqrt{\frac{c_{11}}{\rho}}, \quad \tau = \frac{\eta_{11}}{c_{11}} = \frac{\eta_{11}}{\rho V_{\mathrm{L}}^2}$$

and

$$\alpha_{\mathrm{L}} \cong \frac{\omega^2 \tau}{2V_{\mathrm{L}}} = \frac{\omega^2 \eta_{11}}{2\rho V_{\mathrm{L}}^3} = 2\pi^2 \frac{\eta_{11}}{\rho V_{\mathrm{L}}^3} f^2 . \tag{4.65}$$

For a plane transverse wave ($S = 0$) propagating along x_1 ($j = 1$) and polarized along x_2 ($i = 2$), we have $\partial/\partial x_2 = 0$ and the wave equation (4.64) leads to

$$\rho \frac{\partial^2 u_2}{\partial t^2} = c_{44} \frac{\partial^2 u_2}{\partial x_1^2} + \eta_{44} \frac{\partial^3 u_2}{\partial x_1^2 \partial t} .$$

This has the solution $u_2 = {}^{\circ}u_2 \exp(-\alpha_{\mathrm{T}} x_1) \exp[\mathrm{i}\omega(t - x_1/V_{\mathrm{T}})]$, where

$$V_{\mathrm{T}} \cong \sqrt{\frac{c_{44}}{\rho}}, \quad \tau = \frac{\eta_{44}}{\rho V_{\mathrm{T}}^2} \quad \text{and} \quad \alpha_{\mathrm{T}} \cong \frac{\omega^2 \tau}{2V_{\mathrm{T}}} = 2\pi^2 \frac{\eta_{44}}{\rho V_{\mathrm{T}}^3} f^2 . \tag{4.66}$$

For a path length equal to the wavelength, the attenuation is inversely proportional to the quality factor, since

$$\alpha\lambda \cong \pi\omega\tau = \frac{\pi}{Q} .$$

For a crystal, it is often possible to make use of the results established earlier for no attenuation. For example, in a cubic crystal (Sect. 4.2.5.2a) we know that for any propagation direction in the (001) plane there is a purely transverse wave with polarization along Ox_3 and with velocity $V_{\mathrm{T}} = (c_{44}/\rho)^{1/2}$. To allow for attenuation, it is enough to associate the elastic constant c_{44} with the coefficient η_{44}.

The term 'viscosity' applied to solids refers to physical mechanisms which apply partly in the case of a gas. The losses arise, firstly, from thermal conduction between zones of compression with raised temperature and zones of dilatation with lowered temperature; and, secondly, because the wave perturbs the distribution of incoherent thermal waves (Sect. 4.1) which exist spontaneously (the Akhieser effect). Theoretically, the first mechanism has no role in a transverse wave. The presence of free carriers increases the attenuation, so a wave of a given frequency generally has higher attenuation in a metal than in an insulator. In a polycrystalline material the attenuation is larger than for a monocrystal because scattering by the grain boundaries is an additional source of loss.

Figure 4.31 shows the frequency dependence of the attenuation coefficient α_μ, in decibels per microsecond, for several insulating materials used for delay lines operating in the GHz region. This increases approximately as the square of the frequency.

From the slopes α_μ/f^2 of these lines, it is possible to deduce the order of magnitude of the coefficients η, τ and Q. For example, we can find the coefficient η_{33} and the quality factor Q in single-crystal alumina (sapphire, Al_2O_3), for a longitudinal wave propagating along the Z-axis. From the graph,

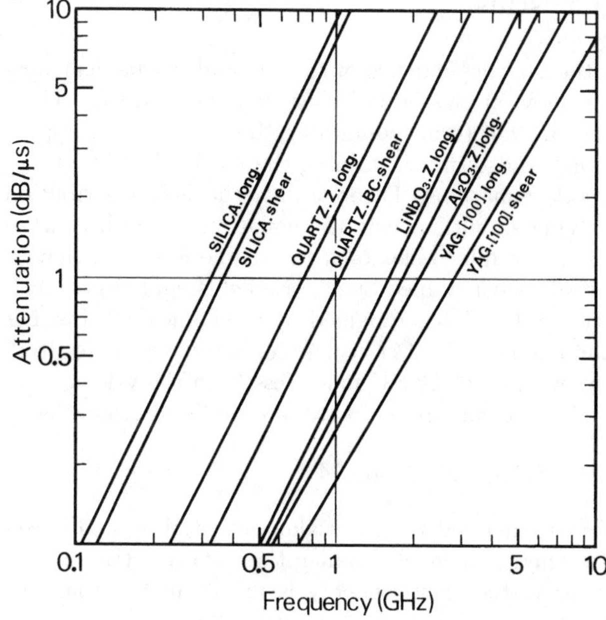

Fig. 4.31. Attenuation of elastic waves, as a function of frequency, for several insulating solids (from [4.2])

$$\alpha_\mu = 1\,\text{dB}/\mu\text{s at } 2\,\text{GHz}\,, \quad \text{so} \quad \alpha_\mu/f^2 = 0.25\times10^{-18}\,\text{dB}/(\mu\text{s}\cdot\text{Hz}^2)$$

Since $\alpha(\text{dB/m}) \approx 8.7\,\alpha(\text{Neper/m})$ and $\rho V_\text{L}^2 = c_{33}$, we have

$$\alpha(\text{dB/m}) = \frac{10^6}{V_\text{L}}\alpha_\mu(\text{dB}/\mu\text{s}) = 8.7\frac{\eta_{33}}{\rho V_\text{L}^3}2\pi^2 f^2$$

giving

$$\eta_{33} = 5\,823\,c_{33}\frac{\alpha_\mu}{f^2}\,.$$

The table in Fig. 3.10 gives $c_{33} \approx 50\times10^{10}\,\text{N/m}^2$, and hence

$$\eta_{33} \cong 730\times10^{-6}\,\text{N·s/m}^2 \quad \text{and} \quad Q = \frac{c_{33}}{\omega\eta_{33}} \cong 54\,500 \text{ at } 2\text{ GHz}\,.$$

These figures are characteristic of a material with excellent quality. For more ordinary materials, polycrystalline or amorphous insulators, the picture is different – the line for silica gives $\alpha_\mu = 10\,\text{dB}/\mu\text{s}$ at a frequency of 1 GHz. To illustrate the range of variation, the attenuation coefficient α (in dB/m) at 1 GHz is 23 for sapphire, a hard insulating material, about 50 times larger for a semiconductor (it is 1000 for silicon), and much larger for a metal – about 1000 times for copper. The attenuation coefficient for several materials used in acousto-optics is given in Chap. 3 of Vol. II.

4.3 Piezoelectric Crystals

In a piezoelectric solid, the interdependence of electric and mechanical variables implies a coupling between elastic and electromagnetic waves. Thus, as already stated in Sect. 3.1.2.4, terms containing the electric field appear in the dynamical equations, and terms containing the mechanical strain appear in the electromagnetic equations. Theoretically, the field distribution can be found only by solving simultaneously the equations of both Newton and Maxwell. The solutions are mixed elasto-electromagnetic waves, that is, elastic waves with velocity V accompanied by electric fields, and electromagnetic waves with velocity $c \approx 10^5\,V$ accompanied by mechanical strains. For the first type of wave the magnetic field is negligible, because it arises from an electric field travelling with a velocity V much less than the velocity c of electromagnetic waves. Thus we can approximate Maxwell's equations as

$$\operatorname{curl}\boldsymbol{E} = -\frac{\partial \boldsymbol{B}}{\partial t} \cong 0 \quad \text{giving} \quad \boldsymbol{E} = -\operatorname{grad}\Phi\,.$$

The field $\boldsymbol{E}$ can therefore be derived from a scalar potential Φ, as in electrostatics. The magnetic energy is much smaller than the electric energy. For the second type of wave, the elastic energy is much smaller than the electromagnetic energy.

Thus, even in a strongly piezoelectric material, the interaction between the three elastic waves and the two electromagnetic waves is weak, because the velocities are so different. The reader familiar with travelling-wave tubes will recognize this effect – to obtain amplification by coupling to the electron beam, the electromagnetic wave has to be 'slowed down' by the helix, so that its velocity becomes comparable. Consequently, the propagation of elastic and electromagnetic waves can be treated separately. The propagation of elastic waves in a piezoelectric (and insulating) material is therefore studied here with the assumption that the electric field appears static in comparison with electromagnetic waves, that is, using the *quasi-static approximation.*

4.3.1 Wave Equation

From (3.94), the stress is given by $T_{ij} = c^{\mathrm{E}}_{ijkl} S_{kl} - e_{kij} E_k$ and on substituting for the strain $S_{kl} = \dfrac{1}{2}\left(\dfrac{\partial u_l}{\partial x_k} + \dfrac{\partial u_k}{\partial x_l}\right)$ and the electric field $E_k = -\dfrac{\partial \Phi}{\partial x_k}$ this becomes

$$T_{ij} = c^{\mathrm{E}}_{ijkl}\frac{\partial u_l}{\partial x_k} + e_{kij}\frac{\partial \Phi}{\partial x_k}\,.$$

Hence Newton's law, in the form of (4.4), becomes

$$\boxed{\rho\frac{\partial^2 u_i}{\partial t^2} = c^{\mathrm{E}}_{ijkl}\frac{\partial^2 u_l}{\partial x_j \partial x_k} + e_{kij}\frac{\partial^2 \Phi}{\partial x_j \partial x_k}}\,. \tag{4.67}$$

Moreover, the electric displacement is

$$D_j = e_{jkl} S_{kl} + \varepsilon^{\mathrm{S}}_{jk} E_k = e_{jkl} \frac{\partial u_l}{\partial x_k} - \varepsilon^{\mathrm{S}}_{jk} \frac{\partial \Phi}{\partial x_k} \tag{4.68}$$

and for an insulating solid this must satisfy Poisson's equation $\partial D_j / \partial x_j = 0$, so that

$$\boxed{e_{jkl} \frac{\partial^2 u_l}{\partial x_j \partial x_k} - \varepsilon^{\mathrm{S}}_{jk} \frac{\partial^2 \Phi}{\partial x_j \partial x_k} = 0} . \tag{4.69}$$

The wave equation for the displacement u_i is obtained by eliminating the potential Φ between (4.67) and (4.69). For a *plane* wave propagating along the direction n_j, solutions for u_i and Φ have the forms

$$u_i = {}^{\circ}u_i \, F\left(t - \frac{n_j x_j}{V}\right), \quad \Phi = \Phi_0 \, F\left(t - \frac{n_j x_j}{V}\right) . \tag{4.70}$$

Each wavefront is an equipotential, so that *the electric field is longitudinal,* being given by

$$E_j = -\frac{\partial \Phi}{\partial x_j} = \frac{n_j}{V} \Phi_0 F' ,$$

where F' is the derivative of the function F. Substituting (4.70) into (4.67) and (4.69) gives the relations

$$\frac{\partial^2 u_i}{\partial t^2} = {}^{\circ}u_i \, F'', \qquad \frac{\partial^2 u_l}{\partial x_j \partial x_k} = \frac{n_j n_k}{V^2} \, {}^{\circ}u_l F'', \qquad \frac{\partial^2 \Phi}{\partial x_j \partial x_k} = \frac{n_j n_k}{V^2} \Phi_0 \, F'' .$$

Defining the quantities

$$\boxed{\Gamma_{il} = c^{\mathrm{E}}_{ijkl} n_j n_k, \quad \gamma_i = e_{kij} n_j n_k, \quad \varepsilon = \varepsilon^{\mathrm{S}}_{jk} n_j n_k} \tag{4.71}$$

the above leads to the system of equations

$$\boxed{\begin{aligned} &\rho V^2 \, {}^{\circ}u_i = \Gamma_{il} \, {}^{\circ}u_l + \gamma_i \Phi_0 \\ &\gamma_l \, {}^{\circ}u_l - \varepsilon \Phi_0 = 0 \end{aligned}} . \tag{4.72}$$

On eliminating the electric potential Φ_0, this leads to

$$\rho V^2 \, {}^{\circ}u_i = \left(\Gamma_{il} + \frac{\gamma_i \gamma_l}{\varepsilon}\right) {}^{\circ}u_l . \tag{4.73}$$

As for a non-piezoelectric solid, the polarizations ${}^{\circ}u_i$ of the plane elastic waves propagating in a chosen direction are the eigenvectors of a tensor of rank two, defined in this case by

$$\bar{\Gamma}_{il} = \Gamma_{il} + \frac{\gamma_i \gamma_l}{\varepsilon} , \tag{4.74}$$

and the eigenvalues $\bar{\gamma} = \rho V^2$ give the phase velocities. The *polarizations* of the three waves are always *mutually orthogonal,* since the tensor $\bar{\Gamma}_{il}$ is symmetric. To show the additional term of piezoelectric origin, dependent on propagation direction, it is appropriate to complete the expansions of (4.11) by adding the components γ_l, which are

$$\begin{aligned}\gamma_l = e_{11l}n_1^2 + e_{22l}n_2^2 + e_{33l}n_3^2 + (e_{12l} + e_{21l})n_1n_2 \\ +(e_{13l} + e_{31l})n_1n_3 + (e_{23l} + e_{32l})n_2n_3\end{aligned} \; .$$

For particular values of the index l, we have

$$\begin{aligned}\gamma_1 &= e_{11}n_1^2 + e_{26}n_2^2 + e_{35}n_3^2 + (e_{16} + e_{21})n_1n_2 \\ &\quad +(e_{15} + e_{31})n_1n_3 + (e_{25} + e_{36})n_2n_3 \\ \gamma_2 &= e_{16}n_1^2 + e_{22}n_2^2 + e_{34}n_3^2 + (e_{12} + e_{26})n_1n_2 \\ &\quad +(e_{14} + e_{36})n_1n_3 + (e_{24} + e_{32})n_2n_3 \\ \gamma_3 &= e_{15}n_1^2 + e_{24}n_2^2 + e_{33}n_3^2 + (e_{14} + e_{25})n_1n_2 \\ &\quad +(e_{13} + e_{35})n_1n_3 + (e_{23} + e_{34})n_2n_3\end{aligned} \tag{4.75}$$

The influence of piezoelectricity on the phase velocity can be expressed in terms of modified stiffness constants of the material. As for a non-piezoelectric solid, the Christoffel tensor can be written in the form

$$\bar{\Gamma}_{il} = \bar{c}_{ijkl}n_jn_k \, ,$$

where the $\bar{c}_{ijkl}$ are effective constants defined by

$$\bar{c}_{ijkl} = c_{ijkl}^{\mathrm{E}} + \frac{(e_{pij}n_p)\,(e_{qkl}n_q)}{\varepsilon_{jk}^{\mathrm{S}}n_jn_k} \, . \tag{4.76}$$

The constants $\bar{c}_{ijkl}$, called 'stiffened' constants, are not true elastic constants since they are defined only for plane waves and they depend on the propagation direction. Moreover, it is preferable to use the expression (4.74) and the expansions (4.75) when treating the examples below.

4.3.2 Examples of Slowness Curves

We consider here bismuth germanium oxide ($Bi_{12}GeO_{20}$), lithium niobate ($LiNbO_3$) and zinc oxide (ZnO). The first of these is used much less frequently than the others, but is notable for its unusually low velocities, less than 3 800 m/s; it has optical applications as a source of non-linear effects. Lithium niobate is commonly available and has large electromechanical coupling. Zinc oxide is often used, in the form of a crystallographically oriented film, to generate bulk waves at high frequencies, above 1 GHz.

(a) Propagation in the (001) Plane of a Cubic Crystal ($Bi_{12}GeO_{20}$, class 23). The matrix of the moduli $e_{i\alpha}$, given in (3.114), has the same form for the two classes $\bar{4}3m$ and 23 in the cubic system whose symmetry is compatible with the presence of piezoelectricity. Since the only non-zero constants are $e_{14} = e_{25} = e_{36}$, the expansions of (4.75) reduce to

$$\gamma_1 = 2e_{14}n_2n_3 \, , \quad \gamma_2 = 2e_{14}n_1n_3 \, , \quad \gamma_3 = 2e_{14}n_1n_2 \, .$$

In the (001) plane, for which $n_1 = \cos\varphi, n_2 = \sin\varphi$ and $n_3 = 0$, this gives

$$\gamma_1 = \gamma_2 = 0 \quad \text{and} \quad \gamma_3 = e_{14}\sin 2\varphi$$

and the components $\bar{\Gamma}_{13}$ and $\bar{\Gamma}_{23}$ are zero, as in the case of a non-piezoelectric cubic crystal (Sect. 4.2.5.2a). Moreover,

$$\bar{\Gamma}_{11} = \Gamma_{11}\,, \quad \bar{\Gamma}_{22} = \Gamma_{22}\,, \quad \bar{\Gamma}_{12} = \Gamma_{12}\,,$$

and the only modified component, $\bar{\Gamma}_{33}$, is

$$\bar{\Gamma}_{33} = \Gamma_{33} + \frac{\gamma_3^2}{\varepsilon_{11}^{\mathrm{S}}} = c_{44}^{\mathrm{E}} + \frac{e_{14}^2}{\varepsilon_{11}^{\mathrm{S}}}\sin^2 2\varphi\,.$$

Equation (4.33) for the velocities V_1 and V_2 of the quasi-longitudinal and quasi-tranverse waves, polarized in the (001) plane, remains valid since only the components $\bar{\Gamma}_{11}$, $\bar{\Gamma}_{22}$ and $\bar{\Gamma}_{12}$ are used. Hence V_1 and V_2 are still given by (4.34) and (4.35), taking the stiffnesses to be those for constant electric field, c_{ijkl}^{E}. These waves are not affected by piezoelectricity since the accompanying strains S_{11}, S_{22}, S_{12} do not generate any longitudinal electric field – this would need moduli e_{ijk} with no index equal to three. In fact, these waves are not coupled to any electric field. However, the transverse wave with polarization along [001] is affected; its velocity V_3, dependent on the propagation direction, is

$$V_3 = \sqrt{\frac{\bar{\Gamma}_{33}}{\rho}} = \sqrt{\frac{c_{44}}{\rho}}\sqrt{1 + \frac{e_{14}^2}{\varepsilon_{11}^{\mathrm{S}} c_{44}^{\mathrm{E}}}\sin^2 2\varphi}\,.$$

The amount of this velocity change is directly related to the value of the *electromechanical coupling coefficient* K, defined by

$$K = \frac{e_{14}}{\sqrt{\varepsilon_{11}^{\mathrm{S}} c_{44}^{\mathrm{E}} + e_{14}^2}} \quad \text{so that} \quad V_3 = \sqrt{\frac{c_{44}}{\rho}}\cdot\sqrt{1 + \frac{K^2}{1-K^2}\sin^2 2\varphi}\,.$$

For bismuth germanium oxide ($Bi_{12}GeO_{20}$, class 23), the constants are given in Figs. 3.10 and 3.15, and the slowness curves are shown in Fig. 4.32. In this case $K = 0.32$ and the largest change of V_3, obtained for propagation along [110], is

$$\frac{\Delta V_3}{V_3} \cong \frac{K^2}{2} \cong 5\,\%\,.$$

(b) Propagation in the YZ Plane of Lithium Niobate ($LiNbO_3$, class $3m$). Lithium niobate is a ferroelectric crystal (Sect. 2.4.2d) belonging to the trigonal class $3m$. For a propagation direction in the symmetry plane YZ ($n_1 = 0, n_2 = \sin\theta, n_3 = \cos\theta$), the piezoelectric tensor of (3.123) and the formulae of (4.75) lead to

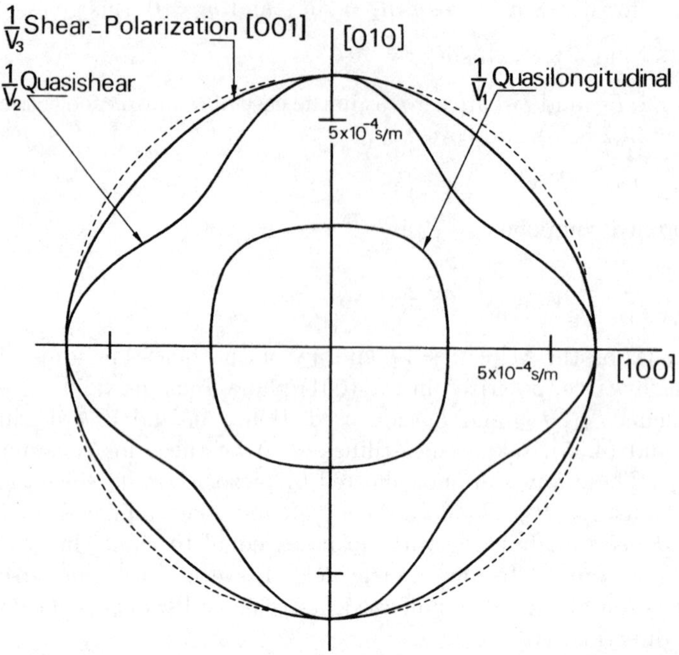

Fig. 4.32. Section of the slowness surfaces of bismuth germanium oxide (class 23) in the (001) plane. Only the transverse wave is piezoelectric, as shown by the gap between its curve and the dashed circle obtained when piezoelectricity is ignored

$$\begin{aligned}
\gamma_1 &= 0 \\
\gamma_2 &= e_{22}\sin^2\theta + (e_{15}+e_{31})\sin\theta\cos\theta \\
\gamma_3 &= e_{15}\sin^2\theta + e_{33}\cos^2\theta \\
\varepsilon &= \varepsilon_{11}\sin^2\theta + \varepsilon_{33}\cos^2\theta\,.
\end{aligned}$$

In Sect. 4.2.5.2d it was established that $\Gamma_{12} = \Gamma_{13} = 0$ for this section of the slowness surfaces, the other components of the tensor Γ_{il} being given by (4.54). Consequently, the only non-zero components of the stiffened Christoffel tensor are

$$\begin{aligned}
\bar{\Gamma}_{11} &= \Gamma_{11} & \bar{\Gamma}_{22} &= \Gamma_{22} + \frac{\gamma_2^2}{\varepsilon} \\
\bar{\Gamma}_{23} &= \Gamma_{23} + \frac{\gamma_2\gamma_3}{\varepsilon} & \bar{\Gamma}_{33} &= \Gamma_{33} + \frac{\gamma_3^2}{\varepsilon}\,.
\end{aligned}$$

There is therefore a transverse wave, polarized normal to the YZ plane, with velocity $V_3 = (\Gamma_{11}/\rho)^{1/2}$, unaffected by piezoelectricity. Different behaviour is shown by the quasi-longitudinal and quasi-transverse waves, polarized in the YZ plane, which respectively have velocities V_1 and V_2 given by

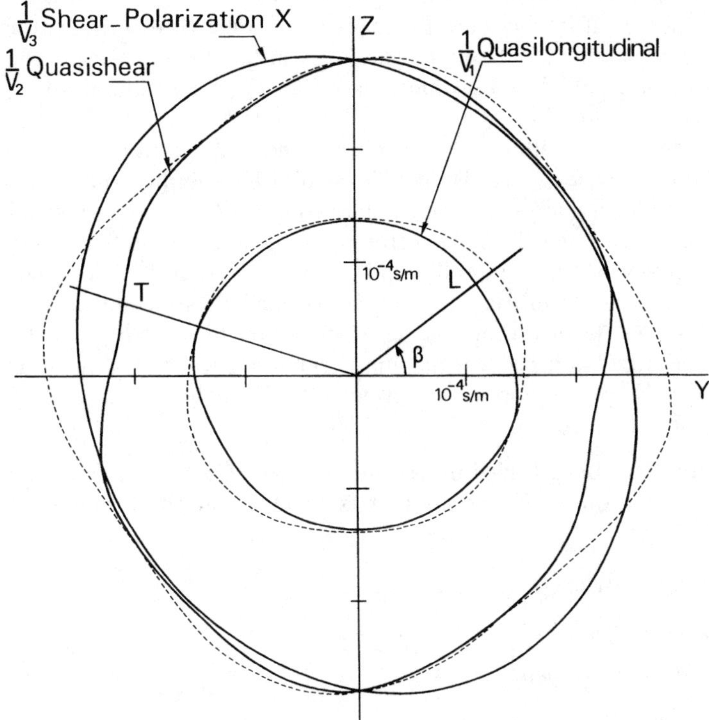

Fig. 4.33. Section of the slowness surfaces of lithium niobate (class $3m$) in the YZ plane. Broken lines are calculated ignoring piezoelectricity. The transverse wave is piezoelectrically inactive

$$2\rho V_{\substack{1\\2}}^2 = \bar{\Gamma}_{22} + \bar{\Gamma}_{33} \pm \sqrt{(\bar{\Gamma}_{22} - \bar{\Gamma}_{33})^2 + 4\bar{\Gamma}_{23}^2} \, . \tag{4.77}$$

Figure 4.33 shows curves calculated using the constants of Figs. 3.10 and 3.15 (full lines), and curves calculated with piezoelectricity ignored (broken lines), demonstrating the significance, dependent on direction, of piezoelectricity in this crystal.

For a particular wave, the magnitude of the gap between the two curves, one taking account of piezoelectricity and the other ignoring it, indicates directly the crystal cuts suitable for a transducer to excite this wave preferentially. The simplest transducer is just a monocrystal in the form of a plate with its major faces metallized. An electric field, generated by a voltage applied between these faces, will cause the crystal to vibrate provided the crystal cut is such that there is coupling between electric fields and strains. The curves in Fig. 4.33 show that a longitudinal wave in lithium niobate, propagating in the OL direction ($\beta = \pi/2 - \theta \approx 36°$), has an associated electric field $\boldsymbol{E}_{\mathrm{L}}$ also directed along OL. Since the piezoelectric effect is reciprocal, an electric field $\boldsymbol{E}_{\mathrm{L}}$ applied across a single-crystal plate with major faces normal to OL (i.e. with cut $Y + 36°$) will generate a longitudinal wave propagating along OL. If its faces are free, this plate constitutes a resonator which vibrates longitudinally. If one of the faces is bonded to a solid the loaded resonator, now called a transducer, launches the longitudinal vibrations into the solid. Note that, if the transducer is to vibrate only in the longitudinal mode, the coupling between electric field and strain needs to be small for the two other modes. Here, this is valid for the direction OL – the coupling is zero for a transverse wave polarized along OX and very small for the quasi-transverse wave. To make a transducer generating preferentially the quasi-transverse wave, the crystalline plate may be chosen to have a cut $Y + 163°$, as can be seen from Fig. 4.33. These conclusions are confirmed by Fig. 4.34, which shows the electromechanical coupling coefficients K_L (for the quasi-longitudinal wave) and K_T (for the quasi-transverse wave), as functions of direction in the YZ plane. For the direction +36° from the Y-axis we find $K_L = 0.49$, and for the +166° direction $K_T = 0.62$.

(c) Propagation in the Meridian Plane of Zinc Oxide (ZnO). Zinc oxide belongs to class $6mm$. Using the matrix (3.119) and the formulae of (4.75), we have

$$\begin{aligned} \gamma_1 &= (e_{15} + e_{31})n_1 n_3 \\ \gamma_2 &= (e_{15} + e_{31})n_2 n_3 \\ \gamma_3 &= e_{15}(n_1^2 + n_2^2) + e_{33}n_3^2 \\ \varepsilon &= \varepsilon_{11}(n_1^2 + n_2^2) + \varepsilon_{33}n_3^2 \,. \end{aligned} \tag{4.78}$$

In the meridian plane YZ we have $n_1 = 0$, $n_2 = \sin\theta$ and $n_3 = \cos\theta$, giving

$$\gamma_1 = 0\,, \quad \gamma_2 = \frac{e_{15} + e_{31}}{2}\sin 2\theta\,, \quad \gamma_3 = e_{15}\sin^2\theta + e_{33}\cos^2\theta\,,$$

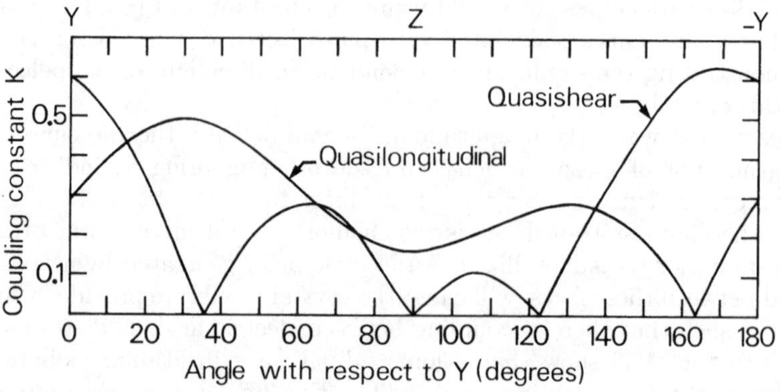

Fig. 4.34. Electromechanical coupling coefficients K_L and K_T as functions of direction in the YZ plane of lithium niobate (Fig. 13 of [4.3])

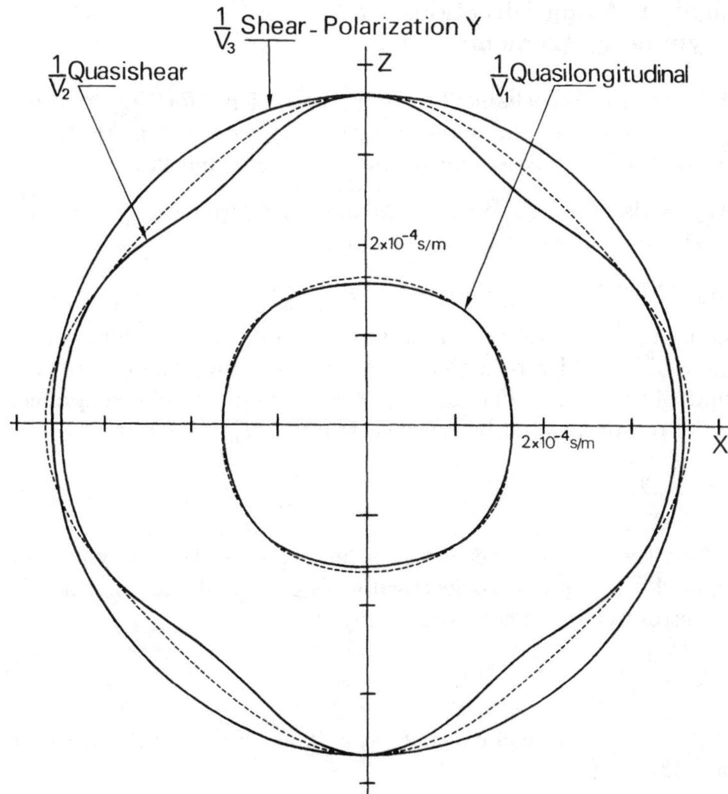

Fig. 4.35. Section of the slowness surfaces of zinc oxide (class $6mm$) in a meridian plane. The broken lines are calculated with piezoelectricity ignored. The transverse wave is piezoelectrically inactive

$$\varepsilon = \varepsilon_{11} \sin^2 \theta + \varepsilon_{33} \cos^2 \theta \,.$$

Using Eq. (4.74), the $\bar{\Gamma}$ tensor is

$$\bar{\Gamma}_{11} = \Gamma_{11} \,, \qquad \bar{\Gamma}_{12} = \Gamma_{12} = 0 \,, \qquad \bar{\Gamma}_{13} = \Gamma_{13} = 0 \,,$$

$$\bar{\Gamma}_{22} = \Gamma_{22} + \frac{\gamma_2^2}{\varepsilon} \,, \quad \bar{\Gamma}_{23} = \Gamma_{23} + \frac{\gamma_2 \gamma_3}{\varepsilon} \,, \quad \bar{\Gamma}_{33} = \Gamma_{33} + \frac{\gamma_3^2}{\varepsilon} \,.$$

Hence there is a transverse wave polarized normal to the meridian plane, with velocity V_3 given by (4.47), and two piezoelectric waves, quasilongitudinal and quasi-transverse, with respective velocities V_1 and V_2 given by (4.77). These results are illustrated in Fig. 4.35.

4.3.3 Propagation Along Directions Related to Symmetry Elements

Here we derive the electromechanical coupling coefficients for the three important cases of propagation (a) along a symmetry axis, (b) in the plane normal to a tetrad or hexad axis, and (c) in a plane of symmetry.

(a) Symmetry Axis. Taking the axis to be along Ox_3 we have $n_1 = n_2 = 0$ and $n_3 = 1$, and the expressions in (4.75) reduce to

$$\gamma_1 = e_{35}\,, \quad \gamma_2 = e_{34}\,, \quad \gamma_3 = e_{33}\,.$$

Inspection of the $e_{i\alpha}$ matrices shows that e_{34} and e_{35} are zero for crystals with a direct axis of any order, or with an inverse axis of any order except two. From the results of Sect. 4.2.2, $\bar{\Gamma}_{13}$ and $\bar{\Gamma}_{23}$ are zero, and the only component of the Christoffel tensor affected by piezoelectricity is $\bar{\Gamma}_{33}$, given by

$$\bar{\Gamma}_{33} = c_{33}^{\mathrm{E}} + \frac{e_{33}^2}{\varepsilon_{33}^{\mathrm{S}}}\,.$$

It follows that the two transverse waves propagating along a symmetry axis (other than $\bar{A}_2$) are not piezoelectrically coupled, and the longitudinal wave has an electromechanical coupling coefficient

$$K = \frac{e_{33}}{\sqrt{\varepsilon_{33}^{\mathrm{S}} c_{33}^{\mathrm{E}} + e_{33}^2}} \qquad A_n \| Ox_3\,.$$

This is zero if the symmetry requires that $e_{33} = 0$, as for example in the case of quartz (class 32).

(b) Plane Normal to a Tetrad or Hexad Axis. The direct tetrad or hexad axis is along Ox_3, so $n_3 = 0$ and the formulae in (4.75) lead to

$$\gamma_1 = e_{11}n_1^2 + e_{26}n_2^2 + (e_{16} + e_{21})n_1 n_2$$

$$\gamma_2 = e_{16}n_1^2 + e_{22}n_2^2 + (e_{12} + e_{26})n_1 n_2$$

$$\gamma_3 = e_{15}n_1^2 + e_{24}n_2^2 + (e_{14} + e_{25})n_1 n_2\,.$$

The $e_{i\alpha}$ matrix of (3.117) shows that

$$e_{11} = e_{26} = e_{16} = e_{21} = e_{22} = 0$$

and hence γ_1 and γ_2 are zero; also, from Sect. 4.2.2 we have $\Gamma_{13} = \Gamma_{23} = 0$. Thus there is a transverse wave, polarized along Ox_3, for all propagation directions in the plane. Moreover, since $e_{25} = -e_{14}$ and $e_{24} = e_{15}$ the component γ_3 is

$$\gamma_3 = e_{15}(n_1^2 + n_2^2) = e_{15}$$

and is thus a constant. Hence the quasi-longitudinal and quasi-transverse waves, polarized in the propagation plane Ox_1x_2 normal to the direct tetrad or hexad axis, are not piezoelectrically coupled.

Propagation direction	Mode	Coupling coefficient K	$Bi_{12}GeO_{20}$ (23)	ZnO (6mm)	CdS (6mm)	Ceramic PZT-4 (6mm)	$LiNbO_3$ (3m)	$LiTaO_3$ (3m)	Quartz (32)
along an axis A_2, A_n or $\bar{A}_n$ ($n \geq 2$) contained by Ox_3	L	$\frac{e_{33}}{\sqrt{\varepsilon_{33}^S c_{33}^E + e_{33}^2}}$	0	0.27	0.155	0.51	0.16	0.18	0
	T_1, T_2	0	0	0	0	0	0	0	0
in a plane $x_1 x_2$ $\perp A_4$ or A_6	T // Ox_3	$\frac{e_{15}}{\sqrt{\varepsilon_{11}^S c_{44}^E + e_{15}^2}}$	×	0.32	0.19	0.70	×	×	×
	QL, QT	0	×	0	0	0	×	×	×
in a mirror	T	0	×	meridian plane	meridian plane	meridian plane	YZ plane	YZ plane	×
A_2 // Ox_1	L	$\frac{e_{11}}{\sqrt{\varepsilon_{11}^S c_{11}^E + e_{11}^2}}$	0	×	×	×	×	×	0.093
	T_1, T_2	0	0	×	×	×	×	×	0

Fig. 4.36. Formulae and values for the electromechanical coupling coefficients K for propagation along symmetry directions, for several materials. The entries shown as crosses are of no interest

The velocity V_3 of the transverse wave, polarized along the symmetry axis, is

$$V_3 = \sqrt{\frac{c_{44}^E + e_{15}^2/\varepsilon_{11}^S}{\rho}}$$

and is independent of the propagation direction. The electromechanical coupling coefficient of this wave is

$$K = \frac{e_{15}}{\sqrt{\varepsilon_{11}^S c_{44}^E + e_{15}^2}}\,.$$

Thus the Ox_1x_2 plane is isotropic for the transverse wave, as in a non- piezoelectric solid. However, this is no longer true if the axis is of type $\bar{A}_4$, as shown in Problem 4.9.

(c) **Symmetry Plane.** The plane is assumed to be normal to Ox_3, so that $n_3 = 0$. The component γ_3 is zero because $e_{15} = e_{24} = e_{14} = e_{25} = 0$, from (3.110). Hence there is a transverse wave with polarization perpendicular to the mirror plane (since $\bar{\Gamma}_{13} = \bar{\Gamma}_{23} = 0$), and it is non-piezoelectric since its velocity $V_3 = (\bar{\Gamma}_{33}/\rho)^{1/2} = (\Gamma_{33}/\rho)^{1/2}$ is unchanged.

These results are summarized in the table of Fig. 4.36, which gives the coupling coefficients for the piezoelectric solids used most frequently.

4.3.4 Bulk Piezoelectric Permittivity

The notion of piezoelectric permittivity, introduced by K.A. Ingebrigtsen [4.4], serves to express the electrical boundary conditions that elastic waves must satisfy, particularly for surface waves (Sect. 5.3.4 below). It is also suitable, as shown by M. Feldmann and J. Hénaff [4.5], to characterize the bulk behaviour of a piezoelectric solid. For example, it can describe the reaction to the introduction of electric charges, that is, to the application of a voltage between electrodes either in the interior of the medium (for an infinite solid, giving bulk waves), or arranged on a surface (for a semi-infinite solid, giving surface waves, Chap. 2 of Vol. II). Following the above-mentioned authors, we define the bulk piezoelectric permittivity starting from a scalar one-dimensional model. Each variable becomes a scalar dependent on only one spatial coordinate, thus simplifying the equations. Although the solid is assumed to be an insulator, we allow for the presence of electric charges, with volume density ρ_{e}, so that charges on the electrodes can be included.

The solid behaves according to (4.67) and (4.69), with ρ_{e} added on the right side of the latter, so we can write, using scalar variables,

$$\begin{cases} c^{\mathrm{E}}\dfrac{\partial^2 u}{\partial x^2} + e\dfrac{\partial^2 \Phi}{\partial x^2} = \rho\dfrac{\partial^2 u}{\partial t^2} & (4.79) \\[2ex] e\dfrac{\partial^2 u}{\partial x^2} - \varepsilon^{\mathrm{S}}\dfrac{\partial^2 \Phi}{\partial x^2} = \rho_{\mathrm{e}}\,. & (4.80) \end{cases}$$

For sinusoidal variations, all variables have the form

$$g(x,t) = \bar{g}\exp[\mathrm{i}(\omega t - kx)]\,,$$

including the propagation phase which is to be understood in the expressions below. Thus,

$$\frac{\partial}{\partial t} = \mathrm{i}\omega \quad \text{and} \quad \frac{\partial}{\partial x} = -\mathrm{i}k\,.$$

From (4.79) the amplitude of the displacement is, in terms of the electric potential,

$$\bar{u} = \frac{ek^2}{\rho\omega^2 - c^{\mathrm{E}}k^2}\bar{\Phi}\,.$$

Substituting into (4.80) leads to the following relation between the two electrical variables, charge density and potential:

$$\bar{\rho}_{\mathrm{e}} = \varepsilon^{\mathrm{S}}k^2\bar{\Phi} - ek^2\bar{u} = \left(\varepsilon^{\mathrm{S}} + \frac{e^2k^2}{c^{\mathrm{E}}k^2 - \rho\omega^2}\right)k^2\bar{\Phi}\,.$$

Here the quantity in brackets, which has the dimension of permittivity, is the *piezoelectric permittivity*, defined by

$$\bar{\varepsilon}(\omega,k) = \frac{\bar{\rho}_{\mathrm{e}}}{k^2\bar{\Phi}} = \varepsilon^{\mathrm{S}}\left(1 + \frac{e^2k^2/\varepsilon^{\mathrm{S}}}{c^{\mathrm{E}}k^2 - \rho\omega^2}\right)\,. \tag{4.81}$$

This is dependent on ω and k. Introducing the two phase velocities $V_\infty = (c^{\mathrm{E}}/\rho)^{1/2}$ and $V_0 = [(c^{\mathrm{E}} + e^2/\varepsilon^{\mathrm{S}})/\rho]^{1/2}$ for elastic waves without and with piezoelectricity, the permittivity becomes

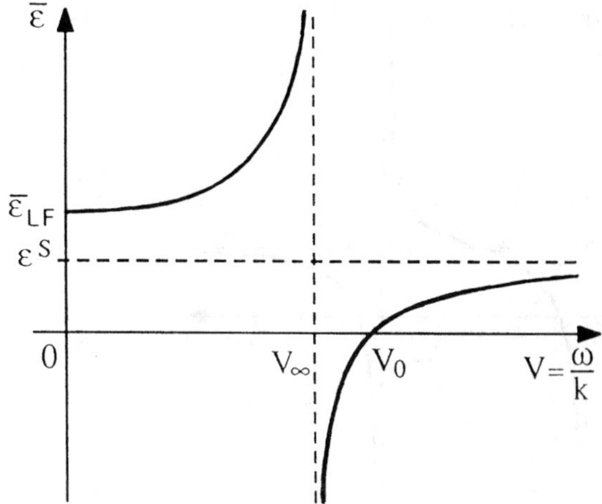

Fig. 4.37. Piezoelectric permittivity, for the one-dimensional case without attenuation, as a function of the phase velocity $V = \omega/k$. The low-frequency value, $\bar{\varepsilon}_{\mathrm{LF}} = \varepsilon^{\mathrm{S}} V_0^2/V_\infty^2$, is larger than the high-frequency value $\bar{\varepsilon}_{\mathrm{HF}} = \varepsilon^{\mathrm{S}}$

$$\bar{\varepsilon}(\omega, k) = \varepsilon^{\mathrm{S}} \frac{\left(c^{\mathrm{E}} + e^2/\varepsilon^{\mathrm{S}}\right) k^2 - \rho\omega^2}{c^{\mathrm{E}} k^2 - \rho\omega^2} = \varepsilon^{\mathrm{S}} \frac{V_0^2 k^2 - \omega^2}{V_\infty^2 k^2 - \omega^2}$$

or, in terms of the phase velocity $V = \omega/k$,

$$\boxed{\bar{\varepsilon}(V) = \varepsilon^{\mathrm{S}} \frac{V_0^2 - V^2}{V_\infty^2 - V^2}} . \tag{4.82}$$

Figure 4.37 shows the variation of the permittivity $\bar{\varepsilon}$ with V. The resonance with infinite amplitude at $V = V_\infty$ expresses the existence of an elastic wave propagating with no attenuation. The low-frequency value, $\bar{\varepsilon}_{\mathrm{LF}} = \varepsilon^{\mathrm{S}} V_0^2/V_\infty^2$, is larger than the high-frequency value $\bar{\varepsilon}_{\mathrm{HF}} = \varepsilon^{\mathrm{S}}$.

In the three-dimensional case there are three resonances, one for each type of elastic wave, with the phase velocity V taking values $V_\infty^{(1)}$, $V_\infty^{(2)}$ and $V_\infty^{(3)}$. Figure 4.38 shows the appearance of the curve for $\bar{\varepsilon}$ in terms of V for this case.

The significance of the piezoelectric permittivity is that it relates electrical variables in such a way that mechanical phenomena, associated with elastic wave propagation, are included. If the dimensions of the medium are finite, as in the case of a resonator of thickness L (Sect. 1.1.2.1), the wavenumber is restricted to values $k = n\pi/L$ and the resonances occur at the fundamental frequency $f = V_\infty/2L$ and its harmonics.

When attenuation is taken into account, writing $c(1+\mathrm{i}\alpha)$ instead of c, this leads of course to a finite amplitude, as shown in Problem 4.11. The variation of $\bar{\varepsilon}/\varepsilon^{\mathrm{S}}$ with V/V_∞, with a resonance, is as shown by the curves in Fig. 4.39.

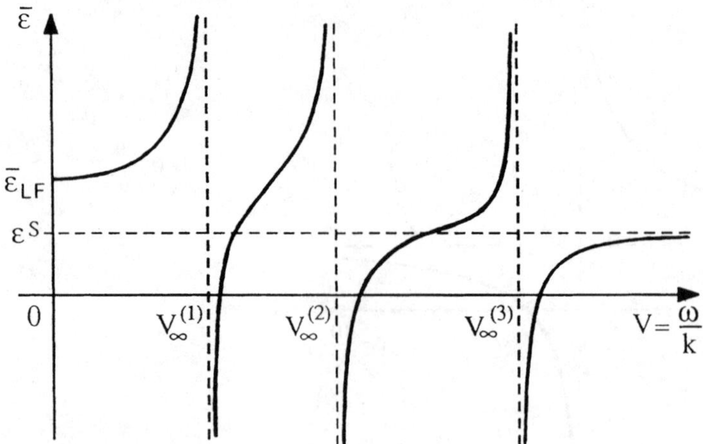

Fig. 4.38. Piezoelectric permittivity for the three-dimensional case

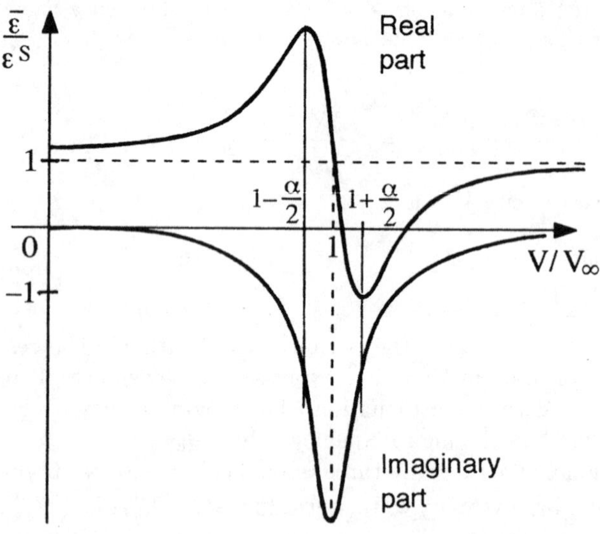

Fig. 4.39. Variation of the piezoelectric permittivity in a solid, in the one-dimensional case, when the elastic waves are attenuated

The significance of the formula (4.81) is more general than it seems at first. It can also be found from the partial differential equations (4.79) and (4.80), on replacing the variables by their Fourier transforms defined by the relations

$$g(x,t) = \frac{1}{(2\pi)^2} \int_{-\infty}^{+\infty} \int_{-\infty}^{+\infty} \overline{\overline{g}}(\omega, k)\, \mathrm{e}^{\mathrm{i}(\omega t - kx)} \mathrm{d}\omega\, \mathrm{d}k \tag{4.83a}$$

$$\overline{\overline{g}}(\omega, k) = \int_{-\infty}^{+\infty} \int_{-\infty}^{+\infty} g(x, t)\, \mathrm{e}^{-\mathrm{i}(\omega t - kx)} \mathrm{d}t\, \mathrm{d}x\,. \tag{4.83b}$$

4.4 Reflection and Refraction

The topic of reflection and refraction has already been considered, in two places, in Chap. 1. The first occasion, Sect. 1.1.2, was concerned with a wave that is totally reflected without changing its nature. This introduced the ideas of stationary waves, eigenmodes of resonators and dispersive propagation of modes in waveguides. Later, Sect. 1.2.4 described a wave partly transmitted across the surface separating two fluids, the wave being necessarily longitudinal and unable to change its nature; this led to the transformed impedance referred to a point in front of the boundary, to matching of impedances, and to the definitions of reflection and transmission coefficients and a study of their variation with the angle of incidence. A total (or partial) reflection, giving a reflected (and refracted) wave of the same type as the incident wave, can also be encountered when the propagation media are solids. However, the situation is generally more complex in solids. In fact, a plane monochromatic wave with a defined polarization can, when incident on the boundary between two (non-piezoelectric) crystals, give rise to three waves propagating on each side of the boundary. Thus a quasi-longitudinal wave can generate three reflected waves in the crystal that it travels in – one quasi-longitudinal and two quasi-transverse (one fast, one slow); and in the other crystal there can be three refracted waves, again one quasi-longitudinal and two quasi-transverse. The case of two isotropic materials is simpler – the incident wave generates at most two waves in each material, and the polarizations are longitudinal or transverse.

For the general problem here the starting point is the propagation direction, polarization and amplitude of the incident wave and the properties of the two crystals; the problem is to determine the propagation directions, polarizations and amplitudes of the reflected and refracted waves and the directions of the energy velocities (that is, of the acoustic rays). Theoretically, the solution can be found from the wave equation in each crystal and the boundary conditions at the interface. In general this can only be done numerically, so here we shall develop the equations only for some particular cases. However, first we shall consider the polarizations and propagation directions of the waves, which can be deduced without solving the equations.

4.4.1 Polarizations and Propagation Directions of Reflected and Refracted Waves

We assume that the two solids, taken to be non-piezoelectric, are firmly bonded at their plane of contact, as in Fig. 4.40. The plane has unit normal $\boldsymbol{l}(= l_1, l_2, l_3)$ and, with the origin in the plane, its equation is

$$\boldsymbol{l} \cdot \boldsymbol{x} = 0 .$$

The waves are taken to be plane waves, described by expressions of the form

$$\mathrm{e}^{\mathrm{i}(\omega t - \boldsymbol{k} \cdot \boldsymbol{x})} = \mathrm{e}^{\mathrm{i}(\omega t - k_1 x_1 - k_2 x_2 - k_3 x_3)} .$$

Continuity of the displacements u_i and mechanical tractions T_i are expressed by

$$\begin{cases} u_i^{\mathrm{I}} + \sum_R u_i^{\mathrm{R}} = \sum_{\mathrm{T}} u_i^{\mathrm{T}} & (4.84a) \\ T_i^{\mathrm{I}} + \sum_R T_i^{\mathrm{R}} = \sum_{\mathrm{T}} T_i^{\mathrm{T}} , & (4.84b) \end{cases}$$

where the superscripts I, R and T refer to incident, reflected and transmitted waves, respectively. This imposes the condition that, at each instant, $\omega_{\mathrm{R}} = \omega_{\mathrm{T}} = \omega_{\mathrm{I}}$, so that the frequencies of the reflected and transmitted waves are the same as that of the incident wave. It also requires, for each point in the plane $\boldsymbol{l} \cdot \boldsymbol{x} = 0$,

$$\boldsymbol{k}_{\mathrm{R}} \cdot \boldsymbol{x} = \boldsymbol{k}_{\mathrm{T}} \cdot \boldsymbol{x} = \boldsymbol{k}_{\mathrm{I}} \cdot \boldsymbol{x} \quad \text{or} \quad (\boldsymbol{k}_{\mathrm{R}} - \boldsymbol{k}_{\mathrm{I}}) \cdot \boldsymbol{x} = 0 , \quad (\boldsymbol{k}_{\mathrm{T}} - \boldsymbol{k}_{\mathrm{I}}) \cdot \boldsymbol{x} = 0 .$$

Thus the vectors $\boldsymbol{k}_{\mathrm{R}} - \boldsymbol{k}_{\mathrm{I}}$ and $\boldsymbol{k}_{\mathrm{T}} - \boldsymbol{k}_{\mathrm{I}}$ are, like $\boldsymbol{l}$, perpendicular to the boundary plane. Thus, *the wave vectors $\boldsymbol{k}_{\mathrm{R}}$ and $\boldsymbol{k}_{\mathrm{T}}$ of the reflected and refracted waves are contained in the plane of incidence, defined by a vector normal to the interface and the wave vector of the incident wave. The projections of $\boldsymbol{k}_{\mathrm{R}}$*

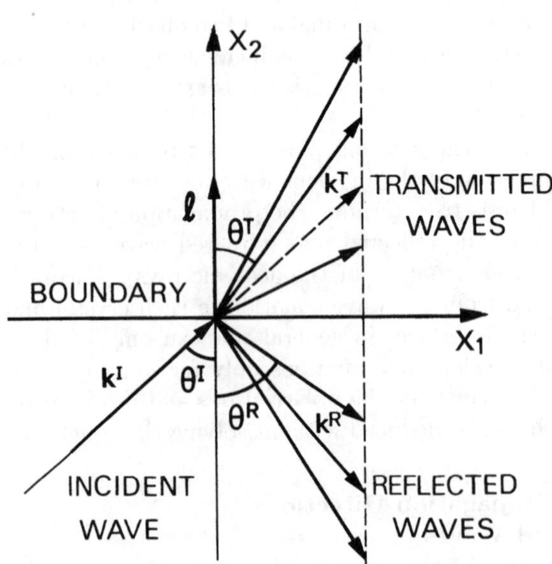

Fig. 4.40. The Snell–Descartes law. For two crystals bound together at a plane, the projections of the wave vectors $\boldsymbol{k}_{\mathrm{R}}$ and $\boldsymbol{k}_{\mathrm{T}}$ of the reflected and refracted waves on to this plane are equal to that of the wave vector of the incident wave

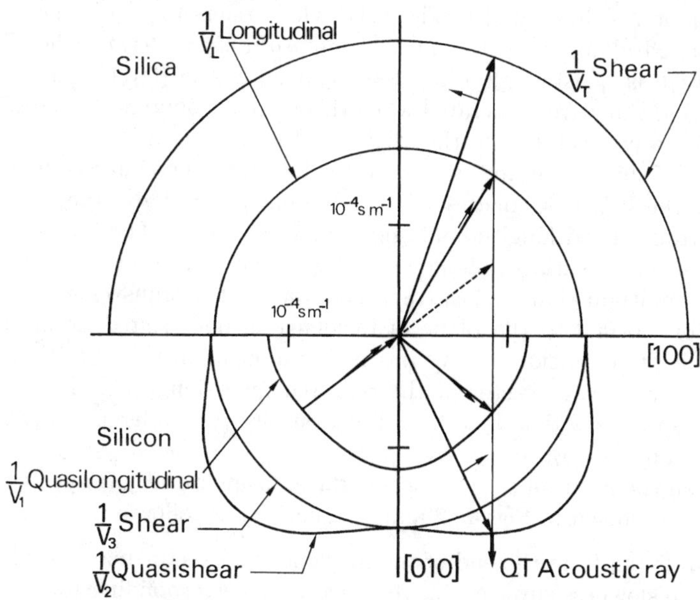

Fig. 4.41. Geometrical construction for the wave vectors and acoustic rays of the reflected and refracted waves generated by an incident longitudinal wave at a silicon-silica interface. The incident wave, which propagates in the silicon with mechanical displacement in the (001) plane, gives rise to a longitudinal wave and a transverse wave in the silica, and to a quasi-longitudinal wave and a quasi-transverse wave in the silicon. The purely transverse wave that can propagate in the silicon is not excited because its polarization is perpendicular to that of the incident wave. For the reflected quasi-transverse wave, the wave vector and the acoustic ray have different directions

and $\boldsymbol{k}_{\mathrm{T}}$ *on to the interface are equal to that of the wave vector of the incident wave*, as in Fig. 4.40.

This result expresses the Snell–Descartes law (Sect. 1.2.4.3), which can be written

$$k_{\mathrm{R}} \sin\theta_{\mathrm{R}} = k_{\mathrm{T}} \sin\theta_{\mathrm{T}} = k_{\mathrm{I}} \sin\theta_{\mathrm{I}}\,, \tag{4.85}$$

where θ_{I}, θ_{R} and θ_{T} denote the angles of incidence, reflection and refraction.

We now show that the *propagation directions* of the waves can be found graphically.

4.4.1.1 Graphical Construction. The slowness surfaces of a material provide the solutions of the wave equation for all propagation directions. Thus, once the wave vector of the incident wave is known, the Snell–Descartes law and the slowness surfaces of the two materials are sufficient to obtain, without calculation, the wave vectors for waves that can propagate in each solid, and their polarizations and the acoustic ray directions (the energy flow directions).

The construction is illustrated in Fig. 4.41, which refers to a *cubic crystal* (silicon) bonded along the (010) plane to an *isotropic material* (silica). The incident wave is quasi-longitudinal, propagating in the (001) plane of the crystal. To find the waves generated when this wave encounters the interface, we draw its wave vector from the origin, and then draw a line passing through the end of this vector and normal to the interface. The intersections of this line with the slowness 'spheres' of the isotropic material give the wave vectors of the transmitted longitudinal and transverse waves. The intersections with the slowness surfaces of the crystal give the wave vectors of the reflected quasi-longitudinal and quasi-transverse waves. The transverse wave with polarization normal to the plane of incidence is not excited because the displacement of the incident wave has no component in this direction. The acoustic ray and wave vector of the reflected quasi-longitudinal wave have almost the same direction, but these directions are very different for the reflected quasi-transverse wave.

Depending on the forms of the slowness surfaces, some rather unexpected cases can occur, as shown in Fig. 4.42 and summarized as follows:

(a) The normal drawn from the end of the incident wave vector may not intersect one of the slowness surfaces, in which case the corresponding wave is evanescent or inhomogeneous. This is expressed mathematically by an imaginary or complex solution of the wave equation for the transmitted wave, as described in Sect. 4.4.1.2 below. The wave amplitude decreases exponentially in the second medium. This situation occurs only beyond a critical angle of incidence, as in Fig. 4.42a.

(b) There can be several critical angles for reflected waves or for refracted waves. This is true even for isotropic solids, as illustrated in Fig. 4.42b.

(c) A wave may be evanescent only between two angles of incidence. As the angle of incidence increases, the wave disappears at one angle and reappears at the other.

(d) The acoustic ray vector (normal to the slowness surface) and the wave vector can be directed on different sides of the interface.

4.4.1.2 Mathematical Analysis. Considering plane waves with displacements of the form

$$u_l = {}^\circ u_l \exp\left[\mathrm{i}\omega\left(t - n_i x_i / V\right)\right]$$

the required results can of course be obtained by applying Christoffel's equation (4.7)

$$\left(c_{ijkl} n_j n_k - \rho V^2 \delta_{il}\right) {}^\circ u_l = 0 \quad i = 1, 2, 3 \tag{4.86}$$

in each of the two media. Introducing the slowness vector

$$\boldsymbol{m} = \frac{\boldsymbol{n}}{V} = \frac{\boldsymbol{k}}{\omega}, \tag{4.87}$$

we obtain

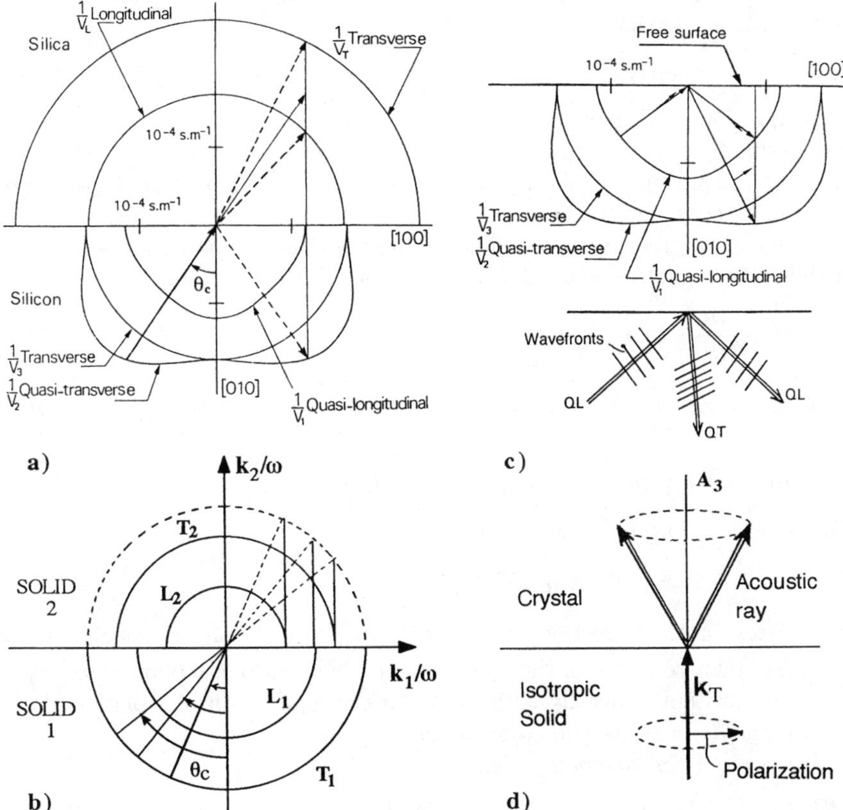

Fig. 4.42. The forms of the slowness curves can give rise to somewhat unexpected results. (**a**) Interface between a cubic crystal and an isotropic solid (silicon–silica). The incident quasi-transverse wave stops generating the quasi-longitudinal wave at the critical angle θ_c. (**b**) A junction between two isotropic solids with three critical angles for an incident transverse wave. (**c**) Wavefronts and acoustic rays at the free surface of a crystal. (**d**) A triad axis

$$u_l = \, ^\circ u_l \, \mathrm{e}^{\mathrm{i}\omega(t - m_i x_i)} ,$$

and (4.86) becomes

$$(G_{il} - \rho \, \delta_{il}) \, ^\circ u_l = 0 \quad \text{with } G_{il} = c_{ijkl} m_j m_k \, . \tag{4.88}$$

From the Snell–Descartes law, the slowness vectors of the incident, reflected and refracted waves, all in the plane of incidence, have the same projection m_1^{I} on the interface (Fig. 4.40), so

$$\boldsymbol{m} = (m_1^{\mathrm{I}}, m_2, 0) , \quad m_1^{\mathrm{I}} = \frac{n_1^{\mathrm{I}}}{V} = \frac{\sin \theta_{\mathrm{I}}}{V(\theta_{\mathrm{I}})} \, .$$

The unknown constant m_2 is given by the characteristic equation of the system (4.88), so that

$$||G_{il}-\rho\delta_{il}|| = ||c_{i11l}(m_1^{\mathrm{I}})^2-\rho\delta_{il}+(c_{i12l}+c_{i21l})m_1^{\mathrm{I}}m_2+c_{i22l}(m_2)^2|| = 0\,. \quad (4.89)$$

This equation, of degree six in m_2, must be solved separately for each of the two media. There are, a priori, six reflected waves and six transmitted waves. Since the coefficients in the equation are real, the roots m_2 must be real or in complex conjugate pairs.

If m_2 is *real*, then $\boldsymbol{m}$ is *real*. The propagation direction and phase velocity of the wave, sometimes called a homogeneous wave, are respectively

$$\boldsymbol{n} = \frac{\boldsymbol{m}}{|\boldsymbol{m}|} \quad \text{and} \quad V = \frac{1}{|\boldsymbol{m}|}\,.$$

If m_2 is *complex* we write $m_2 = m_2^{(\mathrm{r})} + \mathrm{i}m_2^{(\mathrm{i})}$, so that $\boldsymbol{m} = \boldsymbol{m}^{(\mathrm{r})} + \mathrm{i}\boldsymbol{m}^{(\mathrm{i})}$, where

$$\boldsymbol{m}^{(\mathrm{r})} = (m_1^{\mathrm{I}}, m_2^{(\mathrm{r})}, 0) \quad \text{and} \quad \boldsymbol{m}^{(\mathrm{i})} = (0, m_2^{(\mathrm{i})}, 0)\,.$$

The corresponding plane wave, given by

$$u_k = {}^{\circ}u_k\, \mathrm{e}^{\mathrm{i}\omega(t-m_i^{(\mathrm{r})}x_i)}\mathrm{e}^{\omega m_2^{(\mathrm{i})}x_2}$$

propagates in the direction $\boldsymbol{n} = \boldsymbol{m}^{(\mathrm{r})}/|\boldsymbol{m}^{(\mathrm{r})}|$, with phase velocity $V = 1/|\boldsymbol{m}^{(\mathrm{r})}|$ and attenuates in the x_2 direction. The equi-phase planes ($m_i^{(\mathrm{r})}x_i =$ constant) are not generally orthogonal to the equi-amplitude planes ($x_2 =$ constant). The wave is said to be *inhomogeneous*.

If m_2 is *purely imaginary*, then

$$m_2^{(\mathrm{r})} = 0, \quad m_2^{(\mathrm{i})} \neq 0 \quad \text{giving} \quad \boldsymbol{m}^{(\mathrm{r})} = (m_1^{\mathrm{I}}, 0, 0) \quad \text{and} \quad \boldsymbol{m}^{(\mathrm{i})} = (0, m_2^{(\mathrm{i})}, 0)\,.$$

This inhomogeneous wave is *evanescent*, with amplitude

$$u_k = {}^{\circ}u_k\, \mathrm{e}^{\mathrm{i}\omega(t-m_1^{\mathrm{I}}x_1)}\mathrm{e}^{\omega m_2^{(\mathrm{i})}x_2}\,.$$

The planes of constant phase ($x_1 =$ constant) and of constant amplitude (x_2 = constant) are orthogonal.

To be an acceptable solution, a reflected wave must also satisfy the condition that the component of the Poynting vector normal to the interface must be directed into the incident medium ($P_2^{\mathrm{R}} \leq 0$). For a refracted wave, this component must be directed into the adjacent medium ($P_2^{\mathrm{T}} \geq 0$). For an inhomogeneous wave this component is always zero [4.6] – this wave, accompanying the reflection and transmission of a homogeneous wave, carries energy along the interface. Its amplitude must decrease exponentially. These conditions require that reflected waves have $m_2^{(\mathrm{i})} > 0$ and transmitted waves have $m_2^{(\mathrm{i})} < 0$.

Since the tensor $G_{il} = c_{ijkl}m_jm_k$ is known for a given reflected or transmitted wave, the solution of (4.88) yields the polarization, i.e. the eigenvector of G_{il} with associated eigenvalue ρ determined by

$$(G_{il} - \rho\delta_{il})\,{}^{\circ}u_l = 0\,.$$

When there is a double root for the eigenvalue, the propagation direction defined by $\boldsymbol{m}$ is an acoustic axis.

4.4.2 Amplitudes. Continuity Equations

The amplitudes of the reflected and transmitted waves are found from the continuity equations (4.84a), simplified by omitting the propagation terms that are the same at all points on the interface. Continuity of displacements gives

$$\boxed{{}^{\circ}u_i^{\mathrm{I}} + \sum_{\mathrm{R}} {}^{\circ}u_i^{\mathrm{R}} = \sum_{\mathrm{T}} {}^{\circ}u_i^{\mathrm{T}}}\,. \tag{4.90}$$

For the mechanical tractions $T_i = T_{ij}l_j$, we use Hooke's law to obtain

$$T_{ij} = c_{ijkl}\frac{\partial u_k}{\partial x_l} = -\mathrm{i}c_{ijkl}k_l\,{}^{\circ}u_k\,\mathrm{e}^{\mathrm{i}(\omega t - \boldsymbol{k}\cdot\boldsymbol{x})} \tag{4.91}$$

and this gives

$$\boxed{c_{ijkl}l_j\left(k_l^{\mathrm{I}}\,{}^{\circ}u_k^{\mathrm{I}} + \sum_{\mathrm{R}} k_l^{\mathrm{R}}\,{}^{\circ}u_k^{\mathrm{R}}\right) = c'_{ijkl}l_j\sum_{\mathrm{T}} k_l^{\mathrm{T}}\,{}^{\circ}u_k^{\mathrm{T}}}\,, \tag{4.92}$$

where c'_{ijkl} denotes the stiffness constants of the adjacent crystal.

In Sect. 1.2.4.3, the angular dependences of the transmission and reflection coefficients were derived for the case of two homogeneous fluids, assuming the wave to be longitudinal, the only type of wave that can propagate in this case. These expressions are valid in the case of two isotropic solids, and also for two crystals having a symmetry axis such that the propagation is isotropic in a plane, and for transverse horizontal (TH) waves. The last case is considered in the first example below. The second example describes the reflection of a longitudinal wave at the free surface of an isotropic solid or, equivalently, of a crystal with a hexad axis normal to the plane of incidence. The other two examples refer to a solid–liquid interface, often encountered in practice.

4.4.2.1 Solid–Solid Interface. Incident TH Wave. As shown in Sect. 4.2.2, the plane normal to a tetrad or hexad axis is isotropic for transverse waves polarized along the axis, so that velocity $V_3 = (c_{44}/\rho)^{1/2}$ is independent of the propagation direction. This case, shown in Fig. 4.43, therefore gives the simplest behaviour for reflection and transmission at the boundary between two crystals. Here each crystal has a tetrad or hexad axis along $\mathrm{O}x_3$, parallel to the interface, and the incident TH wave is transverse, polarized along $\mathrm{O}x_3$. The plane of incidence is $\mathrm{O}x_1x_2$. Since the allowed displacements in the two crystals are parallel or perpendicular to the symmetry axis $\mathrm{O}x_3$,

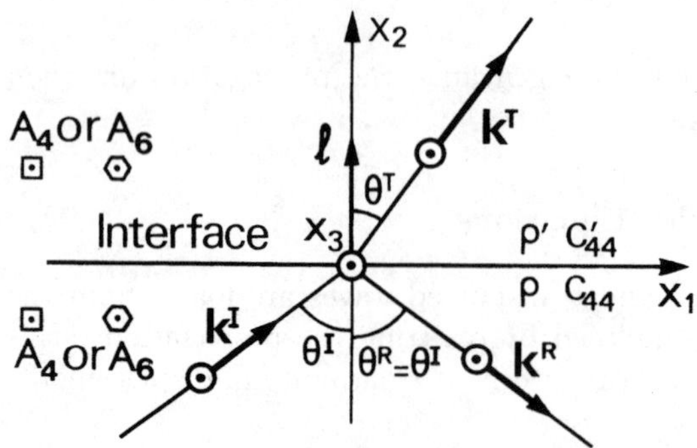

Fig. 4.43. Reflection and refraction of a transverse horizontal (TH) wave at the interface of two crystals with tetrad or hexad axes perpendicular to the plane of incidence (or at the interface of two isotropic solids)

the polarization of the incident wave is maintained – the discontinuity gives rise to only two waves, one reflected and one transmitted, with transverse polarization as for the incident wave. This observation is of course valid also for two isotropic solids. From the result in Sect. 1.2.4.3, we can write the reflection and transmission coefficients directly. However, as an example, we develop it here from (4.90) and (4.92).

Each wave has ${}^\circ u_i = {}^\circ u\, \delta_{i3}$, and continuity of displacements requires

$$ {}^\circ u_\mathrm{I} + {}^\circ u_\mathrm{R} = {}^\circ u_\mathrm{T} \,. \tag{4.93}$$

Using $l_j = \delta_{j2}$ and $u_k = {}^\circ u\, \delta_{k3}$, continuity of the mechanical tractions requires

$$ c_{i23l}\left(k_l^\mathrm{I}\ {}^\circ u_\mathrm{I} + k_l^\mathrm{R}\ {}^\circ u_\mathrm{R}\right) = c'_{i23l} k_l^\mathrm{T}\ {}^\circ u_\mathrm{T}\,, \quad l = 1, 2\,.$$

From (3.75) and (3.76), it can be seen that the tetrad or hexad axis causes all constants to be zero except for those with $i = 3$ and $l = 2$, that is, $c_{3232} = c_{44}$ and $c'_{3232} = c'_{44}$. Hence,

$$ c_{44}\left(k_2^\mathrm{I}\ {}^\circ u_\mathrm{I} + k_2^\mathrm{R}\ {}^\circ u_\mathrm{R}\right) = c'_{44} k_2^\mathrm{T}\ {}^\circ u_\mathrm{T}\,. \tag{4.94}$$

Solving (4.93) and (4.94) yields the ratios $r = {}^\circ u_\mathrm{R}/{}^\circ u_\mathrm{I}$ and $t = {}^\circ u_\mathrm{T}/{}^\circ u_\mathrm{I}$. The Snell–Descartes law demands that $\theta_\mathrm{R} = \theta_\mathrm{I}$ and $(\sin\theta_\mathrm{T})/(\sin\theta_\mathrm{I}) = V'_3/V_3$. Using the relations $k_2^\mathrm{R} = -k_2^\mathrm{I} = -k_\mathrm{I}\cos\theta_\mathrm{I}$ and $k_2^\mathrm{T} = k_\mathrm{T}\cos\theta_\mathrm{T}$, the reflection coefficient r becomes the following function of the angles of incidence and reflection:

$$ r = \frac{c_{44}k_\mathrm{I}\cos\theta_\mathrm{I} - c'_{44}k_\mathrm{T}\cos\theta_\mathrm{T}}{c_{44}k_\mathrm{I}\cos\theta_\mathrm{I} + c'_{44}k_\mathrm{T}\cos\theta_\mathrm{T}}\,. \tag{4.95}$$

Introducing the elastic impedances $Z_3 = \rho V_3$ and $Z'_3 = \rho' V'_3$ for the transverse waves in the two crystals, we have

$$ k_\mathrm{I} c_{44} = \frac{\omega}{V_3} c_{44} = \frac{\omega\rho V_3^2}{V_3} = \omega Z_3 \quad \text{and} \quad k_\mathrm{T} c'_{44} = \omega Z'_3\,.$$

These give

$$r = \frac{Z_3 \cos\theta_\mathrm{I} - Z_3' \cos\theta_\mathrm{T}}{Z_3 \cos\theta_\mathrm{I} + Z_3' \cos\theta_\mathrm{T}} \tag{4.96a}$$

and we also find

$$t = r + 1 = \frac{2Z_3 \cos\theta_\mathrm{I}}{Z_3 \cos\theta_\mathrm{I} + Z_3' \cos\theta_\mathrm{T}}\,, \tag{4.96b}$$

where θ_T is related to θ_I by the Snell–Descartes law.

These formulae resemble those of (1.74) and (1.75), and similar remarks apply – the variation of r and t with the angle of incidence θ_I depends on the impedance ratio Z_3'/Z_3 and the velocity ratio V_3'/V_3. If V_3'/V_3 is less than unity, then θ_T is less than θ_I and there is no critical angle for the transmitted wave. Moreover, if $Z_3'/Z_3 < 1$, the reflection coefficient r is zero for an angle θ_I such that $(\cos\theta_\mathrm{I})/(\cos\theta_\mathrm{T}) = Z_3'/Z_3$; for this direction, all of the energy is transmitted into the second medium. The curves in Fig. 4.44a refer to silica ($V_3' = 3\,763\,\mathrm{m/s}$, $Z_3' = 8.29\,\mathrm{MRayl}$) on silicon ($V_3 = 5\,843\,\mathrm{m/s}$, $Z_3 = 13.6\,\mathrm{MRayl}$).

If $V_3' > V_3$, the transmitted wave is evanescent when the angle of incidence θ_I exceeds the critical value θ_c, defined by $\sin\theta_\mathrm{c} = V_3/V_3'$. The component of the wave vector normal to the surface, k_2^T, is then imaginary, so that $k_2^\mathrm{T} = \mathrm{i}\chi_\mathrm{T}$, and (4.95) shows that the reflection coefficient r is complex. Since the magnitude of r is unity, there is total reflection of the incident wave for any angle of incidence larger than the critical angle. The quantity t, the ratio of the transmitted and incident wave amplitudes at the interface, no longer represents the transmission coefficient, since the transmitted wave amplitude is not constant – it decreases exponentially in the second medium. Figure 4.44b illustrates this case with the above materials interchanged, that is, for silicon on silica.

4.4.2.2 Reflection at a Free Surface. In the case of a free boundary surface, well approximated by a solid in contact with air or another gas under low pressure, there is no transmitted wave. The only boundary condition to be satisfied is that the mechanical traction must be zero at all points on the surface. Thus $T_{ij}l_j = 0$, giving

$$c_{ijkl}l_j\left(k_l^\mathrm{I}\,{}^\circ u_k^\mathrm{I} + \sum_\mathrm{R} k_l^\mathrm{R}\,{}^\circ u_k^\mathrm{R}\right) = 0\,. \tag{4.97}$$

Comparison with (4.92) shows that this condition is equivalent to setting the stiffnesses c'_{ijkl} in the second solid equal to zero. From the above example, $r = 1$ when $Z_3' = 0$, and hence *a transverse wave with polarization parallel to the free surface is totally reflected.*

At the free surface of an isotropic solid, or a hexagonal crystal with the hexad axis normal to the plane of incidence, a longitudinal wave is reflected into both a longitudinal wave and a transverse wave, as shown in Fig. 4.45. These waves are denoted here by indices L and T (instead of RL and RT,

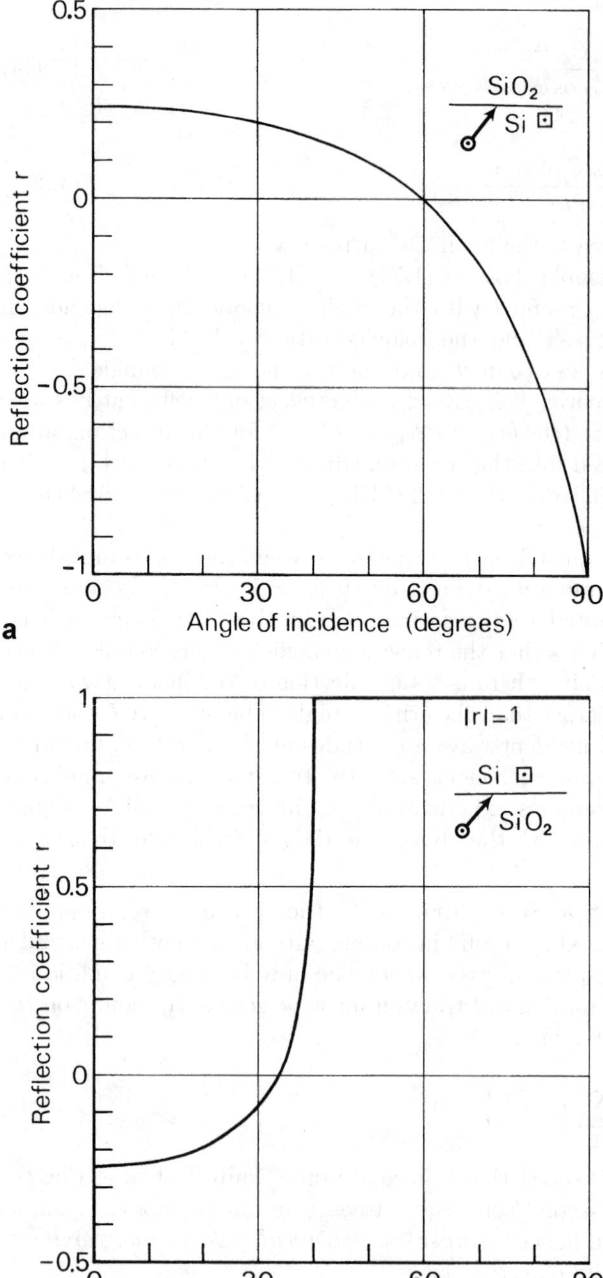

Fig. 4.44. Reflection coefficient for a TH wave as a function of angle of incidence. (**a**) A Si-SiO_2 interface. (**b**) A SiO_2-Si interface. Beyond the critical angle $\theta_c = 40°$ there is total reflection, so $|r| = 1$

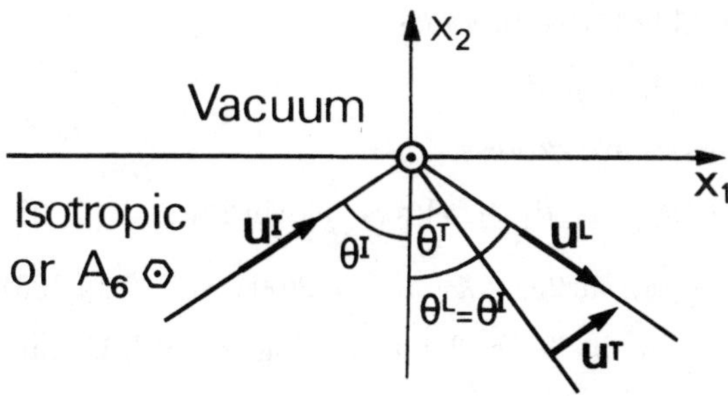

Fig. 4.45. Reflection of a plane longitudinal wave at the free surface of either an isotropic solid or a hexagonal crystal with the hexad axis normal to the plane of incidence

the R being unnecessary in the absence of a transmitted wave). Since the material is isotropic in the plane of incidence, we have

$$k_{\mathrm{L}} = k_{\mathrm{I}} = \frac{\omega}{V_{\mathrm{L}}} \quad \text{and} \quad \frac{k_{\mathrm{T}}}{k_{\mathrm{I}}} = \frac{V_{\mathrm{L}}}{V_{\mathrm{T}}}$$

and the Snell–Descartes law gives

$$\theta_{\mathrm{L}} = \theta_{\mathrm{I}} \quad \text{and} \quad V_{\mathrm{T}} \sin\theta_{\mathrm{I}} = V_{\mathrm{L}} \sin\theta_{\mathrm{T}} \,. \tag{4.98}$$

Expanding (4.91), the mechanical stresses in the x_1x_3 plane are

$$T_{i2}(x_2 = 0) = -\mathrm{i}c_{i2kl}\left(k_l^{\mathrm{I}}\,{}^\circ u_k^{\mathrm{I}} + k_l^{\mathrm{L}}\,{}^\circ u_k^{\mathrm{L}} + k_l^{\mathrm{T}}\,{}^\circ u_k^{\mathrm{T}}\right) \quad i = 1, 2, 3 \tag{4.99}$$

where the indices l and k can only take values 1 or 2 because the vectors $\boldsymbol{k}$ and ${}^\circ\boldsymbol{u}$ are confined to the x_1x_2 plane; their components are

$$\begin{array}{lll} \boldsymbol{k}_{\mathrm{I}} = k_{\mathrm{I}}(\sin\theta_{\mathrm{I}}, \cos\theta_{\mathrm{I}}) & \boldsymbol{k}_{\mathrm{L}} = k_{\mathrm{I}}(\sin\theta_{\mathrm{I}}, -\cos\theta_{\mathrm{I}}) & \boldsymbol{k}_{\mathrm{T}} = k_{\mathrm{T}}(\sin\theta_{\mathrm{T}}, -\cos\theta_{\mathrm{T}}) \\ {}^\circ\boldsymbol{u}_{\mathrm{I}} = {}^\circ u_{\mathrm{I}}(\sin\theta_{\mathrm{I}}, \cos\theta_{\mathrm{I}}) & {}^\circ\boldsymbol{u}_{\mathrm{L}} = {}^\circ u_{\mathrm{L}}(\sin\theta_{\mathrm{I}}, -\cos\theta_{\mathrm{I}}) & {}^\circ\boldsymbol{u}_{\mathrm{T}} = {}^\circ u_{\mathrm{T}}(\cos\theta_{\mathrm{T}}, \sin\theta_{\mathrm{T}}) \end{array} .$$

For $i = 1$ the only non-zero stiffness constants are $c_{1212} = c_{1221} = c_{66}$, so

$$T_{12} = -\mathrm{i}c_{66}\left[k_{\mathrm{I}}\left({}^\circ u_{\mathrm{I}} - {}^\circ u_{\mathrm{L}}\right)\sin 2\theta_{\mathrm{I}} - k_{\mathrm{T}}\,{}^\circ u_{\mathrm{T}} \cos 2\theta_{\mathrm{T}}\right] .$$

For $i = 2$ the only constants involved in (4.99) are $c_{2222} = c_{11}$ and $c_{2211} = c_{12}$, so

$$\begin{aligned} T_{22} = &- \mathrm{i}c_{11}\left[k_{\mathrm{I}}({}^\circ u_{\mathrm{I}} + {}^\circ u_{\mathrm{L}})\cos^2\theta_{\mathrm{I}} - (k_{\mathrm{T}}\,{}^\circ u_{\mathrm{T}}/2)\sin 2\theta_{\mathrm{T}}\right] \\ &- \mathrm{i}c_{12}\left[k_{\mathrm{I}}({}^\circ u_{\mathrm{I}} + {}^\circ u_{\mathrm{L}})\sin^2\theta_{\mathrm{I}} + (k_{\mathrm{T}}\,{}^\circ u_{\mathrm{T}}/2)\sin 2\theta_{\mathrm{T}}\right] . \end{aligned} \tag{4.100}$$

For $i = 3$, all the terms in (4.99) are zero.

In (4.100), the multiplier of $-\mathrm{i}k_{\mathrm{I}}({}^\circ u_{\mathrm{I}} + {}^\circ u_{\mathrm{L}})$, denoted F, can be written

$$F = c_{11}\cos^2\theta_{\mathrm{I}} + c_{12}\sin^2\theta_{\mathrm{I}} = c_{11} + (c_{12} - c_{11})\sin^2\theta_{\mathrm{I}}$$

and, introducing the velocities $V_{\mathrm{L}} = (c_{11}/\rho)^{1/2}$ and $V_{\mathrm{T}} = [(c_{11} - c_{12})/2\rho]^{1/2}$ of longitudinal and shear waves,

$$F = \rho V_{\mathrm{L}}^2 - 2\rho V_{\mathrm{T}}^2 \sin^2\theta_{\mathrm{I}} \,. \tag{4.101}$$

With the Snell–Descartes law (4.98), this becomes

$$F = \rho V_{\rm L}^2 (1 - 2\sin^2\theta_{\rm T}) = \rho V_{\rm L}^2 \cos 2\theta_{\rm T} \,.$$

At the surface $x_2 = 0$, the stresses are expressed as

$$\left\{ \begin{array}{ll} T_{22} = -{\rm i}\rho\, V_{\rm L}^2 k_{\rm I}({}^\circ u_{\rm I} + {}^\circ u_{\rm L}) \cos 2\theta_{\rm T} + {\rm i}\rho\, V_{\rm T}^2 k_{\rm T}\, {}^\circ u_{\rm T} \sin 2\theta_{\rm T} & (4.102) \\ T_{12} = -{\rm i}\rho\, V_{\rm T}^2 [k_{\rm I}({}^\circ u_{\rm I} - {}^\circ u_{\rm L}) \sin 2\theta_{\rm I} - k_{\rm T}\, {}^\circ u_{\rm T} \cos 2\theta_{\rm T}] \,. & (4.103) \end{array} \right.$$

Replacing $k_{\rm I}$ by $\omega/V_{\rm L}$ and $k_{\rm T}$ by $\omega/V_{\rm T}$, and introducing $Z_{\rm L} = \rho V_{\rm L}$ and $Z_{\rm T} = \rho\, V_{\rm T}$, these become

$$\left\{ \begin{array}{ll} T_{22} = -{\rm i}\omega[({}^\circ u_{\rm I} + {}^\circ u_{\rm L}) Z_{\rm L} \cos 2\theta_{\rm T} - \; {}^\circ u_{\rm T} Z_{\rm T} \sin 2\theta_{\rm T}] & (4.104) \\ T_{12} = -{\rm i}\omega \dfrac{V_{\rm T}}{V_{\rm L}} [({}^\circ u_{\rm I} - {}^\circ u_{\rm L}) Z_{\rm T} \sin 2\theta_{\rm I} - \; {}^\circ u_{\rm T} Z_{\rm L} \cos 2\theta_{\rm T}] \,. & (4.105) \end{array} \right.$$

The two unknowns are (with the first index L indicating that the incident wave is longitudinal)

$$r_{\rm LL} = \frac{{}^\circ u_{\rm L}}{{}^\circ u_{\rm I}} \quad \text{and} \quad r_{\rm LT} = \frac{{}^\circ u_{\rm T}}{{}^\circ u_{\rm I}} \,. \tag{4.106}$$

Applying the condition of zero surface stresses, these need to satisfy the equations

$$\left\{ \begin{array}{l} r_{\rm LL}\, Z_{\rm L} \cos 2\theta_{\rm T} - r_{\rm LT}\, Z_{\rm T} \sin 2\theta_{\rm T} = -Z_{\rm L} \cos 2\theta_{\rm T} \\ r_{\rm LL}\, Z_{\rm T} \sin 2\theta_{\rm I} + r_{\rm LT}\, Z_{\rm L} \cos 2\theta_{\rm T} = Z_{\rm T} \sin 2\theta_{\rm I} \,. \end{array} \right.$$

Solving these, the reflection coefficient for the longitudinal wave is

$$r_{\rm LL} = \frac{{}^\circ u_{\rm L}}{{}^\circ u_{\rm I}} = \frac{v^2 \sin 2\theta_{\rm I} \sin 2\theta_{\rm T} - \cos^2 2\theta_{\rm T}}{v^2 \sin 2\theta_{\rm I} \sin 2\theta_{\rm T} + \cos^2 2\theta_{\rm T}} \quad \text{with} \quad v = \frac{V_{\rm T}}{V_{\rm L}} \,, \tag{4.107}$$

and the coefficient for conversion into the transverse wave is

$$r_{\rm LT} = \frac{{}^\circ u_{\rm T}}{{}^\circ u_{\rm I}} = \frac{2v \sin 2\theta_{\rm I} \cos 2\theta_{\rm T}}{v^2 \sin 2\theta_{\rm I} \sin 2\theta_{\rm T} + \cos^2 2\theta_{\rm T}} \quad \text{with} \quad \sin\theta_{\rm T} = v \sin\theta_{\rm I} \,. \tag{4.108}$$

For $\theta_{\rm I} = 0$ we find $r_{\rm LT} = 0$ and $r_{\rm LL} = -1$, so that ${}^\circ\boldsymbol{u}_{\rm L} = {}^\circ\boldsymbol{u}_{\rm I}$ with the vectors oriented as in Fig. 4.45. Thus, for normal incidence, the longitudinal wave is totally reflected. The mechanical displacements are added in phase at the free surface. Since $V_{\rm T} < V_{\rm L}/\sqrt{2}$ the angle $\theta_{\rm T}$ is less than $\pi/4$, and the coefficient $r_{\rm LT}$ is positive for any angle of incidence $\theta_{\rm I}$. In contrast, $r_{\rm LL}$ is zero for two values of $\theta_{\rm I}$ if $V_{\rm T}/V_{\rm L}$ exceeds 0.565, as shown by the family of curves in Fig. 4.46. At these points, the incident longitudinal wave is reflected into a single transversely-polarized wave. *This property can be exploited to convert a longitudinal wave into a transverse wave.*

In a similar way, an incident *transverse* wave with amplitude ${}^\circ u_{\rm I}$, polarized in the plane of incidence (i.e. a TV wave), is reflected into a transverse wave with amplitude ${}^\circ u_{\rm T}$ and converted into a longitudinal wave with amplitude ${}^\circ u_{\rm L}$, as in Fig. 4.47. The reflection and conversion coefficients are given

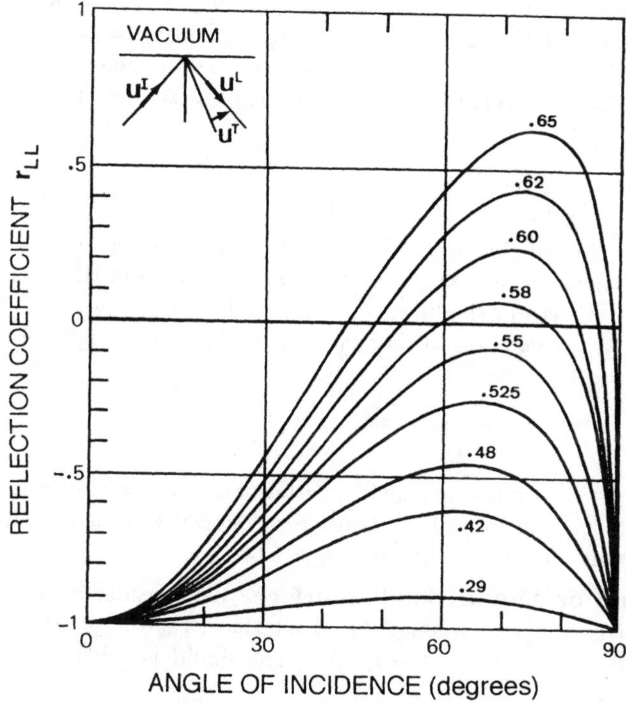

Fig. 4.46. Reflection coefficient of a longitudinal wave at the free surface of an isotropic solid, as a function of the angle of incidence, for various values of the parameter V_T/V_L. For silica, $V_T/V_L = 0.63$. (Fig. 3 of [4.7])

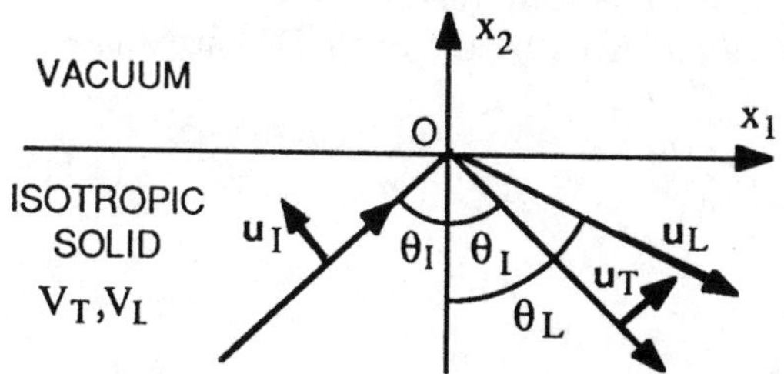

Fig. 4.47. Reflection of a plane transverse vertical wave at the free surface of an isotropic solid

by

$$r_{TT} = \frac{{}^{\circ}u_T}{{}^{\circ}u_I} = \frac{v^2 \sin 2\theta_I \sin 2\theta_L - \cos^2 2\theta_I}{v^2 \sin 2\theta_I \sin 2\theta_L + \cos^2 2\theta_I} \quad \text{with} \quad v = \frac{V_T}{V_L} \tag{4.109}$$

$$r_{TL} = \frac{{}^{\circ}u_L}{{}^{\circ}u_I} = \frac{-v \sin 4\theta_I}{v^2 \sin 2\theta_I \sin 2\theta_L + \cos^2 2\theta_I} \quad \text{with} \quad \sin\theta_L = \frac{1}{v}\sin\theta_I \,. \tag{4.110}$$

For $\theta_{\mathrm{I}} = 0$ these give $r_{\mathrm{TL}} = 0$ and $r_{\mathrm{TT}} = -1$, so that ${}^{\circ}\boldsymbol{u}_{\mathrm{T}} = {}^{\circ}\boldsymbol{u}_{\mathrm{I}}$ with the vectors oriented as in Fig. 4.47. For normal incidence, the TV displacement is totally reflected, without a change of sign. As the angle of incidence is increased, θ_{L} reaches $\pi/2$ for a critical angle θ_{c}, always less than $\pi/4$, given by

$$\sin\theta_{\mathrm{c}} = v = \frac{V_{\mathrm{T}}}{V_{\mathrm{L}}} < \frac{1}{\sqrt{2}} .$$

For example, $\theta_{\mathrm{c}} = 30°$ for an isotropic solid with $V_{\mathrm{L}} = 2V_{\mathrm{T}}$. As θ_{I} increases from 0 to θ_{c}, the coefficient r_{TL} is always negative, since $\sin 4\theta_{\mathrm{I}} > 0$. However, r_{TT}, which equals -1 at both of the limits $\theta_{\mathrm{I}} = 0, \pi/2$, is zero for two values of θ_{I} if the ratio $V_{\mathrm{T}}/V_{\mathrm{L}}$ exceeds 0.565. For $\theta_{\mathrm{I}} = \theta_{\mathrm{c}}$ we have $\theta_{\mathrm{L}} = \pi/2$, and (4.109) and (4.110) lead to

$$r_{\mathrm{TT}} = -1 \quad \text{and} \quad r_{\mathrm{TL}} = -\frac{4v^2\sqrt{1-v^2}}{1-2v^2} .$$

Beyond the critical angle, the reflection coefficient becomes complex but gives $|r_{\mathrm{TT}}| = 1$; thus, the total reflection of the transverse vertical wave is accompanied by a phase change, equal to $-\pi$ when $\theta_{\mathrm{I}} = \pi/4$.

4.4.2.3 Solid–Liquid or Liquid–Solid Interface. We consider an incident longitudinal wave, firstly at a solid–liquid interface, then at a liquid–solid interface, taking the solid to be isotropic. The liquid is taken to be non-viscous.

(a) Interface between isotropic solid and liquid, with incident longitudinal wave. The variables for the longitudinal wave in the liquid, Fig. 4.48, are written without indices, and its wavenumber is $k = \omega/c$. As before, it is not necessary to use the subscript R to indicate reflected waves.
The directions of the wave vectors are given by the Snell–Descartes law, so that

$$\theta_{\mathrm{L}} = \theta_{\mathrm{I}} , \qquad \frac{\sin\theta}{c} = \frac{\sin\theta_{\mathrm{T}}}{V_{\mathrm{T}}} = \frac{\sin\theta_{\mathrm{I}}}{V_{\mathrm{L}}} . \tag{4.111}$$

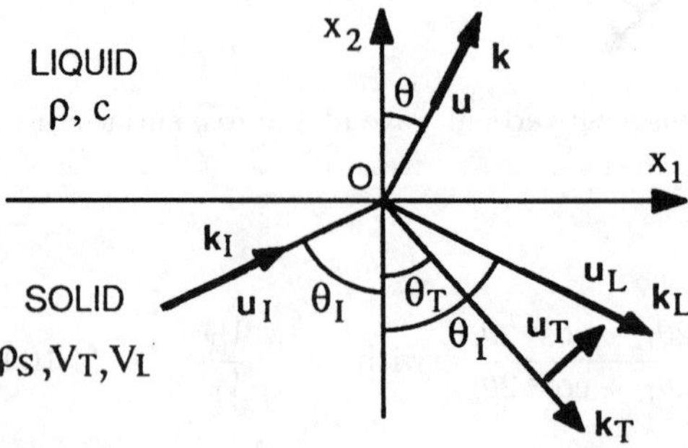

Fig. 4.48. Interface between an isotropic solid and a liquid. The incident wave is longitudinal

Comparing with the previous case, we note that

(1) there is a transmitted longitudinal wave in the liquid, with wave vector and displacement

$$\boldsymbol{k} = (k\sin\theta, k\cos\theta) \quad \boldsymbol{u} = (u\sin\theta, u\cos\theta)\,; \tag{4.112}$$

(2) the normal stress at $x_2 = 0$ is no longer zero – it is equal and opposite to the acoustic overpressure δp in the liquid, so that

$$T_{22} = -\delta p = -Z\dot{u} = -\mathrm{i}\omega Z u\,, \tag{4.113}$$

where $Z = \rho c$ is the acoustic impedance of the liquid;

(3) the tangential stresses are zero because the liquid is non-viscous, so that $T_{12} = 0$; also, T_{32} is necessarily zero;

(4) the normal component of displacement is continuous, so

$$\,^{\circ}u_{\mathrm{I}}\cos\theta_{\mathrm{I}} - \,^{\circ}u_{\mathrm{L}}\cos\theta_{\mathrm{I}} + \,^{\circ}u_{\mathrm{T}}\sin\theta_{\mathrm{T}} = \,^{\circ}u\cos\theta\,. \tag{4.114}$$

Using the expression (4.104) for T_{22} in the solid, (4.113) gives

$$(\,^{\circ}u_{\mathrm{I}} + \,^{\circ}u_{\mathrm{L}})Z_{\mathrm{L}}\cos 2\theta_{\mathrm{T}} - \,^{\circ}u_{\mathrm{T}}Z_{\mathrm{T}}\sin 2\theta_{\mathrm{T}} = Z\,^{\circ}u\,. \tag{4.115}$$

Setting the tangential stress (4.105) to zero gives

$$(\,^{\circ}u_{\mathrm{I}} - \,^{\circ}u_{\mathrm{L}})Z_{\mathrm{T}}\sin 2\theta_{\mathrm{I}} - \,^{\circ}u_{\mathrm{T}}Z_{\mathrm{L}}\cos 2\theta_{\mathrm{T}} = 0\,. \tag{4.116}$$

The coefficients for reflection into a longitudinal wave, r_{LL}, and for conversion into a transverse wave, r_{LT}, were defined in (4.106). From (4.116), these are related by

$$r_{\mathrm{LT}} = \frac{\,^{\circ}u_{\mathrm{T}}}{\,^{\circ}u_{\mathrm{I}}} = \frac{Z_{\mathrm{T}}\sin 2\theta_{\mathrm{I}}}{Z_{\mathrm{L}}\cos 2\theta_{\mathrm{T}}}(1 - r_{\mathrm{LL}})\,. \tag{4.117}$$

Substituting this into (4.114) leads to

$$\left(\cos\theta_{\mathrm{I}} + \frac{Z_{\mathrm{T}}\sin 2\theta_{\mathrm{I}}}{Z_{\mathrm{L}}\cos 2\theta_{\mathrm{T}}}\sin\theta_{\mathrm{T}}\right)(1 - r_{\mathrm{LL}}) = t\cos\theta\,,$$

where $t = \,^{\circ}u/\,^{\circ}u_{\mathrm{I}}$ is the transmission coefficient for the longitudinal wave in the liquid. The Snell–Descartes law gives

$$Z_{\mathrm{T}}\sin 2\theta_{\mathrm{I}} = 2\rho_{\mathrm{S}}V_{\mathrm{T}}\sin\theta_{\mathrm{I}}\cos\theta_{\mathrm{I}} = 2Z_{\mathrm{L}}\sin\theta_{\mathrm{T}}\cos\theta_{\mathrm{I}}\,, \tag{4.118}$$

and with $\cos 2\theta_{\mathrm{T}} = 1 - 2\sin^2\theta_{\mathrm{T}}$ we find

$$(1 - r_{\mathrm{LL}})\cos\theta_{\mathrm{I}} = t\cos\theta\cos 2\theta_{\mathrm{T}}\,. \tag{4.119}$$

We also have, eliminating $\,^{\circ}u_{\mathrm{T}}$ from (4.115) and (4.116),

$$Z_{\mathrm{L}}[(1 + r_{\mathrm{LL}})\cos^2 2\theta_{\mathrm{T}} - (1 - r_{\mathrm{LL}})v^2\sin 2\theta_{\mathrm{I}}\sin 2\theta_{\mathrm{T}}] = Zt\cos 2\theta_{\mathrm{T}}\,.$$

The left side of this equation can be written in terms of the numerator N and denominator D of (4.107), which is the reflection coefficient r_{LL} for a longitudinal wave reflected at a free surface. Thus,

$$(Dr_{\mathrm{LL}} - N)Z_{\mathrm{L}} = Zt\cos 2\theta_{\mathrm{T}}\,. \tag{4.120}$$

Dividing corresponding sides of this by (4.119) gives

$$\frac{Dr_{\mathrm{LL}} - N}{1 - r_{\mathrm{LL}}} = \frac{Z \cos\theta_{\mathrm{I}}}{Z_{\mathrm{L}} \cos\theta} = R \quad \text{so} \quad r_{\mathrm{LL}} = \frac{N + R}{D + R} .$$

Thus the *reflection coefficient* for the longitudinal wave is

$$r_{\mathrm{LL}} = \frac{v^2 \sin 2\theta_{\mathrm{I}} \sin 2\theta_{\mathrm{T}} - \cos^2 2\theta_{\mathrm{T}} + \dfrac{Z \cos\theta_{\mathrm{I}}}{Z_{\mathrm{L}} \cos\theta}}{v^2 \sin 2\theta_{\mathrm{I}} \sin 2\theta_{\mathrm{T}} + \cos^2 2\theta_{\mathrm{T}} + \dfrac{Z \cos\theta_{\mathrm{I}}}{Z_{\mathrm{L}} \cos\theta}} \quad \text{with} \quad v = \frac{V_{\mathrm{T}}}{V_{\mathrm{L}}} . \tag{4.121}$$

Using (4.117), the *conversion coefficient*, giving the transverse wave amplitude, is

$$r_{\mathrm{LT}} = \frac{2v \sin 2\theta_{\mathrm{I}} \cos 2\theta_{\mathrm{T}}}{v^2 \sin 2\theta_{\mathrm{I}} \sin 2\theta_{\mathrm{T}} + \cos^2 2\theta_{\mathrm{T}} + \dfrac{Z \cos\theta_{\mathrm{I}}}{Z_{\mathrm{L}} \cos\theta}} \quad \text{with } \sin\theta_{\mathrm{T}} = v \sin\theta_{\mathrm{I}} . \tag{4.122}$$

From (4.119) and (4.121), the *transmission coefficient*, giving the longitudinal wave in the liquid, is

$$t = \frac{{}^{\circ}u}{{}^{\circ}u_{\mathrm{I}}} = \frac{2 \cos 2\theta_{\mathrm{T}} \dfrac{\cos\theta_{\mathrm{I}}}{\cos\theta}}{v^2 \sin 2\theta_{\mathrm{I}} \sin 2\theta_{\mathrm{T}} + \cos^2 2\theta_{\mathrm{T}} + \dfrac{Z \cos\theta_{\mathrm{I}}}{Z_{\mathrm{L}} \cos\theta}} .$$

The reader can verify that, for $Z \to 0$, the formulae (4.121) and (4.122) lead back to the corresponding expressions (4.107) and (4.108) for a free surface. Although there is no wave transmitted into the liquid in this case, the transmission coefficient is not zero because the displacement of the free surface is finite. It is more appropriate to consider the transmission coefficient for the acoustic pressure, given by

$$t_{\mathrm{p}} = \frac{Z}{Z_{\mathrm{L}}} t = \frac{2 \cos 2\theta_{\mathrm{T}} \dfrac{Z \cos\theta_{\mathrm{I}}}{Z_{\mathrm{L}} \cos\theta}}{v^2 \sin 2\theta_{\mathrm{I}} \sin 2\theta_{\mathrm{T}} + \cos^2 2\theta_{\mathrm{T}} + \dfrac{Z \cos\theta_{\mathrm{I}}}{Z_{\mathrm{L}} \cos\theta}} . \tag{4.123}$$

Note that for normal incidence this relation leads to the simple expression (1.61) in Sect. 1.2.4.2. Figure 4.49 shows the three coefficients r_{LL}, r_{LT} and t_{p} as functions of the angle of incidence in the solid, for a liquid and solid such that

$$v = \frac{V_{\mathrm{T}}}{V_{\mathrm{L}}} = 0.5 , \quad c = 0.25\, V_{\mathrm{L}} \quad \text{and} \quad Z = 0.09 Z_{\mathrm{L}} .$$

These correspond approximately to the case of duralumin and water.

(b) Interface between liquid and isotropic solid. The incident wave must of course be a longitudinal wave here, as in Fig. 4.50. This situation is encountered in non-destructive evaluation of metals (or other materials). The sample, immersed in water, is irradiated by an ultrasonic beam which can

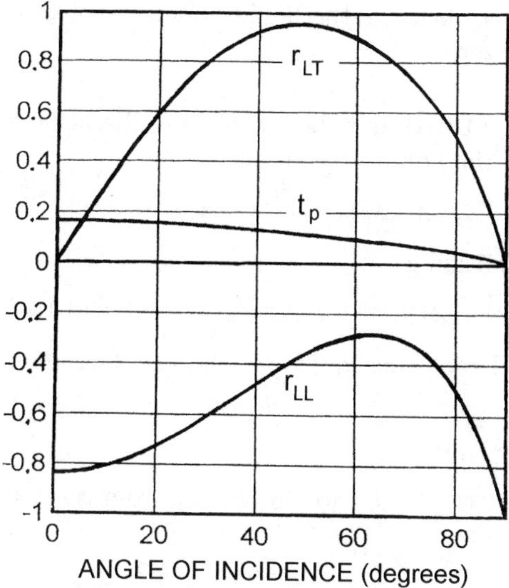

Fig. 4.49. Interface between an isotropic solid and liquid, with longitudinal wave incident in the solid. The coefficients for reflection r_{LL}, conversion into transverse waves, r_{LT}, and for transmission into the liquid, t_{p}

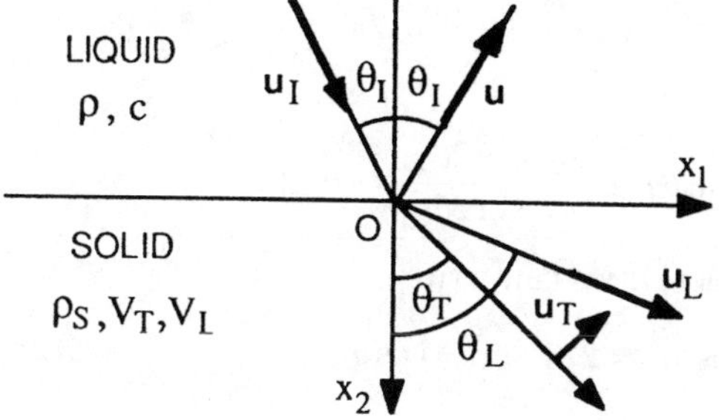

Fig. 4.50. Interface between a liquid and an isotropic solid

generate longitudinal and transverse waves in the metal. The proportions of these waves depend on the angle of incidence. The notation is simplified here by omitting the index T for transmitted waves.

The boundary conditions at the plane $x_2 = 0$ impose the following:

(1) Continuity of the normal component of displacement requires

$$^{\circ}u_{\mathrm{L}} \cos\theta_{\mathrm{L}} - {^{\circ}u_{\mathrm{T}}} \sin\theta_{\mathrm{T}} = ({^{\circ}u_{\mathrm{I}}} - {^{\circ}u}) \cos\theta_{\mathrm{I}} \,. \tag{4.124}$$

(2) Continuity of normal stress requires

$$T_{22} = -\delta p = -\rho c({^{\circ}\dot{u}_{\mathrm{I}}} + {^{\circ}\dot{u}}) = -\mathrm{i}\omega Z({^{\circ}u_{\mathrm{I}}} + {^{\circ}u}) \,.$$

(3) The tangential stress must be zero because the liquid is non-viscous, so $T_{12} = 0$. Also, T_{32} is necessarily zero.

For an isotropic solid T_{22} is given by (4.104), but the $°u_\mathrm{I}$ term is absent here because the incident wave is now in the liquid, so that

$$T_{22} = -\mathrm{i}\omega(°u_\mathrm{L} Z_\mathrm{L} \cos 2\theta_\mathrm{T} - °u_\mathrm{T} Z_\mathrm{T} \sin 2\theta_\mathrm{T}),$$

and hence the second boundary condition gives

$$°u_\mathrm{L} Z_\mathrm{L} \cos 2\theta_\mathrm{T} - °u_\mathrm{T} Z_\mathrm{T} \sin 2\theta_\mathrm{T} = Z(°u_\mathrm{I} + °u). \tag{4.125}$$

Similarly, using (4.105) for T_{12} and replacing θ_I by θ_L, the last boundary condition becomes

$$°u_\mathrm{L} Z_\mathrm{T} \sin 2\theta_\mathrm{L} + °u_\mathrm{T} Z_\mathrm{L} \cos 2\theta_\mathrm{T} = 0. \tag{4.126}$$

To solve the three equations (4.124)–(4.126), we derive $°u_\mathrm{L}$ from the last equation, giving

$$°u_\mathrm{L} = -\,°u_\mathrm{T} \frac{Z_\mathrm{L}}{Z_\mathrm{T}} \frac{\cos 2\theta_\mathrm{T}}{\sin 2\theta_\mathrm{L}} \tag{4.127}$$

and substitute this into (4.125) to obtain

$$Z(°u_\mathrm{I} + °u) = -\,°u_\mathrm{T} Z_\mathrm{T} \left(\sin 2\theta_\mathrm{T} + \frac{Z_\mathrm{L}^2}{Z_\mathrm{T}^2} \frac{\cos^2 2\theta_\mathrm{T}}{\sin 2\theta_\mathrm{L}} \right)$$

and into (4.124) to give

$$(°u_\mathrm{I} - °u) \cos\theta_\mathrm{I} = -\,°u_\mathrm{T} \left(\sin\theta_\mathrm{T} + \frac{Z_\mathrm{L}}{Z_\mathrm{T}} \frac{\cos 2\theta_\mathrm{T}}{\sin 2\theta_\mathrm{L}} \cos\theta_\mathrm{L} \right).$$

Using the Snell–Descartes law (4.111), we can write

$$Z_\mathrm{T} \sin 2\theta_\mathrm{L} = 2\rho_\mathrm{S} V_\mathrm{T} \sin\theta_\mathrm{L} \cos\theta_\mathrm{L} = 2 Z_\mathrm{L} \sin\theta_\mathrm{T} \cos\theta_\mathrm{L}, \tag{4.128}$$

and this enables the first of the above two equations to be expressed as

$$°u_\mathrm{I} + °u = -\,°u_\mathrm{T} \frac{Z_\mathrm{L}^2}{Z} \frac{N}{Z_\mathrm{T} \sin 2\theta_\mathrm{L}} = -\,°u_\mathrm{T} \frac{Z_\mathrm{L}}{Z} \frac{N}{2 \sin\theta_\mathrm{T} \cos\theta_\mathrm{L}}, \tag{4.129}$$

where

$$N = v^2 \sin 2\theta_\mathrm{T} \sin 2\theta_\mathrm{L} + \cos^2 2\theta_\mathrm{T}. \tag{4.130}$$

We also find

$$°u_\mathrm{I} - °u = -\,°u_\mathrm{T} \frac{2\sin^2\theta_\mathrm{T} + \cos 2\theta_\mathrm{T}}{2 \sin\theta_\mathrm{T} \cos\theta_\mathrm{I}} = -\frac{°u_\mathrm{T}}{2 \sin\theta_\mathrm{T} \cos\theta_\mathrm{I}}.$$

The coefficients for reflection of the longitudinal wave in the liquid, for transmission into the solid, and for conversion into a transverse wave in the solid, are obtained from

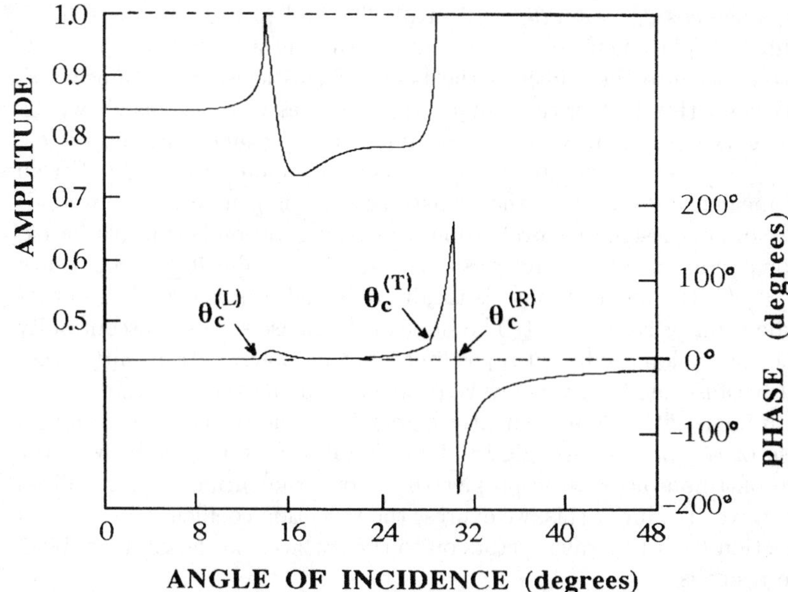

Fig. 4.51. Water–duralumin interface. Amplitude and phase of the reflection coefficient r_{LL} as a function of the angle of incidence of the longitudinal wave in the water

$$\frac{1-r_{LL}}{1+r_{LL}} = \frac{R}{N} \quad \text{with} \quad R = \frac{Z\cos\theta_L}{Z_L\cos\theta_I}, \tag{4.131}$$

which gives

$$r_{LL} = \frac{{}^\circ u}{{}^\circ u_I} = \frac{N-R}{N+R} = \frac{v^2 \sin 2\theta_T \sin 2\theta_L + \cos^2 2\theta_T - \dfrac{Z\cos\theta_L}{Z_L\cos\theta_I}}{v^2 \sin 2\theta_T \sin 2\theta_L + \cos^2 2\theta_T + \dfrac{Z\cos\theta_L}{Z_L\cos\theta_I}}. \tag{4.132}$$

From (4.129) we have

$$t_{LT} = \frac{{}^\circ u_T}{{}^\circ u_I} = -2\frac{ZZ_T}{Z_L^2} \cdot \frac{\sin 2\theta_L}{N+R} \tag{4.133}$$

and from (4.127),

$$t_{LL} = \frac{{}^\circ u_L}{{}^\circ u_I} = 2\frac{Z}{Z_L} \cdot \frac{\cos 2\theta_T}{N+R}. \tag{4.134}$$

Figure 4.51 illustrates the case in which the wave in the liquid is slower than the transverse wave in the solid – the liquid is water, with $c = 1\,480\,\mathrm{m/s}$, and the solid is duralumin, with $V_T = 3\,100\,\mathrm{m/s}$ and $V_L = 6\,300\,\mathrm{m/s}$. As θ_I increases, the coefficient r_{LL} is almost constant and remains real up to the critical angle $\theta_c^L (= 13.6°)$ for which $\theta_L = \pi/2$. At this point $r_{LL} = 1$ and

$t_{LT} = 0$, since $\cos\theta_L = \sin 2\theta_L = 0$ in (4.132) and (4.133). In the solid, a longitudinal displacement propagates along the surface, but no power is transmitted because the width of the beam of elastic waves diminishes as $\cos\theta_L$. Beyond this first critical angle, r_{LL} becomes complex, as shown by the phase curve in Fig. 4.51; the longitudinal wave is evanescent and the coefficient t_{LT} for conversion into the transverse wave increases rapidly. For an angle of incidence $\theta_I \approx 19.7°$, the transverse wave is generated at an angle $\theta_T = 45°$. In non-destructive evaluation, this configuration is suitable for examination of cracks. When θ_I reaches a second critical value $\theta_c^T (= 28.5°)$ such that $\theta_T = \pi/2$, the incident wave is totally reflected; r_{LL} is complex, but its magnitude is unity, so $|r_{LL}| = 1$. The slope of the phase shows a discontinuity here. The conversion coefficient t_{LT} is finite, but no power is transmitted into the solid because the transverse wave propagates along the surface.

For an angle $\theta_c^R (= 30.6°)$, slightly higher than the second critical angle, the phase of r_{LL} passes through 2π. This singularity is associated with the existence of a non-plane wave propagating along the surface, a generalized Rayleigh wave. To clarify this, we express the reflection coefficient in terms of the projection k_1 of the wave vectors on to the surface. As shown in Problem 4.13, the result is

$$r_{LL}(k_1) = \frac{R(k_1) - \mathrm{i}S(k_1)}{R(k_1) + \mathrm{i}S(k_1)}, \tag{4.135}$$

where

$$R(k_1) = (2k_1^2 - k_T^2)^2 - 4k_1^2(k_1^2 - k_T^2)^{1/2}(k_1^2 - k_L^2)^{1/2} \tag{4.136}$$

and

$$S(k_1) = \frac{\rho}{\rho_S} k_T^4 \left(\frac{k_1^2 - k_L^2}{k^2 - k_1^2} \right)^{1/2} . \tag{4.137}$$

It is shown in Problem 4.13 that the equation $R(k_1) = 0$ is the same as (5.43) in Chap. 5 below, giving the velocity V_R of Rayleigh waves on the surface of an isotropic solid. A real solution is obtained only for $V_R < V_T < V_L$, that is, for $k_R > k_T > k_L$ (Sect. 5.3.2). In the presence of a liquid, the roots of the numerator and denominator of (4.135) are complex conjugate of each other [4.9], giving

$$k_{1N} = \beta - \mathrm{i}\alpha \quad \text{and} \quad k_{1D} = \beta + \mathrm{i}\alpha, \quad \text{with} \quad \beta \approx k_R \quad \text{and} \quad \alpha \ll k_R .$$

When k_1 is in the region of k_R, the magnitude of $r_{LL}(k_1)$ is proportional to $(k_1 - k_{1N})/(k_1 - k_{1D})$. As k_1 passes through k_R, the magnitude is unity and the phase changes from π to $-\pi$; this occurs when the angle of incidence θ_I equals the Rayleigh angle θ_c^R defined by

$$k_1 = k \sin\theta_c^R = k_R \quad \text{giving} \quad \sin\theta_c^R = \frac{k_R}{k} = \frac{c}{V_R} .$$

The fact that the roots become complex expresses the attenuation of the Rayleigh wave due to radiation into the liquid, as shown in Problem 5.7.

Problems

4.1. Show that the equation of motion for a particle, (4.1) reduces to the wave equation (1.25) for a continuous medium when $\lambda \gg a$.

Solution. If $\lambda \gg a$, the difference $u_{n+1} - u_n$ becomes an infinitesimal quantity du, so that

$$\frac{u_{n+1} - u_n}{a} = \left(\frac{\partial u}{\partial x}\right)_{x_{n+1}} \quad \text{and} \quad \frac{u_n - u_{n-1}}{a} = \left(\frac{\partial u}{\partial x}\right)_{x_n} .$$

Equation (4.1) becomes

$$M\frac{\partial^2 u}{\partial t^2} = Ka\left[\left(\frac{\partial u}{\partial x}\right)_{x_{n+1}} - \left(\frac{\partial u}{\partial x}\right)_{x_n}\right] .$$

Applying the same reasoning again gives

$$\frac{\partial^2 u}{\partial t^2} = \frac{Ka^2}{M}\frac{\partial^2 u}{\partial x^2} = V_0^2\frac{\partial^2 u}{\partial x^2} ,$$

which is the same as (1.25), since $V_0 = a\sqrt{\frac{K}{M}}$ is the low-frequency velocity of elastic waves.

4.2. Consider a chain of equidistant atoms, with even numbered atoms having mass M_1 and odd numbered atoms having mass $M_2 > M_1$, as in Fig. 4.52. Write the equations of motion for atoms $2n$ and $2n+1$. Assuming $u_{2n} = A_1 \exp[\mathrm{i}(\omega t - 2nka)]$ and $u_{2n+1} = A_2 \exp \mathrm{i}[\omega t - (2n+1)ka]$, derive the $\omega(k)$ relation and sketch the dispersion curves. Study the cases $ka \ll 1$ and $ka = \pi/2$.

Solution. The equations of motion of atoms $2n$ and $2n+1$, analogous to (4.1), are

$$\begin{cases} M_1\dfrac{\mathrm{d}^2 u_{2n}}{\mathrm{d}t^2} = K\left(u_{2n+1} + u_{2n-1} - 2u_{2n}\right) \\[2ex] M_2\dfrac{\mathrm{d}^2 u_{2n+1}}{\mathrm{d}t^2} = K\left(u_{2n+2} + u_{2n} - 2u_{2n+1}\right) . \end{cases}$$

Substituting the expressions for the displacements gives

$$\begin{cases} (2K - M_1\omega^2)A_1 - (2K\cos ka)A_2 = 0 \\ -(2K\cos ka)A_1 + (2K - M_2\omega^2)A_2 = 0 . \end{cases}$$

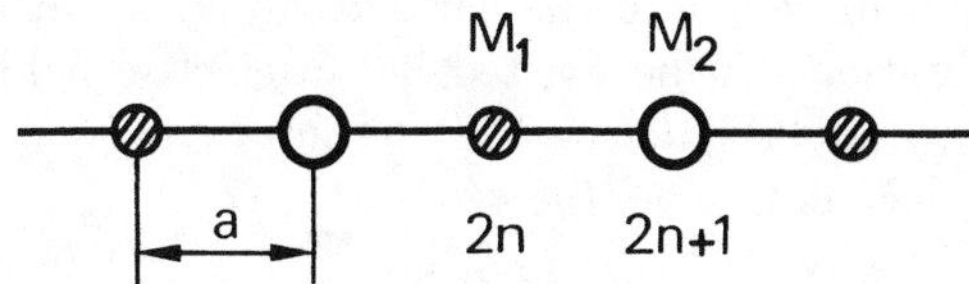

Fig. 4.52. Diatomic chain

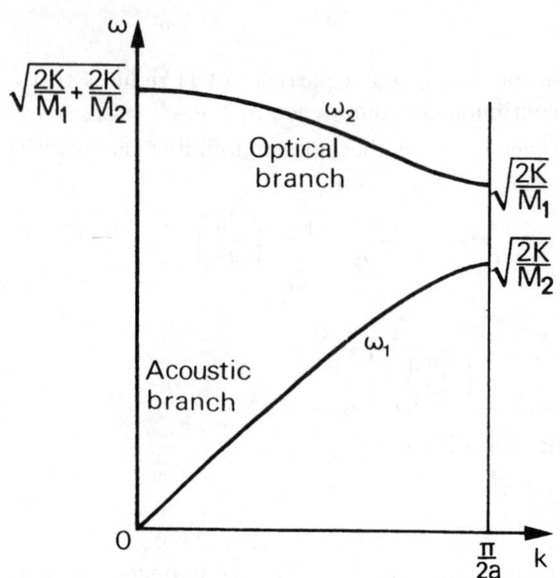

Fig. 4.53. Dispersion curves for longitudinal waves on a diatomic chain of atoms

The condition for compatibility of these homogeneous equations is

$$\begin{vmatrix} 2K - M_1\omega^2 & -2K\cos ka \\ -2K\cos ka & 2K - M_2\omega^2 \end{vmatrix} = 0\,,$$

and this gives the dispersion relations

$$\left(\omega_{\substack{2\\1}}\right)^2 = K\left(\frac{1}{M_1}+\frac{1}{M_2}\right) \pm K\left[\left(\frac{1}{M_1}+\frac{1}{M_2}\right)^2 - 4\frac{\sin^2 ka}{M_1M_2}\right]^{\frac{1}{2}}\,.$$

The dispersion curve $\omega_1(k)$ has the same appearance (Fig. 4.53) as that for a monatomic chain, Fig. 4.3. For $ka \ll 1$, the solution $\omega_1 \cong ak\sqrt{2K/(M_1+M_2)}$ with $A_1 = A_2$ corresponds to propagation of an elastic wave with velocity $V_0 \cong a\sqrt{2K/(M_1+M_2)}$, hence the name *acoustic branch* given to the curve $\omega_1(k)$. The other solution gives, for $ka \ll 1$, $\omega_2 = \sqrt{2K\left(\frac{1}{M_1}+\frac{1}{M_2}\right)}$ and $M_1A_1 + M_2A_2 = 0$. In this case, two adjacent atoms vibrate in phase opposition and with amplitudes A_1 and A_2 such that the centre of mass of the molecule is immobile. If the particles are ions, with opposite electric charges, this vibration can be excited by an electric field with frequency $\omega_2(0)/2\pi(\cong 10^{12}$ to 10^{13} Hz) located in the infrared, hence the name *optical branch* given to the solution $\omega_2(k)$.

For $ka = \pi/2$,

$$\omega_1 = \sqrt{\frac{2K}{M_2}} \text{ and } A_1 = 0, \text{so the light atoms are immobile;}$$

$$\omega_2 = \sqrt{\frac{2K}{M_1}} \text{ and } A_2 = 0, \text{so the heavy atoms are immobile.}$$

4.3. Show that a longitudinal wave is always pure.

Solution. Since ${}^{\circ}u_j = n_j$, the energy velocity (4.24) is

$$V_i^{\mathrm{e}} = c_{ijkl}\frac{n_j n_k\ {}^{\circ}u_l}{\rho V} = \frac{\Gamma_{il}\ {}^{\circ}u_l}{\rho V}\,.$$

Using Christoffel's equation $\Gamma_{il}\ {}^{\circ}u_l = \rho V^2\ {}^{\circ}u_i$, we find $V_i^{\mathrm{e}} = V\ {}^{\circ}u_i = Vn_i$. Thus the energy velocity is parallel to the propagation direction.

4.4. Show that, if there exists a pure mode propagating along $\boldsymbol{n}$ with polarization ${}^{\circ}\boldsymbol{u}$ and velocity V, then there is also a pure mode propagating in the direction ${}^{\circ}\boldsymbol{u}$ and polarized along $\boldsymbol{n}$ having the same velocity. Application: show that, in the YZ plane of a trigonal crystal, there are only two directions for pure transverse waves polarized along X (designated by AC and BC in Fig. 4.28).

Solution. From (4.24), the condition $V_i^{\mathrm{e}} = Vn_i$ for a pure mode can be written

$$c_{ijkl}\ {}^{\circ}u_j\ {}^{\circ}u_l n_k = \gamma n_i \quad \text{with} \quad \gamma = \rho V^2$$

or, on interchanging the indices k and l and noting that $c_{ijlk} = c_{ijkl}$,

$$c_{ijkl}\ {}^{\circ}u_j\ {}^{\circ}u_k n_l = \gamma n_i\,.$$

Comparison with the Christoffel equation (4.7) shows that $\boldsymbol{n}$ is the polarization of a wave propagating in the direction ${}^{\circ}\boldsymbol{u}$ with velocity V. This mode is pure since, from (4.7)

$$c_{ijkl}n_j n_l\ {}^{\circ}u_k = \gamma\ {}^{\circ}u_i\,.$$

Thus, *when the mode is pure, the directions of propagation and polarization are interchangeable.* For a crystal in class $\bar{3}m$, $3m$ or 32, the transverse wave polarized along X is pure for *two* perpendicular directions in the YZ plane. These are the polarization directions of the two *pure, non-degenerate* transverse waves propagating along the dyad axis A_2 or $\bar{A}_2$, which is along X.

4.5. Relation between the energy velocity and the phase velocity for a wave propagating in a plane of symmetry.

(a) Considering a section of the slowness surface, show that $\tan(\theta^{\mathrm{e}} - \theta) = \frac{1}{V}\frac{\mathrm{d}V}{\mathrm{d}\theta}$. Deduce the projection of the vector $\boldsymbol{V}^{\mathrm{e}}$ onto the wavefront, and the relation between V^{e} and $V(\theta)$.

(b) Considering a section of the wave surface, show that $\tan(\theta^{\mathrm{e}} - \theta) = \frac{1}{V^{\mathrm{e}}}\frac{\mathrm{d}V^{\mathrm{e}}}{\mathrm{d}\theta^{\mathrm{e}}}$. Deduce a simple expression for V in terms of V^{e}.

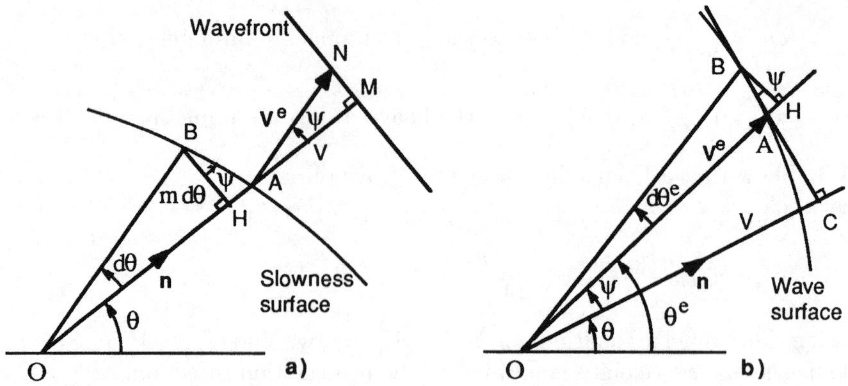

Fig. 4.54a,b. Relation between phase and energy velocities

Solution.
(a) From (4.25), the projection of $\boldsymbol{V}^e$ onto the propagation direction is equal to V, as in Fig. 4.54a. In the triangle ABH we have $HB \cong AB \cong m\mathrm{d}\theta$ and $AH \cong \mathrm{d}m$, and the vertex angle $\psi = \theta^e - \theta$ is given by

$$\tan\psi = -\frac{\mathrm{d}m}{m\mathrm{d}\theta} = \frac{1}{V}\frac{\mathrm{d}V}{\mathrm{d}\theta} \quad \text{since} \quad m(\theta) = \frac{1}{V(\theta)}\,.$$

The minus sign shows that $\theta^e > \theta$ if $m(\theta)$ decreases with θ. The triangle AMN has

$$MN = V\tan\psi = \frac{\mathrm{d}V}{\mathrm{d}\theta} \quad \text{and hence} \quad V^e = \sqrt{V^2(\theta) + \left(\frac{\mathrm{d}V}{\mathrm{d}\theta}\right)^2}\,.$$

(b) In Fig. 4.54b, the triangle ABH has $AH = \mathrm{d}V^e$ and $HB = V^e\mathrm{d}\theta^e$, and the angle ψ is defined by

$$\tan\psi = \frac{AH}{HB} = \frac{1}{V^e}\frac{\mathrm{d}V^e}{\mathrm{d}\theta^e}\,.$$

Since $V = V^e\cos\psi$, triangle OAC gives

$$V(\theta) = \frac{V^e(\theta^e)}{\sqrt{1 + \left(\frac{1}{V^e}\frac{\mathrm{d}V^e}{\mathrm{d}\theta^e}\right)^2}}\,.$$

4.6. Determine the direction θ^e of the energy velocity for a transverse wave polarized perpendicular to the meridian plane of a hexagonal crystal. What value of θ maximizes the deviation $\theta - \theta^e$?

Solution. With the axes arranged as in Sect. 4.2.5.2c, we have

$${}^{\circ}u_i = \delta_{i1} \quad n_1 = 0 \quad n_2 = \sin\theta \quad n_3 = \cos\theta\,,$$

and (4.24) gives

$$V_i^e = c_{i1k1}\frac{n_k}{\rho V}\,.$$

$$\text{Thus,}\quad V_1^{\mathrm{e}} = 0\,, \quad V_2^{\mathrm{e}} = c_{66}\frac{\sin\theta}{\rho V}\,, \quad V_3^{\mathrm{e}} = c_{44}\frac{\cos\theta}{\rho V}$$

and hence

$$\tan\theta^{\mathrm{e}} = \frac{V_2^{\mathrm{e}}}{V_3^{\mathrm{e}}} = \frac{c_{66}}{c_{44}}\tan\theta = k\tan\theta \quad \text{where} \quad k = \frac{c_{66}}{c_{44}}\,.$$

The deviation $\theta^{\mathrm{e}} - \theta$ is maximized when $\mathrm{d}\theta^{\mathrm{e}}/\mathrm{d}\theta = 1$, that is, from the above equation, when

$$\frac{\cos\theta_{\mathrm{M}}}{\cos\theta_{\mathrm{M}}^{\mathrm{e}}} = \sqrt{k}\,.$$

Substituting

$$\tan\theta_{\mathrm{M}}^{\mathrm{e}} = \sqrt{\frac{1}{\cos^2\theta_{\mathrm{M}}^{\mathrm{e}}} - 1} = \sqrt{\frac{k}{\cos^2\theta_{\mathrm{M}}} - 1} = \sqrt{k - 1 + k\tan^2\theta_{\mathrm{M}}}$$

into the equation $\tan\,\theta_{\mathrm{M}}^{\mathrm{e}} = k\tan\theta_{\mathrm{M}}$ gives

$$\tan\theta_{\mathrm{M}} = \frac{1}{\sqrt{k}} = \sqrt{\frac{c_{44}}{c_{66}}}\,.$$

4.7. For a crystal with symmetry $4, \bar{4}$ or $4/m$, evaluate the angle φ_0 such that the constant c_{16} becomes zero when the $\mathrm{O}x_1x_2x_3$ axes are rotated about $\mathrm{O}x_3$ through this angle.

Solution. The matrix α_i^j for rotation through an angle φ about $\mathrm{O}x_3$ is given by (2.14), and the new modulus c_{16}' is

$$c_{16}' = c_{1112}' = \alpha_1^i\alpha_1^j\alpha_1^k\alpha_2^l c_{ijkl}\,.$$

The only original moduli concerned are those in the array of (3.75) with no index equal to 3, that is

$$c_{1111} = c_{2222} = c_{11}, \quad c_{1122} = c_{12}, \quad c_{1212} = c_{66}, \quad c_{1112} = c_{16}, \quad c_{2212} = -c_{16}\,.$$

The coefficients of c_{11}, c_{12} and c_{66} are, respectively,

$$a_{11} = \alpha_1^1\alpha_1^1\alpha_1^1\alpha_2^1 + \alpha_1^2\alpha_1^2\alpha_1^2\alpha_2^2 = -\cos^3\varphi\sin\varphi + \sin^3\varphi\cos\varphi = -\frac{\sin 4\varphi}{4}\,,$$

$$a_{12} = \alpha_1^1\alpha_1^1\alpha_1^2\alpha_2^2 + \alpha_1^2\alpha_1^2\alpha_1^1\alpha_2^1 = \cos^3\varphi\sin\varphi - \sin^3\varphi\cos\varphi = \frac{\sin 4\varphi}{4}\,,$$

$$a_{66} = (\alpha_1^1\alpha_1^2 + \alpha_1^2\alpha_1^1) + (\alpha_1^1\alpha_2^2 + \alpha_1^2\alpha_2^1) = \sin 2\varphi\cos 2\varphi = \frac{\sin 4\varphi}{2}\,,$$

and the coefficient of c_{16} is

$$a_{16} = \alpha_1^1\alpha_1^1(\alpha_1^1\alpha_2^2 + \alpha_1^2\alpha_2^1) + \alpha_1^1\alpha_2^1(\alpha_1^1\alpha_1^2 + \alpha_1^2\alpha_1^1)$$
$$-\alpha_1^2\alpha_1^2(\alpha_1^1\alpha_2^2 + \alpha_1^2\alpha_2^1) - \alpha_1^2\alpha_2^2(\alpha_1^1\alpha_1^2 + \alpha_1^2\alpha_1^1)$$

$$a_{16} = \cos^2\varphi\cos 2\varphi - \frac{\sin^2 2\varphi}{2} - \sin^2\varphi\cos 2\varphi - \frac{\sin^2 2\varphi}{2}$$
$$= \cos^2 2\varphi - \sin^2 2\varphi = \cos 4\varphi\,.$$

Using these, the new modulus is

$$c'_{16} = \frac{1}{2}\left(c_{66} - \frac{c_{11} - c_{12}}{2}\right) \sin 4\varphi + c_{16} \cos 4\varphi$$

and this is zero for an angle φ_0 such that

$$\boxed{\tan 4\varphi_0 = \frac{4c_{16}}{c_{11} - c_{12} - 2c_{66}}} \, .$$

4.8. For a crystal with symmetry 3 or $\bar{3}$, evaluate the angle φ_0 such that the constant c_{25} becomes zero when the $Ox_1x_2x_3$ axes are rotated about Ox_3 through this angle.

Solution. Since $\alpha_3^l = \delta_{3l}$, the new modulus

$$c'_{25} = c'_{2213} = \alpha_2^i \alpha_2^j \alpha_1^k c_{ijk3}$$

contains only the original constants c_{ijk3} in which i, j and k are unequal to 3, and from the array of (3.74) these are

$$c_{1113} = -c_{25} \, , \quad c_{2213} = c_{25} \, , \quad c_{1213} = c_{2113} = c_{14} \, ,$$
$$c_{1123} = c_{14} \, , \quad c_{2223} = -c_{14} \, , \quad c_{1223} = c_{2123} = c_{25} \, .$$

Collecting the terms in c_{25} and c_{14} leads to

$$c'_{25} = c_{25}(\cos^3 \varphi - 3 \sin^2 \varphi \cos \varphi) + c_{14}(\sin^3 \varphi - 3 \cos^2 \varphi \sin \varphi) \, .$$

The new modulus is

$$c'_{25} = c_{25} \cos 3\varphi - c_{14} \sin 3\varphi$$

and this is zero for an angle φ_0 such that

$$\boxed{\tan 3\varphi_0 = \frac{c_{25}}{c_{14}}} \, .$$

4.9. Calculate the velocity V_3 of a transverse wave polarized along an $\bar{A}_4$ axis.

Solution. Using (3.126), with $n_1 = \cos \varphi, n_2 = \sin \varphi$ and $n_3 = 0$, the γ's of (4.75) become

$$\gamma_1 = \gamma_2 = 0, \quad \gamma_3 = e_{15} \cos 2\varphi + e_{14} \sin 2\varphi \, .$$

From the results of Sect. 4.2.2, this gives

$$\rho V_3^2 = \bar{\Gamma}_{33} = \Gamma_{33} + \frac{\gamma_3^2}{\varepsilon} = c_{44}^{\mathrm{E}} + \frac{(e_{15} \cos 2\varphi + e_{14} \sin 2\varphi)^2}{\varepsilon_{11}^{\mathrm{S}}} \, .$$

4.10. For a solid, a suitable relation between stresses and strains is Zener's model

$$T + \tau_2 \frac{\partial T}{\partial t} = K\left(S + \tau_1 \frac{\partial S}{\partial t}\right) \, .$$

(a) Write the Laplace transform of this equation, defined by $\bar{F}(s) = \int_0^\infty f(t)\mathrm{e}^{-st}\mathrm{d}t$ and deduce the transfer function $\bar{S}/\bar{T}$. Here s is the Laplace variable.

(b) Deduce the response $S(t)$ to a stress $T = T_0\Gamma(t)$ in the form of a step function, the latter having the transform $\bar{T} = T_0/s$.

(c) For the harmonic case, derive the delay of the strain with respect to the stress.

(d) Establish the wave equation for a longitudinal wave.

Solution.

(a) The Laplace transform gives $\bar{T}(1+\tau_2 s) = K(1+\tau_1 s)\bar{S}$. The transfer function is

$$\frac{\bar{S}}{\bar{T}} = \frac{1}{K}\left(\frac{1+\tau_2 s}{1+\tau_1 s}\right) = \frac{1}{K}\left(\frac{1}{1+\tau_1 s} + \tau_2\frac{s}{1+\tau_1 s}\right).$$

(b) The step function $T_0\Gamma(t)$ has transform $\bar{T} = T_0/s$, so the response has transform

$$\bar{S} = \frac{T_0}{K}\left(\frac{1}{s} - \frac{1}{s+1/\tau_1} + \frac{\tau_2}{\tau_1}\frac{1}{s+1/\tau_1}\right)$$

and hence

$$S(t) = \frac{T_0}{K}\left(1 - \mathrm{e}^{-t/\tau_1} + \frac{\tau_2}{\tau_1}\mathrm{e}^{-t/\tau_1}\right)\Gamma(t)\,.$$

The step response of this solid has the form of the response of a first-order system to a step function plus the response to a Dirac impulse, as shown in Fig. 4.55.

(c) In the harmonic case, the delay expressed as the phase change δ is found by writing $s = \mathrm{i}\omega$, so that

$$S(\mathrm{i}\omega) = \frac{1}{K}\left(\frac{1+\mathrm{i}\omega\tau_2}{1+\mathrm{i}\omega\tau_1}\right)T(\mathrm{i}\omega) = \frac{1+\omega^2\tau_1\tau_2 - \mathrm{i}\omega(\tau_1-\tau_2)}{K(1+\omega^2\tau_1^2)}T(\mathrm{i}\omega)$$

and this gives

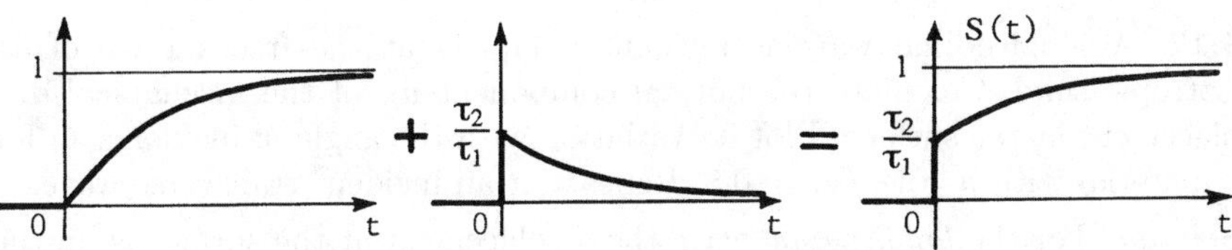

Fig. 4.55. Strain in a solid, obeying Zener's model, in response to a step function of stress

$$\delta = \arctan \frac{\omega(\tau_1 - \tau_2)}{1 + \omega^2 \tau_1 \tau_2} .$$

(d) For the equation of motion for a longitudinal wave, we write

$$\rho \frac{\partial^2 u}{\partial t^2} = \frac{\partial T}{\partial x}, \quad S = \frac{\partial u}{\partial x} \quad \text{giving} \quad \rho \frac{\partial^2 S}{\partial t^2} = \frac{\partial^2 T}{\partial x^2} .$$

Differentiating the stress–strain relation gives

$$\frac{\partial^2 T}{\partial x^2} + \tau_2 \frac{\partial}{\partial t}\left(\frac{\partial^2 T}{\partial x^2}\right) = K \left[\frac{\partial^2 S}{\partial x^2} + \tau_1 \frac{\partial}{\partial t}\left(\frac{\partial^2 S}{\partial x^2}\right)\right]$$

and hence

$$\frac{\partial^2 S}{\partial t^2} + \tau_2 \frac{\partial^3 S}{\partial t^3} = \frac{K}{\rho}\left(\frac{\partial^2 S}{\partial x^2} + \tau_1 \frac{\partial^3 S}{\partial t \partial x^2}\right) .$$

4.11. Derive an expression for the piezoelectric permittivity taking account of wave attenuation as in the one-dimensional model of Sect. 4.3.4 (that is, write $T = cS + \eta \partial S/\partial t + e\partial \Phi/\partial x$). Sketch the curve for $\bar{\varepsilon}/\varepsilon^{\mathrm{S}}$ as a function of V/V_∞.

Solution. In this case, $T = (c^{\mathrm{E}} + \mathrm{i}\omega\eta)S - \mathrm{i}ek\Phi$, so the constant c^{E} for the loss-less case is to be replaced by $c^{\mathrm{E}} + \mathrm{i}\omega\eta = c^{\mathrm{E}}(1 + \mathrm{i}\alpha)$. Here $\alpha = \omega\eta/c^{\mathrm{E}} = 1/Q$, and $Q \gg 1$ is the quality factor. For the loss-less case, (4.81) gives

$$\frac{\bar{\varepsilon}}{\varepsilon^{\mathrm{S}}} = 1 + \frac{e^2}{\varepsilon^{\mathrm{S}}} \frac{1}{c^{\mathrm{E}} - \rho V^2} .$$

Making the above substitution, and using $c^{\mathrm{E}} = \rho V_\infty^2$, gives

$$\frac{\bar{\varepsilon}}{\varepsilon^{\mathrm{S}}} = 1 + \frac{e^2}{\varepsilon^{\mathrm{S}}} \frac{1}{c^{\mathrm{E}}(1 + \mathrm{i}\alpha) - \rho V^2} = 1 + \frac{e^2}{\varepsilon^{\mathrm{S}} c^{\mathrm{E}}} \frac{1}{1 + \mathrm{i}\alpha - V^2/V_\infty^2} ,$$

and, on defining $v = V/V_\infty$ and $K^2 = e^2/(\varepsilon^{\mathrm{S}} c^{\mathrm{E}})$, this becomes

$$\begin{aligned} \frac{\bar{\varepsilon}}{\varepsilon^{\mathrm{S}}} &= 1 + K^2 \frac{1}{1 - v^2 + \mathrm{i}\alpha} = 1 + K^2 \frac{(1 - v^2 - \mathrm{i}\alpha)}{(1 - v^2)^2 + \alpha^2} \\ &= \mathrm{Re}\left[\frac{\bar{\varepsilon}}{\varepsilon^{\mathrm{S}}}\right] + \mathrm{i}\,\mathrm{Im}\left[\frac{\bar{\varepsilon}}{\varepsilon^{\mathrm{S}}}\right] . \end{aligned}$$

Thus $\bar{\varepsilon}$ is complex. The real part $\mathrm{Re}\left[\bar{\varepsilon}/\varepsilon^{\mathrm{S}}\right]$ is extremal for $1 - v^2 = \pm\alpha$, and the imaginary part $\mathrm{Im}\left[\bar{\varepsilon}/\varepsilon^{\mathrm{S}}\right]$ is always negative, as shown in Fig. 4.39.

4.12. A longitudinal wave is incident obliquely at the free surface of an isotropic solid. Calculate the normal component u_{N} of the mechanical displacement at the surface. Plot its variation with the angle of incidence θ_{I} for a material with $v = V_{\mathrm{T}}/V_{\mathrm{L}} = 0.5$. Repeat for an incident transverse wave.

Solution. For the *longitudinal* wave the displacement at the surface is, in the notation of Sect. 4.4.2.2,

$$\boldsymbol{u} = {}^{\circ}\boldsymbol{u}_{\mathrm{I}} + {}^{\circ}\boldsymbol{u}_{\mathrm{L}} + {}^{\circ}\boldsymbol{u}_{\mathrm{T}}$$

and the normal component is

$$\begin{aligned} {}^\circ u_{\rm N} = {}^\circ u_2 &= {}^\circ u_{\rm I}\cos\theta_{\rm I} - {}^\circ u_{\rm L}\cos\theta_{\rm I} + {}^\circ u_{\rm T}\sin\theta_{\rm T} \\ &= {}^\circ u_{\rm I}[(1-r_{\rm LL})\cos\theta_{\rm I} + r_{\rm LT}\sin\theta_{\rm T}]\,. \end{aligned}$$

Equations (4.107) and (4.108) lead to

$$\frac{{}^\circ u_2}{{}^\circ u_{\rm I}} = \frac{2\cos 2\theta_{\rm T}\cos\theta_{\rm I}}{v^2\sin 2\theta_{\rm I}\sin 2\theta_{\rm T} + \cos^2 2\theta_{\rm T}} \quad \text{with} \quad \sin\theta_{\rm T} = v\sin\theta_{\rm I}\,,$$

and hence, replacing $\theta_{\rm I}$ by θ,

$$\frac{{}^\circ u_2}{{}^\circ u_{\rm I}} = \frac{2\cos\theta\,(1-2v^2\sin^2\theta)}{4v^3\sin^2\theta\cos\theta\,(1-v^2\sin^2\theta)^{1/2} + (1-2v^2\sin^2\theta)^2}\,. \tag{4.138}$$

This ratio is plotted in Fig. 4.56.

For an incident *transverse* wave the normal component of displacement at the surface is, from Fig. 4.47,

$$\begin{aligned} {}^\circ u_2 &= {}^\circ u_{\rm I}\sin\theta_{\rm I} + {}^\circ u_{\rm T}\sin\theta_{\rm I} - {}^\circ u_{\rm L}\cos\theta_{\rm L} \\ &= {}^\circ u_{\rm I}[(1+r_{\rm TT})\sin\theta_{\rm I} - r_{\rm TL}\cos\theta_{\rm L}]\,. \end{aligned}$$

From (4.109) and (4.110) this gives

$$\frac{{}^\circ u_2}{{}^\circ u_{\rm I}} = \frac{2v\sin 2\theta_{\rm I}\,(2v\sin\theta_{\rm L}\sin\theta_{\rm I} + \cos 2\theta_{\rm I})\cos\theta_{\rm L}}{v^2\sin 2\theta_{\rm I}\sin 2\theta_{\rm L} + \cos^2 2\theta_{\rm I}} \quad \text{with } v\sin\theta_{\rm L} = \sin\theta_{\rm I}\,,$$

and, writing θ for $\theta_{\rm I}$,

$$\frac{{}^\circ u_2}{{}^\circ u_{\rm I}} = \frac{2v\sin 2\theta}{2v\sin 2\theta\sin\theta + (1-v^{-2}\sin\theta)^{-1/2}\cos^2 2\theta} \tag{4.139}$$

for

$$\theta < \theta_c = \arcsin v\,.$$

When the angle of incidence exceeds the critical angle θ_c (here 30°), it is necessary to take the modulus of the denominator of the right side of (4.139), as done in Fig. 4.56. For an angle of 45° the transverse wave is totally reflected and the normal displacement at the surface is $\sqrt{2}$ times that of the incident wave.

4.13. For the case of total reflection at a liquid–solid interface, express the reflection coefficient for the longitudinal wave incident in the liquid (formula 4.132) in terms of the projection k_1 of the wave vectors onto the surface. Show that this agrees with the relation (5.43) for the velocity of Rayleigh waves on the surface of an isotropic solid.

Solution. The projection k_1 of the wave vectors is given by $k_1 = k_{\rm T}\sin\theta_{\rm T} = k_{\rm L}\sin\theta_{\rm L} = k\sin\theta$, where $k = k_{\rm I}$ and $\theta = \theta_{\rm I}$ refer to the incident wave in the liquid. In terms of k_1, the functions $\sin 2\theta_{\rm T}$, $\sin 2\theta_{\rm L}$ and $\cos 2\theta_{\rm T}$ can be written as

$$\sin 2\theta_{\rm T} = 2\sin\theta_{\rm T}\cos\theta_{\rm T} = 2\frac{k_1}{k_{\rm T}}\left(1-\frac{k_1^2}{k_{\rm T}^2}\right)^{1/2} = 2\mathrm{i}\frac{k_1}{k_{\rm T}^2}(k_1^2-k_{\rm T}^2)^{1/2}$$

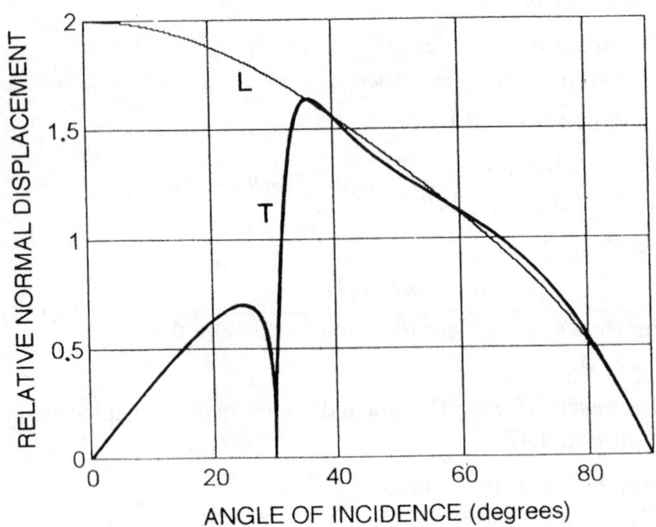

Fig. 4.56. Reflection at the free surface of an isotropic solid with $v = V_\mathrm{T}/V_\mathrm{L} = 0.5$. Amplitude of the normal displacement at the surface, divided by the amplitude of the longitudinal (L) or transverse (T) incident wave, as a function of the angle of incidence

since $k_\mathrm{T} < k_1$,

$$\sin 2\theta_\mathrm{L} = 2\sin\theta_\mathrm{L}\cos\theta_\mathrm{L} = 2\mathrm{i}\frac{k_1}{k_\mathrm{L}^2}(k_1^2 - k_\mathrm{L}^2)^{1/2} \quad \text{since} \quad k_\mathrm{L} < k_1\,,$$

$$\cos 2\theta_\mathrm{T} = 1 - 2\sin^2\theta_\mathrm{T} = (k_\mathrm{T}^2 - 2k_1^2)/k_\mathrm{T}^2\,.$$

Substituting these into (4.132) gives, with $v = V_\mathrm{T}/V_\mathrm{L} = k_\mathrm{L}/k_\mathrm{T}$,

$$r_\mathrm{LL}(k_1) = \frac{\left(2k_1^2 - k_\mathrm{T}^2\right)^2 - 4k_1^2(k_1^2 - k_\mathrm{T}^2)^{1/2}(k_1^2 - k_\mathrm{L}^2)^{1/2} - \mathrm{i}\dfrac{\rho}{\rho_\mathrm{S}}k_\mathrm{T}^4\left(\dfrac{k_1^2 - k_\mathrm{L}^2}{k^2 - k_1^2}\right)^{1/2}}{\left(2k_1^2 - k_\mathrm{T}^2\right)^2 - 4k_1^2(k_1^2 - k_\mathrm{T}^2)^{1/2}(k_1^2 - k_\mathrm{L}^2)^{1/2} + \mathrm{i}\dfrac{\rho}{\rho_\mathrm{S}}k_\mathrm{T}^4\left(\dfrac{k_1^2 - k_\mathrm{L}^2}{k^2 - k_1^2}\right)^{1/2}}$$

and hence

$$r_\mathrm{LL}(k_1) = \frac{R(k_1) - \mathrm{i}S(k_1)}{R(k_1) + \mathrm{i}S(k_1)}\,,$$

where $R(k_1)$ and $S(k_1)$ are defined in (4.136) and (4.137). We introduce the decay constants χ_1 and χ_2 defined, as in (5.39) below, such that

$$\chi_1^2 = 1 - \frac{V^2}{V_\mathrm{L}^2} = 1 - \frac{k_\mathrm{L}^2}{k_1^2}\,, \quad \chi_2^2 = 1 - \frac{V^2}{V_\mathrm{T}^2} = 1 - \frac{k_\mathrm{T}^2}{k_1^2}\,,$$

and these give

$$k_1^2 - k_L^2 = k_1^2 \chi_1^2 \quad \text{and} \quad k_1^2 - k_T^2 = k_1^2 \chi_2^2 \, .$$

The function $R(k_1)$ can then be written as

$$R(k_1) = k_1^4 \left[\left(1 + \chi_2^2\right)^2 - 4\chi_1\chi_2 \right] \, . \tag{4.140}$$

The equation $R(k_1) = 0$ is equivalent to (5.43), which gives the velocity V_R of Rayleigh waves in an isotropic solid. It has the solution $k_1 = k_R = \omega / V_R$.

5. Guided Waves

As explained in Chap. 4, the free propagation of elastic waves in a solid is characterized by *slowness surfaces.* For an isotropic solid these are two spheres, while a crystal generally has three surfaces. For a given direction and for one of the three types of wave, each surface shows the inverse of velocity and, in particular, the energy flow direction. The *pure mode* directions, for which the energy vector and the wave vector are parallel to each other and normal to the slowness surface, are directly seen. In a piezoelectric material, the variation with direction of the electromechanical coupling is also seen if the actual slowness surface is accompanied by another one calculated without taking account of piezoelectricity. For a given crystal, the slowness surfaces represent the plane-wave solutions of three equations of motion plus, for a piezoelectric insulator, Poisson's equation. This analysis assumes that the dimensions of the 'beam' of waves are small in comparison with those of the solid, so that these bulk waves are assumed to propagate without encountering any boundaries.

When the solid is not infinite, for example if it has a free plane surface (as in a semi-infinite solid in contact with air), the boundary imposes conditions on the mechanical and electrical variables. This causes the waves to be reflected there, often with a change of wave type, as seen in Sect. 4.4. Their directions are changed. If the crystal has the form of a plate, with two plane and parallel free surfaces, the wave can progress by reflection at the two surfaces alternately. This is a guided wave.

These observations, based on reflection and refraction of a given type of incident wave, lead to the idea of elastic waveguides. They can be developed further, but this approach is not sufficiently general. In fact, to derive the various propagation modes of a structure, it is necessary to start by writing equations, firstly for the propagation, and secondly for the constraints imposed by the structure, that is, the mechanical boundary conditions at the surfaces and, if the material is piezoelectric, the electrical boundary conditions. These two groups of equations must then be solved simultaneously. In this way, it can be shown that even a simple free surface can act as a waveguide.

In general, the investigation of wave propagation in a crystal with one surface or two parallel surfaces, for an arbitrary crystallographic orientation, is only possible numerically. However, there are two situations in which a crystal symmetry element causes the equations to split into two independent parts. This decomposition highlights two important families of modes – modes with displacement components contained in the sagittal plane, and those with a single transverse component. The presence of piezoelectricity does not affect this conclusion. The boundary conditions concerning surface stresses can be applied separately for the two mode types.

This chapter therefore begins by studying these two situations, relating to a plane waveguide. The main types of wave guided along one or more parallel planes are then described qualitatively. Section 5.2 considers the elastic power transported by a guide, irrespective of the type of wave.

Section 5.3, the most important part, treats Rayleigh waves, which are guided by a simple surface. The general method for finding solutions is explained, and the propagation is analysed in the most accessible cases, isotropic solids and crystals in which the symmetry reduces the number of displacement components to two. For a piezoelectric crystal, the need to satisfy the electrical boundary conditions leads to the introduction of the fruitful ideas of surface permittivity and electromechanical coupling, already introduced for bulk waves in Sects. 4.3.2b and 4.3.4. The propagation medium for Rayleigh waves is generally a piezoelectric material which, for convenience, also plays a part in generating the waves. The characteristics of the commonest materials – quartz, lithium niobate and several others – are presented in a table.

Waves with a single transverse component, parallel to the surface, are considered in the Sect. 5.4. These are the simple non-piezoelectric wave propagating in a plate, the Love wave propagating in a film on a substrate, and the piezoelectric Bleustein–Gulyaev–Shimizu wave which is confined to the region near a surface of a half-space.

The subject of Sect. 5.5 is the propagation of Lamb waves in an isotropic plate. The two components of these waves are in the sagittal (vertical) plane. The dispersion relation is derived and is used to clarify the behaviour of the two mode types, symmetric and antisymmetric, and the associated displacements of the plate surfaces.

Lastly, Sect. 5.6 treats waves in an isotropic cylinder. These modes are classified into the three families of torsion, compression and flexion.

One reason for the interest in guided waves lies in the reduction of losses due to diffraction. Guidance in one of the two lateral directions leads to an appreciable improvement since, far from the source, the energy decreases as r^{-1} instead of r^{-2}. Another advantage of guidance is the accessibility, at all points on a free surface, of the mechanical displacement or the elasto-electric field. This explains, in particular, the widespread development of Rayleigh-wave devices. These waves are not dispersive – for a given crystallographic direction the velocity is independent of frequency. Nevertheless, the dispersion required for pulse compression (Chap. 4 of Vol. II) can be obtained by exploiting the accessibility of the piezoelectric field, generating and detecting the waves in a dispersive manner.

5.1 Planar Waveguides

The crystal is assumed to be semi-infinite or bounded by two parallel planes, as in Fig. 5.1, and the waves propagate along x_1. The (x_1, x_2) plane, containing the surface normal and the propagation direction, is called the *sagittal plane.*

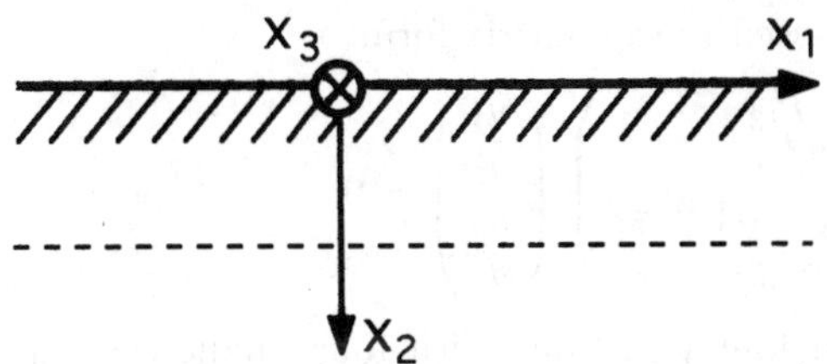

Fig. 5.1. A semi-infinite crystal occupying the region $x_2 > 0$, with surface plane (x_1, x_3). Also, a crystal bounded by two parallel planes, with one plane shown by the broken line

If the guide has properties invariant with time and with x_1, and if the material behaves linearly and without attenuation, the *normal modes* have the form

$$g(x_i, t) = g_0(x_2, x_3)\mathrm{e}^{\mathrm{i}(\omega t - kx_1)}\,, \tag{5.1}$$

where g represents one of the components u_i of the mechanical displacement, or the electric potential Φ.

5.1.1 Decomposition of Equations

The general equations for propagation in a piezoelectric insulator were derived in Sect. 4.3.1. For the harmonic case, they are written

$$\begin{cases} c^{\mathrm{E}}_{ijkl}\dfrac{\partial^2 u_l}{\partial x_j \partial x_k} + e_{kij}\dfrac{\partial^2 \Phi}{\partial x_j \partial x_k} = -\rho\omega^2 u_i\,, \quad i = 1, 2, 3 \\ e_{jkl}\dfrac{\partial^2 u_l}{\partial x_j \partial x_k} - \varepsilon^{\mathrm{S}}_{jk}\dfrac{\partial^2 \Phi}{\partial x_j \partial x_k} = 0\,. \end{cases} \tag{5.2}$$

We define the dimensionless variable $X_2 = kx_2$ and neglect the effects of diffraction in the x_3 direction, so that

$$\frac{\partial}{\partial x_1} = -\mathrm{i}k\,, \quad \frac{\partial}{\partial x_2} = k\frac{\partial}{\partial X_2}\,, \quad \frac{\partial}{\partial x_3} = 0\,.$$

We also take $u_4 \equiv \Phi$, and introduce the phase velocity $V = \omega/k$ and the quantities

$$\begin{cases} \Gamma_{il} = c^{\mathrm{E}}_{i11l} - c^{\mathrm{E}}_{i22l}\dfrac{\partial^2}{\partial X_2^2} + \mathrm{i}(c^{\mathrm{E}}_{i12l} + c^{\mathrm{E}}_{i21l})\dfrac{\partial}{\partial X_2} \\ \gamma_l \ = e_{11l} - e_{22l}\dfrac{\partial^2}{\partial X_2^2} + \mathrm{i}(e_{12l} + e_{21l})\dfrac{\partial}{\partial X_2} \\ \varepsilon \ = \varepsilon^{\mathrm{S}}_{11} - \varepsilon^{\mathrm{S}}_{22}\dfrac{\partial^2}{\partial X_2^2} + 2\mathrm{i}\,\varepsilon^{\mathrm{S}}_{12}\dfrac{\partial}{\partial X_2}\,. \end{cases} \tag{5.3}$$

Equations (5.2) can then be expressed in the matrix form

$$\begin{pmatrix} \Gamma_{11} - \rho V^2 & \Gamma_{12} & \Gamma_{13} & \gamma_1 \\ \Gamma_{12} & \Gamma_{22} - \rho V^2 & \Gamma_{23} & \gamma_2 \\ \Gamma_{13} & \Gamma_{23} & \Gamma_{33} - \rho V^2 & \gamma_3 \\ \gamma_1 & \gamma_2 & \gamma_3 & -\varepsilon \end{pmatrix} \begin{pmatrix} u_1 \\ u_2 \\ u_3 \\ u_4 \end{pmatrix} = 0\,. \tag{5.4}$$

This is a homogeneous system of four equations with four unknowns, the three components of mechanical displacement and the electric potential. It simplifies if the sagittal plane (x_1, x_2) is normal to a dyad axis, which is thus along x_3. In this case, *the elastic constants with one index equal to 3 are zero*, as shown in (3.63), and this gives $\Gamma_{13} = \Gamma_{23} = 0$.

We consider the two cases in which there is a direct or inverse axis normal to the sagittal plane, the latter being equivalent to a mirror plane parallel to the sagittal plane.

(a) **Sagittal Plane Normal to a Direct Axis of Even Order.** This case includes a dyad axis. The piezoelectric constants with no index equal to 3 are zero, as shown in (3.111), so that $\gamma_1 = \gamma_2 = 0$. The system (5.4) divides into two independent subsystems. The first,

$$\begin{pmatrix} \Gamma_{11} - \rho V^2 & \Gamma_{12} \\ \Gamma_{12} & \Gamma_{22} - \rho V^2 \end{pmatrix} \begin{pmatrix} u_1 \\ u_2 \end{pmatrix} = 0 \tag{5.5}$$

has as solution a non-piezoelectric wave *polarized in the sagittal plane* (x_1, x_2). This is designated P_2. The second subsystem is

$$\begin{pmatrix} \Gamma_{33} - \rho V^2 & \gamma_3 \\ \gamma_3 & -\varepsilon \end{pmatrix} \begin{pmatrix} u_3 \\ u_4 \end{pmatrix} = 0 \tag{5.6}$$

and the solution of this is a *transverse horizontal* wave with polarization u_3. This wave, which is *piezoelectric*, was discovered in 1968 by Bleustein and Gulyaev. It is designated $\overline{\mathrm{TH}}$, where the bar indicates that it is piezoelectric.

(b) **Sagittal Plane Parallel to a Mirror Plane.** In this case the piezoelectric constants with one index equal to 3 are zero, as shown by (3.110), so $\gamma_3 = 0$. The system again splits into two parts. The first,

$$\begin{pmatrix} \Gamma_{11} - \rho V^2 & \Gamma_{12} & \gamma_1 \\ \Gamma_{12} & \Gamma_{22} - \rho V^2 & \gamma_2 \\ \gamma_1 & \gamma_2 & -\varepsilon \end{pmatrix} \begin{pmatrix} u_1 \\ u_2 \\ u_4 \end{pmatrix} = 0 \tag{5.7}$$

gives a wave with *sagittal polarization* and an associated *electric* field, denoted $\overline{\mathrm{P}}_2$. The electric field is always contained in the sagittal plane, since $E_3 = -\partial\Phi/\partial x_3 = 0$. The second part,

$$(\Gamma_{33} - \rho V^2)u_3 = 0 \tag{5.8}$$

corresponds to a *transverse horizontal* (TH) wave, in this case non-piezoelectric.
Summarizing, a direct or inverse dyad axis normal to the sagittal plane decouples waves polarized in the sagittal plane from transverse horizontal waves. We now consider whether the boundary conditions maintain this decoupling.

5.1.2 Mechanical Boundary Conditions

On the surface normal to x_2, the mechanical boundary conditions concern the surface stresses T_{i2}. In the abbreviated notation, these are $T_{12} = T_6$, $T_{22} = T_2$ and $T_{32} = T_4$. The general expression

$$T_{ij} = \frac{\partial}{\partial x_k}\left(c^{\mathrm{E}}_{ijkl}u_l + e_{kij}\Phi\right) \tag{5.9}$$

leads to the following, with $\alpha = 2$, 4 or 6 and with the superscript E on c^{E}_{ijkl} omitted for convenience:

$$\begin{aligned} T_\alpha &= \frac{\partial}{\partial x_1}(c_{\alpha 1}u_1 + c_{\alpha 6}u_2 + c_{\alpha 5}u_3 + e_{1\alpha}u_4) \\ &+ \frac{\partial}{\partial x_2}(c_{\alpha 6}u_1 + c_{\alpha 2}u_2 + c_{\alpha 4}u_3 + e_{2\alpha}u_4)\,. \end{aligned} \tag{5.10}$$

For $\alpha = 4$ the constants $c_{\alpha 1}$, $c_{\alpha 2}$ and $c_{\alpha 6}$ are zero because they have only one index equal to 3, and hence

$$T_4 = T_{32} = \frac{\partial}{\partial x_1}(c_{45}u_3 + e_{14}u_4) + \frac{\partial}{\partial x_2}(c_{44}u_3 + e_{24}u_4)\,.$$

For any wave *polarized in the sagittal plane* ($u_3 = 0$), this stress is zero because

(a) when $A_2 \| x_3$ the wave is not piezoelectric, so $u_4 = 0$;
(b) when $M \perp x_3$ the constants $e_{14} = e_{123}$ and $e_{24} = e_{223}$ are zero, because these constants have only one index equal to 3.

For $\alpha = 2$ or 6, the constants $c_{\alpha 4}$ and $c_{\alpha 5}$ are zero (one index equal to 3), so

$$T_\alpha = e_{1\alpha}\frac{\partial u_4}{\partial x_1} + e_{2\alpha}\frac{\partial u_4}{\partial x_2} \quad \text{for } \alpha = 2 \text{ or } 6\,.$$

For a *transverse horizontal* wave, TH ($u_1 = u_2 = u_4 = 0$) or $\overline{\mathrm{TH}}$ ($u_4 \neq 0$), these stresses are zero because

(a) when $A_2 \| x_3$ the piezoelectric constants with no index equal to 3 are zero, so $e_{1\alpha} = e_{2\alpha} = 0$;

Wave polarized in the sagittal plane, P_2 or $\overline{P}_2$, $u_3 = 0$:

$$\left\{\begin{matrix} T_{22}\ (\alpha=2) \\ T_{12}\ (\alpha=6) \end{matrix}\right\} = \frac{\partial}{\partial x_1}(c^E_{\alpha 1} u_1 + c^E_{\alpha 6} u_2 + e_{1\alpha} u_4) + \frac{\partial}{\partial x_2}(c^E_{\alpha 6} u_1 + c^E_{\alpha 2} u_2 + e_{2\alpha} u_4)$$

$$T_{32} = 0$$

Transverse horizontal wave, TH or $\overline{\text{TH}}$, $u_1 = u_2 = 0$:

$$T_{32} = \frac{\partial}{\partial x_1}(c^E_{45}\, u_3 + e_{14}\, u_4) + \frac{\partial}{\partial x_2}(c^E_{44}\, u_3 + e_{24}\, u_4), \qquad T_{22} = T_{12} = 0.$$

Fig. 5.2. Expressions for the stresses associated with waves of type P_2 standard, $\overline{P}_2$ or TH, $\overline{\text{TH}}$ guided along a plane normal to x_2 (or between two such planes) and propagating along x_1

(b) when $M \perp x_3$ the TH wave is non-piezoelectric, so $u_4 = 0$.

We conclude that the stresses on the horizontal surface, associated with waves polarized in the sagittal plane, become decoupled from those associated with transverse horizontal waves when there is a dyad axis normal to the sagittal plane. These results are summarized in the table of Fig. 5.2.

5.1.3 Main Types of Guided Wave

The above results relate to a variety of structures. The simplest is an isotropic semi-infinite solid with a free surface. For a wave polarized in the sagittal plane to be guided along the surface, so that it remains close to the surface as it propagates, the component u_2 must decrease with depth. Noting that the mechanical stresses T_{12} and T_{22} are zero at the free surface and couple the two components u_1 and u_2, it follows that the decrease of u_2 with depth implies a decrease of u_1. This wave, with two displacement components, is called a *Rayleigh wave* since Lord Rayleigh discovered it in 1885, showing that it explained the last component of an earth tremor, observed by geophysicists but not previously understood. The two displacement components have a phase difference of $\pi/2$, so that the polarization is elliptical, and their amplitudes decrease with depth at different rates. The undulation of the surface, associated with the wave motion, extends to a depth of the order of the wavelength λ, as illustrated in Fig. 5.3. The Rayleigh-wave velocity V_R is slightly less than the velocity V_T of bulk transverse waves, because the solid behaves less rigidly in the absence of material above the surface. In silica, for example, $V_T = 3\,760$ m/s and $V_R = 3\,410$ m/s, giving $\lambda = 34\ \mu$m at 100 MHz. For this reason, this wave is the last type to reach a seismograph following a seismic shock, in which all types of wave are generated. We denote this Rayleigh wave by R_2.

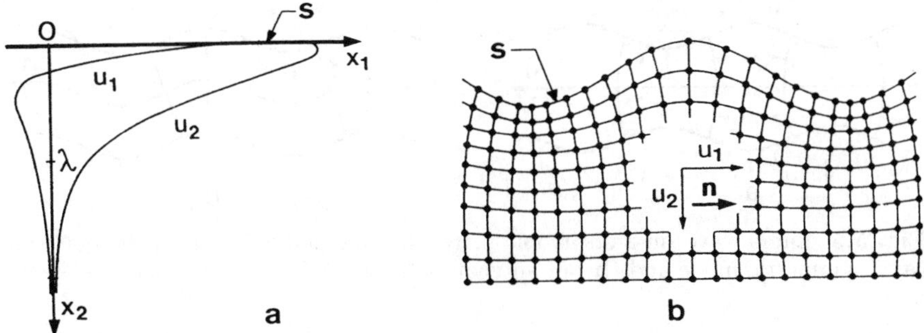

Fig. 5.3. Rayleigh wave R_2 in an isotropic semi-infinite solid, where s is the free surface and $\boldsymbol{n}$ is the propagation direction. (**a**) Decrease of the longitudinal and transverse components u_1 and u_2 with depth. (**b**) Displacement of the surface

If the semi-infinite solid is a piezoelectric crystal, with one of the above symmetry cases $A_2 \| x_3$ or $M \perp x_3$, the propagating wave is a Rayleigh wave R_2 (with the $\overline{\mathrm{TH}}$ wave) or a similar piezoelectric wave $\overline{\mathrm{R}}_2$ (with the TH wave). A transverse horizontal wave $\overline{\mathrm{TH}}$, with amplitude decreasing with depth, is also a solution to (5.6). This wave penetrates less deeply into the material as the latter becomes more strongly piezoelectric, but its penetration depth, some tens of wavelengths, is always larger than that of Rayleigh waves (Fig. 5.4).

If the crystal does not have one of the symmetry elements (A_2 or M) oriented as above, numerical calculations show that there is still a wave propagating near the surface, but in general it involves *three* displacement components. This is also called a generalized Rayleigh wave, designated by R_3 or $\overline{\mathrm{R}}_3$. The polarization vector describes an ellipse in a plane inclined to the sagittal plane. The velocity depends on the crystal cut and the propagation direction. The mode is pure only for particular sagittal plane orientations

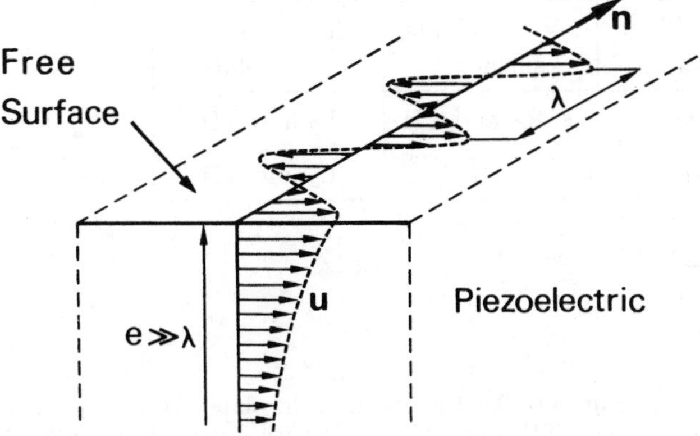

Fig. 5.4. $\overline{\mathrm{TH}}$ type surface wave with transverse horizontal polarization, a Bleustein–Gulyaev–Shimizu wave. The surface of the solid remains flat

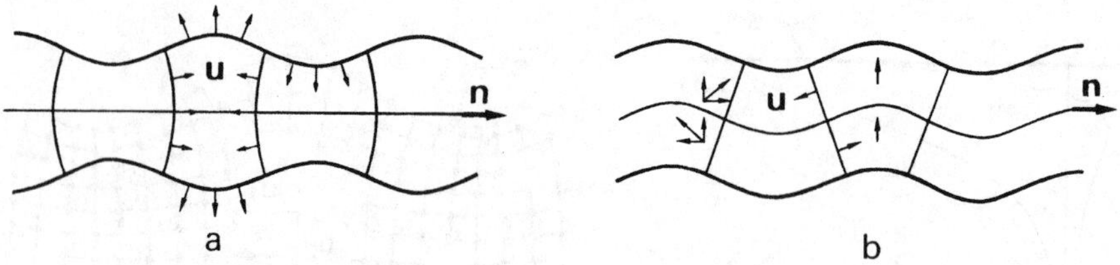

Fig. 5.5. Lamb waves in an isotropic plate, showing deformations of the plate for (**a**) a symmetric mode and (**b**) an antisymmetric mode

and propagation directions. In quartz, which lacks a mirror plane, a pure $\overline{\mathrm{R}}_3$ mode with velocity 3 154 m/s can propagate along the (X, Z) plane (i.e. for a Y-cut crystal) in the direction of the dyad axis X. The plane of polarization is inclined at 43° to the sagittal plane.

For an isotropic solid bounded by two parallel free surfaces, i.e. a free plate, the normal guided wave with two components is called the *Lamb wave*, denoted L_2. One way of interpreting the formation of a Lamb wave and its relation to the Rayleigh wave is to note that a Rayleigh wave (R_2) can propagate independently on each free surface if the thickness is much larger than the wavelength λ. When the plate thickness h becomes comparable to λ, the longitudinal and transverse components of the R_2 waves are coupled, giving rise to symmetric or antisymmetric deformations of the plate, as in Fig. 5.5. The isotropic plate is also a waveguide for the transverse TH wave, Fig. 5.6. A crystalline piezoelectric plate with free surfaces, with one of the symmetries considered above (Sect. 5.1.1), can guide waves of type L_2 and $\overline{\mathrm{TH}}$ (*Bleustein–Gulyaev waves*), or of type $\overline{\mathrm{L}}_2$ and TH, plus, if $h \gg \lambda$, the R_2 or $\overline{\mathrm{R}}_2$ waves.

The conclusions for a guide with one free surface x_1x_3 (valid in practice when in contact with air), and for a plate with two free surfaces, are summarized below for the two symmetry cases.

Sagittal Plane orientation		Guidance Structure surface	plate
$\perp A_2$	$\Rightarrow$	R_2 and $\overline{\mathrm{TH}}$	L_2 and $\overline{\mathrm{TH}}$
$\parallel M$	$\Rightarrow$	$\overline{\mathrm{R}}_2$	$\overline{\mathrm{L}}_2$ and TH

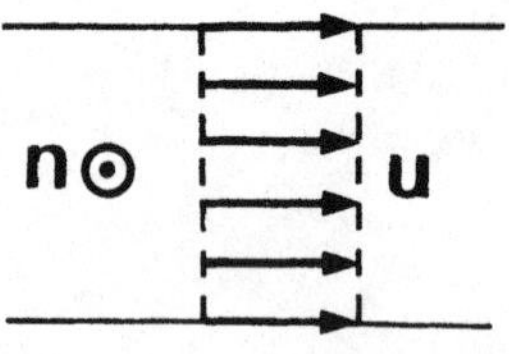

Fig. 5.6. Another mode of the plate, the purely transverse TH wave. The mode with order zero is shown

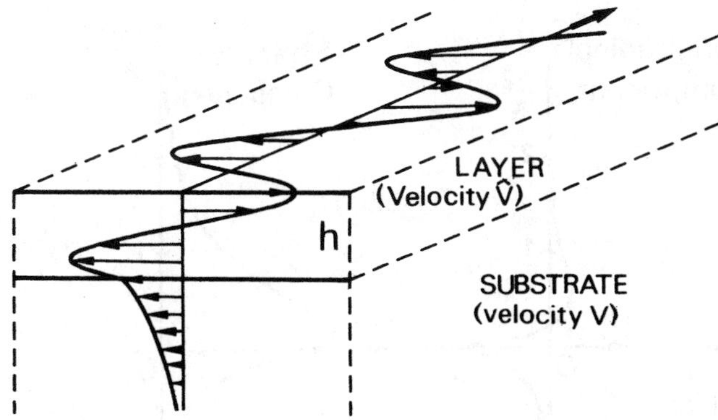

Fig. 5.7. The Love wave, a transverse wave propagating in a layer and its supporting substrate, provided $V_{\mathrm{T}} > \widehat{V}_{\mathrm{T}}$. The penetration depth into the substrate is large at low frequencies, when $\lambda \gg h$

Between the most complex structure (a plate between two solids) and the simplest structure (a free surface) lies the intermediate case of a plate on a semi-infinite solid, that is, a layer on a substrate. Here, the energy propagates partly in the substrate. In the light of the above discussion, the reader can envisage that various waves may propagate, depending on the nature of the two solids. The easiest case to consider is that of a transverse TH wave in an isotropic film and substrate, called the *Love wave*, shown in Fig. 5.7. This wave is guided only if the velocity V_{T} of transverse bulk waves in the substrate is higher than $\widehat{V}_{\mathrm{T}}$, that of the same waves in the film. At high frequencies, such that the wavelength λ is less than the film thickness h, the wave is concentrated in the film and has velocity close to $\widehat{V}_{\mathrm{T}}$; at low frequencies where $\lambda \gg h$, the velocity is close to V_{T}.

Another guiding structure is the boundary joining two semi-infinite solids, Fig. 5.8. For a suitable choice of materials, a Rayleigh-like wave can propagate on either side of the boundary. This interface wave is called a *Stoneley* wave.

Elastic waves can of course propagate in more complex structures such as anisotropic piezoelectric plates or layered media. For these cases it is more difficult to anticipate the propagation characteristics, and numerical methods are needed to investigate them [5.1].

Rayleigh waves ($\overline{\mathrm{R}}_2$ or $\overline{\mathrm{R}}_3$) are much exploited because, for piezo-electric crystals, they can be readily generated and detected by means of comb-shaped electrodes made by photolithography (Chap. 2 of Vol. II). This method of generation, introduced in 1965 by White and Voltmer [5.2], has been the main reason for the usage of the wave in electronics, 80 years after its discovery (see Historical Survey at the beginning of the book). The lengths and the spacings of the fingers in the combs can be chosen at will, and consequently a variety of functions can be synthesized. Some hundreds of millions of Rayleigh-wave filters are produced each year (Chap. 4 of Vol. II), justifying the emphasis of this chapter on this topic.

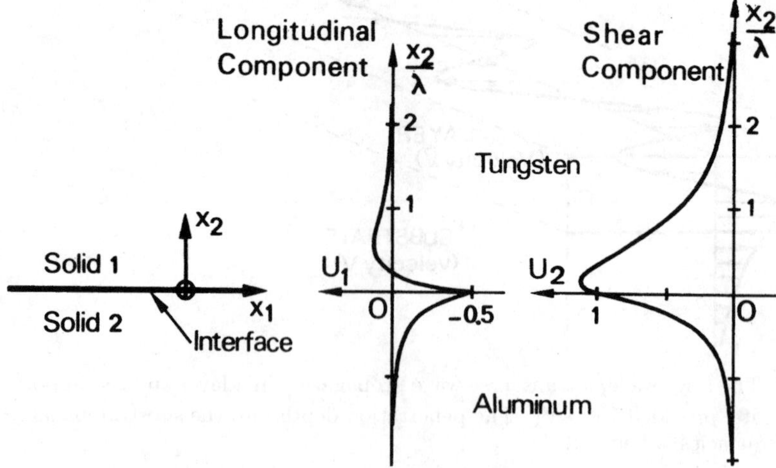

Fig. 5.8. The Stoneley wave, which can propagate along the interface between particular pairs of solids. Example showing the decrease of the longitudinal and transverse components for the pair aluminium and tungsten [5.3]

Around the 1960s, there were attempts to exploit the dispersive propagation of Love waves in matched filters for radar signals, but this method was outweighed by the ability of simple Rayleigh-wave comb transducers to produce dispersion by employing pitch variations.

Lamb waves are used in sensors.

5.2 Power Flow

Figure 5.9 represents the general structure of a waveguide, infinite in the x_1 direction, limited laterally by an 'acoustic' material boundary Σ_a and by an electrical boundary Σ_e enclosing Σ_a. External mechanical forces, with density p_i per unit area, are exerted on the boundary Σ_a. The electrical boundary carries electric charges with density σ.

It was shown in Sect. 3.1.4 that the power per unit volume generated by volume mechanical and electrical sources is

$$\frac{\mathrm{d}w}{\mathrm{d}t} = \frac{\mathrm{d}}{\mathrm{d}t}(e_\text{k} + e_\text{p}) + \frac{\partial P_j}{\partial x_j}, \tag{5.11}$$

where P_j is the Poynting vector given by

$$P_j(x_i, t) = -T_{ij}\frac{\partial u_i}{\partial t} + \Phi\frac{\partial D_j}{\partial t}, \tag{5.12}$$

and e_k and e_p are the volume densities of kinetic and potential energy.

In the case of a waveguide, the powers of the mechanical and electrical sources at the surfaces, $p_i \mathrm{d}u_i/\mathrm{d}t$ and $\Phi\,\mathrm{d}\sigma/\mathrm{d}t$, are involved in the energy balance. Designating by S a normal section of the guide, the densities E_k and E_p of kinetic energy and potential energy per unit length in the x_1 direction are

$$E_\mathrm{k}(x_1,t) = \int_S e_\mathrm{k}(x_i,t)\mathrm{d}x_2\,\mathrm{d}x_3 \quad \text{and} \quad E_\mathrm{p}(x_1,t) = \int_S e_\mathrm{p}(x_i,t)\mathrm{d}x_2\,\mathrm{d}x_3\,. \tag{5.13}$$

Integrating (5.11) over S, the total power produced per unit length is

$$\frac{\mathrm{d}W}{\mathrm{d}t} = \frac{\mathrm{d}}{\mathrm{d}t}(E_\mathrm{k}+E_\mathrm{p}) + \int_S \frac{\partial P_j}{\partial x_j}\mathrm{d}x_2\,\mathrm{d}x_3 + \int_{C_\mathrm{a}} p_i \frac{\partial u_i}{\partial t}\mathrm{d}c + \int_{C_\mathrm{e}} \Phi\,\frac{\partial \sigma}{\partial t}\mathrm{d}c\,. \tag{5.14}$$

The last two integrals represent the power produced by the mechanical and electrical surface sources; dc is the length element along the acoustic contour C_a or electric contour C_e of the normal Section S of the guide. The power P crossing the section of the guide is

$$P(x_1,t) = \int_S P_1(x_1,x_2,x_3,t)\mathrm{d}x_2\mathrm{d}x_3\,. \tag{5.15}$$

Applying the divergence theorem in two dimensions, the first integral of (5.14) becomes

$$\begin{aligned}\int_S \frac{\partial P_j}{\partial x_j}\mathrm{d}x_2\mathrm{d}x_3 &= \frac{\partial P(x_1,t)}{\partial x_1} + \int_S \left(\frac{\partial P_2}{\partial x_2} + \frac{\partial P_3}{\partial x_3}\right)\mathrm{d}x_2\mathrm{d}x_3 \\ &= \frac{\partial P}{\partial x_1} + \int_C P_\mathrm{n}\mathrm{d}c\,.\end{aligned}$$

Here P_n is the component of the Poynting vector normal to the contour C, taken to be C_a or C_e according to the variable being considered. Using (5.12) this is given by

$$P_\mathrm{n} = P_i l_i = -T_i(\boldsymbol{l})\frac{\partial u_i}{\partial t} + \Phi\frac{\partial D_\mathrm{n}}{\partial t}\,, \tag{5.16}$$

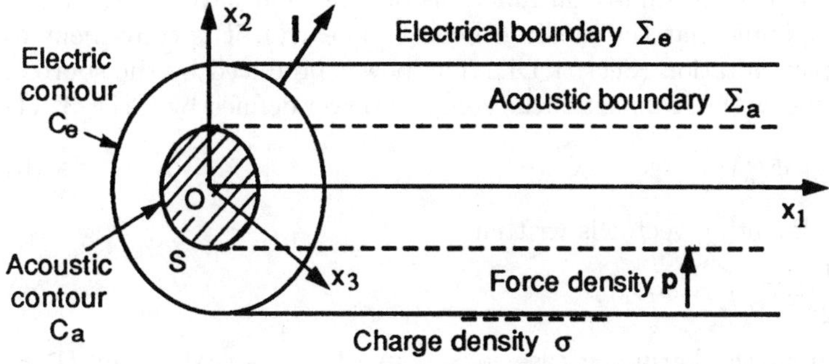

Fig. 5.9. General structure of a guide. The waves propagate along x_1

where $T_i(\boldsymbol{l})$ is the mechanical traction and $D_{\rm n}$ the normal component of the electric displacement. Substituting into (5.14) gives

$$\frac{\mathrm{d}W}{\mathrm{d}t} = \frac{\mathrm{d}}{\mathrm{d}t}(E_{\rm k} + E_{\rm p}) + \frac{\partial P}{\partial x_1} + \int_{C_{\rm a}} [p_i - T_i(\boldsymbol{l})] \frac{\partial u_i}{\partial t}\mathrm{d}c + \int_{C_{\rm e}} \Phi \frac{\partial(\sigma + D_{\rm n})}{\partial t}\mathrm{d}c\,.$$

Using the boundary conditions (3.23) to (3.27), the integrals over the contours $C_{\rm a}$ and $C_{\rm e}$ are expressed in terms of variables relating to the external medium, which are indicated by primes. This gives

$$\frac{\mathrm{d}W}{\mathrm{d}t} = \frac{\mathrm{d}}{\mathrm{d}t}(E_{\rm k} + E_{\rm p}) + \frac{\partial P}{\partial x_1} + \int_{C_{\rm a}} -T_i'(\boldsymbol{l}) \frac{\partial u_i'}{\partial t}\mathrm{d}c + \int_{C_{\rm e}} \Phi' \frac{\partial D_{\rm n}'}{\partial t}\mathrm{d}c\,. \quad (5.17)$$

Here the first integral, which gives the radiation of mechanical energy into the external medium, is zero if either

(a) the mechanical surface $\Sigma_{\rm a}$ is free, so that $T_i'(\boldsymbol{l}) = 0$; or
(b) the external medium is infinitely rigid, so that $u_i' = 0$.

The second integral gives the flow of electrical energy into the external medium, and is zero if either

(a) the surface $\Sigma_{\rm e}$ is metallized (short-circuit condition), so that $\Phi' = 0$; or
(b) the external medium has zero permittivity (open-circuit condition), so that $D_{\rm n}' = 0$.

Thus, the law of energy conservation is expressed, for any section at position x_1 and for any time t, by

$$\frac{\mathrm{d}W}{\mathrm{d}t} = \frac{\mathrm{d}}{\mathrm{d}t}(E_{\rm k} + E_{\rm p}) + \frac{\partial P}{\partial x_1}\,. \quad (5.18)$$

These non-dissipative boundary conditions are in practice very important; in the harmonic case, they lead to conservation of the average power carried by the wave in source-free regions of the guide, as shown in Problem 5.1.

5.2.1 Harmonic Case

When the variables are sinusoidal functions of time, and have the same frequency ω (implying that the medium behaves linearly), it is convenient to use the complex notation (Sect. 1.1.1). The power produced by the sources, per unit volume, is the real part of the complex power defined by (Sect. 3.1.4)

$$\frac{1}{2}(f_i \dot{u}_i^* + \Phi \dot{\rho}_{\rm e}^*)\,. \quad (5.19)$$

The complex Poynting vector is written

$$\overline{P}_j = \frac{1}{2}(-T_{ij}\dot{u}_i^* + \Phi \dot{D}_j^*)\,. \quad (5.20)$$

Knowing that in the harmonic case $\ddot{u}_i = \mathrm{i}\omega\dot{u}_i$, $\dot{S}_{ij}^* = -\mathrm{i}\omega^* S_{ij}^*$ and $\dot{D}_j^* = -\mathrm{i}\omega^* D_j^*$, the relation (3.30) becomes, after simplification by the factor $\exp[\mathrm{i}(\omega - \omega^*)t]$,

$$\frac{1}{2}(f_i\dot{u}_i^* + \Phi\dot{\rho}_e^*) = \frac{i\omega}{2}\rho\dot{u}_i\dot{u}_i^* - \frac{i\omega^*}{2}(T_{ij}S_{ij}^* + E_jD_j^*) + \frac{\partial\overline{P}_j}{\partial x_j}.$$

The quantity $\rho\dot{u}_i\dot{u}_i^*$ is equal to four times the mean value of the kinetic energy, which is

$$\langle e_k\rangle = \frac{1}{2}\mathrm{Re}\left[\frac{1}{2}\rho\dot{u}_i\dot{u}_i^*\right] = \frac{1}{4}\rho\dot{u}_i\dot{u}_i^*.$$

If the medium is *linear and non-dissipative*, so that the constants are real, the quantity $T_{ij}S_{ij}^* + E_jD_j^*$ is real and equal to four times the internal energy density, given by

$$\langle e_p\rangle = \frac{1}{2}\mathrm{Re}\left[\frac{1}{2}T_{ij}S_{ij}^* + E_jD_j^*\right] = \frac{1}{4}(T_{ij}S_{ij}^* + E_jD_j^*).$$

Thus the complex Poynting theorem takes the form

$$\frac{1}{2}(f_i\dot{u}_i^* + \Phi\dot{\rho}_e^*) = 2i\left(\omega\langle e_k\rangle - \omega^*\langle e_p\rangle\right) + \frac{\partial\overline{P}_j}{\partial x_j}$$

or, on taking $\omega = \omega' + i\omega''$,

$$\frac{1}{2}(f_i\dot{u}_i^* + \Phi\dot{\rho}_e^*) = 2i\,\omega'\langle e_k - e_p\rangle - 2\omega''\langle e_k + e_p\rangle + \frac{\partial\overline{P}_j}{\partial x_j}. \quad (5.21)$$

In the following, we assume that the sources are only of electrical origin. For a waveguide consisting of an insulating medium, the electric charges can only be on the surface. Using (3.28), the power delivered by the electric sources at the surface is, per unit length along x_1,

$$\int_{C_e}\Phi\dot{\sigma}\,\mathrm{d}c = \int_{C_e}\Phi(J_n - J_n')\mathrm{d}c. \quad (5.22)$$

Since the potential Φ on the surface Σ_e is constant on the contour C_e at position x_1, and the conduction current density J_i is zero inside the insulating guide, we have

$$\int_{C_e}\Phi\dot{\sigma}\,\mathrm{d}c = \int_{C_e}-\Phi J_n'\mathrm{d}c = \Phi j\,, \quad \text{where} \quad j = \int_{C_e}-J_n'\mathrm{d}c \quad (5.23)$$

is the density of current entering per unit length, the minus sign being present because the normal is directed outwards.

We now derive the complex power per unit length along the guide. If there is no dissipation at the lateral surfaces Σ_a and Σ_e, this is obtained, as in the real case, by integrating over a normal section, so that

$$\frac{1}{2}\Phi j^* = 2i\,\omega'\langle E_k - E_p\rangle - 2\omega''\langle E_k + E_p\rangle + \frac{\partial\overline{P}}{\partial x_1}, \quad (5.24)$$

where E_k and E_p are the densities of kinetic and potential energy per unit length, and

$$\overline{P}(x_1,t) = \int_S \overline{P}_1(x_i,t)\mathrm{d}x_2\mathrm{d}x_3 \tag{5.25}$$

is the complex power transported along the guide axis.

All variables associated with normal modes of a guide *homogeneous* along x_1 have the form of (5.1). Thus, if we take $k = k' + \mathrm{i}k''$ then products of the type ab^* give

$$ab^* = a_0 b_0^*\, \mathrm{e}^{\mathrm{i}(k^*-k)x_1} = a_0 b_0^*\, \mathrm{e}^{2k''x_1} .$$

Thus,

$$\Phi j^* = \Phi_0 j_0^*\, \mathrm{e}^{2k''x_1} \quad \text{and} \quad \overline{P} = \frac{1}{2}\int_S \left(-T_{i1}\dot{u}_i^* + \Phi \dot{D}_1^*\right) \mathrm{d}x_2\mathrm{d}x_3 = \overline{P}_0\, \mathrm{e}^{2k''x_1} .$$

After simplifying by removing the factor $\exp(2k''x_1)$, (5.24) becomes

$$\boxed{\frac{1}{2}\Phi_0 j_0^* = 2\mathrm{i}\,\omega'\langle E_\mathrm{k} - E_\mathrm{p}\rangle - 2\omega''\langle E_\mathrm{k} + E_\mathrm{p}\rangle + 2k''\overline{P}_0} . \tag{5.26}$$

Thus the complex power supplied by the electrical sources is given by the complex Poynting vector and the sum and difference of the kinetic and potential energies per unit length of the guide.

5.2.2 Susceptance

In the steady-state case, and in the absence of dissipation, ω and k are real. The above equation then shows that the complex power is purely imaginary, since $\omega'' = k'' = 0$. The potential Φ_0 and the current density j_0 are therefore in phase quadrature. The most general linear relation between these quantities is

$$j_0 = \mathrm{i}B(\omega,k)\Phi_0 , \tag{5.27}$$

where $B(\omega,k)$ is called the susceptance per unit length of the guide. This is a real function when ω and k are real.

If ω'' and k'' are vanishingly small, but not zero, we write $\omega'' = \delta\omega$ and $k'' = \delta k$ and (5.26) becomes

$$\begin{aligned} &\Phi_0 \left[\mathrm{i}B(\omega,k) - \delta\omega\frac{\partial B}{\partial \omega} - \delta k\frac{\partial B}{\partial k}\right]^* \Phi_0^* \\ &\quad = 4\mathrm{i}\omega\langle E_\mathrm{k} - E_\mathrm{p}\rangle - 4\delta\omega\langle E_\mathrm{k} + E_\mathrm{p}\rangle + 4\delta k\left(\langle P\rangle + \mathrm{i}\,\mathrm{Im}\left[\mathrm{P}_0\right]\right) , \end{aligned}$$

where $\langle P\rangle = \mathrm{Re}[\overline{P}_0]$ is the mean value of the power crossing the guide section. Equating the real parts of the two sides gives

$$\langle E_\mathrm{k} + E_\mathrm{p}\rangle = \frac{1}{4}\frac{\partial B}{\partial \omega}\,|\Phi_0|^2 \quad \text{and} \quad \langle P\rangle = -\frac{1}{4}\frac{\partial B}{\partial k}\,|\Phi_0|^2 \tag{5.28}$$

and the imaginary parts give, noting that δk is indefinitely small,

$$\langle E_\mathrm{k} - E_\mathrm{p}\rangle = -\frac{1}{4}\frac{B}{\omega}\,|\Phi_0|^2 . \tag{5.29}$$

Finally, in the sinusoidal case the mean values of the kinetic and potential energies and the power transported are given by

$$\boxed{\langle E_\mathrm{k}\rangle = \frac{1}{8}\left(\frac{\partial B}{\partial\omega} - \frac{B}{\omega}\right)|\Phi_0|^2,\ \langle E_\mathrm{p}\rangle = \frac{1}{8}\left(\frac{\partial B}{\partial\omega} + \frac{B}{\omega}\right)|\Phi_0|^2,\ \langle P\rangle = -\frac{1}{4}\frac{\partial B}{\partial k}|\Phi_0|^2}\,. \tag{5.30}$$

Since the kinetic and potential energies, E_k and E_p, are positive, we see that

$$\frac{\partial B}{\partial\omega} \geq \left|\frac{B}{\omega}\right| > 0\,.$$

This inequality, equivalent to Foster's inequality for electrical impedances, shows that the waveguide susceptance is always an increasing function of frequency [5.4].

5.2.3 Free Modes

When there are no sources ($j = 0$), the normal modes of the guide, also called free modes, satisfy the equation

$$B(\omega, k) = 0\,, \tag{5.31}$$

which expresses the dispersion relation between ω and k. To obtain the group velocity we differentiate this relation, so that

$$\mathrm{d}B = \frac{\partial B}{\partial\omega}\mathrm{d}\omega + \frac{\partial B}{\partial k}\mathrm{d}k = 0 \quad \text{giving} \quad V_\mathrm{g} = \frac{\mathrm{d}\omega}{\mathrm{d}k} = -\frac{\partial B}{\partial k}\bigg/\frac{\partial B}{\partial\omega}\,.$$

From (5.28), the energy velocity is

$$V_\mathrm{e} = \frac{\langle P\rangle}{\langle E_\mathrm{k} + E_\mathrm{p}\rangle} = -\frac{\partial B}{\partial k}\bigg/\frac{\partial B}{\partial\omega} = V_\mathrm{g} \tag{5.32}$$

and is thus equal to the group velocity V_g. For these free modes the mean values of the kinetic and potential energies are equal since, from (5.30) with $B = 0$,

$$\langle E_\mathrm{k}\rangle = \langle E_\mathrm{p}\rangle = \frac{1}{8}\left(\frac{\partial B}{\partial\omega}\right)|\Phi_0|^2\,.$$

Note that this property is not generally true for the volume energy densities e_k and e_p. The equality $e_\mathrm{k} = e_\mathrm{p}$ holds only for plane waves (Sect. 4.2.4). In a wave-guide, these densities vary from point to point over the guide cross-section S, but the equality is obtained when they are integrated over the cross-section. It follows that the mean power transported along the guide axis can be expressed in terms of only the displacement u_i of the wave, so that

$$\langle P\rangle = V_\mathrm{g}\int_S \frac{1}{2}\rho\,|\dot{u}_i|^2\,\mathrm{d}x_2\mathrm{d}x_3 = \frac{1}{2}\rho\,\omega^2 V_\mathrm{g}\int_S |u_i|^2\,\mathrm{d}x_2\mathrm{d}x_3\,. \tag{5.33}$$

In summary, for no dissipation we have the following:

(a) the energy velocity is equal to the group velocity;
(b) per unit length, the mean kinetic energy is equal to the mean potential energy; and
(c) the mean power transported by the guided wave can be expressed in terms of the displacement alone.

5.3 Rayleigh Waves

In this section we first explain the general search method for these surface waves and deduce the modes for the symmetries considered in Sect. 5.1, and then we treat the analytically accessible case of an isotropic solid. The application to anisotropic materials rapidly leads to equations which can only be solved numerically, except for some particular symmetries. For a piezoelectric solid, the electrical boundary conditions are formulated using the idea of *surface permittivity*. Consideration of different cases – a free surface, a metallized surface, or a surface in contact with a hypothetical medium of zero permittivity – leads to the definition of a coupling coefficient applicable to any material. We relate the amplitudes of the mechanical displacement and electric potential of the wave to the power density, and we tabulate the characteristics of the commonest materials.

5.3.1 Search Procedure

The coordinate axes are taken to be as in Fig. 5.1, so that Ox_2 is normal to the surface and directed into the solid, and Ox_1 is the propagation direction. The function

$$u_l = {}^{\circ}u_l \, \mathrm{e}^{-\chi k x_2} \, \mathrm{e}^{\mathrm{i}(\omega t - k x_1)}, \quad \text{with} \quad \mathrm{Re}[\chi k] > 0 \tag{5.34}$$

represents an inhomogeneous wave whose displacement components u_1, u_2, u_3 and electric potential $\Phi = u_4$ decrease with the depth x_2. Substituting this into the wave equation (5.2) yields a linear and homogeneous system of equations similar to that of (5.4). In the expansions of (5.3) for Γ_{il}, γ_l and ε, the derivative with respect to $X_2 = kx_2$ is replaced by $-\chi$ and the second derivative by $\chi^2 = -(\mathrm{i}\chi)^2$, giving

$$\begin{cases} \Gamma_{il} = c^{\mathrm{E}}_{i11l} + c^{\mathrm{E}}_{i22l}(\mathrm{i}\chi)^2 - (c^{\mathrm{E}}_{i12l} + c^{\mathrm{E}}_{i21l})(\mathrm{i}\chi) \\ \gamma_l = e_{11l} + e_{22l}(\mathrm{i}\chi)^2 - (e_{12l} + e_{21l})(\mathrm{i}\chi) \\ \varepsilon = \varepsilon^{\mathrm{S}}_{11} + \varepsilon^{\mathrm{S}}_{22}(\mathrm{i}\chi)^2 - 2\varepsilon^{\mathrm{S}}_{12}(\mathrm{i}\chi) \,. \end{cases} \tag{5.35}$$

Taking the velocity V as the variable to be solved for, the condition for compatibility of the system (5.4) is an eighth-order equation in $(\mathrm{i}\chi)$. The coefficients of the powers of $(\mathrm{i}\chi)$ are real, so the roots will be four pairs of complex conjugates. The only acceptable solutions are the four $\mathrm{i}\chi_r$ with

$r = 1, 2, 3, 4$, such that $\mathrm{Re}[\chi_r k] > 0$. The others, with $\mathrm{Re}[\chi_r k] < 0$, lead to a wave whose amplitude increases with x_2. For each of the values χ_r, the equations determine the mechanical displacements ${}^{\circ}u_l^{(r)}$ and the potential $\Phi_0^{(r)} = {}^{\circ}u_4^{(r)}$, apart from an arbitrary constant multiplier. The general solution is a combination of these four partial waves, with the same velocity V, so that

$$u_l = \left(\sum_{r=1}^{4} A_r \, {}^{\circ}u_l^{(r)} \mathrm{e}^{-\chi_r X_2} \right) \mathrm{e}^{\mathrm{i}(\omega t - k x_1)}, \quad X_2 = k x_2 \,. \tag{5.36}$$

The *mechanical* boundary conditions at the free surface $x_2 = 0$ are (Fig. 5.2)

$$T_{i2} = 0, \quad \text{that is,} \quad T_\alpha = 0, \quad \text{with} \quad \alpha = 2, 4, 6 \quad \text{and} \quad x_2 = 0 \,, \tag{5.37}$$

and these are written as

$$\sum_{r=1}^{3} t_\alpha^{(r)} A_r = -t_\alpha^{(4)} A_4 \,. \tag{5.38}$$

Here the coefficients $t_\alpha^{(r)}$ do not depend on the amplitudes A_r. If the solid is not piezoelectric, A_4 is zero and the velocity V is determined by the compatibility condition for the linear homogeneous system (5.38) (see Appendix B). If the solid is *piezoelectric*, then (5.38) gives the coefficients A_1, A_2 and A_3 in terms of A_4. The velocity is then affected by the *electrical* boundary conditions, as shown in Sect. 5.3.4.2 below. Generally there is only one root $V = V_\mathrm{R}$ that is acceptable, that is, such that $\mathrm{Re}[\chi_r k] > 0$. This gives a Rayleigh wave designated here by $\overline{\mathrm{R}}_3$; its mechanical displacement has three components and it is accompanied by an electric field, which must be in the sagittal plane $x_1 x_2$ since $E_3 = -\partial\Phi/\partial x_3 = 0$.

There are however two solutions in the two cases for which the wave equations and the boundary conditions cause decoupling, discussed in Sect. 5.1. These cases are:

(a) *The sagittal plane is normal to a direct binary axis* (A_2, A_4 or A_6). This gives a non-piezoelectric Rayleigh wave R_2 polarized in this plane (so $u_3 = 0$), and a piezoelectric transverse horizontal wave $\overline{\mathrm{TH}}$ (or B) which can propagate independently.

(b) *The sagittal plane is parallel to a mirror plane.* This gives a piezoelectric Rayleigh wave $\overline{\mathrm{R}}_2$, polarized in the sagittal plane, and a TH wave which can propagate independently. Of course, the TH wave is not a surface wave.

The application of these results to the various classes of crystal symmetry leads to the table of Fig. 5.10.

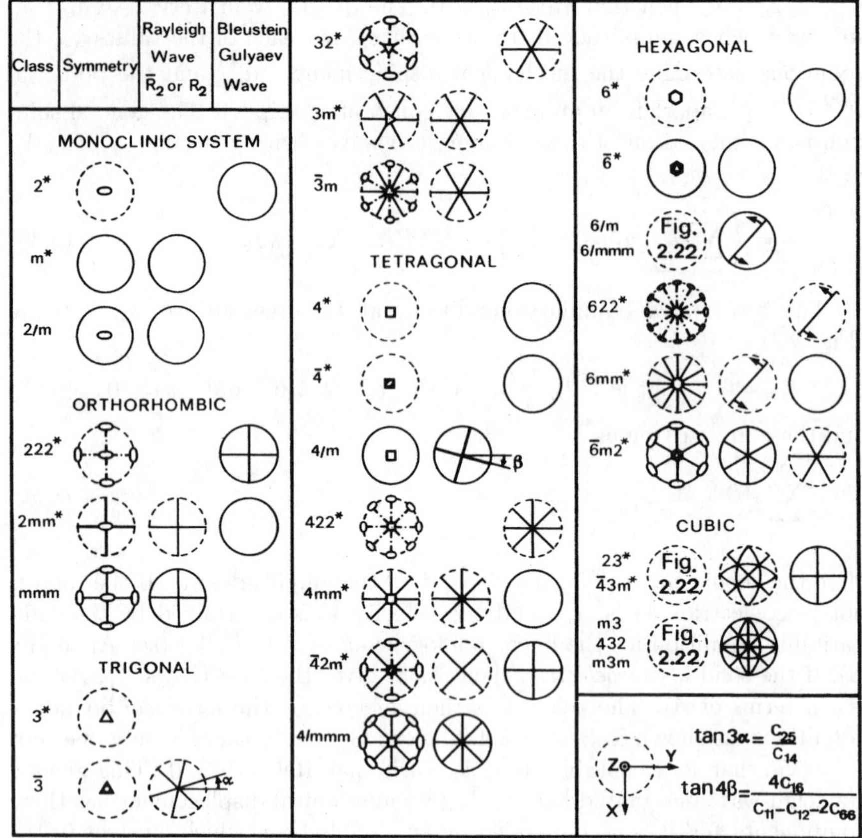

Fig. 5.10. Sagittal planes for piezoelectric transverse waves $\overline{\mathrm{TH}}$, and for two-component Rayleigh waves, $\overline{\mathrm{R}}_2$ or R_2, depending on whether the crystal is or is not piezoelectric. Piezoelectric classes are indicated by asterisks. α and β are the angles through which the coordinate axes must be rotated to cancel c_{25} and c_{16} (Problems 4.7, 4.8)

5.3.2 Isotropic Solid. Solutions for Components

Equation (5.34), which represents an inhomogeneous wave (Sect. 4.4.1.2) because its amplitude varies with x_2, can be put into the following form similar to that of a plane wave:

$$u_l = {}^{\circ}u_l \, \mathrm{e}^{\mathrm{i}(\omega t - k_1 x_1 - k_2 x_2 - k_3 x_3)} .$$

Here, the component k_2 of the wave vector $\boldsymbol{k}$ is complex, so that $k_1 = k$, $k_2 = -\mathrm{i}\chi k$ and $k_3 = 0$. The results of Sect. 4.2.3 can be applied here.

Since a Rayleigh wave in an isotropic solid is polarized in the sagittal plane, the decay constants χ_1 and χ_2 can be obtained from the propagation equation (5.5). However, this is unnecessary. For an isotropic solid we know that the velocity of the bulk longitudinal wave, $V_L = (c_{11}/\rho)^{1/2}$ with polarization ${}^\circ\boldsymbol{u}^{(1)}$, and that of the bulk transverse wave, $V_T = (c_{66}/\rho)^{1/2}$ with polarization ${}^\circ\boldsymbol{u}^{(2)}$, are independent of the direction of the wave vector. This property can be expressed in terms of the dispersion relation $\omega^2 = V^2(k_1^2+k_2^2+k_3^2)$, giving

$$\begin{cases} \omega^2 = V_L^2 k^2(1-\chi_1^2) & \text{for the partial wave} \quad {}^\circ\boldsymbol{u}^{(1)} = (1, -\mathrm{i}\chi_1, 0)\|\boldsymbol{k} \\ \omega^2 = V_T^2 k^2(1-\chi_2^2) & \text{for the partial wave} \quad {}^\circ\boldsymbol{u}^{(2)} = (\mathrm{i}\chi_2, 1, 0)\perp\boldsymbol{k}\,. \end{cases}$$

The Rayleigh-wave phase velocity $V = \omega/k$ must be less than V_L and V_T so that the real parts of the decay constants

$$\chi_1 = (1 - V^2/V_L^2)^{1/2} \quad \text{and} \quad \chi_2 = (1 - V^2/V_T^2)^{1/2} \tag{5.39}$$

are positive.

The two components of the total wave are, from (5.36),

$$\begin{cases} u_1 = \left(A_1\,\mathrm{e}^{-\chi_1 X_2} + \mathrm{i}\chi_2\,A_2\mathrm{e}^{-\chi_2 X_2}\right)\mathrm{e}^{\mathrm{i}(\omega t - kx_1)} \\ u_2 = \left(-\mathrm{i}\chi_1 A_1\,\mathrm{e}^{-\chi_1 X_2} + A_2\,\mathrm{e}^{-\chi_2 X_2}\right)\mathrm{e}^{\mathrm{i}(\omega t - kx_1)}\,, \end{cases} \tag{5.40}$$

in which the coefficients A_1 and A_2 are determined by the mechanical boundary conditions, Fig. 5.2. Using $\partial/\partial x_1 = -ik$ and $\partial/\partial x_2 = k\partial/\partial X_2$, we have

$$T_\alpha = k\left[-\mathrm{i}c_{\alpha 1}u_1 + c_{\alpha 6}\left(-\mathrm{i}u_2 + \frac{\partial u_1}{\partial X_2}\right) + c_{\alpha 2}\frac{\partial u_2}{\partial X_2}\right]_{X_2=0} = 0\,,$$

for $\alpha = 6, 2$. With $c_{61} = c_{62} = 0$, we obtain

$$\begin{cases} 2\chi_1 A_1 + \mathrm{i}(1+\chi_2^2)A_2 = 0 & \alpha = 6 \quad (5.41) \\ \mathrm{i}(-c_{12} + c_{22}\chi_1^2)A_1 + \chi_2(c_{12} - c_{22})A_2 = 0 & \alpha = 2\,. \quad (5.42) \end{cases}$$

Making use of the relations

$$c_{22} - c_{12} = 2c_{66} \quad \text{and} \quad c_{22}\chi_1^2 - c_{12} = 2c_{66} - \rho V^2 = c_{66}(1+\chi_2^2)\,,$$

the compatibility condition is

$$(1+\chi_2^2)^2 - 4\chi_1\chi_2 = 0 \tag{5.43}$$

and, on substituting for χ_1 and χ_2 using (5.39), this leads to the following equation for the Rayleigh wave velocity V_R:

$$16\left(1 - \frac{V^2}{V_L^2}\right)\left(1 - \frac{V^2}{V_T^2}\right) - \left(2 - \frac{V^2}{V_T^2}\right)^4 = 0\,.$$

Here the constant term cancels, so the equation is of third degree in V^2. Defining $R = V^2/V_T^2$, it can be put into the form

$$R^3 - 8(R-1)(R - 1 - c_{12}/c_{11}) = 0 \tag{5.44}$$

as established by Lord Rayleigh in 1885. When c_{12}/c_{11} is between 0 to 1, the permitted range for an isotropic solid, this equation has only one positive

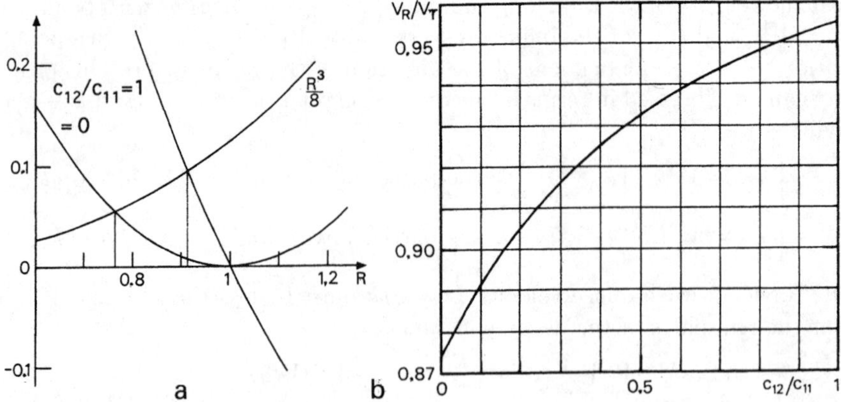

Fig. 5.11. Rayleigh wave in an isotropic solid. (**a**) Graphical solution to (5.44) in the limiting cases $c_{12}/c_{11} = 0$ and $c_{12}/c_{11} = 1$. (**b**) Variation of V_R/V_T, the Rayleigh-wave velocity normalized to the transverse-wave velocity, with the stiffness ratio c_{12}/c_{11}

root less than unity, i.e. with $0 < V_R < V_T$. This is shown by the graphical solution of Fig. 5.11a. The unknown, R, is determined by the intersection of the cubic $y_1 = R^3/8$ with the parabola $y_2 = (R-1)(R-1-c_{12}/c_{11})$. The curves, drawn for the extreme cases $c_{12} = 0$ and $c_{12} = c_{11}$, show that only one of the roots for R is positive and less than unity, and its value is between 0.764 and 0.912. The ratio $V_R/V_T = \sqrt{R}$ is between 0.874 and 0.955, as shown in Fig. 5.11b. Viktorov [5.5] gives the formula

$$\frac{V_R}{V_T} = \frac{0.718 - (V_T/V_L)^2}{0.75 - (V_T/V_L)^2} = \frac{0.436 + c_{12}/c_{11}}{0.50 + c_{12}/c_{11}}$$

as a useful approximation for this ratio.

The fact that V_R is always less than V_T can be explained by the absence of material above the free surface, which is equivalent to reducing the stiffness constants (Problem 5.2).

Equations (5.41) and (5.43) give the ratio of the coefficients A_2 and A_1 as

$$\frac{A_2}{A_1} = \frac{2\mathrm{i}\chi_1}{1+\chi_2^2} = \mathrm{i}\left(\frac{\chi_1}{\chi_2}\right)^{1/2} .$$

From this, the longitudinal and transverse displacement components, u_1 and u_2, can be written as

$$\begin{cases} u_1 = A_1\left(\mathrm{e}^{-k\chi_1 x_2} - \sqrt{\chi_1\chi_2}\,\mathrm{e}^{-k\chi_2 x_2}\right)\mathrm{e}^{\mathrm{i}(\omega t - kx_1)} \\ u_2 = \mathrm{i}\sqrt{\dfrac{\chi_1}{\chi_2}}A_1\left(\mathrm{e}^{-k\chi_2 x_2} - \sqrt{\chi_1\chi_2}\,\mathrm{e}^{-k\chi_1 x_2}\right)\mathrm{e}^{\mathrm{i}(\omega t - kx_1)} \end{cases} . \tag{5.45}$$

These are in *phase quadrature.* The particles therefore describe an ellipse, which changes form with the depth x_2 because the amplitudes vary differ-

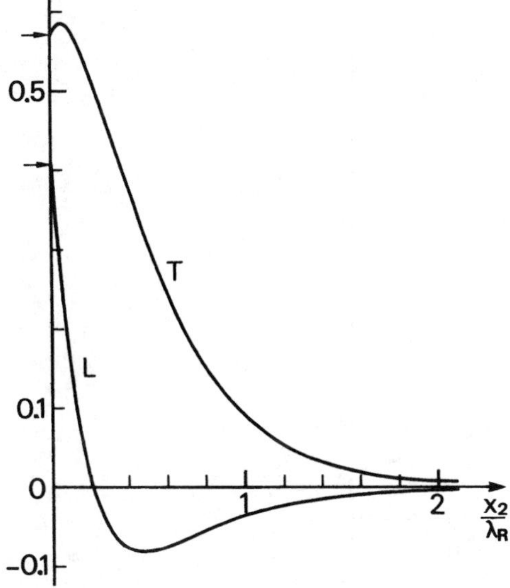

Fig. 5.12. Depth variation of the amplitudes of the longitudinal (L) and transverse (T) components of the Rayleigh wave in silica

ently with x_2. At the surface the motion is retrograde. From a depth of about $0.2\,\lambda_\mathrm{R}$, where the longitudinal component changes sign, the motion reverses and becomes prograde. The curves in Fig. 5.12, drawn for silica ($V_\mathrm{R} = 3\,410\,\mathrm{m/s}$), are typical of many isotropic materials. At the surface, the transverse component is about 1.5 times the longitudinal component.

Displacement Amplitude and Power. The magnitude of the Rayleigh-wave displacement is related to the density of power transported. From (5.33), the mean value of the elastic power depends only on the amplitudes of the displacement components u_i. Since u_i depends only on $X_2 = k_\mathrm{R} x_2$ and there is no dispersion, we have, for a beam of width w in the x_3 direction,

$$\langle P \rangle = \frac{1}{2} \rho\, \omega^2 V_\mathrm{R} w \int_0^\infty |u_i|^2 \mathrm{d}x_2 \,.$$

The power density per unit width of the beam, $P_\mathrm{R} = \langle P \rangle / w$, is expressed as

$$P_\mathrm{R} = \frac{1}{2} \rho\, \omega V_\mathrm{R}^2 \int_0^\infty |u_i(x_1, x_2, t)|^2 \mathrm{d}X_2 \,.$$

We write the displacements as

$$u_i(x_1, x_2, t) = \left(\frac{2P_\mathrm{R}}{\rho \omega V_\mathrm{R}^2} \right)^{1/2} U_i(X_2)\, \mathrm{e}^{\mathrm{i}(\omega t - k x_1)} \,, \tag{5.46}$$

where $U_i(X_2)$ is a non-dimensional function of the non-dimensional variable X_2, expressing the variation of the component u_i with depth. This function satisfies the condition

$$\int_0^\infty |U_i(X_2)|^2 \mathrm{d}X_2 = 1 \,, \tag{5.47}$$

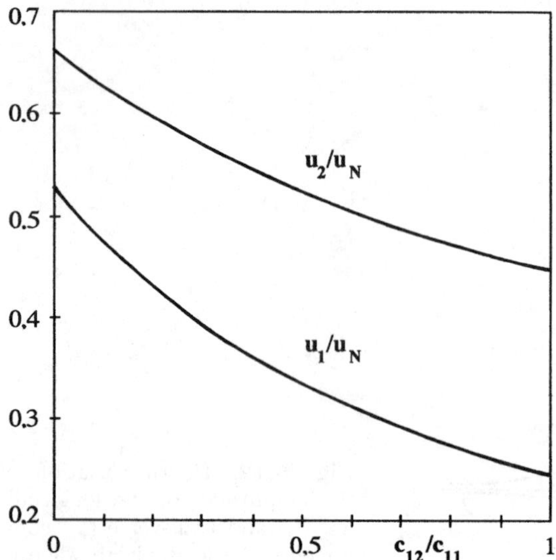

Fig. 5.13. Amplitudes of the longitudinal and transverse components u_1 and u_2 at the free surface, as functions of the ratio c_{12}/c_{11}, normalized to the parameter $u_\mathrm{N} = \sqrt{2P_\mathrm{R}/\rho\omega V_\mathrm{R}^2}$

which facilitates the calculation of the absolute displacement of the material, as shown in Problem 5.3.

Figure 5.13 shows the amplitudes of the displacements u_1 and u_2 at the surface, as functions of the ratio c_{12}/c_{11}, normalized with respect to the parameter

$$u_\mathrm{N} = \left(\frac{2P_\mathrm{R}}{\rho\omega V_\mathrm{R}^2}\right)^{1/2} . \tag{5.48}$$

Numerical example. For silica, Fig. 3.10 gives $\rho = 2\,200\,\mathrm{kg/m^3}$, $V_\mathrm{R} = 3\,410\,\mathrm{m/s}$ and $c_{12}/c_{11} = 0.205$, and Fig. 5.13 shows that $u_2/u_\mathrm{N} = 0.59$. For a frequency $f = 16\,\mathrm{MHz}$ ($\omega = 10^8$) and a power density $P_\mathrm{R} = 1\,\mathrm{W/cm}$, the normal displacement at the free surface has amplitude $u_2 = 5.2\,\mathrm{nm}$.

5.3.3 Crystals

Generally, the anisotropy of crystals can show itself in the following ways:

(a) The velocity of the Rayleigh wave depends on the direction of propagation in the plane of the free surface, though not when the surface is perpendicular to an A_6 axis (the latter case being elastically isotropic).
(b) The energy flow is no longer parallel to the propagation direction.
(c) If the sagittal plane is not a plane of elastic symmetry (i.e. mirror or a plane normal to a dyad axis), the plane of the elliptical particle displacement is *inclined* to the sagittal plane.

(d) The decay constants χ_r are no longer real – they become *complex*. The displacements therefore oscillate with depth as well as decreasing.

(e) In particular directions on particular planes, there is a solution of the Rayleigh type with velocity higher than that of the slow transverse wave. However, on moving away from this special direction, this wave is no longer strictly a surface wave; it has a component which gives rise to radiation of energy into the bulk of the material. This solution, whose displacement remains finite at infinite depth, is called a *pseudo-surface wave*.

As stated earlier, the propagation of Rayleigh waves in crystals is most frequently investigated using numerical techniques. An algebraic analysis is difficult because the mechanical displacement needs to satisfy three coupled equations of motion and three boundary conditions. However, the problem is simplified in cases where the wave has only two components and needs to satisfy only two boundary conditions. As shown in Sect. 5.1, this simplification occurs when the sagittal plane (x_1x_2) is normal to a dyad axis. It is also found that the expressions for the two boundary conditions, shown in Fig. 5.2, are simplified if one of the axes x_1 or x_2 is along a dyad axis (direct or inverse). It is then possible to derive the equation for the velocity and to deduce expressions for the displacement components. This is the purpose of the following section. The existence of pseudo-surface waves is explained in a later section.

5.3.3.1 Equation for Velocity. Mechanical Displacements. The coordinate axes are, as usual, taken to be as in Fig. 5.1, and we seek a solution with the form of (5.34) but here with $\chi = iq$. The material is assumed to be non-piezoelectric. The system of equations (5.4) (with $u_4 = 0$) splits into two parts under the conditions considered in Sect. 5.1.1, that is, when the sagittal plane x_1x_2 is perpendicular to an A_2 axis or parallel to a mirror plane M, implying that

$$c_{14} = c_{24} = c_{15} = c_{25} = c_{64} = c_{65} = 0$$

and hence

$$\Gamma_{13} = \Gamma_{23} = 0\,. \tag{5.49}$$

If, in addition, one of the axes x_1 or x_2 is along a symmetry axis, also of order two, then

$$c_{1211} = c_{61} = 0 \quad \text{and} \quad c_{1222} = c_{62} = 0\,, \tag{5.50}$$

and the two boundary conditions of Fig. 5.2 can be written as

$$T_{12} = c_{66}\left(\frac{\partial u_2}{\partial x_1} + \frac{\partial u_1}{\partial x_2}\right) = 0 \quad \text{for } x_2 = 0 \tag{5.51a}$$

and

$$T_{22} = c_{21}\frac{\partial u_1}{\partial x_1} + c_{22}\frac{\partial u_2}{\partial x_2} = 0 \quad \text{for} \quad x_2 = 0\,. \tag{5.51b}$$

Symmetry	ORTHORHOMBIC	TETRAGONAL (4, $\bar{4}$, 4/m)					
Propagation direction	A_2	$\perp Z$		$//Z$		$X+\varphi_0$	$X+\varphi_0+\pi/4$
Normal to the free surface	A'_2	$X+\varphi_0$	$X+\varphi_0+\pi/4$	$X+\varphi_0$	$X+\varphi_0+\pi/4$	$//Z$	

Symmetry	HEXAGONAL			TETRAGONAL (422, 4*mm*, $\bar{4}$2*m*, 4/*mmm*) or CUBIC					
Direction	$\perp Z$	$//Z$	$\perp Z$	$\perp Z$		$//Z$		X	$X+\pi/4$
Normal	$\perp Z$	$\perp Z$	$//Z$	X	$X+\pi/4$	X	$X+\pi/4$	$//Z$	

Fig. 5.14. The sixteen configurations for which the Rayleigh wave (R_2) is decoupled from the transverse horizontal wave (TH), and also the boundary conditions are simplified. The angle φ_0 is given by $\tan 4\varphi_0 = 4c_{16}/(c_{11} - c_{12} - 2c_{66})$

The symmetries of the two cases in (5.49) and (5.50) are satisfied only for orthorhombic, tetragonal, cubic and hexagonal crystals. With respect to the crystallographic axes, there are sixteen possible combinations, that is, orientations of the propagation direction x_1 and the surface normal x_2 [5.6]. The table of Fig. 5.14 shows that these include the three cases studied by Stoneley [5.7, 5.8] – propagation along the [100] or [110] directions in the (001) plane of a cubic crystal, and propagation in the basal plane of a hexagonal crystal.

We therefore seek a wave with the form

$$u_i = {}^\circ u_i\, \mathrm{e}^{-iqkx_2}\, \mathrm{e}^{\mathrm{i}(\omega t - kx_1)} \quad \text{with} \quad \mathrm{Im}[\mathrm{q}] < 0\,.$$

With the conditions (5.49) and (5.50) satisfied, we define $\zeta = \rho V^2$ and $\chi = \mathrm{i}q$ and the system (5.5) becomes

$$\begin{pmatrix} c_{11} + c_{66}\, q^2 - \zeta & (c_{12} + c_{66})q \\ (c_{12} + c_{66})q & c_{66} + c_{22}q^2 - \zeta \end{pmatrix} \begin{pmatrix} {}^\circ u_1 \\ {}^\circ u_2 \end{pmatrix} = 0\,. \tag{5.52}$$

The secular equation reduces to

$$c_{22}\, c_{66}\, q^4 + [c_{11}\, c_{22} - c_{12}^2 - 2c_{12}c_{66} - (c_{22} + c_{66})\zeta]q^2 + (c_{11} - \zeta)(c_{66} - \zeta) = 0$$

and hence

$$q^4 - Sq^2 + P = 0\,, \tag{5.53}$$

where P and S are the product and sum of the roots q_1^2 and q_2^2 of the quadratic equation, so that

$$\left\{ \begin{aligned} P &= q_1^2 q_2^2 = (c_{11} - \zeta)(c_{66} - \zeta)/c_{22}c_{66} && \text{(5.54a)} \\ S &= q_1^2 + q_2^2 = [2c_{12}c_{66} + c_{66}\zeta - c_{22}(c - \zeta)]/c_{22}c_{66}\,, && \text{(5.54b)} \end{aligned} \right.$$

in which $c = c_{11} - c_{12}^2/c_{22}$.

For the two acceptable solutions q_r $(r = 1, 2)$, which have negative imaginary parts, the components ${}^{\circ}u_i^{(r)}$ of the eigenvector are given by (5.52). Taking ${}^{\circ}u_1^{(r)} = 1$ and ${}^{\circ}u_3^{(r)} = 0$, and defining p_r to be equal to ${}^{\circ}u_2^{(r)}$ in this situation, we have

$$p_r = {}^{\circ}u_2^{(r)} = -\frac{c_{11} - \zeta + c_{66}\, q_r^2}{(c_{12} + c_{66}) q_r}\,, \quad \text{for} \quad r = 1, 2\,. \tag{5.55}$$

The general solution is a linear combination of these two displacements, with the same velocity $V = \omega/k$, so that

$$\begin{cases} U_1(x_2) = A_1 \mathrm{e}^{-\mathrm{i}q_1 k x_2} + A_2\, \mathrm{e}^{-\mathrm{i}q_2 k x_2} & \text{(5.56a)} \\ U_2(x_2) = A_1 p_1\, \mathrm{e}^{-\mathrm{i}q_1 k x_2} + A_2 p_2\, \mathrm{e}^{-\mathrm{i}q_2 k x_2}\,. & \text{(5.56b)} \end{cases}$$

Here the weighting factors A_1 and A_2 and the velocity V are determined by the mechanical boundary conditions (5.51a,b), which give

$$\begin{cases} (q_1 + p_1) A_1 + (q_2 + p_2) A_2 = 0 & \text{(5.57a)} \\ (c_{21} + c_{22}\, q_1 p_1) A_1 + (c_{21} + c_{22} q_2 p_2) A_2 = 0\,. & \text{(5.57b)} \end{cases}$$

For compatibility, the determinant of these two equations is set to zero, giving

$$(p_1 - p_2)(c_{12} - c_{22} q_1 q_2) + (q_1 - q_2)(c_{12} - c_{22} p_1 p_2) = 0\,. \tag{5.58}$$

From the sum S and product P of the roots, we have

$$c_{22} p_1 p_2 q_1 q_2 = c_{11} - \zeta \tag{5.59}$$

and with the value of $p_1 - p_2$ deduced from (5.55), we find that (5.58) reduces to

$$q_1 q_2 = -\zeta (c_{11} - \zeta)/c_{22}(c - \zeta)\,. \tag{5.60}$$

Substituting this expression into (5.54a) leads to the velocity equation

$$f(\zeta) = c_{66}(c_{11} - \zeta)\zeta^2 - c_{22}(c_{66} - \zeta)(c - \zeta)^2 = 0\,, \tag{5.61}$$

which resembles that established by Stoneley [5.8] for the particular case of a Rayleigh wave propagating along the basal plane of a hexagonal crystal. However, the above equation is more general, being applicable to all sixteen cases of Fig. 5.14.

$f(\zeta)$ is negative when $\zeta = 0$, and $f(\zeta)$ is positive for $\zeta = \min[c_{66}, c]$ because $c < c_{11}$. Thus there is one and only one root ζ_{R} such that $0 < \zeta_{\mathrm{R}} < \min[c_{66}, c]$. The Rayleigh-wave velocity is given by $\zeta_{\mathrm{R}} = \rho V_{\mathrm{R}}^2$.

With the velocity known, the components of the mechanical displacement can be deduced. Substituting $c_{11} - \zeta$ from (5.59) into (5.55) gives

$$p_r (c_{12} + c_{22} p_s q_s) = -c_{66}(p_r + q_r)\,, \quad r \neq s = 1, 2$$

and, using (5.57a),

$$p_1/p_2 = (A_2/A_1)^2\,. \tag{5.62}$$

If the roots q_1^2 and q_2^2 of (5.53) are real and negative, that is, if $S < -2\sqrt{P}$, the two physically acceptable solutions are $q_r = -\mathrm{i}\chi_r$, with $\chi_r > 0$. The coefficients p_r given by (5.55) are pure imaginary, so that $p_r = -\mathrm{i}a_r$. From (5.54b) we have

$$a_r = (c_{11} - \zeta - c_{66}\chi_r^2)/(c_{12} + c_{66})\chi_r \tag{5.63}$$

showing that a_r is positive. From (5.57a), the ratio A_2/A_1 is real and negative. With the relation (5.62) we have

$$A_2/A_1 = -(a_1/a_2)^{1/2} \quad \text{and} \quad p_2 A_2 = \mathrm{i}aA_1\,,$$

where a is defined, using (5.59) and (5.60), as

$$a = (-p_1 p_2)^{1/2} = (a_1 a_2)^{1/2} = (c/\zeta_\mathrm{R} - 1)^{1/2}\,.$$

The longitudinal and transverse components of the mechanical displacement, given by (5.56), are

$$\begin{cases} U_1(x_2) = A_1 \left[\mathrm{e}^{-\chi_1 k x_2} - \left(\dfrac{a_1}{a_2}\right)^{1/2} \mathrm{e}^{-\chi_2 k x_2} \right] \\ U_2(x_2) = \mathrm{i}aA_1 \left[\mathrm{e}^{-\chi_2 k x_2} - \left(\dfrac{a_1}{a_2}\right)^{1/2} \mathrm{e}^{-\chi_1 k x_2} \right] \end{cases} \tag{5.64}$$

and are thus in phase quadrature. Their decrease with depth is the same as that of the components of a Rayleigh wave in an isotropic solid, for which $c_{66} = (c_{11} - c_{12})/2$. In particular, since a_r is a decreasing function of χ_r, the transverse component does not change sign, while the longitudinal component passes through zero at a certain depth.

If $|S| < 2\sqrt{P}$ the roots of (5.53) are complex, and therefore conjugate, and the two acceptable solutions are

$$q_r = (-1)^r g - \mathrm{i}h \quad \text{with} \quad \begin{Bmatrix} g \\ h \end{Bmatrix} = \frac{1}{2}\left(2\sqrt{P} \pm S\right)^{1/2}. \tag{5.65}$$

Equation (5.55) shows that $p_1 = -p_2^* = \mathrm{i}a\exp(-\mathrm{i}2\alpha)$, and with (5.62) and (5.57a) we find

$$\mathrm{e}^{-\mathrm{i}2\alpha} = A_2/A_1 = (q_2 + p_2)^*/(q_2 + p_2)\,.$$

The two components of the Rayleigh wave, given by (5.56), are still in phase quadrature. However, their amplitudes, given by

$$\begin{cases} U_1(x_2) = 2A_1\,\mathrm{e}^{-hkx_2}\cos(gkx_2 + \alpha) \\ U_2(x_2) = 2\mathrm{i}aA_1\,\mathrm{e}^{-hkx_2}\cos(gkx_2 - \alpha) \end{cases} \tag{5.66}$$

are oscillatory functions of depth with a spatial period λ_R/g, the ratio of the wavelength λ_R to the magnitude g of the real parts of the roots ($q_2 = -q_1^*$). On the other hand, the depth variation of the mechanical displacement, dependent on the imaginary part h of these roots, can be much slower than in the previous case. The difference arises from the anisotropy of the solid in the sagittal plane.

Influence of the Anisotropy Factor. Cubic and tetragonal crystals have $c_{22} = c_{11}$ if the tetrad symmetry axis is along the x_3 axis. The anisotropy factor in the sagittal plane, $A = 2c_{66}/(c_{11} - c_{12})$, is much larger than unity if c_{12} is only a little less than c_{11}. The solution ζ_R of (5.61), necessarily less than $c = c_{11} - (c_{12})^2/c_{11}$, is then small compared with c_{11} and c_{66}, giving $\zeta_R \approx c/2$. Expanding the sum S and product P of the roots to second order in ζ_R yields the imaginary part, defined in (5.65), in the form

$$h \approx (1 + c_{66}/c_{11})/2A\,, \quad A \gg 1\,. \tag{5.67}$$

Consequently, *the oscillations decay more slowly if the anisotropy factor is increased.* This result is illustrated by the curves in Fig. 5.15, referring to propagation along the dyad axis [100] of crystals of YAG (yttrium aluminium garnet, class $m3m$), silicon (Si, $m3m$), gallium arsenide (GaAs, $\bar{4}3m$), rutile (TiO_2, $4/mmm$) and tellurium dioxide (TeO_2, 422). The stiffness constants used are from the table of Fig. 3.10. Two of these crystals, GaAs and TeO_2, are piezoelectric, but this effect was not taken into account because the sagittal plane is normal to a direct dyad axis (Sect. 5.1.1a).

It is interesting to note that, as A increases, the ratio of the Rayleigh wave velocity V_R to the transverse bulk wave velocity $V_T = (c_{66}/\rho)^{1/2}$ decreases, as can be predicted for $A \gg 1$. Thus,

$$\rho V_R^2 \cong \frac{c_{11}^2 - c_{12}^2}{2c_{11}} \quad \text{giving} \quad \frac{V_R}{V_T} \cong \left[\frac{2}{A}\left(1 - \frac{c_{66}}{Ac_{11}}\right)\right]^{1/2}. \tag{5.68}$$

For example, in tellurium dioxide V_R is almost a quarter of V_T. For an isotropic solid, $0.874 < V_R/V_T < 0.955$. In practice, (5.67) and (5.68) are good approximations if $A > 4$.

5.3.3.2 Pseudo-Surface Waves. A pseudo-surface wave appears in certain crystals when, because of the anisotropy, the Rayleigh wave velocity is greater than that of one of the bulk transverse waves.

For example, in a cubic crystal, characterized by the three constants c_{11}, c_{12} and c_{66}, a vertically-polarized transverse wave (TV) and a horizontally-polarized quasi-transverse wave (TH) can propagate in a plane such as (001), as shown in Fig. 5.16. The transverse TV wave has velocity $V_{TV} = (c_{66}/\rho)^{1/2}$, independent of direction. In contrast, the quasi-transverse TH wave has velocity $V_{TH}(\theta)$, dependent on the angle θ between the propagation direction and [100], with $V_{TH}(0) = V_{TV}$. Since the (010) and (1$\bar{1}$0) planes are planes of symmetry, the transverse-horizontal TH wave is not coupled to the displacement of a Rayleigh wave propagating along [100] or [110]; these Rayleigh waves have displacements confined to the sagittal plane. The Rayleigh wave velocity may therefore be higher than V_{TH}.

Apart from these two particular directions, there are two cases to be considered, depending on whether the anisotropy factor

$$A = \frac{2c_{66}}{c_{11} - c_{12}} = \left(\frac{V_{TV}}{V_{TH}[110]}\right)^2 \tag{5.69}$$

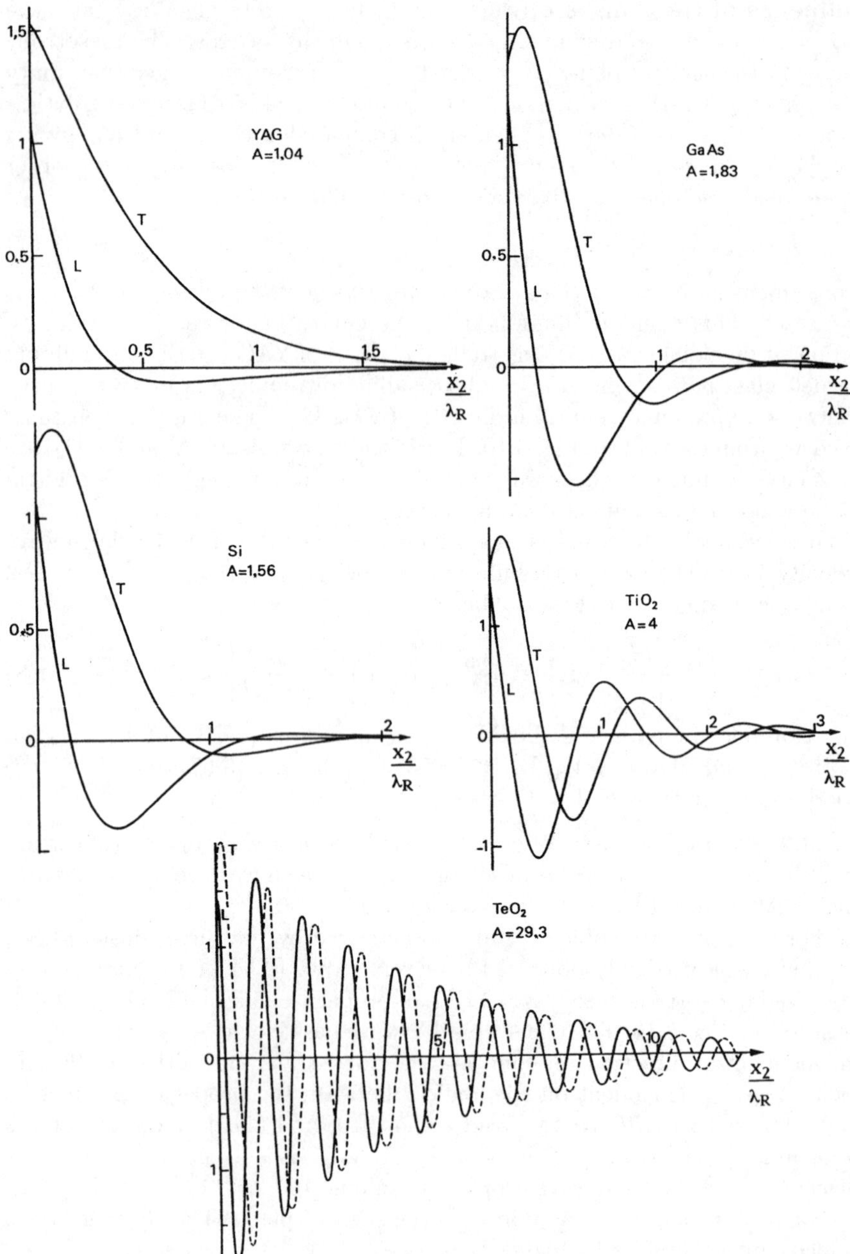

Fig. 5.15. Variation with depth of the longitudinal (L) and transverse (T) components of mechanical displacement in a Rayleigh wave propagating along the (010) surface in the [100] direction, for crystals of YAG, Si, GaAs, TiO_2 and TeO_2. The decay of the oscillations becomes slower as the anisotropy factor $A > 1$ increases

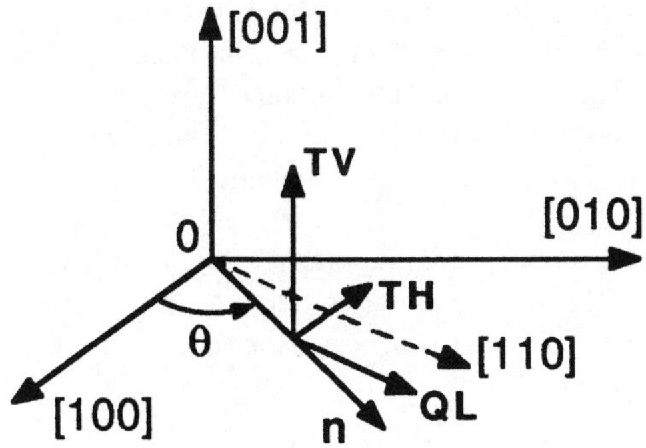

Fig. 5.16. TV and quasi-TH waves propagating in the (001) plane of a cubic crystal

is less than or greater than unity. The first case, with $A < 1$ and $V_{TH}[110] > V_{TV}$, is shown in Fig. 5.17a. The TH-wave velocity increases with θ and is maximized in the [110] direction. The coupling between this wave and the Rayleigh wave is weak, so the velocity $V_R(\theta)$ of the latter remains almost constant, and is slightly below V_{TV}.

The second case, with $A > 1$ and $V_{TV} > V_{TH}[110]$, is shown in Fig. 5.17b. The velocity $V_{TH}(\theta)$ decreases continuously with θ, starting from its value along [100], and is minimized along [110] where it is less than the Rayleigh wave velocity $V_R(\pi/4)$. As θ increases, the coupling between the TH wave and the Rayleigh wave becomes progressively stronger, and the Rayleigh wave de-

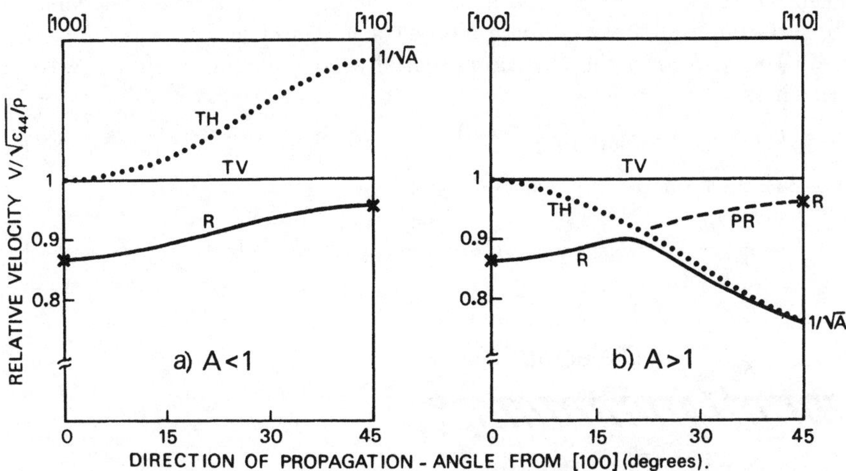

Fig. 5.17. Propagation of surface waves, of type Rayleigh (R) or pseudo-Rayleigh (PR), along the (001) plane of a cubic crystal. The anisotropy factor A is (**a**) less than, or (**b**) greater than, unity. Where the Rayleigh wave is denoted by a cross it is not coupled to the bulk TH wave, so its velocity can be greater than V_{TH}

generates into a TH wave where the two curves meet. In the reverse direction, starting with the velocity $V_{\mathrm{R}}(\pi/4)$ along [110], there is also coupling between the Rayleigh wave and the TH wave, of course. The surface wave, with velocity $V_{\mathrm{R}} > V_{\mathrm{TH}}$, generates an increasing amount of the TH wave, and so is no longer confined near the surface. The wave is now a pseudo-surface wave, or leaky surface wave.

These results for $A > 1$ and for the (001) plane, can also be found when $A < 1$ and for a plane other than (001), if the velocities behave similarly. For example, Farnell [5.9] has shown that the above observations apply qualitatively for $A < 1$ and for the (110) plane.

5.3.4 Piezoelectric Solid

The search for surface-wave solutions in a piezoelectric solid involves the consideration of electrical boundary conditions in addition to mechanical ones. However, numerical solutions show that, regarding the wave velocity and the decay of the displacement components with depth, the character of the Rayleigh wave is hardly affected by piezoelectricity. The significance of piezoelectricity lies mainly in the possibility of electrical coupling between the crystal and an external medium. The method chosen to express this electrical coupling via the surface is analogous to that of Sect. 4.3.4. We introduce the surface piezoelectric permittivity, which relates the normal component of electric displacement and the electric potential at the surface, taking account of mechanical coupling.

5.3.4.1 Surface Piezoelectric Permittivity. The piezoelectric solid is taken to be in contact with a dielectric, as in Fig. 5.18. We first consider the surface permittivity of a non-piezoelectric dielectric.

In the *dielectric*, the electric potential accompanying a surface wave is written as

$$\Phi(x_1, x_2, t) = \Phi_0(x_2)\mathrm{e}^{\mathrm{i}(\omega t - kx_1)} \tag{5.70}$$

and this must satisfy Laplace's equation

$$\varepsilon_{ij}\frac{\partial^2 \Phi}{\partial x_i \partial x_j} = \varepsilon_{22}\frac{\partial^2 \Phi_0}{\partial x_2^2} - 2\mathrm{i}k\varepsilon_{12}\frac{\partial \Phi_0}{\partial x_2} - \varepsilon_{11}k^2\Phi_0 = 0\,.$$

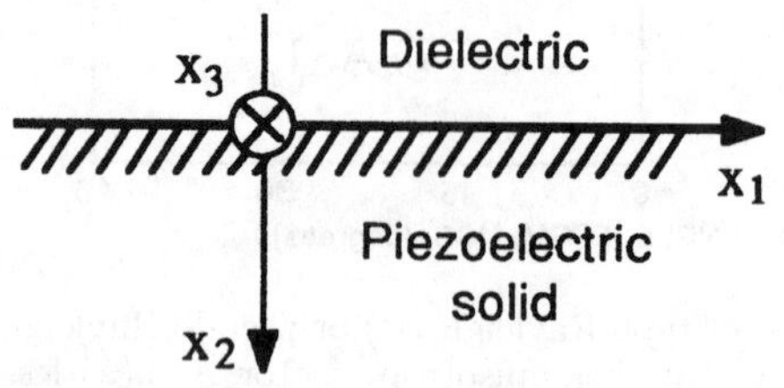

Fig. 5.18. A piezoelectric insulating crystal in contact with a non-piezoelectric material

The solution, vanishing at $x_2 = -\infty$, is $\Phi_0 = A\exp(\chi x_2)$ with $\mathrm{Re}[\chi] > 0$, and hence

$$\varepsilon_{22}\chi^2 - 2\mathrm{i}k\varepsilon_{12}\chi - \varepsilon_{11}k^2 = 0\,,$$

and since $\varepsilon_{11}\varepsilon_{22} > \varepsilon_{12}^2$ we have

$$\varepsilon_{22}\chi = \mathrm{i}k\varepsilon_{12} + |k|(\varepsilon_{11}\varepsilon_{22} - \varepsilon_{12}^2)^{1/2}\,.$$

The normal component of electric displacement is $D_2 = \varepsilon_{2i}E_i$, and this is proportional to the potential Φ so that, for any x_2,

$$-D_2 = \varepsilon_{21}\frac{\partial\Phi}{\partial x_1} + \varepsilon_{22}\frac{\partial\Phi}{\partial x_2} = (-\mathrm{i}k\varepsilon_{12} + \varepsilon_{22}\chi)\Phi\,.$$

The *surface permittivity* of the dielectric is defined by

$$\varepsilon_{\mathrm{d}} = -\frac{D_2}{|k|\Phi} = (\varepsilon_{11}\varepsilon_{22} - \varepsilon_{12}^2)^{1/2}\,. \tag{5.71}$$

This is a constant, independent of ω and k. It is necessary to use $|k|$ so that ε_{d} does not change sign with the direction of wave propagation.

For a *piezoelectric* material, the electric displacement and the potential $\Phi = u_4$ accompanying the surface elastic wave are in a ratio that is no longer independent of x_2, because the wave is a linear combination of partial waves decreasing at different rates; thus, from (5.36),

$$u_l = \sum_{r=1}^{4} A_r u_l^{(r)} \quad \text{and} \quad u_l^{(r)} = {}^{\circ}u_l^{(r)}\,\mathrm{e}^{-k\chi_r x_2}\,\mathrm{e}^{\mathrm{i}(\omega t - kx_1)}\,.$$

However, the mechanical boundary conditions (5.38) at the free surface $x_2 = 0$ show that the coefficients A_1, A_2 and A_3 are proportional to A_4. It follows that, at the surface, the electric potential

$$\Phi(0) = \sum_{r=1}^{4} A_r\,{}^{\circ}u_4^{(r)}$$

is proportional to the normal component of electric displacement (from 4.68)

$$D_2(0_+) = \frac{\partial}{\partial x_k}(e_{2kl}u_l - \varepsilon_{2k}\Phi)_{x_2=0_+}\,.$$

The *piezoelectric surface permittivity* $\overline{\varepsilon}$ is therefore defined by the relation

$$\overline{\varepsilon} = \frac{D_2(0_+)}{|k|\Phi(0)}\,. \tag{5.72}$$

Bearing in mind that $\partial/\partial x_1 = -\mathrm{i}k, \partial u_l^{(r)}/\partial x_2 = -k\chi_r$ and $\partial/\partial x_3 = 0$, we find

$$D_2(0_+) = k\sum_{r=1}^{4} A_r\left[\mathrm{i}\left(\varepsilon_{21}\,{}^{\circ}u_4^{(r)} - e_{21l}\,{}^{\circ}u_l^{(r)}\right) + \chi_r\left(\varepsilon_{22}\,{}^{\circ}u_4^{(r)} - e_{22l}\,{}^{\circ}u_l^{(r)}\right)\right]$$

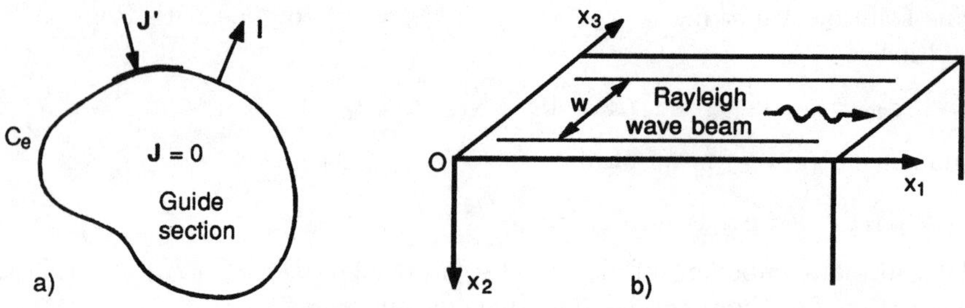

Fig. 5.19. Piezoelectric waveguide. (**a**) General case. (**b**) Plane waveguide

and, on taking ${}^{\circ}u_4^{(r)} = 1$, we have

$$\overline{\varepsilon} = \frac{k}{|k|} \frac{\sum_{r=1}^{4} A_r \left[\mathrm{i}\,\varepsilon_{21} + \chi_r \varepsilon_{22} - (\mathrm{i} e_{21l} + \chi_r e_{22l})\ {}^{\circ}u_l^{(r)} \right]}{\sum_{r=1}^{4} A_r} .$$

For a given crystal and orientation, the surface piezoelectric permittivity $\overline{\varepsilon}$ depends only on the phase velocity $V = \omega/k$.

Relation to Susceptance. In Sect. 5.2.2 the susceptance $B(\omega, k)$ was defined by the relation (5.27) between the surface potential Φ and the density j of current entering from external sources, so that (Fig. 5.19a)

$$j = \mathrm{i}B(\omega, k)\Phi \quad \text{with} \quad j = \int_{C_e} -J_n' \mathrm{d}c\,.$$

In the present case we have a plane waveguide, Fig. 5.19b, using an insulating material, so $J_n = 0$. For a Rayleigh wave whose beam width along x_3 is w, we have, noting that $J_n - J_n' = \dot{\sigma}$,

$$j = w\dot{\sigma} = \mathrm{i}B(\omega, k)\Phi\,.$$

Using the definition (5.72) for the surface permittivity, and noting that the x_2 axis is directed outwards for the dielectric, we have $\varepsilon_{\mathrm{d}} = -D_2(0_-)/|k|\Phi(0_-)$, and this gives

$$\sigma = D_2(0_+) - D_2(0_-) = (\overline{\varepsilon} + \varepsilon_{\mathrm{d}})|k|\Phi\,,$$

where we have also used the fact that the potential is continuous, so that $\Phi(0_+) = \Phi(0_-)$. The surface susceptance is therefore

$$B(\omega, k) = w\,\omega|k|[\overline{\varepsilon}(\omega/k) + \varepsilon_{\mathrm{d}}]\,. \tag{5.73}$$

Since this is real when ω and k are real, the same can be said for $\overline{\varepsilon}(V)$, assuming ε_{d} to be real.

5.3.4.2 Electrical Boundary Conditions. Velocity of Rayleigh Waves. For freely propagating waves, that is, in the absence of sources ($j = \dot{\sigma} = 0$), the dispersion relation (5.31) becomes

$$\bar{\varepsilon}(V) + \varepsilon_d = 0\,. \tag{5.74}$$

This shows that the phase velocity $V = \omega/k$ depends on the adjacent medium through its permittivity ε_d. We distinguish three important cases:

(a) If the surface is covered by a very thin *metallic film* we have the short circuit condition with $\Phi = 0$, or equivalently $\varepsilon_d = \infty$, giving

$$V = V_\infty \quad \text{and} \quad \bar{\varepsilon}(V_\infty) = \infty\,. \tag{5.75}$$

(b) If the surface is covered by a hypothetical medium with *zero permittivity* we have the open-circuit condition, with $D_2(0) = 0$, giving

$$V = V_0 \quad \text{and} \quad \bar{\varepsilon}(V_0) = 0\,. \tag{5.76}$$

(c) If the adjacent 'medium' is a *vacuum* ($\varepsilon_d = \varepsilon_0$), the Rayleigh-wave velocity V_R is between the two extremes V_∞ and V_0, so that

$$V_\infty < V_R < V_0 \quad \text{and} \quad \bar{\varepsilon}(V_R) = -\varepsilon_0\,. \tag{5.77}$$

The velocity V_∞ is less than V_0 because a metal film deposited on the surface eliminates the tangential electric field, and thus partially suppresses the piezoelectricity of the material. For a material with relatively high permittivity, such as lithium niobate, the velocities V_R and V_0 are very similar. The fractional change of velocity, defined by

$$\frac{\Delta V}{V} = \frac{V_0 - V_\infty}{V_0} \tag{5.78}$$

is always small, at most a few percent.

From the above comments, we can anticipate the form of $\bar{\varepsilon}$ as a function of the velocity $V = \omega/k$, as shown in Fig. 5.20. The resonance at velocity V_∞, lower than the bulk wave velocities, corresponds to a surface wave, and this has no attenuation because the imaginary part of $\bar{\varepsilon}$ is zero at these low velocities. The pseudo-resonances at higher velocities correspond to 'pseudo-surface waves' whose attenuation, expressed by the negative imaginary part of $\bar{\varepsilon}$, is due to radiation of elastic energy into the bulk of the material. For very high velocities, that is at high frequencies, the limiting value $\bar{\varepsilon}_{HF}$ is equal to the effective permittivity of the piezoelectric material as given by (5.71), so that

$$\bar{\varepsilon}_{HF} = \varepsilon_p^T = \left[\varepsilon_{11}^T \varepsilon_{22}^T - \left(\varepsilon_{12}^T\right)^2\right]^{1/2}, \tag{5.79}$$

here the elastic phenomena are effectively no longer coupled to electrical phenomena because they act much more slowly. The constants ε_{ij}^T must be chosen because the mechanical traction is zero at the surface. At low frequencies, the permittivity $\bar{\varepsilon}_{LF}$ is greater than the high-frequency value ε_p^T.

Given the symmetry about the axis $\omega/k = 0$, the surface permittivity can be expressed, near a resonance, by a formula analogous to that established for bulk waves in Sect. 4.3.4, so that

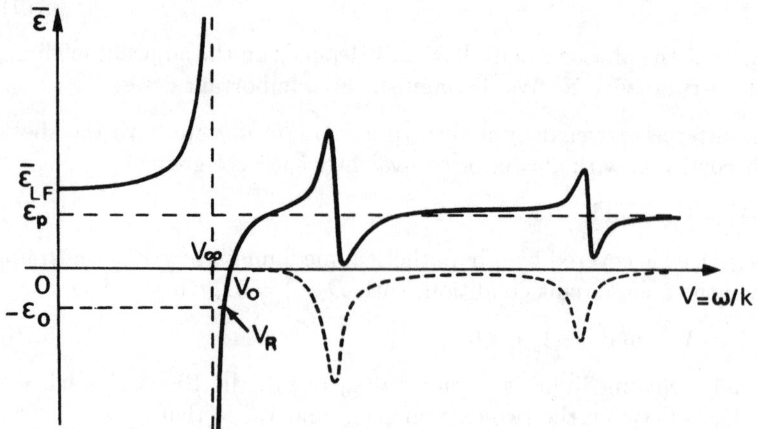

Fig. 5.20. Variation of the real (*solid line*) and imaginary (*broken line*) parts of the surface permittivity of a piezoelectric solid, as functions of the phase velocity $V = \omega/k$

$$\boxed{\bar{\varepsilon}(\omega, k) = \varepsilon_{\mathrm{p}}^{\mathrm{T}} \frac{(\omega/k)^2 - V_0^2}{(\omega/k)^2 - V_\infty^2}} . \tag{5.80a}$$

If the propagation direction of the wave is specified as $x_1 > 0$, for example, this formula can be simplified to

$$\bar{\varepsilon}(\omega, k) = \varepsilon_{\mathrm{p}}^{\mathrm{T}} \frac{V - V_0}{V - V_\infty} = \varepsilon_{\mathrm{p}}^{\mathrm{T}} \frac{\omega - kV_0}{\omega - kV_\infty} \quad k > 0 . \tag{5.80b}$$

5.3.4.3 Electromechanical Coupling Coefficient. As for bulk waves, it is useful to define a factor of merit K_{S}, called the electromechanical coupling coefficient. This factor expresses the fact that a piezoelectric material transforms a fraction K_{S}^2 of the applied electrical energy into mechanical energy at the surface; it is in the range $0 < K_{\mathrm{S}} < 1$. It is related to the fact that the surface piezoelectric permittivity is greater at low frequencies than at high frequencies.

Imagine a continuous voltage v_0 to be applied between two electrodes situated on the free surface. With the solid, these electrodes constitute a capacitor, with capacitance C. The energy supplied by the electrical source and stored in the capacitor is

$$U_{\mathrm{s}} = \frac{1}{2} C_{\mathrm{LF}} v_0^2 .$$

The static capacitance C_{LF} is proportional to $\bar{\varepsilon}_{\mathrm{LF}} + \varepsilon_{\mathrm{d}}$. If the electrodes are short-circuited at $t = 0$, as in Fig. 5.21a, the capacitor returns an electrical energy

$$U_{\mathrm{r}} = \frac{1}{2} C_{\mathrm{HF}} v_0^2$$

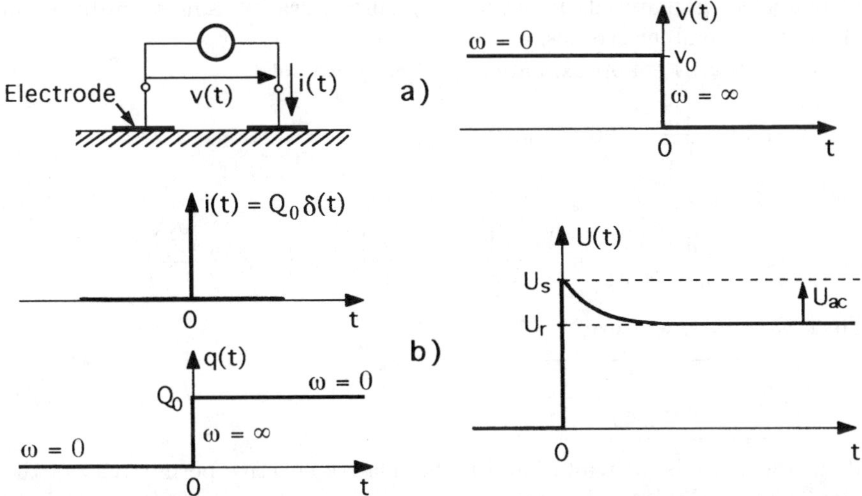

Fig. 5.21. A capacitor formed from two electrodes on the surface of a piezoelectric solid. (**a**) Instantaneous discharge of a capacitor previously charged to a voltage v_0. (**b**) Charging by a current impulse $Q_0\delta(t)$. The mechanical energy U_{ac} generated is, in both cases, a fraction K_S^2 of the supplied electrical energy U_s

because the spectrum of the discontinuity at $t = 0$, a negative step of voltage, consists of high frequency components. The difference $U_s - U_r$ represents the mechanical energy U_{ac} generated, so

$$U_{ac} = \frac{1}{2}(C_{LF} - C_{HF})v_0^2 = \frac{C_{LF} - C_{HF}}{C_{LF}} U_s\,. \tag{5.81}$$

The ratio U_{ac}/U_s is, by definition, the square of the electromechanical coupling coefficient, so that

$$K_S^2 = \frac{U_{ac}}{U_s}\,, \quad \text{giving} \quad \boxed{K_S^2 = \frac{\bar{\varepsilon}_{LF} - \bar{\varepsilon}_{HF}}{\bar{\varepsilon}_{LF} + \varepsilon_d}}\,. \tag{5.82}$$

The same result is obtained if we imagine that charges $\pm Q_0$ are injected into the electrodes at $t = 0$, as in Fig. 5.21b. The energy supplied by the generator, in the form of a current impulse $i(t) = Q_0\delta(t)$, is $U_s = Q_0^2/2C_{HF}$, since the high-frequency permittivity determines the reaction of the material to the discontinuity. In the steady state the stored energy is $Q_0^2/2C_{LF}$, which is smaller. The difference, giving the mechanical energy produced, is

$$U_{ac} = \left(\frac{1}{C_{HF}} - \frac{1}{C_{LF}}\right)\frac{Q_0^2}{2} = \left(1 - \frac{C_{HF}}{C_{LF}}\right)U_s$$

and this is identical to the previous value of (5.81).

These two arguments show that, whether the source is a voltage or current, the coupling coefficient K_S is very useful for comparing the effectiveness of piezoelectric materials for generating surface waves using transducers in the form of comb electrodes.

When $\overline{\varepsilon}(\omega, k)$ is approximated by (5.80a), we have

$$\overline{\varepsilon}_{\mathrm{LF}} = \varepsilon_{\mathrm{p}} \frac{V_0^2}{V_\infty^2} \quad \text{and} \quad \overline{\varepsilon}_{\mathrm{HF}} = \varepsilon_{\mathrm{p}} ,$$

and this gives

$$K_{\mathrm{S}}^2 = \frac{V_0^2 - V_\infty^2}{V_0^2 + \frac{\varepsilon_{\mathrm{d}}}{\varepsilon_{\mathrm{p}}} V_\infty^2} = \frac{V_0 - V_\infty}{V_0} \cdot \frac{V_0 + V_\infty}{V_0} \Big/ \left(1 + \frac{\varepsilon_{\mathrm{d}}}{\varepsilon_{\mathrm{p}}} \frac{V_\infty^2}{V_0^2}\right) .$$

Since $V_0 \approx V_\infty$, we have

$$\boxed{K_{\mathrm{S}}^2 \cong \frac{2}{1 + \varepsilon_{\mathrm{d}}/\varepsilon_{\mathrm{p}}} \cdot \frac{\Delta V}{V}} . \tag{5.83}$$

If the dielectric is a vacuum, and if the relative effective permittivity of the piezoelectric solid is much greater than unity (as in the case of lithium niobate or tantalate for example), the above formula can be approximated by

$$K_{\mathrm{S}}^2 \cong 2 \frac{\Delta V}{V} = 2 \frac{V_0 - V_\infty}{V_0} . \tag{5.84}$$

5.3.4.4 Magnitude of Electric Potential and Power Flow. The mechanical displacement has its amplitude related to power flow by the formula (5.48).

The *electric potential* at the surface, Φ, is related to the mean power $\langle P \rangle$ and the power density $P_{\mathrm{R}} = \langle P \rangle / w$ by (5.30), so that

$$\langle P \rangle = w P_{\mathrm{R}} = -\frac{1}{4} \frac{\partial B}{\partial k} |\Phi|^2 .$$

Using (5.73) for the susceptance gives

$$\frac{\partial B}{\partial k} = \frac{B(\omega, k)}{k} + w\omega |k| \frac{\partial \overline{\varepsilon}}{\partial k} .$$

A freely-propagating mode gives $B(\omega, k) = 0$, so the first term is zero. Thus, with $k = \omega / V$, we have

$$P_{\mathrm{R}} = -\frac{\omega}{4} k_{\mathrm{R}} \left(\frac{\partial \overline{\varepsilon}}{\partial k}\right)_{k_{\mathrm{R}}} |\Phi|^2 = \frac{\omega}{4} V_{\mathrm{R}} \left(\frac{\partial \overline{\varepsilon}}{\partial V}\right)_{V_{\mathrm{R}}} |\Phi|^2 . \tag{5.85}$$

From (5.80a) we have

$$V_{\mathrm{R}} \left(\frac{\partial \overline{\varepsilon}}{\partial V}\right)_{V_{\mathrm{R}}} = 2 V_{\mathrm{R}}^2 \left(\frac{\partial \overline{\varepsilon}}{\partial V^2}\right)_{V_{\mathrm{R}}} = 2 \varepsilon_{\mathrm{p}} \frac{V_{\mathrm{R}}^2}{V_0^2 - V_\infty^2} \left(\frac{V_0^2 - V_\infty^2}{V_R^2 - V_\infty^2}\right)^2 .$$

Here $\overline{\varepsilon}(V_{\mathrm{R}}) + \varepsilon_{\mathrm{d}} = 0$ so

$$\frac{V_{\mathrm{R}}^2 - V_0^2}{V_{\mathrm{R}}^2 - V_\infty^2} = -\frac{\varepsilon_{\mathrm{d}}}{\varepsilon_{\mathrm{p}}} ,$$

leading to

$$\frac{V_0^2 - V_\infty^2}{V_\mathrm{R}^2 - V_\infty^2} = 1 - \frac{V_\mathrm{R}^2 - V_0^2}{V_\mathrm{R}^2 - V_\infty^2} = 1 + \frac{\varepsilon_\mathrm{d}}{\varepsilon_\mathrm{p}} .$$

Introducing the Rayleigh-wave electromechanical coupling coefficient K_R, from (5.83), gives

$$\frac{V_0^2 - V_\infty^2}{V_\mathrm{R}^2} \cong 2\frac{\Delta V}{V} = \left(1 + \frac{\varepsilon_\mathrm{d}}{\varepsilon_\mathrm{p}}\right) K_\mathrm{R}^2$$

giving

$$V_\mathrm{R} \left(\frac{\partial \bar{\varepsilon}}{\partial V}\right)_{V_\mathrm{R}} = 2\frac{\varepsilon_\mathrm{p} + \varepsilon_\mathrm{d}}{K_\mathrm{R}^2} . \tag{5.86}$$

Substituting into (5.85), the surface electric potential accompanying the Rayleigh wave has amplitude

$$\boxed{|\Phi| = K_\mathrm{R} \left(\frac{2}{\varepsilon_\mathrm{p} + \varepsilon_\mathrm{d}} \cdot \frac{P_\mathrm{R}}{\omega}\right)^{1/2}} . \tag{5.87}$$

Like the mechanical displacement, the potential is proportional to $(P_\mathrm{R}/\omega)^{1/2}$.

The tangential component of electric field is $E_1 = \mathrm{i}k_\mathrm{R}\Phi$, so

$$|E_1| = \frac{K_\mathrm{R}}{V_\mathrm{R}} \left(\frac{2\omega P_\mathrm{R}}{\varepsilon_\mathrm{p} + \varepsilon_\mathrm{d}}\right)^{1/2} .$$

The normal component of electric displacement is

$$D_2(0_+) = |k_\mathrm{R}|\bar{\varepsilon}(V_\mathrm{R})\Phi = -|k_\mathrm{R}|\varepsilon_\mathrm{d}\Phi ,$$

and hence

$$|D_2(0_+)| = \varepsilon_\mathrm{d}\frac{K_\mathrm{R}}{V_\mathrm{R}} \left(\frac{2\omega P_\mathrm{R}}{\varepsilon_\mathrm{p} + \varepsilon_\mathrm{d}}\right)^{1/2} . \tag{5.88}$$

Numerical Example. For $Y + 131.5°$ lithium niobate, with propagation along X, Fig. 5.27 shows that $K_\mathrm{R}^2 = 5.6\%$, $V_\mathrm{R} = 4\,000\,\mathrm{m/s}$ and $\varepsilon_\mathrm{p} + \varepsilon_0 = 6 \times 10^{-10}\,\mathrm{F/m}$. Taking $P_\mathrm{R} = 1\,\mathrm{W/cm}$ and $\omega = 10^9$, i.e. $f = 160\,\mathrm{MHz}$, we find $|\Phi| = 4.3\,\mathrm{V}$ and $|E_1| = 1.1 \times 10^6\,\mathrm{V/m}$.

5.3.4.5 Characteristics of Main Materials. Once the velocity V has been found by solving (5.74), the amplitudes $°u_l^{(r)}$ obtained from the system (5.4) for each of the four values of the decay constants χ_r satisfy the characteristic equation of this system. The coefficients A_1, A_2 and A_3 are deduced from the three mechanical boundary conditions (5.38). Equation (5.36) then gives the displacement components u_1, u_2 and u_3 and the electric potential $\Phi = u_4$.

In practice, the equations are too complicated to be solved by hand – they need to be solved numerically, using for example the flow chart of Fig. 5.22. The complete characterization of a material requires a considerable number of operations, even if a step as coarse as 10° is used for the rotations of the propagation direction and the plane of the cut. It is therefore advisable to take maximum advantage of the crystal symmetry.

The most obvious scheme of investigation is to rotate the propagation direction $\boldsymbol{n}$ in the plane P of the free surface, with the latter fixed, as in Fig. 5.23a. This method is not very informative in view of the fact that only pure modes, whose energy velocity $\boldsymbol{V}^{\mathrm{e}}$ is parallel to $\boldsymbol{n}$, are normally used. The pure mode directions are in fact symmetry axes, apart from a few cases where the angle between $\boldsymbol{V}^{\mathrm{e}}$ and $\boldsymbol{n}$ is zero incidentally.

An alternative scheme is to rotate the free surface P about the propagation direction $\boldsymbol{n}$ with the latter fixed, as in Fig. 5.23b. For any such orientation, the mode is pure if $\boldsymbol{n}$ is along a symmetry axis that leaves P unchanged, that is, an A_2, A_4, $\overline{A}_4$ or A_6 axis which contains a dyad axis, or an $\overline{A}_2$ or $\overline{A}_6$

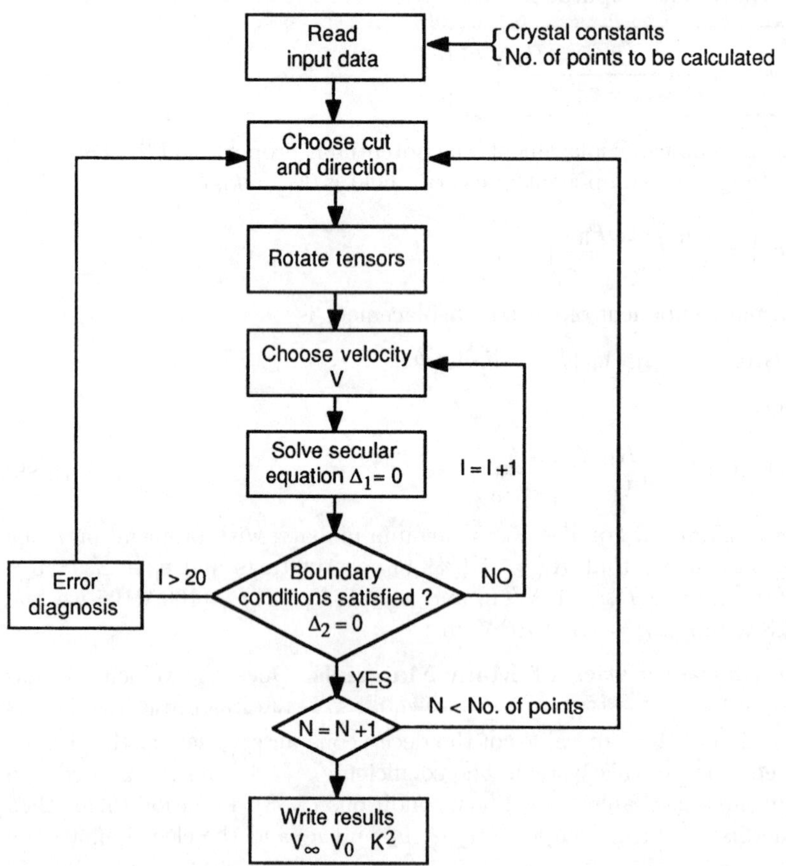

Fig. 5.22. Flow chart for calculation of surface wave velocities, from [5.10]. Δ_1 is the determinant of the propagation equations (5.4), and Δ_2 is the determinant of the boundary condition equations (5.38)

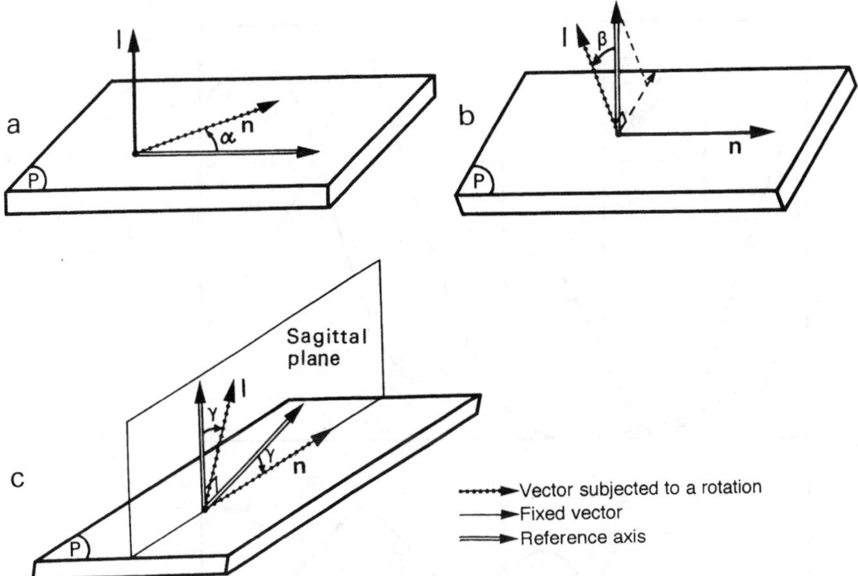

Fig. 5.23. Schemes for investigating surface-wave properties of crystals. (**a**) Rotation of the propagation direction in the plane of the cut. (**b**) Rotation of the cut about the propagation direction. (**c**) Simultaneous rotation of the cut and propagation direction

axis which includes a perpendicular mirror plane. With this scheme, it is possible to choose the cut angle so as to optimize some particular characteristic. For example, the coupling coefficient can be maximized, as in the $Y + 131.5°$ cut of lithium niobate, with propagation along the X axis which is perpendicular to a mirror plane. Alternatively, the temperature coefficient can be minimized, giving zero for the ST-cut of quartz with propagation along the dyad axis X.

A third scheme is to fix the sagittal plane and rotate the propagation direction in this plane, as in Fig. 5.23c. Here the propagation direction and the cut are both varying. For any propagation direction $\boldsymbol{n}$, the mode is pure if the sagittal plane is a mirror plane because the energy velocity, contained in the free surface P and in the mirror, must be along $\boldsymbol{n}$.

Among the main piezoelectric solids that have been studied, we note zinc oxide (ZnO, class $6mm$), gallium arsenide (GaAs, $\bar{4}3m$), bismuth germanium oxide ($Bi_{12}GeO_{20}$, 23), berlinite ($AlPO_4$, 32), lithium niobate ($LiNbO_3$, $3m$), lithium tantalate ($LiTaO_3$, $3m$) and quartz (SiO_2, 32). The last three are commonest.

Lithium niobate (class $3m$) is a strongly piezoelectric material with large crystals available – cylinders of diameter 10 cm and length more than 10 cm. Since the YZ plane is a mirror, the Rayleigh waves are pure for any free surface containing the X axis, provided the propagation is along or normal

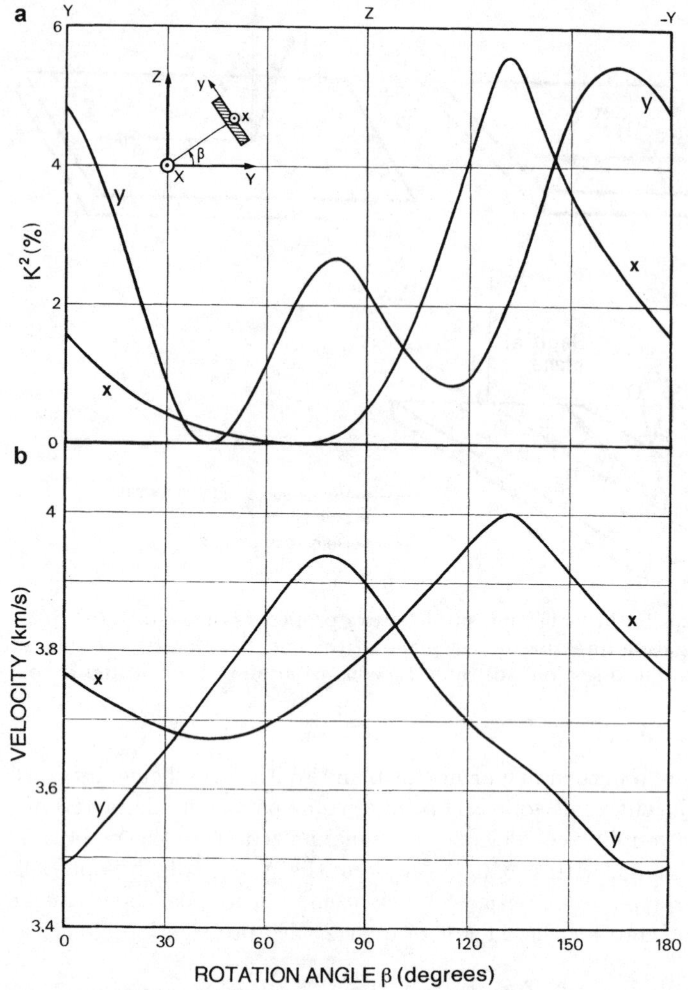

Fig. 5.24. Lithium niobate. Rayleigh wave propagation along a direction x parallel to X, and a direction y normal to X. The curves show (**a**) the coupling coefficient and (**b**) the velocity V_0, as functions of the cut angle β. From [5.11]

to this axis. Figure 5.24 shows the phase velocity and the square of the coupling coefficient as functions of the cut angle β. The largest coupling, $K_{\mathrm{R}}^2 = 5.6\,\%$, occurs for $\beta = 131.5°$ ($Y + 131.5°$ cut), with propagation along X. For propagation along Z in the XZ plane, i.e. for the Y-cut, the coupling is $K_{\mathrm{R}}^2 = 4.8\,\%$. Figure 5.25 shows the two displacement components and the electric potential as functions of depth, for the two classic electrical conditions at the surface.

Quartz (class 32) lacks a mirror plane, and therefore does not give waves of type $\overline{\mathrm{R}}_2$. However, waves of type $\overline{\mathrm{R}}_3$ propagating along the dyad axis X

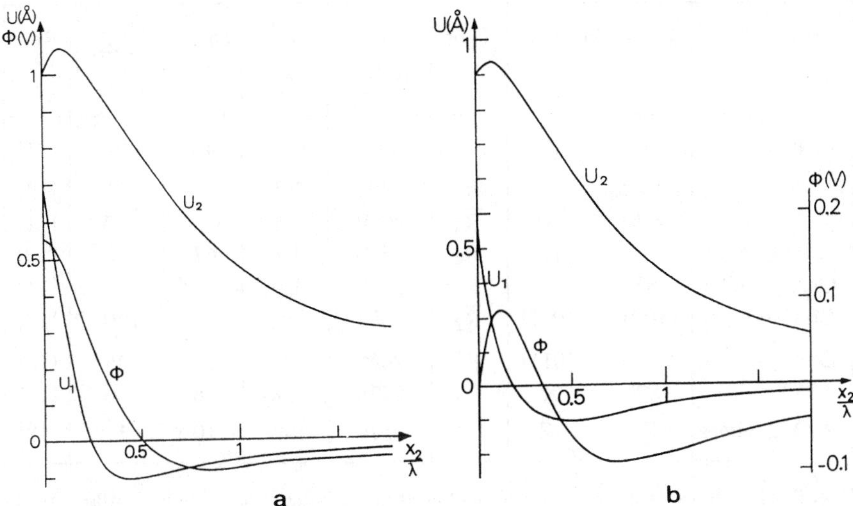

Fig. 5.25. Y-cut lithium niobate, with propagation along Z. Variation of mechanical displacements and electric potential with distance from the surface. (**a**) Free surface. (**b**) Metallized. From [5.12]

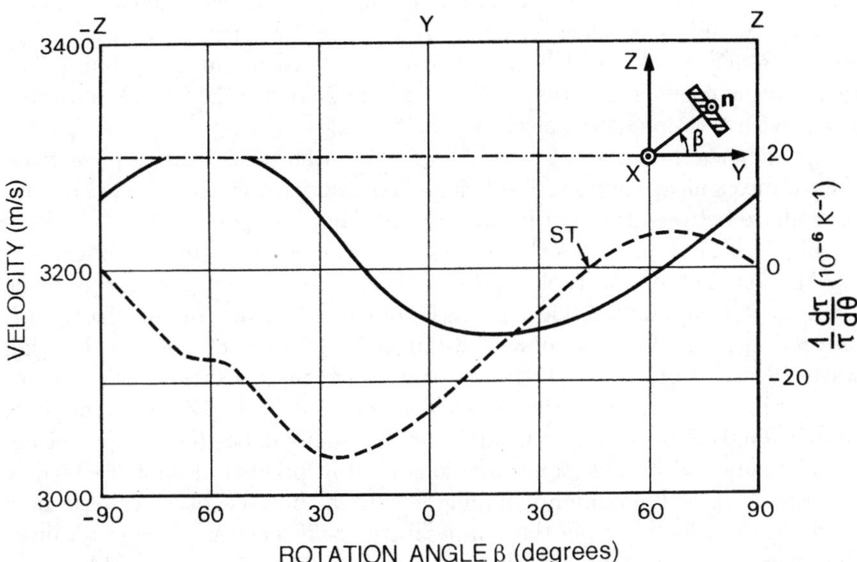

Fig. 5.26. Quartz. Rayleigh-wave propagation along X. Velocity (solid line) and temperature coefficient of delay $(1/\tau)\mathrm{d}\tau/\mathrm{d}\theta$ (broken line), as functions of the surface orientation defined by the angle β. From [5.13]

Crystal	Cut - Direction		Type	Velocity (m/s)	K^2 (%)	$\varepsilon/\varepsilon_0$	$\left\lvert\frac{1}{\tau}\frac{d\tau}{d\theta}\right\rvert$ $10^{-6}K^{-1}$	Ref.
$LiNbO_3$	Y	Z	$\overline{R}_2$	3488	4.8	46	94	[11]
"	Y + 128°	X	$\overline{R}_3$	3992	5.3		75	[14]
"	Y + 131.5°	X	$\overline{R}_3$	4000	5.6		72	[11]
$LiTaO_3$	Y	Z	$\overline{R}_2$	3230	0.9	47	38	[10]
Quartz (SiO_2)	ST	X	$\overline{R}_3$	3158	0.16	4.5	0	[11]
$Bi_{12}GeO_{20}$	(100)	[011]	$\overline{R}_2$	1683	1.4	40	140	[15]
GaAs	(100)	[011]	$\overline{R}_2$	2865	0.07	11	49	[16]
$AlPO_4$	ST	X	$\overline{R}_3$	2736	0.56	6.1	0	[17]
ZnO/ E / Glass	Z	⊥Z		2576	1.4	10.8	11	[18]

Fig. 5.27. Characteristics of piezoelectric materials used for generation and propagation of Rayleigh waves

are pure for any orientation of the free surface. For the Y-cut, the displacement describes an ellipse in the plane obtained by rotating the sagittal plane through 43° about the X axis. The velocity is 3 154 m/s. Figure 5.26 shows the velocities of Rayleigh waves propagating along X, on a surface whose normal is in the YZ plane and makes an angle β with the Y axis. Although quartz is weakly piezoelectric, giving $K_R^2 = 0.22\,\%$ for the Y-cut with propagation along X, it is widely used because of its temperature stability. The temperature coefficient of the delay τ is zero at 25 °C for the ST cut, a rotated Y-cut with rotation 42.5°, as shown in Fig. 5.26.

Lithium tantalate is a piezoelectric material whose main properties, electro-mechanical coupling coefficient and temperature coefficient, are intermediate between those of lithium niobate and quartz.

Figure 5.27 summarizes the properties of the crystals quoted above.

The last line of the table, ZnO/E/glass, refers to a layer of zinc oxide on glass, with aluminium electrodes (E) of comb form at the interface. Such electrodes, in the form of connected 'fingers', are used to generate Rayleigh waves (Sect. 2.1 of Vol. II). If the substrate is not piezoelectric, it is necessary to cover it with a piezoelectric film such as ZnO or AlN. The ZnO film, with its principal axis Z normal to the substrate, is usually deposited by sputtering.

The cuts and propagation directions used in practice do not necessarily correspond to the maximum coupling coefficient because other criteria may be important, for example the temperature coefficient. Another criterion is the unwanted bulk waves which the electrodes can generate, in addition to the Rayleigh waves, to a degree dependent on substrate orientation. In this respect, the Y+128° cut of lithium niobate is superior to the Y+131.5° cut.

There is some scatter in the values for the constants given by various authors, because of variations in the quality of the samples and the measurement techniques. However, with the development of practical applications and the simulation of devices, the reproducibility required has called for more and more accurate values of the constants. Measurements are therefore made by manufacturers from time to time, on samples of different origins. The reader should not be surprised to find, in articles or catalogues, figures somewhat different from those quoted here.

The number of available piezoelectric crystals is somewhat limited, so various laboratories are investigating new ones such as $GaPO_4$, $Li_2B_4O_7$ and $La_3Ga_5SiO_{14}$. However, to be useful a new material must satisfy a range of criteria including purity, coupling coefficient, temperature stability, size and cost.

5.4 Transverse Horizontal Waves

As shown in Sect. 5.1.1, a transverse horizontal wave ($\overline{\mathrm{TH}}$ or TH), with polarization u_3, can propagate along x_1 if the sagittal plane x_1x_2 is normal to a direct or inverse dyad axis. This wave satisfies the propagation equations (5.6), so that

$$\begin{pmatrix} \Gamma_{33} - \rho V^2 & \gamma_3 \\ \gamma_3 & -\varepsilon \end{pmatrix} \begin{pmatrix} {}^\circ u_3 \\ {}^\circ u_4 \end{pmatrix} = 0 \quad \text{with} \quad V = \frac{\omega}{k} \quad \text{and} \quad u_4 = \Phi \,. \tag{5.89}$$

With $X_2 = kx_2$, the components of the Christoffel tensor are, from (5.3),

$$\begin{cases} \Gamma_{33} = c_{55}^{\mathrm{E}} - c_{44}^{\mathrm{E}} \dfrac{\partial^2}{\partial X_2^2} + 2\,\mathrm{i}\, c_{45}^{\mathrm{E}} \dfrac{\partial}{\partial X_2} \\ \gamma_3 \;\; = e_{15} - e_{24} \dfrac{\partial^2}{\partial X_2^2} + \mathrm{i}(e_{14} + e_{25}) \dfrac{\partial}{\partial X_2} \\ \varepsilon \;\;\; = \varepsilon_{11}^{\mathrm{S}} - \varepsilon_{22}^{\mathrm{S}} \dfrac{\partial^2}{\partial X_2^2} + 2\,\mathrm{i}\, \varepsilon_{12}^{\mathrm{S}} \dfrac{\partial}{\partial X_2} \,. \end{cases} \tag{5.90}$$

For boundaries normal to the x_2 axis the *mechanical* boundary conditions concern only T_{32}, because T_{12} and T_{22} are necessarily zero (Fig. 5.2), and we have

$$T_{32} = -\mathrm{i}k\,(c_{45}^{\mathrm{E}} u_3 + e_{14} u_4) + \frac{\partial}{\partial x_2}(c_{44} u_3 + e_{24} u_4)\,. \tag{5.91}$$

The *electrical* boundary conditions concern the electric potential $\Phi = u_4$ and the normal component of electric displacement, D_2.

We consider first the TH wave propagating in a plate, already mentioned in Sect. 1.1.2, and then the TH wave propagating in a layer firmly bonded to a semi-infinite solid (the Love wave). Finally, we describe the $\overline{\mathrm{TH}}$ wave (the Bleustein–Gulyaev wave) that can propagate along the surface of a semi-infinite piezoelectric solid.

5.4.1 Non-piezoelectric TH Waves

If the solid is not piezoelectric, or if the sagittal plane is parallel to a mirror plane ($\gamma_3 = 0$), the wave does not have an associated electric field. Using (5.90), the propagation equation (5.89) becomes

$$(c_{55} - \rho V^2)u_3 - c_{44}\frac{\partial^2 u_3}{\partial X_2^2} + 2\,\mathrm{i}\,c_{45}\frac{\partial u_3}{\partial X_2} = 0\,.$$

Substituting $X_2 = kx_2$ and $V = \omega/k$, and defining $u \equiv u_3$, we have

$$c_{44}\frac{\partial^2 u}{\partial x_2^2} - 2\,\mathrm{i}\,kc_{45}\frac{\partial u}{\partial x_2} + (\rho\omega^2 - c_{55}\,k^2)u = 0\,. \tag{5.92}$$

We simplify the calculation by assuming that the medium is isotropic in relation to TH waves. This is the case if the x_3 axis is along a crystal tetrad or hexad axis, or if the medium is totally isotropic so that $c_{45} = 0$ and $c_{55} = c_{44}$. The TH wave velocity is $V_{\mathrm{T}} = (c_{44}/\rho)^{1/2}$, independent of the direction of propagation in the x_1x_2 plane, and the above equation simplifies to

$$\frac{\partial^2 u}{\partial x_2^2} + \left(\frac{\omega^2}{V_{\mathrm{T}}^2} - k^2\right) u = 0\,. \tag{5.93}$$

The solutions depend on the mechanical boundary conditions, involving the stress

$$T_{32} = c_{44}\frac{\partial u}{\partial x_2} \tag{5.94}$$

and therefore depend on the structure of the waveguide.

5.4.1.1 Parallel-Sided Plate. The requirement that $T_{32} = 0$ at the free surfaces $x_2 = 0$ and $x_2 = -h$ is satisfied if the displacement is maximized at these points. Hyperbolic functions are excluded because they cannot have more than one maximum, and the solution is sinusoidal, with the form

$$u(x_2) = u_0 \cos\left[\frac{n\pi}{h}(x_2 + h)\right]\,, \quad n = 0, 1, 2, \ldots\,. \tag{5.95}$$

Figure 5.28 shows the first three modes, with $n = 0, 1, 2$. Along x_2 the wave appears to be stationary, and one can speak of a transverse resonance. For n even (odd), the mode is symmetric (antisymmetric). Substituting (5.95) into (5.93) yields the dispersion relation

$$\frac{\omega^2}{V_{\mathrm{T}}^2} - k^2 = \frac{n^2\pi^2}{h^2}\,, \quad n = 0, 1, 2, \ldots$$

and with normalized variables this becomes

$$\left(\frac{\omega h}{\pi V_{\mathrm{T}}}\right)^2 = \left(\frac{kh}{\pi}\right)^2 + n^2$$

or alternatively

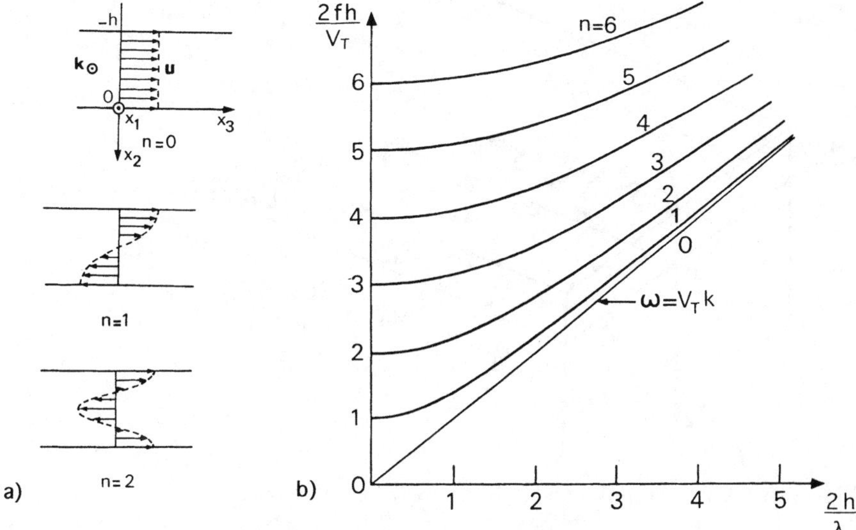

Fig. 5.28. TH wave propagation in a parallel-sided plate. (**a**) Displacement for the first three modes, with maxima at the free surfaces. (**b**) Dispersion curves. The normalized phase velocity V/V_{T} of the nth mode is the slope of the line joining the point on the curve to the origin

$$\frac{2fh}{V_{\mathrm{T}}} = \left[\left(\frac{2h}{\lambda}\right)^2 + n^2\right]^{1/2} . \tag{5.96}$$

The phase velocity $V = \omega/k$ of the guided TH wave depends on frequency, except for the mode of order zero which has velocity V_{T}. This mode, which has no cut-off frequency and can propagate at very low frequencies, has no counterpart in electromagnetism.

5.4.1.2 Layer on a Substrate. Love Waves. This is shown in Fig. 5.29. In the *substrate*, the displacement must vanish at $x_2 = \infty$, so the solution of the wave equation (5.93) has the form

$$u(x_2) = u_0 \, \mathrm{e}^{-k\chi x_2} \quad \text{with} \quad \mathrm{Re}[k\chi] > 0 , \quad x_2 > 0 .$$

This implies that

$$\frac{\omega^2}{V_{\mathrm{T}}^2} - k^2 = -k^2\chi^2 < 0 \quad \text{and hence} \quad V = \frac{\omega}{k} < V_{\mathrm{T}}$$

and

$$\chi = \left(1 - \frac{V^2}{V_{\mathrm{T}}^2}\right)^{1/2} . \tag{5.97}$$

For the *layer*, the variables are distinguished by using a hat. The mechanical stress $T_{32} = \widehat{c}_{44}\partial u_3/\partial x_2$ is zero at the free surface $x_2 = -h$, where the displacement is maximal. The solution is sinusoidal, with the form

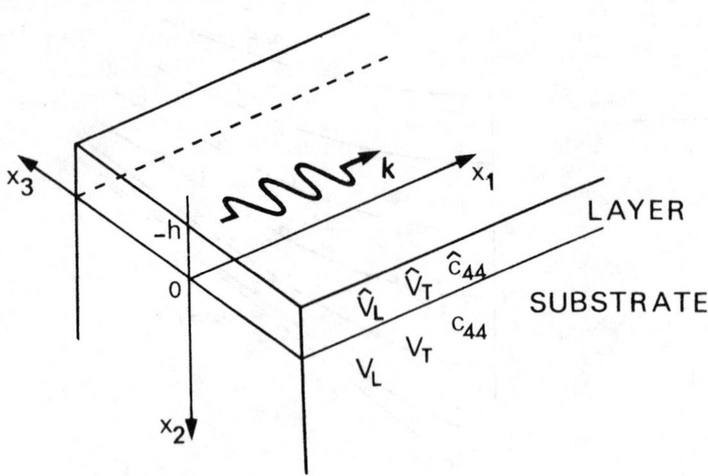

Fig. 5.29. A layer firmly bonded to a substrate

$$u(x_2) = \hat{u}_0 \cos k\hat{\chi}(x_2 + h), \quad -h < x_2 < 0\,,$$

and this implies that

$$\frac{\omega^2}{\hat{V}_T^2} - k^2 = k^2\hat{\chi}^2 > 0 \quad \text{and hence}$$
$$V > \hat{V}_T \quad \text{and} \quad \hat{\chi} = \left(\frac{V^2}{\hat{V}_T^2} - 1\right)^{1/2} . \tag{5.98}$$

The two inequalities above show that the velocity of the bulk TH wave in the substrate needs to be greater than that of the TH wave in the layer, so that $V_T > \hat{V}_T$. The Love wave velocity V is between these, so that $\hat{V}_T < V < V_T$.

At the interface $x_2 = 0$, continuity of displacement requires that

$$\hat{u}_0 = \frac{u_0}{\cos(k\hat{\chi}h)}$$

and continuity of stress requires

$$c_{44}\left(\frac{\partial u_3}{\partial x_2}\right)_{x_2=0_+} = \hat{c}_{44}\left(\frac{\partial u_3}{\partial x_2}\right)_{x_2=0_-}$$

and hence

$$-c_{44}u_0\, k\,\chi = -\hat{c}_{44}\hat{u}_0 k\hat{\chi}\sin(k\hat{\chi}h)\,.$$

These equations yield the dispersion relation, relating ω to k, in terms of the coefficients χ and $\hat{\chi}$, giving

$$\tan(\hat{\chi}kh) = \frac{c_{44}\chi}{\hat{c}_{44}\hat{\chi}}\,. \tag{5.99}$$

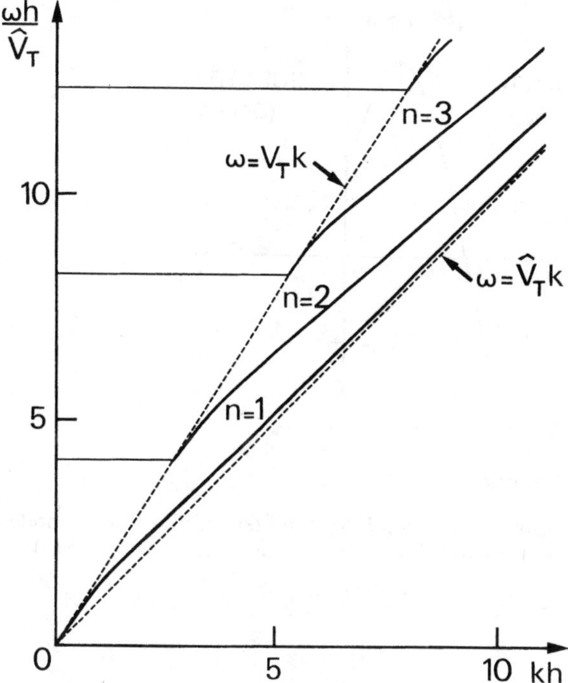

Fig. 5.30. Dispersion curves of the first three Love wave modes, for a silica layer ($\widehat{V}_{\mathrm{T}} = 3\,764\,\mathrm{m/s}$) on a crystalline silicon substrate ($V_{\mathrm{T}} = 5\,843\,\mathrm{m/s}$)

For a particular phase velocity between $\widehat{V}_{\mathrm{T}}$ and V_{T}, χ and $\widehat{\chi}$ have fixed values given by (5.97) and (5.98). The dispersion relation (5.99) has an infinite number of solutions given by

$$(kh)_n = \frac{1}{\widehat{\chi}} \tan^{-1}\left(\frac{c_{44}\chi}{\widehat{c}_{44}\widehat{\chi}}\right) + n\frac{\pi}{\widehat{\chi}}, \quad n = 0, 1, 2, \ldots . \tag{5.100}$$

In the $(\omega h, kh)$ plane, Fig. 5.30, the points representing these modes are equidistant along a line with slope $V = \omega/k$ situated between the two lines representing the extreme values $\widehat{V}_{\mathrm{T}}$ and V_{T}. If the frequencies are taken to be identical, so that the wavenumbers are also the same, (5.100) shows that the points correspond to an increase $\Delta h = \pi/\widehat{\chi}k$ of the layer thickness from one mode to the next. In this situation, the variation $u(x_2)$ of displacement with depth depends only on the materials and the velocity chosen (Fig. 5.31). The free surface is placed at an extremum of the sinusoid, chosen according to the order of the mode, in order to maximize the displacement and therefore cancel the surface stress.

Figure 5.32 shows the phase velocity V and group velocity $V_{\mathrm{g}} = \mathrm{d}\omega/\mathrm{d}k$ as functions of the product (frequency × film thickness), for the first mode in a silica layer on a silicon substrate. At low frequencies ($\omega h < \widehat{V}_{\mathrm{T}}$) the

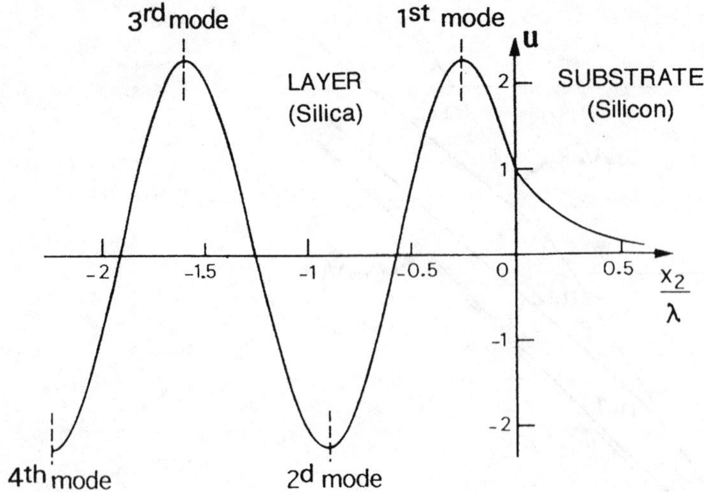

Fig. 5.31. Displacement as a function of depth for the first four Love wave modes, for a silica layer on a silicon substrate. The phase velocity is chosen to be $V = (V_T \widehat{V}_T)^{1/2} = 4\,690\,\text{m/s}$

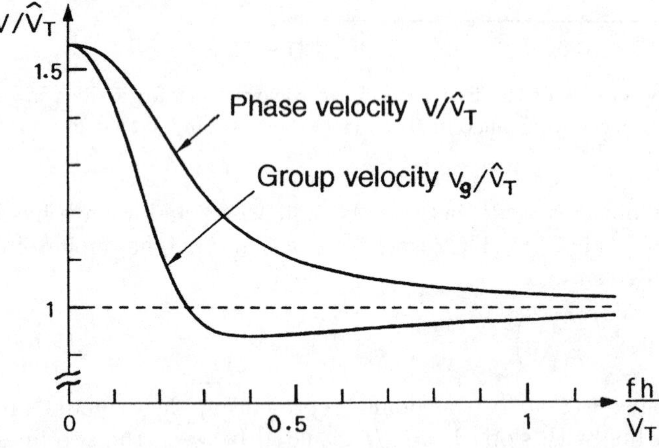

Fig. 5.32. Phase velocity V and group velocity V_g for the first Love wave mode as a function of the product (frequency × thickness). Silica layer on a silicon substrate

energy propagates mainly in the substrate, with velocity V_T, because the layer thickness is much less than the wavelength. As the frequency increases, the fraction of the energy travelling in the layer increases, so V and V_g approach $\widehat{V}_T$. Thus the dispersion of Love waves has a straightforward physical interpretation.

5.4.2 Piezoelectric TH Waves ($\overline{\mathrm{TH}}$ Waves)

In a semi-infinite piezoelectric solid (Fig. 5.1), a $\overline{\mathrm{TH}}$ wave (a Bleustein–Gulyaev wave) can exist if the sagittal plane is normal to a dyad axis, particularly, for example, in a crystal or piezoelectric ceramic with symmetry $6mm$ [5.19, 5.20]. In this case, the orientation of the x_1 and x_2 axes is arbitrary in the plane normal to the A_6 axis. We consider a mechanical displacement and electric potential of the form, for $x_2 > 0$,

$$\begin{Bmatrix} u_3 \\ u_4 \end{Bmatrix} = \begin{Bmatrix} {}^\circ u_3 \\ {}^\circ u_4 \end{Bmatrix} \mathrm{e}^{-\chi X_2} \,\mathrm{e}^{\mathrm{i}(\omega t - k x_1)}\,, \quad \mathrm{Re}[\chi] > 0\,. \tag{5.101}$$

With $c_{45} = 0$, $e_{14} = e_{25} = 0$ and $e_{15} = e_{24}$, the components of the Christoffel tensor are, from (5.90),

$$\Gamma_{33} = c_{44}^{\mathrm{E}}(1-\chi^2)\,, \quad \gamma_3 = e_{15}(1-\chi^2)\,, \quad \varepsilon = \varepsilon_{11}^{\mathrm{S}}(1-\chi^2)\,.$$

Equations (5.89) become

$$\begin{pmatrix} c_{44}^{\mathrm{E}}(1-\chi^2) - \rho V^2 & e_{15}(1-\chi^2) \\ e_{15}(1-\chi^2) & -\varepsilon_{11}^{\mathrm{S}}(1-\chi^2) \end{pmatrix} \begin{pmatrix} {}^\circ u_3 \\ {}^\circ u_4 \end{pmatrix} = 0 \tag{5.102}$$

and setting the determinant to zero gives the equation

$$(1-\chi^2)\left[\rho V^2 \varepsilon_{11}^{\mathrm{S}} - (1-\chi^2)(c_{44}^{\mathrm{E}}\varepsilon_{11}^{\mathrm{S}} + e_{15}^2)\right] = 0\,,$$

which has roots χ_1 and χ_2 given by

$$\begin{cases} \chi_1 = 1 & \text{giving } {}^\circ u_3^{(1)} = 0 \quad \text{and} \quad {}^\circ u_4^{(1)} = 1 \\ \chi_2 = \sqrt{1 - V^2/V_{\mathrm{T}}^2} & \text{giving } {}^\circ u_3^{(2)} = \varepsilon_{11}^{\mathrm{S}}/e_{15} \text{ on taking } {}^\circ u_4^{(2)} = 1\,. \end{cases} \tag{5.103}$$

Here

$$V_{\mathrm{T}} = \sqrt{\frac{c_{44}^{\mathrm{E}} + e_{15}^2/\varepsilon_{11}^{\mathrm{S}}}{\rho}} > V$$

is the velocity of a bulk transverse wave with the same polarization, propagating in the piezoelectric solid.

The general solution, a linear combination of partial waves of the form (5.101) with the same phase velocity $V = \omega/k$, is written as

$$\begin{cases} u_3 = u = \dfrac{\varepsilon_{11}^{\mathrm{S}}}{e_{15}} A_2 \,\mathrm{e}^{-k\chi_2 x_2} \,\mathrm{e}^{\mathrm{i}(\omega t - k x_1)} \\ u_4 = \Phi = \left(A_1 \,\mathrm{e}^{-k x_2} + A_2 \,\mathrm{e}^{-k\chi_2 x_2}\right) \mathrm{e}^{\mathrm{i}(\omega t - k x_1)}\,. \end{cases}$$

The coefficients A_1 and A_2 are determined by the *mechanical boundary* conditions, given by (5.91). Since c_{45} and e_{14} are zero, we have

$$T_{32} = \left[\frac{\partial}{\partial x_2}(c_{44} u_3 + e_{15} u_4)\right]_{x_2=0} = -k\left[\chi_2\left(c_{44}\frac{\varepsilon_{11}^{\mathrm{S}}}{e_{15}} + e_{15}\right) A_2 + e_{15} A_1\right] = 0\,.$$

We introduce, from Sect. 4.3.3b, the transverse-wave electromechanical coupling coefficient

$$K_{\mathrm{T}} = \frac{e_{15}}{\sqrt{c_{44}^{\mathrm{E}}\varepsilon_{11}^{\mathrm{S}} + e_{15}^2}}$$

and this leads to

$$A_2 = -\frac{K_{\mathrm{T}}^2}{\chi_2}A_1 \quad \text{and hence} \quad \Phi = A_1\left(\mathrm{e}^{-kx_2} - \frac{K_{\mathrm{T}}^2}{\chi_2}\,\mathrm{e}^{-k\chi_2 x_2}\right)\mathrm{e}^{\mathrm{i}(\omega t - kx_1)}.$$

The phase velocity is determined by the electrical boundary conditions, which can be expressed with the aid of the surface piezoelectric permittivity $\overline{\varepsilon}$ defined by (5.72). Since $e_{24} = e_{15}$ and $\varepsilon_{22} = \varepsilon_{11}$, the normal component of electric displacement can be written

$$D_2 = \frac{\partial}{\partial x_2}\left(e_{15}u_3 - \varepsilon_{11}^{\mathrm{S}}u_4\right) = \frac{\partial}{\partial x_2}\left(-\varepsilon_{11}^{\mathrm{S}}\,A_1\,\mathrm{e}^{-kx_2}\right) = k\varepsilon_{11}^{\mathrm{S}}A_1\,\mathrm{e}^{-kx_2}$$

and hence

$$\overline{\varepsilon} = \frac{D_2(0)}{|k|\Phi(0)} = \frac{\varepsilon_{11}^{\mathrm{S}}}{1 - K_{\mathrm{T}}^2/\chi_2} \quad \text{giving} \quad \frac{1}{\overline{\varepsilon}} = \frac{1}{\varepsilon_{11}^{\mathrm{S}}}\left(1 - \frac{K_{\mathrm{T}}^2}{\sqrt{1 - V^2/V_{\mathrm{T}}^2}}\right). \tag{5.104}$$

We consider two limiting electrical conditions, in a manner similar to Sect. 5.3.4.2.

(a) *Metallized Surface.* The surface is covered by a metallic film with thickness much less than the wavelength and with zero potential, so that $\overline{\varepsilon} = \infty$ and hence $\chi_2 = K_{\mathrm{T}}^2$. From (5.104) the Bleustein–Gulyaev wave velocity is

$$V_\infty = V_{\mathrm{T}}\sqrt{1 - K_{\mathrm{T}}^4} \tag{5.105}$$

and is therefore very close to the transverse-wave velocity V_{T}. The decay of the displacement u and potential Φ with depth becomes more rapid as the coupling coefficient K_{T} increases. Figure 5.33 shows curves for cadmium sulphide (CdS), for which $K_{\mathrm{T}} = 0.19$, giving $\chi_2 = 0.036$ and $V_\infty = 0.9994\,V_{\mathrm{T}} = 1\,788\,\mathrm{m/s}$.

(b) *Free Surface.* Here the value of the piezoelectric permittivity is $\overline{\varepsilon} = -\varepsilon_0$, which gives $\chi_2 = K_T^2/(1 + \varepsilon_{11}/\varepsilon_0)$, and the TH wave velocity is

$$V_{\mathrm{B}} = V_{\mathrm{T}}\sqrt{1 - \frac{K_{\mathrm{T}}^4}{(1 + \varepsilon_{11}/\varepsilon_0)^2}}. \tag{5.106}$$

The wave penetrates more deeply into the solid when the surface is not metallized. In both cases the depth involved (10 – 100 λ) is much greater than the wavelength, and hence the velocity is little different from that of the bulk TH wave. The penetration depth $1/\chi_2$ of the mechanical displacement depends on the permittivity ε_{d} of the external medium since, with $\overline{\varepsilon} = -\varepsilon_{\mathrm{d}}$, we have

$$\frac{1}{\chi_2} = \left(1 - \frac{\varepsilon_{11}^{\mathrm{S}}}{\overline{\varepsilon}}\right)\Big/K_{\mathrm{T}}^2 = \left(1 + \frac{\varepsilon_{11}^{\mathrm{S}}}{\varepsilon_{\mathrm{d}}}\right)\Big/K_{\mathrm{T}}^2. \tag{5.107}$$

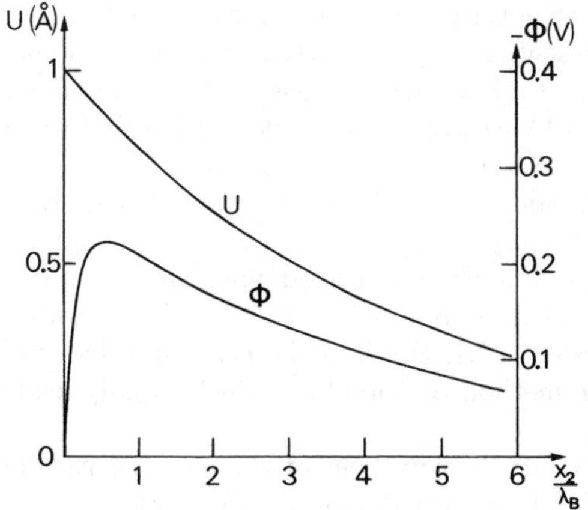

Fig. 5.33. $\overline{\text{TH}}$ wave in cadmium sulphide. Variation of mechanical displacement and electric potential with depth, for a metallized surface

Defining $A = (\varepsilon_{11}^{S}/e_{15})A_2$, we find

$$\begin{cases} u = A\,\mathrm{e}^{-k\chi_2 x_2}\,\mathrm{e}^{\mathrm{i}(\omega t - kx_1)} \\ \Phi = A\dfrac{e_{15}}{\varepsilon_{11}^{S}}\left(\mathrm{e}^{-k\chi_2 x_2} - \dfrac{\varepsilon_d}{\varepsilon_d + \varepsilon_{11}^{S}}\,\mathrm{e}^{-kx_2}\right)\mathrm{e}^{\mathrm{i}(\omega t - kx_1)}\,. \end{cases} \tag{5.108}$$

The absolute amplitude A of the displacement at the surface determines the power density per unit width of the beam. Using (5.33) with $k = \omega/V_B$ we have

$$P_B = \frac{\langle P\rangle}{w} = \frac{1}{2}\rho\omega\,V_B^2\int_0^\infty |u|^2\,\mathrm{d}(kx_2) = \frac{A^2}{2}\rho\omega\,V_B^2\int_0^\infty \mathrm{e}^{-2\chi_2 X_2}\,\mathrm{d}X_2$$

and hence

$$A = \left(\frac{4\chi_2 P_B}{\rho\omega\,V_B^2}\right)^{1/2} = \frac{2K_T}{V_B}\left(\frac{\varepsilon_d}{\varepsilon_d + \varepsilon_{11}^{S}}\right)^{1/2}\left(\frac{P_B}{\rho\omega}\right)^{1/2}.$$

The magnitude of the potential at the surface is

$$|\Phi(0)| = \frac{e_{15}}{\varepsilon_d + \varepsilon_{11}^{S}}A\,.$$

Numerical Example. For cadmium sulphide, $\varepsilon_{11}^{S} = 9\,\varepsilon_0 = 8\times 10^{-11}$ F/m and $\rho = 4\,824$ kg/m^3. For $P_B = 1$ W/cm and $f = 16$ MHz we find $A = 0.97$ nm, and with $e_{15} = -0.21$ C/m^2 we have $|\Phi(0)| = 2.55$ V.

5.5 Lamb Waves in an Isotropic Plate

As shown in Sect. 5.4.1.2, a transverse horizontal (TH) wave can propagate alone in a plane stratified structure, because its polarization is unchanged on reflection or refraction. This is not the case for a longitudinal (L) or transverse vertical (TV) wave – these waves are coupled at an interface. Thus, Sect. 4.4.2.2 showed that reflection at a free surface is associated with partial conversion, so that an incident L (TV) wave gives a reflected L (TV) wave and a converted TV (L) wave (Fig. 5.34). The analysis of wave propagation in a plate is thus more complicated for a sagittally-polarized wave than for a TH wave.

Decomposition into partial waves is a general method, applicable for propagation in an anisotropic plate as well as an isotropic one [5.21]. We have used it elsewhere for Rayleigh waves, Sect. 5.3.1. However, here we consider only an isotropic plate, and we use the method of potentials which rapidly leads to the dispersion relation.

From the results of Sect. 4.2.3, the displacement of the material can be derived from a scalar potential ϕ and a vector potential Ψ, so that

$$\boldsymbol{u} = \nabla\phi + \nabla \wedge \Psi \tag{5.109}$$

and the potentials satisfy the wave equations

$$\nabla^2\phi - \frac{1}{V_{\mathrm{L}}^2}\frac{\partial^2\phi}{\partial t^2} = 0 \quad \text{and} \quad \nabla^2\Psi - \frac{1}{V_{\mathrm{T}}^2}\frac{\partial^2\Psi}{\partial t^2} = 0\,. \qquad (5.110\text{a}, 5.110\text{b})$$

Here $V_{\mathrm{L}} = (c_{11}/\rho)^{1/2}$ and $V_{\mathrm{T}} = (c_{66}/\rho)^{1/2}$ are the phase velocities of bulk longitudinal and transverse waves, respectively. Strains associated with dilatation, causing volume changes, are expressed by ϕ; and shear strains, causing no volume changes, are expressed by Ψ. In a plate with faces normal to x_2, as in Fig. 5.35, the wave polarized in the sagittal plane x_1x_2 (i.e. the L_2 Lamb wave of Sect. 5.1.3) is decoupled from the TH wave polarized along x_3.

The Lamb wave is taken to propagate along x_1, and we ignore diffraction in the x_3 direction, so that $\partial_3 \equiv \partial/\partial x_3 = 0$. For the sinusoidal case we have $\partial/\partial x_1 = \partial_1 = -\mathrm{i}k$, and the displacement components are

$$\begin{cases} u_1 = \partial_1\phi + \partial_2\Psi_3 = -\mathrm{i}k\phi + \partial_2\Psi_3 \\ u_2 = \partial_2\phi - \partial_1\Psi_3 = \partial_2\phi + \mathrm{i}k\Psi_3\,. \end{cases} \tag{5.111}$$

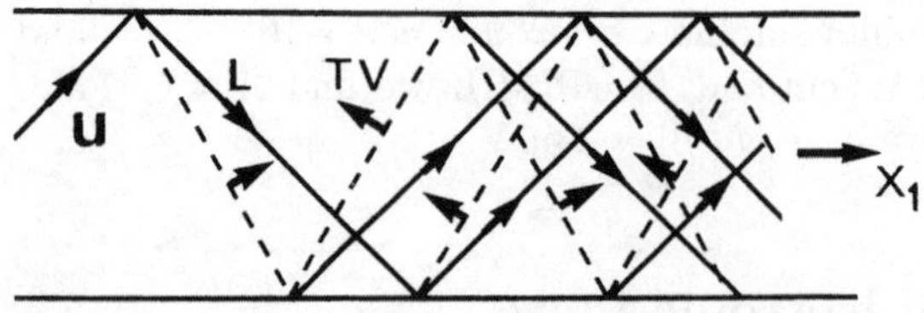

Fig. 5.34. Lamb waves. The L and TV constituents travelling along x_1 are successively reflected at each surface of the plate. At each reflection, the L (TV) constituent generates the TV (L) constituent

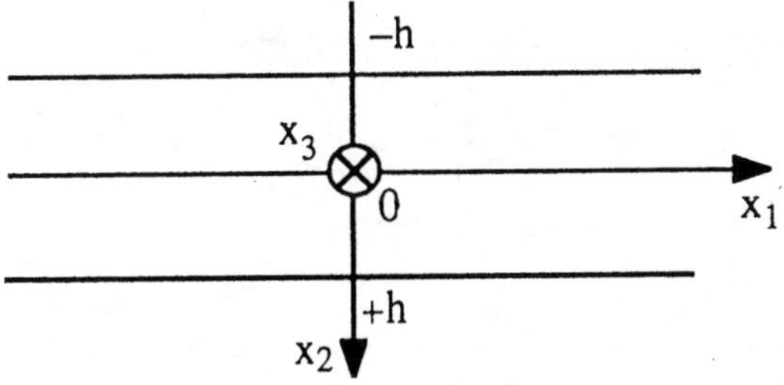

Fig. 5.35. Isotropic plate

The TH wave has displacement $u_3 = -(\mathrm{i}k\Psi_2+\partial_2\Psi_1)$. The Laplacian is written as

$$\nabla^2 = \frac{\partial^2}{\partial x_1^2} + \frac{\partial^2}{\partial x_2^2} = -k^2 + \frac{\partial^2}{\partial x_2^2}.$$

We define

$$p^2 = \frac{\omega^2}{V_\mathrm{L}^2} - k^2 \quad \text{and} \quad q^2 = \frac{\omega^2}{V_\mathrm{T}^2} - k^2, \tag{5.112a,b}$$

and the functions ϕ and $\Psi \equiv \Psi_3$ then satisfy the equations

$$\partial_2^2\phi + p^2\phi = 0 \quad \text{and} \quad \partial_2^2\Psi + q^2\Psi = 0. \tag{5.113}$$

Acceptable solutions are determined by the boundary conditions at the faces $x_2 = \pm h$. Since $T_{32} = 0$, this only concerns the stresses $T_{22} = T_2$ and $T_{12} = T_6$ (see Fig. 5.2). Thus we need to set $T_2 = T_6 = 0$ at the faces.

5.5.1 Dispersion Relation. Modes

For an isotropic solid, the stresses are given by (4.13), with $c_{44} = c_{66}$. Since $S_{33} = \partial_3 u_3 = 0$, the dilatation is

$$S = S_{11} + S_{22} = \partial_1 u_1 + \partial_2 u_2 = \nabla^2\phi$$

and the normal stress $T_2 = T_{22} = c_{11}S - 2c_{66}S_{11}$ is given by

$$T_2 = c_{11}\nabla^2\phi - 2c_{66}\partial_1 u_1 = c_{11}\nabla^2\phi + 2c_{66}(k^2\phi + \mathrm{i}k\partial_2\Psi).$$

Using (5.110a), and (5.112a,b) in the form $\rho\omega^2 = c_{66}(k^2+q^2)$, we find

$$c_{11}\nabla^2\phi = -\rho\omega^2\phi = -c_{66}(k^2+q^2)\phi$$

and this gives

$$T_2 = c_{66}[(k^2 - q^2)\phi + 2\mathrm{i}k\partial_2\Psi]. \tag{5.114}$$

On the other hand, the tangential stress $T_6 = T_{12} = 2\,c_{66}S_{12}$ is expressed as

$$T_6 = c_{66}(\partial_2 u_1 + \partial_1 u_2) = c_{66}(\partial_2^2\Psi + k^2\Psi - 2\mathrm{i}k\,\partial_2\phi)$$

and with $\partial_2^2\Psi = -q^2\Psi$ this becomes

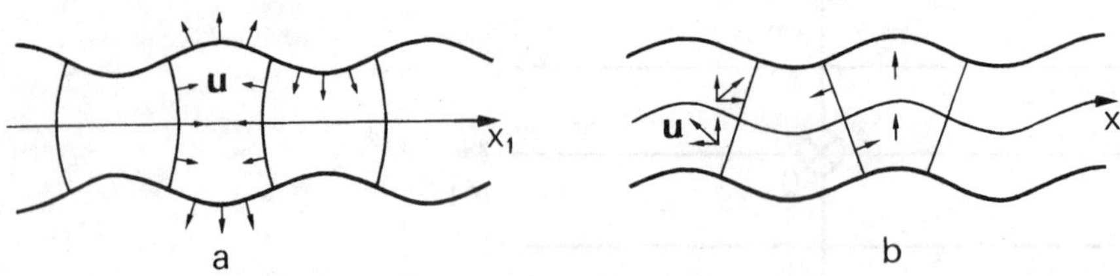

Fig. 5.36. Lamb waves. (**a**) Symmetric – on either side of the median plane, the longitudinal components are equal and the transverse components are opposite. (**b**) Antisymmetric – on either side of the median plane, the transverse components are equal and the longitudinal components are opposite

$$T_6 = c_{66}\left[(k^2 - q^2)\Psi - 2\mathrm{i}k\,\partial_2\phi\right]. \tag{5.115}$$

The requirements $T_2 = T_6 = 0$ at $x_2 = \pm h$ are satisfied simultaneously only if the stresses T_2 and T_6 are *even* or *odd* functions of x_2. The solutions of (5.113) must have different parity, and they can be taken as

$$\phi = B\cos(px_2 + \alpha) \quad \text{and} \quad \Psi = A\sin(qx_2 + \alpha), \tag{5.116}$$

where $\alpha = 0$ (T_2 even, T_6 odd) or $\alpha = \pi/2$ (T_2 odd, T_6 even). The propagation factor $\exp[\mathrm{i}(\omega t - kx_1)]$ has been omitted here. The mechanical displacements are given by

$$\begin{cases} u_1 = -\mathrm{i}kB\cos(px_2 + \alpha) + qA\cos(qx_2 + \alpha) \\ u_2 = -pB\sin(px_2 + \alpha) + \mathrm{i}kA\sin(qx_2 + \alpha). \end{cases} \tag{5.117}$$

Thus, there are two types of Lamb wave (Fig. 5.36):

(a) *Symmetric* modes, with $\alpha = 0$, in which the longitudinal component is an even function of x_2 and the transverse component is an odd function of x_2.
(b) *Antisymmetric* modes, with $\alpha = \pi/2$, in which the longitudinal component is an odd function of x_2 and the transverse component is an even function of x_2.

When $\alpha = 0$ or $\pi/2$, the boundary conditions yield the two equations

$$(k^2 - q^2)B\cos(ph + \alpha) + 2\mathrm{i}kqA\cos(qh + \alpha) = 0 \tag{5.118a}$$

$$2\mathrm{i}kpB\sin(ph + \alpha) + (k^2 - q^2)A\sin(qh + \alpha) = 0. \tag{5.118b}$$

For compatibility of these the determinant of the coefficients is set to zero, giving

$$(k^2 - q^2)^2\cos(ph+\alpha)\sin(qh+\alpha) + 4k^2pq\sin(ph+\alpha)\cos(qh+\alpha) = 0. \tag{5.119}$$

Expanding the initial square, we have

$$(k^2 - q^2)^2 = (k^2 + q^2)^2 - 4k^2q^2 = \frac{\omega^4}{V_{\mathrm{T}}^4} - 4k^2q^2$$

and hence

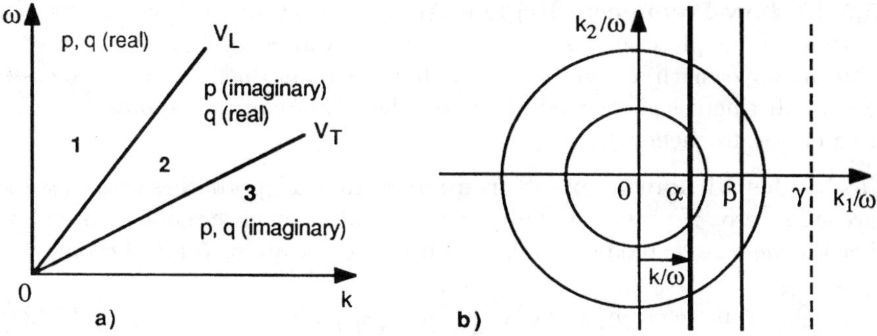

Fig. 5.37. (**a**) The (ω, k) plane is divided into three regions. For $V > V_\mathrm{L}$, p and q are real; for $V_\mathrm{T} < V < V_\mathrm{L}$, p is imaginary and q is real; and for $V < V_\mathrm{T}$, p and q are imaginary. (**b**) The position of the line k/ω with respect to the two circles that are sections of the slowness surfaces

$$\boxed{\frac{\omega^4}{V_\mathrm{T}^4} = 4\,k^2q^2\left[1 - \frac{p\tan(ph+\alpha)}{q\tan(qh+\alpha)}\right]} \quad \text{with} \quad \alpha = 0\,\text{or}\,\pi/2\,. \tag{5.120}$$

This dispersion relation between ω and k, equivalent to (5.119), is called the *Rayleigh–Lamb equation.*

5.5.2 Dispersion Curves. Mechanical Displacement

The displacement components are obtained by substituting B from (5.118a) into (5.117) giving

$$\begin{cases} u_1 = qA\left[\cos\left(qx_2+\alpha\right) - \dfrac{2k^2}{k^2-q^2}\,\dfrac{\cos\left(qh+\alpha\right)}{\cos\left(ph+\alpha\right)}\cos\left(px_2+\alpha\right)\right] \\ u_2 = \mathrm{i}kA\left[\sin\left(qx_2+\alpha\right) + \dfrac{2pq}{k^2-q^2}\,\dfrac{\cos\left(qh+\alpha\right)}{\cos\left(ph+\alpha\right)}\sin\left(px_2+\alpha\right)\right]. \end{cases} \tag{5.121}$$

It is not possible to obtain analytic expressions for the various branches of the dispersion curve from (5.120). In the (ω, k) plane, Fig. 5.37, three regions can be distinguished, depending on whether the phase velocity $V = \omega/k$ exceeds the longitudinal wave velocity $V_\mathrm{L} = (c_{11}/\rho)^{1/2}$ or the transverse wave velocity $V_\mathrm{T} = (c_{66}/\rho)^{1/2}$. From (5.112a,b) we can write

$$p^2 = \omega^2\left(\frac{1}{V_\mathrm{L}^2} - \frac{1}{V^2}\right) \quad \text{and} \quad q^2 = \omega^2\left(\frac{1}{V_\mathrm{T}^2} - \frac{1}{V^2}\right).$$

hence the three cases

(a) $V > V_\mathrm{L} > V_\mathrm{T}$, or $k < \omega/V_\mathrm{L} < \omega/V_\mathrm{T}$ the wavenumbers p and q are real;

(b) $V_\mathrm{T} < V < V_\mathrm{L}$, or $\omega/V_\mathrm{L} < k < \omega/V_\mathrm{T}$ q is real and p is imaginary;

(c) $V < V_\mathrm{T} < V_\mathrm{L}$, or $k > \omega/V_\mathrm{T} > \omega/V_\mathrm{L}$ p and q are imaginary.

5.5.2.1 Low-Frequency Region. An important distinction in the behaviour of the plate modes occurs for small k values, that is, when $kh \ll 1$ and the wavelength λ is much greater than the plate thickness $2h$. Two cases can be distinguished, depending on whether the frequency approaches zero or a cut-off frequency f_c.

(a) **Modes without a cut-off frequency.** Here ω approaches zero as k approaches zero. We consider symmetric and antisymmetric modes separately. For the *symmetric* mode, $\alpha = 0$. The dispersion relation (5.120) becomes

$$\frac{\omega^4}{V_T^4} \cong 4k^2(q^2 - p^2) = 4\,k^2\omega^2\left(\frac{1}{V_T^2} - \frac{1}{V_L^2}\right). \tag{5.122}$$

This shows that the phase velocity $V = \omega/k$ tends to a finite limit V_P, called the *plate velocity*, given by

$$V_P = 2V_T\left(1 - \frac{V_T^2}{V_L^2}\right)^{1/2} = V_T\sqrt{2}\sqrt{1 + \frac{c_{12}}{c_{11}}} = V_L\sqrt{1 - \frac{c_{12}^2}{c_{11}^2}}. \tag{5.123}$$

The plate velocity is always between $V_T\sqrt{2}$ and V_L. Here q and p are given by

$$q^2 = k^2\left(\frac{V_P^2}{V_T^2} - 1\right) = k^2\left(1 + 2\frac{c_{12}}{c_{11}}\right) \quad \text{and} \quad p = \mathrm{i}k\frac{c_{12}}{c_{11}},$$

so that q is real and p is imaginary, as in region 2 of Fig. 5.37. The mechanical displacement follows from the formulae of (5.121) with $\alpha = 0$. Taking $ph \to 0$, $qh \to 0$ and $q^2 - k^2 = 2k^2 c_{12}/c_{11}$, we find

$$\begin{cases} u_1 = qA\left(1 - \dfrac{2k^2}{k^2 - q^2}\right) = qA\left(1 + \dfrac{c_{11}}{c_{12}}\right) \\[2ex] u_2 = \mathrm{i}kqA\left(1 + \dfrac{2p^2}{k^2 - q^2}\right)x_2 = \mathrm{i}qA\left(1 + \dfrac{c_{12}}{c_{11}}\right)kx_2\,. \end{cases} \tag{5.124}$$

The longitudinal component is uniform, while the transverse component, in phase quadrature, has its maximum values at the surfaces $x_2 = \pm h$, given by

$$\frac{|u_2(\pm h)|}{u_1} = \frac{c_{12}}{c_{11}}kh \le kh \ll 1\,. \tag{5.125}$$

When $kh \ll 1$, this symmetric mode with no cut-off, called S_0, is essentially *longitudinal.*

The *antisymmetric* mode has $\alpha = \pi/2$. In the dispersion relation (5.120), the term in brackets tends to zero in a manner such that the phase velocity approaches zero with kh. Expanding the tangents gives

$$\frac{\omega^4}{V_T^4} \cong 4k^2q^2\left(1 - \frac{1 + q^2h^2/3}{1 + p^2h^2/3}\right) \cong \frac{4}{3}q^2(p^2 - q^2)k^2h^2\,.$$

Here ω approaches zero faster than k, and consequently $q^2 \approx -k^2$. The phase velocity is given by

$$V = \frac{\omega}{k} \cong \frac{2V_{\mathrm{T}}}{\sqrt{3}} \sqrt{1 - V_{\mathrm{T}}^2/V_{\mathrm{L}}^2}\, kh = \frac{V_{\mathrm{P}}}{\sqrt{3}} kh$$

and thus varies as the product kh. Near the origin, the dispersion curve $\omega(k)$ becomes the parabola

$$\boxed{\omega = \frac{V_{\mathrm{P}}}{\sqrt{3}} k^2 h} . \tag{5.126}$$

The displacements are given by (5.121), with $\alpha = \pi/2$, $ph \to 0$ and $qh \to 0$, so that

$$\begin{cases} u_1 = q^2 A \dfrac{k^2+q^2}{k^2-q^2} x_2 \cong -kA \dfrac{k^2+q^2}{k^2-q^2} k x_2 \\ u_2 \cong \mathrm{i}kA \dfrac{k^2+q^2}{k^2-q^2} . \end{cases}$$

The transverse component is invariant across the plate thickness, while the longitudinal component is maximized at the free surfaces $x_2 = \pm\, h$, where its values are given by

$$\frac{|u_1(\pm h)|}{u_2} = kh . \tag{5.127}$$

When $kh \ll 1$, this antisymmetric mode with no cut-off, called A_0, is a *flexural* mode.

(b) Modes with a cut-off frequency. For these, ω approaches a cut-off frequency ω_{c} as k approaches zero. The wavenumbers p and q approach $\omega_{\mathrm{c}}/V_{\mathrm{L}}$ and $\omega_{\mathrm{c}}/V_{\mathrm{T}}$, respectively. Equations (5.118) reduce to

$$\begin{cases} B \cos\left(\dfrac{\omega_{\mathrm{c}} h}{V_{\mathrm{L}}} + \alpha\right) = 0 \\ A \sin\left(\dfrac{\omega_{\mathrm{c}} h}{V_{\mathrm{T}}} + \alpha\right) = 0 \end{cases} . \tag{5.128}$$

These equations allow two types of solutions, such that only one of the coefficients A and B is non-zero. We consider successively the two types of Lamb waves.

For *symmetric* modes, $\alpha = 0$. The even solutions $S_{2n} (n = 1, 2, \ldots)$ are such that $A \neq 0$, giving

$$\frac{\omega_{\mathrm{c}} h}{V_{\mathrm{T}}} = n\pi \quad \Rightarrow \quad f_{\mathrm{c}} \times 2h = nV_{\mathrm{T}} , \tag{5.129}$$

where $2h$ is the plate thickness. Since $B = 0$ the displacement is *longitudinal*; from (5.117), with $k \to 0$, it is given by

$$u_1 = qA \cos\left(n\pi \frac{x_2}{h}\right) , \quad u_2 = 0 .$$

The odd solutions, S_{2m+1} with m integer, are such that $B \neq 0$, giving

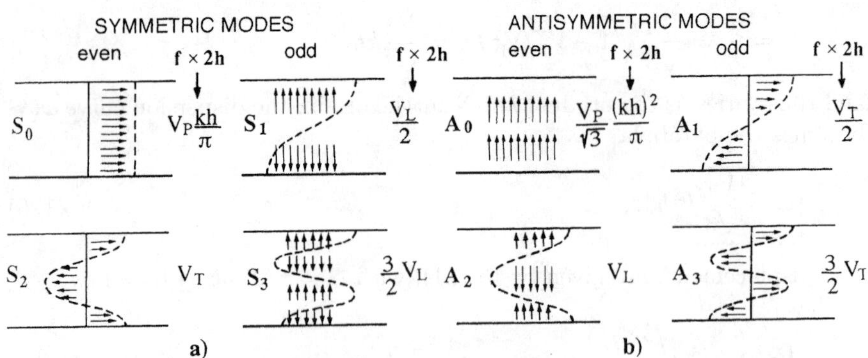

Fig. 5.38. Mechanical displacements of Lamb waves in an isotropic plate, for the first few modes, when $k \approx 0$, i.e. $kh \ll 1$. (**a**) Symmetric modes (dilatational). (**b**) Antisymmetric modes (flexural)

$$\frac{\omega_c h}{V_L} = (2m+1)\frac{\pi}{2}, \quad m = 0, 1, 2, \ldots$$

giving (5.130)

$$f_c \times 2h = \left(m + \tfrac{1}{2}\right) V_L .$$

With $A = 0$ we find that the displacement is *transverse*, giving, with $k \to 0$,

$$u_2 = -pB \sin\left[(2m+1)\frac{\pi}{2}\frac{x_2}{h}\right], \quad u_1 = 0 .$$

For the *antisymmetric* modes, $\alpha = \pi/2$, the even solutions A_{2n} give

$$\frac{\omega_c h}{V_L} = n\pi, \quad n = 1, 2, 3, \ldots \quad \text{and hence} \quad f_c \times 2h = nV_L . \qquad (5.131)$$

Using $A = 0$ the displacement is found to be *transverse*, and given by

$$u_2 = -pB \cos\left(n\pi\frac{x_2}{h}\right), \quad u_1 = 0 .$$

The odd solutions A_{2m+1} give

$$\frac{\omega_c h}{V_T} = (2m+1)\frac{\pi}{2}, \quad m = 0, 1, 2, \ldots$$

giving (5.132)

$$f_c \times 2h = \left(m + \frac{1}{2}\right) V_T .$$

Here $B = 0$ and the displacement is *longitudinal*, given by

$$u_1 = -qA \sin\left[(2m+1)\frac{\pi}{2}\frac{x_2}{h}\right], \quad u_2 = 0 .$$

Figure 5.38 illustrates these results for the first four symmetric and antisymmetric modes. The lowest cut-off frequency is that of mode A_1, occurring when the product (frequency × plate thickness), $f \times 2h$, is equal to half the transverse wave velocity.

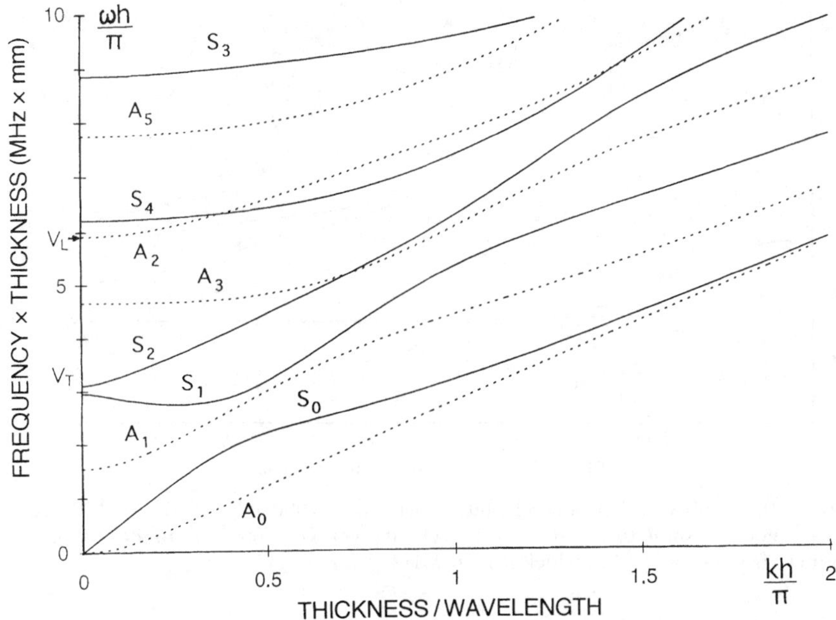

Fig. 5.39. Lamb waves in a steel plate, for which $V_L = 5\,900\,\text{m/s}$ and $V_T = 3\,100\,\text{m/s}$. Dispersion curves for symmetric modes (*solid lines*) and antisymmetric modes (*broken lines*)

From (5.129) – (5.132), the cut-off frequencies of the Lamb modes occur when the plate thickness $2h$ is equal to an even or odd multiple of the half-wavelength $\lambda_L/2$ or $\lambda_T/2$, so that

$$2h = p\lambda/2\,, \quad \text{with} \quad p = 2n \quad \text{or} \quad p = 2m+1 \tag{5.133}$$

and $\lambda = V_T/f_c$ or V_L/f_c.

The distribution of the mechanical displacements in the plate varies substantially with frequency, i.e. with kh. However, the distinction between the modes without a cut-off, S_0 and A_0, and those with a cut-off (S_n and A_n with $n \geq 1$) remains at high frequencies.

5.5.2.2 High-Frequency Region. Dispersion Curves. The high-frequency behaviour is obtained by setting $kh \gg 1$ in (5.119). For example, if p is imaginary ($p = \mathrm{i}k\chi_1$), as in region 2 of Fig. 5.37a, then for $kh \to \infty$ the dispersion relation gives $q_n h = n\pi$ for the symmetric modes ($\alpha = 0$) and $q_n h = (2n+1)\pi/2$ for the antisymmetric modes. This is shown in Problems 5.10 and 5.11. From (5.112a,b), the velocities of the S_n and A_n modes, with $n \geq 1$, approach V_T. If p and q are both imaginary (region 3), so that $p = \mathrm{i}k\chi_1$ and $q = \mathrm{i}k\chi_2$, the dispersion relation becomes that of Rayleigh waves, formula (5.43), when $kh \to \infty$. The velocities of the S_0 and A_0 modes approach V_R, the Rayleigh wave velocity (Problem 5.12).

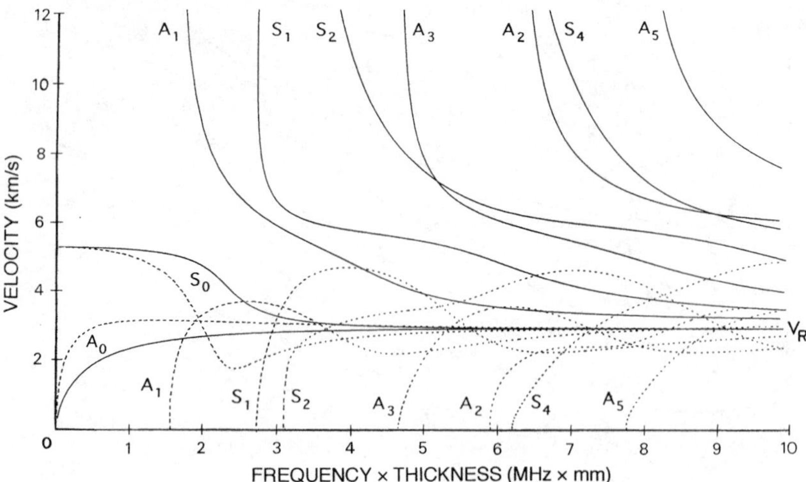

Fig. 5.40. Lamb waves in a steel plate, with $V_L = 5\,900$ m/s and $V_T = 3\,100$ m/s. Phase velocity (*solid lines*) and group velocity (*broken lines*) as functions of the product (frequency × plate thickness in MHz × mm)

Figure 5.39 shows dispersion curves for a steel plate. Their forms show physically the frequency variation of the proportion of the longitudinal and transverse components. The scale for the vertical axis is $\omega h/\pi = f \times 2h =$ (frequency × thickness) product, and that for the horizontal axis is $kh/\pi = 2h/\lambda$, the ratio of plate thickness to wavelength.

The dispersion curves for modes of the same family do not cross each other, though a curve for a symmetric (antisymmetric) mode may cross a curve for an antisymmetric (symmetric) mode. The order of the curves in the same family depends on the ratio V_L/V_T. For example, the curve for mode S_2 is above or below that for S_1 according to whether V_T is less or more than $V_L/2$. For the S_1 mode, and for small values of kh, ω decreases with k; the group velocity $V_g = \mathrm{d}\omega/\mathrm{d}k$ is therefore negative, that is, in the opposite direction to the phase velocity. Thus the acoustic energy is transported in the direction opposite to that of the wave vector.

Figure 5.40 shows the phase velocity $V_\varphi = \omega/k$ and the group velocity $V_g = \mathrm{d}\omega/\mathrm{d}k$ as functions of the product (frequency × plate thickness). This quantity, $f \times 2h$, has the dimensions of velocity and may conveniently be expressed in mm/µs or MHz × mm.

5.5.2.3 Lamé Modes. The particular case $q^2 = k^2$ is of special interest. From (5.112b), this leads to

$$\frac{\omega^2}{V_T^2} = 2k^2 \quad \text{and so} \quad V_\varphi = \frac{\omega}{k} = V_T\sqrt{2}\,.$$

Consequently $V_\varphi < V_L$, so p is imaginary, and we write $p = \mathrm{i}\chi$. Moreover,

$$V_P = V_T\sqrt{2}\sqrt{1 + c_{12}/c_{11}} > V_T\sqrt{2}\,.$$

The equality is therefore possible for all modes except the antisymmetric mode A_0 whose velocity is less than V_R, which in turn is less than V_T.

For these modes, the boundary conditions given by (5.118) can be written as follows:

symmetric modes ($\alpha = 0$)

$$\begin{cases} A \cos qh = 0\,, \text{so}\, qh = (n+1/2)\pi \\ B \sinh \chi h = 0\,, \text{so}\, B = 0 \end{cases}$$

antisymmetric modes ($\alpha = \pi/2$)

$$\begin{cases} A \sin qh = 0\,, \text{so}\, qh = n\pi \\ B \cosh \chi h = 0\,, \text{so}\, B = 0\,. \end{cases}$$

In general, since $q = k$ we can write $kh = m\pi/2$, with m odd for symmetric modes and m even for antisymmetric modes. The Lamé modes occur for equally-spaced values of the frequency–thickness product, given by

$$f \times 2h = m\frac{V_T}{\sqrt{2}}\,, \qquad m = 1, 2, 3 \ldots\,. \tag{5.134}$$

Noting that $B = 0$ implies $\phi = 0$, (5.117) gives the displacements as follows:

symmetric modes

$$\begin{cases} u_1 = kA \cos qx_2 \\ u_2 = \mathrm{i}kA \sin qx_2 \end{cases}$$

antisymmetric modes

$$\begin{cases} u_1 = kA \sin qx_2 \\ u_2 = \mathrm{i}kA \cos qx_2\,. \end{cases}$$

Thus we have

$$|u_1|^2 + |u_2|^2 = k^2 A^2 = \text{constant}\,.$$

Figure 5.41a shows the variations of u_1 and u_2 across the plate for the S_0 mode. Here, u_1 is zero on the faces because the Lamb wave is simply a TV wave, a transverse wave with polarization contained in the vertical plane $x_1 x_2$. As seen in Fig. 5.41b, this propagates at an angle of $\pi/4$ to the plate axis, so that $V_\varphi = V_T\sqrt{2}$. This is confirmed by the values of the stresses, which are, from (5.115) and (5.114),

$$\begin{cases} T_{12} = c_{66}\left[(k^2 - q^2)\Psi - 2\mathrm{i}k\dfrac{\partial \phi}{\partial x_2}\right] \equiv 0 \quad \text{since} \quad q^2 = k^2 \quad \text{and} \quad \phi = 0 \\ T_{22} = 2\mathrm{i}k\, c_{66}\dfrac{\partial \Psi}{\partial x_2} = 2\mathrm{i}kq\, A \cos(qx_2 + \alpha)\,. \end{cases}$$

Thus T_{12} is zero and T_{22} is maximal.

5.6 Cylindrical Waveguides

The waves considered so far propagate in plane waveguides, either near a plane surface (Rayleigh, Bleustein–Gulyaev or Love waves), or between parallel plane surfaces (transverse-horizontal waves or Lamb waves). In fact, these waves also propagate in a similar manner if the surface(s) (parallel if more than one surface is involved) is (are) slightly curved, with radius $R \gg \lambda$. The analysis of these waveguides is relatively simple because they

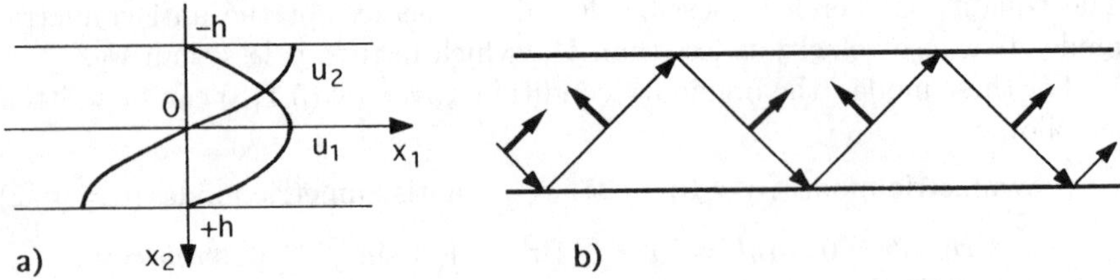

Fig. 5.41. The particular case $q^2 = k^2$. (**a**) Displacement components for the S_0 mode. (**b**) The Lamb wave is a transverse wave (a Lamé mode) propagating at 45° to the plate axis

have an infinite cross-section. They are often used at high frequencies, above for example 50 MHz, for which the material must be crystalline. However, there are also waveguides with finite cross-sections, used in particular for low-frequency waves travelling over large distances. For these, the material is of course isotropic. The classic example is the cylindrical guide [5.22].

The material displacement vector $\boldsymbol{u}$ is derived from a scalar potential ϕ and a vector potential Ψ, as in (5.109), and from (5.110a, 5.110b) these satisfy the wave equations

$$\nabla^2\phi - \frac{1}{V_\mathrm{L}^2}\frac{\partial^2\phi}{\partial t^2} = 0 \quad \text{and} \quad \nabla^2\Psi - \frac{1}{V_\mathrm{T}^2}\frac{\partial^2\Psi}{\partial t^2} = 0\,. \tag{5.135a,b}$$

Considering propagation along the cylinder axis z, it is convenient to adopt cylindrical coordinates r, θ, z, as in Fig. 5.42. In this case, the Laplacian becomes (from Appendix A)

$$\nabla^2 = \frac{\partial^2}{\partial r^2} + \frac{1}{r}\frac{\partial}{\partial r} + \frac{1}{r^2}\frac{\partial^2}{\partial\theta^2} + \frac{\partial^2}{\partial z^2}\,.$$

We assume that the scalar potential can be written in the form

$$\phi = f(r)g(\theta)\,\mathrm{e}^{\mathrm{i}(\omega t - kz)}\,,$$

where the separation of variables is justified *a posteriori*. The equation for the scalar potential then splits into the two parts

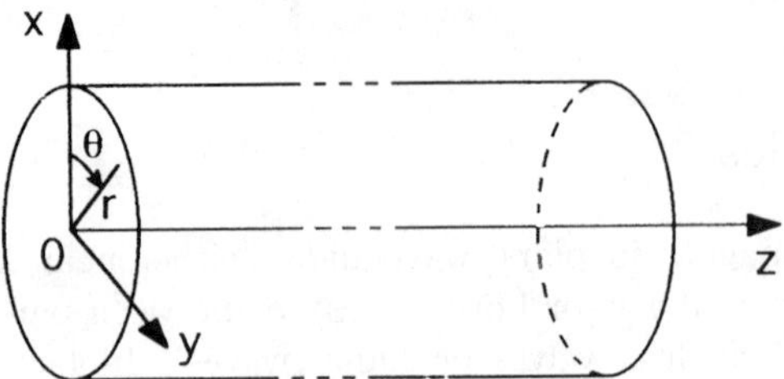

Fig. 5.42. Cylindrical coordinates r, θ, z. The radius of the cylinder is a

$$\begin{cases} \dfrac{d^2 g}{d\theta^2} + n^2 g(\theta) = 0\,, \quad n \,\text{integer} & (5.136\text{a}) \\ \dfrac{d^2 f}{dr^2} + \dfrac{1}{r}\dfrac{df}{dr} + \left(\dfrac{\omega^2}{V_\mathrm{L}^2} - k^2\right) f - \dfrac{n^2}{r^2} f(r) = 0\,. & (5.136\text{b}) \end{cases}$$

The development of the vectorial Laplacian $\nabla^2 \Psi$ is described in Appendix A. For the time-harmonic case it gives the three equations

$$\begin{cases} \nabla^2 \Psi_z + \dfrac{\omega^2}{V_\mathrm{T}^2} \Psi_z = 0 & (5.137\text{a}) \\ \nabla^2 \Psi_r - \dfrac{\Psi_r}{r^2} - \dfrac{2}{r^2}\dfrac{\partial \Psi_\theta}{\partial \theta} + \dfrac{\omega^2}{V_\mathrm{T}^2} \Psi_r = 0 & (5.137\text{b}) \\ \nabla^2 \Psi_\theta - \dfrac{\Psi_\theta}{r^2} + \dfrac{2}{r^2}\dfrac{\partial \Psi_r}{\partial \theta} + \dfrac{\omega^2}{V_\mathrm{T}^2} \Psi_\theta = 0\,. & (5.137\text{c}) \end{cases}$$

Equation (5.137a), which contains only the component Ψ_z, is similar to (5.135a) for the scalar potential, so the solutions are also similar. The components Ψ_r and Ψ_θ are coupled.

The solutions of (5.136a) for $g(\theta)$ are $\sin n\theta$ and $\cos n\theta$, where n must be an integer to ensure continuity of $g(\theta)$ and its derivatives. Equation (5.136b) for $f(r)$ is of course satisfied by Bessel functions. The fact that the several variables involved must be finite at $r = 0$ implies that the Bessel functions must be of the first kind, that is,

$$J_n(pr) \quad \text{with} \quad p^2 = \frac{\omega^2}{V_\mathrm{L}^2} - k^2\,. \tag{5.138}$$

Thus the solution for ϕ is, for example,

$$\phi = A J_n(pr) \cos n\theta \, e^{i(\omega t - kz)}\,, \tag{5.139}$$

and Ψ_z has the similar solution

$$\Psi_z = B J_n(qr) \cos n\theta \, e^{i(\omega t - kz)} \quad \text{with} \quad q^2 = \frac{\omega^2}{V_\mathrm{T}^2} - k^2\,. \tag{5.140}$$

The expressions for Ψ_r and Ψ_θ are, *a priori*, of the form

$$\begin{cases} \Psi_r = \Psi_r(r) \sin n\theta \, e^{i(\omega t - kz)} \\ \Psi_\theta = \Psi_\theta(r) \cos n\theta \, e^{i(\omega t - kz)}\,, \end{cases}$$

where the presence of $\sin n\theta$ in the first expression implies $\cos n\theta$ in the second, because the coupling terms in (5.137b) and (5.137c) have differentials with respect to θ with different signs. Substitution gives

$$\frac{d^2 \Psi_r}{dr^2} + \frac{1}{r}\frac{d\Psi_r}{dr} + \frac{1}{r^2}\left(-n^2 \Psi_r + 2n\,\Psi_\theta - \Psi_r\right) + q^2 \Psi_r = 0$$

and

$$\frac{d^2\Psi_\theta}{dr^2} + \frac{1}{r}\frac{d\Psi_\theta}{dr} + \frac{1}{r^2}\left(-n^2\Psi_\theta + 2n\,\Psi_r - \Psi_\theta\right) + q^2\Psi_\theta = 0\,,$$

where the second equation is the same as the first except for interchanging the indices r and θ. Subtracting and adding these equations give two new equations whose solutions are, respectively,

$$\begin{cases} \Psi_r - \Psi_\theta = 2\,C J_{n+1}(qr) \\ \Psi_r + \Psi_\theta = 2\,C_1 J_{n-1}(qr)\,, \end{cases}$$

from which we have

$$\begin{cases} \Psi_r = C_1\,J_{n-1}(qr) + C J_{n+1}(qr) \\ \Psi_\theta = C_1\,J_{n-1}(qr) - C J_{n+1}(qr)\,. \end{cases}$$

Four unknown constants, A, B, C and C_1 appear in the potentials ϕ, Ψ_z, Ψ_r and Ψ_θ. However, the boundary conditions, expressing the fact that the surface $r = a$ is free, only give the three relations $T_{rr} = 0$, $T_{rz} = 0$ and $T_{r\theta} = 0$. Fortunately, we are free to impose another condition between the components Ψ_r, Ψ_θ and Ψ_z, and the choice $\Psi_r = -\Psi_\theta$ is convenient because it gives $C_1 = 0$ [5.23]. The expressions for ϕ and the components of Ψ then become

$$\phi = AJ_n(pr)\begin{Bmatrix}\cos n\theta \\ \sin n\theta\end{Bmatrix} e^{i(\omega t - kz)} \qquad \Psi_z = BJ_n(qr)\begin{Bmatrix}\sin n\theta \\ \cos n\theta\end{Bmatrix} e^{i(\omega t - kz)} \tag{5.141a,b}$$

$$\Psi_r = CJ_{n+1}(qr)\begin{Bmatrix}\sin n\theta \\ \cos n\theta\end{Bmatrix} e^{i(\omega t - kz)} \quad \Psi_\theta = -CJ_{n+1}(qr)\begin{Bmatrix}\cos n\theta \\ \sin n\theta\end{Bmatrix} e^{i(\omega t - kz)}\,. \tag{5.141c,d}$$

Boundary Conditions. From Appendix A, the components of the displacement (u_r along r, u_θ along θ and u_z along z) are given in terms of the potentials by

$$\begin{cases} u_r = \dfrac{\partial\phi}{\partial r} + \dfrac{1}{r}\dfrac{\partial\Psi_z}{\partial\theta} - \dfrac{\partial\Psi_\theta}{\partial z} & (5.142\text{a}) \\[2ex] u_\theta = \dfrac{1}{r}\dfrac{\partial\phi}{\partial\theta} + \dfrac{\partial\Psi_r}{\partial z} - \dfrac{\partial\Psi_z}{\partial r} & (5.142\text{b}) \\[2ex] u_z = \dfrac{\partial\phi}{\partial z} + \dfrac{1}{r}\dfrac{\partial(r\Psi_\theta)}{\partial r} - \dfrac{1}{r}\dfrac{\partial\Psi r}{\partial\theta}\,. & (5.142\text{c}) \end{cases}$$

The local dilatation S is

$$S = \mathrm{div}\boldsymbol{u} = \frac{1}{r}\frac{\partial(ru_r)}{\partial r} + \frac{1}{r}\frac{\partial u_\theta}{\partial\theta} + \frac{\partial u_z}{\partial z}$$

and the stresses, involving S and the Lamé constants λ and μ, are

$$\begin{cases} T_{rr} = \lambda S + 2\mu \dfrac{\partial u_r}{\partial r} & (5.143\text{a}) \\ T_{r_\theta} = \mu \left[\dfrac{1}{r} \left(\dfrac{\partial u_r}{\partial \theta} - u_\theta \right) + \dfrac{\partial u_\theta}{\partial r} \right] & (5.143\text{b}) \\ T_{rz} = \mu \left(\dfrac{\partial u_r}{\partial z} + \dfrac{\partial u_z}{\partial r} \right) . & (5.143\text{c}) \end{cases}$$

Setting these three stresses to zero at the free surface $r = a$ yields three linear homogeneous equations in the three unknowns A, B and C, giving non-zero solutions if the determinant of the coefficients is zero. Expressing this explicitly gives the dispersion relation relating ω to k and the circumferential order n. This can only be solved numerically, whatever the value of the integer n. For any n and k (with k real) there is an infinite number of roots, which are the frequencies of modes propagating along z. However, the modes can be grouped into three families: (1) *compressional* waves, axially symmetric, with the displacements u_r and u_z independent of θ; (2) *torsional* waves with circumferential displacement u_θ independent of θ; and (3) *flexural* waves with displacements u_r, u_θ, u_z dependent on r, θ and z. Fortunately, expressions (5.141) lead conveniently to the essential features in the simple cases of compressional or torsional waves with $n = 0$, and flexural waves with $n = 1$.

5.6.1 Compressional Waves

In this case, with $n = 0$, the particle motion is described by two displacement components u_r and u_z, which are independent of θ. The expressions for the potentials in (5.141) become

$$\begin{cases} \phi \;\; = A\, J_0(pr)\, \mathrm{e}^{\mathrm{i}(\omega t - kz)} \\ \Psi_\theta = -C J_1(qr)\, \mathrm{e}^{\mathrm{i}(\omega t - kz)} \end{cases} \quad \text{with} \quad \Psi_r = \Psi_z = 0 \, .$$

Using the relations

$$\frac{\mathrm{d}}{\mathrm{d}x}[J_0(x)] = -J_1(x) \quad \text{and} \quad \frac{\mathrm{d}}{\mathrm{d}x}[J_1(x)] = J_0(x) - \frac{J_1(x)}{x} \, , \qquad (5.144\text{a,b})$$

the radial and axial components of displacement in (5.142a) and (5.142c) become

$$\begin{cases} u_r = -\,[pAJ_1(pr) + \mathrm{i}kCJ_1(qr)]\; \mathrm{e}^{\mathrm{i}(\omega t - kz)} & (5.145\text{a}) \\ u_z = -\,[\mathrm{i}kAJ_0(pr) + qCJ_0(qr)]\; \mathrm{e}^{\mathrm{i}(\omega t - kz)} \, . & (5.145\text{b}) \end{cases}$$

These expressions, and the dilatation $S = -(p^2 + k^2)A\, J_0(pr)$, are substituted into (5.143a) and (5.143c) to obtain the stresses T_{rr} and T_{rz}. Setting these stresses to zero at the surface $r = a$, we have

$$T_{rr} = \left\{ \left[-\lambda(p^2 + k^2) - 2\mu p^2 \right] J_0(pa) + 2\mu \tfrac{p}{a} J_1(pa) \right\} A$$
$$+2\mu i k \left[-q J_0(qa) + \tfrac{1}{a} J_1(qa) \right] C = 0 ,$$
$$T_{rz} = -2\mu i k p J_1(pa) A - \mu(q^2 - k^2) J_1(qa) C = 0 .$$

Making use of the equality $\lambda(p^2 + k^2) + 2\mu p^2 = \mu(q^2 - k^2)$, these equations become

$$\begin{cases} \left[-(q^2 - k^2) J_0(pa) + 2\dfrac{p}{a} J_1(pa) \right] A + 2ik \left[-q J_0(qa) + \dfrac{1}{a} J_1(qa) \right] C = 0 \\ 2ikp J_1(pa) A + (q^2 - k^2) J_1(qa) C = 0 \end{cases}$$

and these are satisfied only if the determinant of coefficients is zero, giving

$$\begin{aligned} &\tfrac{2p}{a} (q^2 + k^2) J_1(pa) J_1(qa) - (q^2 - k^2)^2 J_0(pa) J_1(qa) \\ &\quad -4k^2 pq J_1(pa) J_0(qa) = 0 . \end{aligned} \tag{5.146}$$

This dispersion relation is often called the Pochhammer–Chree equation. With p and q given by (5.138) and (5.140), it contains the angular frequency ω, the wavenumber k, the velocities of bulk longitudinal and transverse waves V_{L} and V_{T}, and the cylinder radius a. Depending on whether the phase velocity $V = \omega/k$ is greater than V_{L} or V_{T}, the coefficients p and q are real or purely imaginary. In the latter case, the equations may be rewritten using the identity $J_n(ix) = \mathrm{i}^n I_n(x)$, where $I_n(x)$ denotes the modified Bessel functions of the first kind. Figure 5.43a shows dispersion curves $\omega(k)$ for the first few compressional modes in a cylinder of stainless steel, which has $\rho = 7\,930\,\mathrm{kg/m^3}$, $V_{\mathrm{L}} = 5\,692\,\mathrm{m/s}$ and $V_{\mathrm{T}} = 3\,172\,\mathrm{m/s}$, giving a Poisson's ratio of $\nu = 0.27$. The ordinate and abscissa represent the non-dimensional quantities $\omega a/V_{\mathrm{T}}$ and ka, respectively. All the modes are dispersive. The first mode, L(0, 1), is the only one with no cut-off frequency. Figure 5.43b shows the first 18 modes, and it can be seen that the dispersion curves do not cross.

At high frequencies, where $\lambda \ll a$ and $ka \gg 1$, the phase velocity of the L(0, 1) mode approaches the Rayleigh-wave velocity $V_{\mathrm{R}} = 2\,940\,\mathrm{m/s}$, which is less than V_{T}. The other mode velocities approach V_{T}.

The phase velocity $V_\varphi = \omega/k$ and group velocity $V_{\mathrm{g}} = \mathrm{d}\omega/\mathrm{d}k$, normalized to the bulk transverse wave velocity V_{T}, are shown in Fig. 5.44 as functions of the normalized angular frequency $\omega a/V_{\mathrm{T}}$, for the first two longitudinal modes. At low frequencies, where the wavelength is large so that $\lambda \gg a$, the phase velocity is equal to $V_{\mathrm{b}} = (E/\rho)^{1/2} = 5\,072\,\mathrm{m/s}$. This result can be obtained from the dispersion relation by replacing $J_0(pa)$ and $J_0(qa)$ by unity, and $J_1(pa)$ and $J_1(qa)$ by $pa/2$ and $qa/2$ respectively. These are the initial terms of the expansions

$$J_0(x) = 1 - x^2/4 + \ldots , \qquad J_1(x) = x/2 - x^2/16 + \ldots .$$

Problem 5.13 shows that this result can also be obtained directly by noting that the wave variables must have negligible variation across the width of

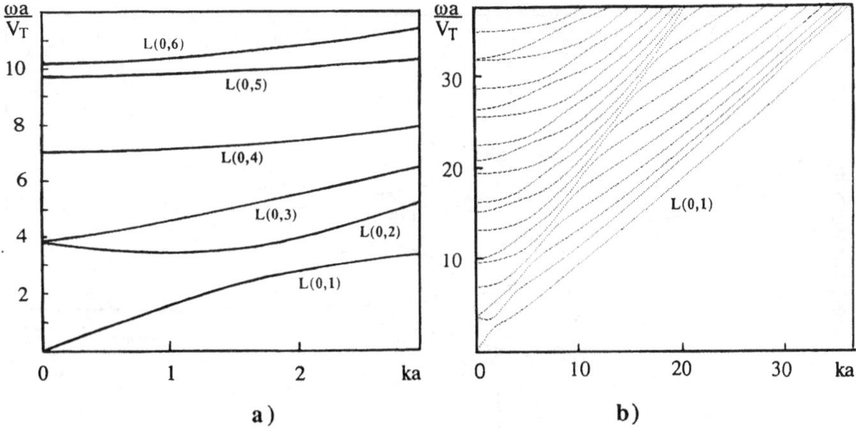

Fig. 5.43. Dispersion curves for the family of longitudinal modes in a steel cylinder, with $\nu = 0.27$. (**a**) Only the first mode L(0, 1) has no cut-off frequency. (**b**) The curves do not cross. A line with slope unity corresponds to a phase velocity equal to V_{T}

the cylinder, because the radius is much less than the wavelength; thus, the wavefronts are planes normal to the cylinder axis. Taking account of the lateral motion of the particles, Rayleigh derived the relation

$$V_{\mathrm{b}} = \sqrt{\frac{E}{\rho}}\left[1 - \nu^2\left(\frac{ka}{2}\right)^2\right]$$

and this is found from the above equations when the x^2 term of $J_0(x)$ is included.

The group velocity V_{g} of the first mode has a region with little dispersion, where $V_{\mathrm{g}} \approx V_{\mathrm{b}} =$ constant, at low frequencies where $\omega < V_{\mathrm{T}}/a$. On the other hand, for an angular frequency $\omega_{\mathrm{m}} \approx 3.17\, V_{\mathrm{T}}/a$ the group velocity passes through a minimum.

5.6.2 Torsional Waves

Here there is only one displacement component $u_\theta(r, z, t)$, and this is independent of the angle θ. It is derived from a single potential component Ψ_z, from (5.141a). Using (5.144a), we find

$$\Psi_{\mathrm{z}} = BJ_0(qr)\,\mathrm{e}^{\mathrm{i}(\omega t - kz)}$$

giving (5.147)

$$u_\theta = -\frac{\partial \Psi_z}{\partial r} = -BqJ_1(qr)\,\mathrm{e}^{\mathrm{i}(\omega t - kz)}\,.$$

The stress $T_{r\theta}$ must be zero at the free surface $r = a$, so

$$T_{r\theta} = \mu\left(\frac{\partial u_\theta}{\partial r} - \frac{u_\theta}{r}\right)_{r=a} = 0$$

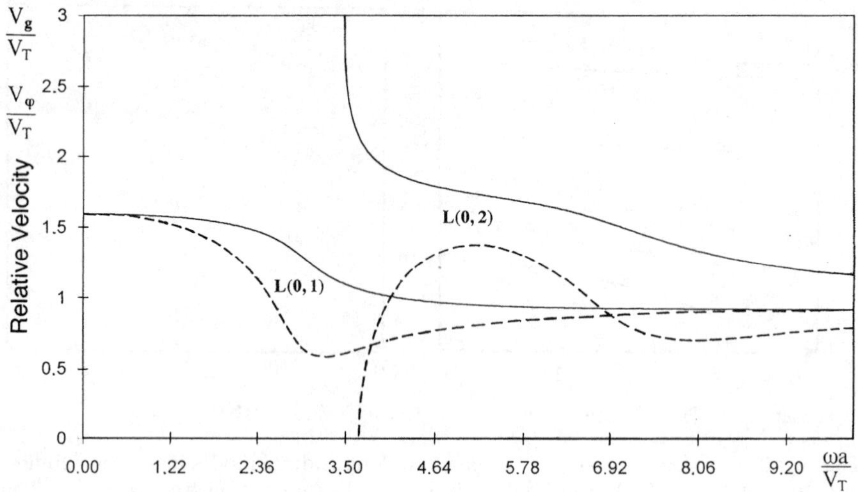

Fig. 5.44. Normalized phase velocity $V_\varphi/V_{\rm T}$ (*solid line*) and group velocity $V_{\rm g}/V_{\rm T}$ (*broken line*) as functions of normalized frequency $\omega a/V_{\rm T}$, for the first two longitudinal modes in a steel cylinder

and with (5.144) this gives

$$qa\, J_0(qa) - 2J_1(qa) = 0\,. \tag{5.148}$$

The simplest solution of this equation is $qa = 0$. Defining $A = -Bq^2/2$ and taking the limit $q \to 0$, the solution for u_θ is

$$u_\theta = Ar\,\mathrm{e}^{\mathrm{i}(\omega t - kz)}\,.$$

This is the lowest order torsional mode. Each disc-shaped section of the cylinder rotates about its centre, since the displacement is proportional to the radial coordinate. From (5.140), the value $q = 0$ also requires the velocity to be a constant, equal to $V_{\rm T}$. Hence this torsional mode is not dispersive. The other modes are dispersive, with velocities given by the relation

$$\frac{\omega a}{V_{\rm T}} = \sqrt{(ka)^2 + (q_n a)^2}$$

where $q_n a$ are the solutions of (5.148), for example $q_1 a \approx 5.14$, $q_2 a \approx 8.42$, $q_3 a \approx 11.62$. These results resemble those of Sect. 1.1.2.2.

Figure 5.45 shows the dispersion curves for the first three torsional modes of the steel cylinder considered earlier. They are similar to the curves in Fig. 1.8. As for other types of wave, the cut-off frequencies can be explained by the fact that the wavefronts make an angle α with the cylinder axis, so that the phase velocity in the axial direction is $V_\varphi = V_{\rm T}/\cos\alpha$. The cut-off occurs when α approaches $\pi/2$, so that V_φ becomes infinite.

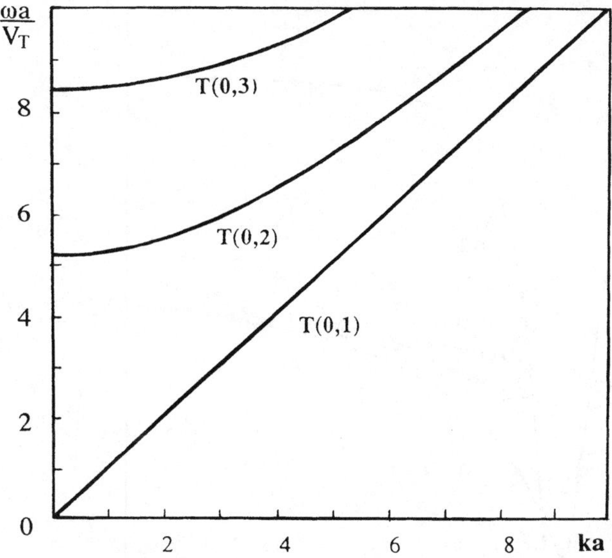

Fig. 5.45. Dispersion curves for the first three torsional modes. The first mode, T(0, 1), is non-dispersive

5.6.3 Flexural Waves

Here the particle motion is described by the three displacement components u_r, u_θ and u_z, which vary with θ as $\sin n\theta$ and $\cos n\theta$. To illustrate the vibration of a mode propagating along the cylinder, it is sufficient to consider the case $n = 1$. The mode is described by the potentials of (5.141), omitting the exponential propagation term for convenience, so that

$$\begin{aligned} \phi &= AJ_1(pr)\cos\theta \quad & \Psi_z &= BJ_1(qr)\sin\theta \\ \Psi_r &= CJ_2(qr)\sin\theta \quad & \Psi_\theta &= -CJ_2(qr)\cos\theta\,. \end{aligned}$$

Using (5.142), these give

$$u_r = U_r(r)\cos\theta\,, \quad u_\theta = U_\theta(r)\sin\theta\,, \quad u_z = U_z(r)\cos\theta\,,$$

where $U_r(r)$, $U_\theta(r)$ and $U_z(r)$ are dependent only on r.

Inspection of the components u_r (along r), u_θ (along θ) and u_z (along z), as functions of the angle θ, explains why these waves are called flexural. Thus, for $\theta = 0$ we have $u_\theta = 0$, so points in the (x, z) plane are displaced in this plane. For $\theta = \pm\pi/2$ only u_θ is non-zero, so the points in the (y, z) plane are displaced parallel to the (x, z) plane.

As for other families of modes, the dispersion relation is obtained by substituting the expressions for the displacement components into the formulae for the stresses T_{rr}, $T_{r\theta}$ and T_{rz}, and then equating these stresses to zero at $r = a$. This gives three equations in A, B and C, which are satisfied only if the determinant of the coefficients is zero. The solutions provide the various modes of the family.

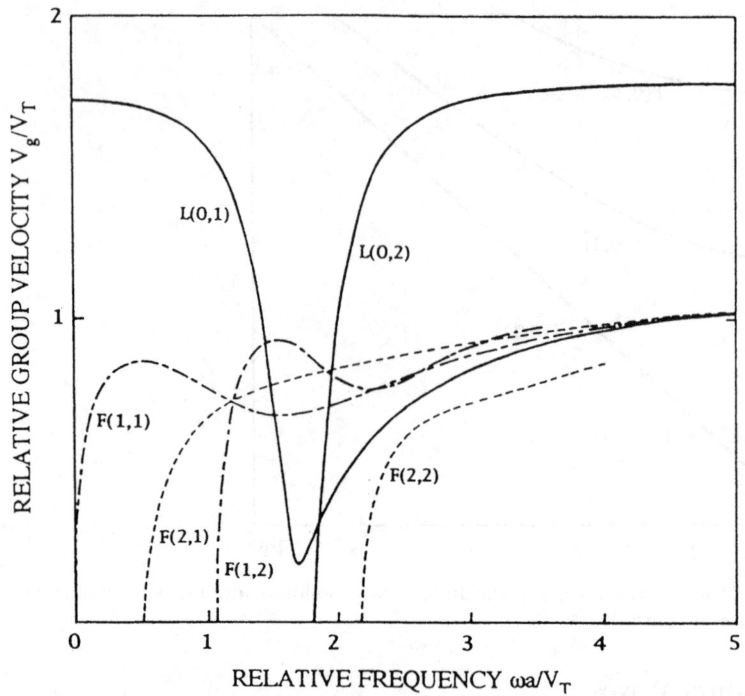

Fig. 5.46. Steel tube, with mean radius a=3.75 mm and thickness 1.5 mm. Group velocity as a function of normalized frequency for the first few compressional (L) and flexural (F) modes. The group velocity of the L(0, 1) mode has a deep minimum for $\omega a/V_T = 1.68$, where the frequency is 227 kHz

5.6.4 Tubular Waveguides

In principle, propagation in a cylindrical tube can be analysed by a method similar to that used for propagation in a cylinder. However, the tube has two free surfaces instead of the one surface of a simple cylinder – the boundary conditions (zero stresses) need to be satisfied on both the external and internal surfaces. This implies that a Bessel function of the second kind needs to be introduced into the scalar potential and into the three components of the vector potential. Six constants appear in the six equations expressing the boundary conditions on the two free surfaces, and these equations can be satisfied only if the determinant is zero. Setting the determinant to zero gives the dispersion relation. The equation can be solved only by numerical techniques [5.24, 5.25].

An approximate model based on the assumption of a *thin shell* was established by Cooper and Naghdi [5.26] and, independently, by Mirsky and Hermann [5.27]. Within the thickness of the shell, the displacements u_z and u_θ are taken to be linear functions of a radial coordinate $\rho = r - a$ referenced to the mean radius a, so that

$$\begin{cases} u_r = U_r(z,\theta,t) \\ u_\theta = U_\theta(z,\theta,t) + \rho\beta_\theta(z,\theta,t) \\ u_z = U_z(z,\theta,t) + \rho\beta_z(z,\theta,t)\,. \end{cases}$$

Here U_r, U_θ and U_z are the displacement components at the mean radius, with the latter taken as the origin. The functions β_θ and β_z are small inclination angles of the surface at the mean radius a, respectively in the rz and $r\theta$ planes.

Figure 5.46 shows the normalized group velocities $V_\mathrm{g}/V_\mathrm{T}$ as functions of normalized frequency $\omega a/V_\mathrm{T}$, for the first few compressional and flexural modes, calculated using the thin shell approximation. These refer to a steel tube with mean radius $a = 3.75\,\mathrm{mm}$ and thickness $h = 1.5\,\mathrm{mm}$.

Problems

5.1. Show that, in the harmonic regime, the mean power $\langle P\rangle$ transported along a waveguide with non-dissipative boundary conditions is conserved if the region does not contain any sources.

Solution. The absence of sources implies that $\mathrm{d}W/\mathrm{d}t = 0$, and integration of (5.18) over one period T gives

$$\frac{1}{T}\int_{-\mathrm{T}/2}^{+\mathrm{T}/2}\frac{\partial P}{\partial x_1}\mathrm{d}t = \frac{\partial\langle P\rangle}{\partial x_1} = -\frac{1}{T}[E_\mathrm{k}+E_\mathrm{p}]_{-\mathrm{T}/2}^{+\mathrm{T}/2}\,.$$

Since, in the periodic case, all variables take the same value after one period, this leads to

$$\frac{\partial\langle P\rangle}{\partial x_1} = 0 \quad \text{and hence} \quad \langle P\rangle = \text{constant}\,.$$

5.2. By comparing the stiffness of a cube A of material adjacent to the surface of a solid with that of a cube B entirely enclosed within it (Fig. 5.47), estimate the ratio $V_\mathrm{R}/V_\mathrm{T}$ of the Rayleigh wave and bulk transverse wave velocities in an isotropic solid.

Solution. The velocity of an elastic bulk wave has the form $(c/\rho)^{1/2}$, where c is a stiffness constant. Since one of the six faces of cube A is free, the stiffness of the material in the region of the surface is $5c/6$. Thus, the Rayleigh wave velocity can be estimated as

$$V_\mathrm{R} = \sqrt{\frac{5c}{6\rho}} = \sqrt{\frac{5}{6}}V_\mathrm{T} = 0.91V_\mathrm{T}\,.$$

This result agrees well with the full calculation of Sect. 5.3.2, which leads to a ratio $V_\mathrm{R}/V_\mathrm{T}$ between 0.89 and 0.94.

5.3. For a Rayleigh wave in an isotropic solid, given by (5.45), express the absolute amplitude of the longitudinal component at the free surface, $A = U_1(0)$, in terms of the ratio $R = (V_\mathrm{R}/V_\mathrm{T})^2$. Deduce from this the amplitude $U_2(0)$ of the transverse component.

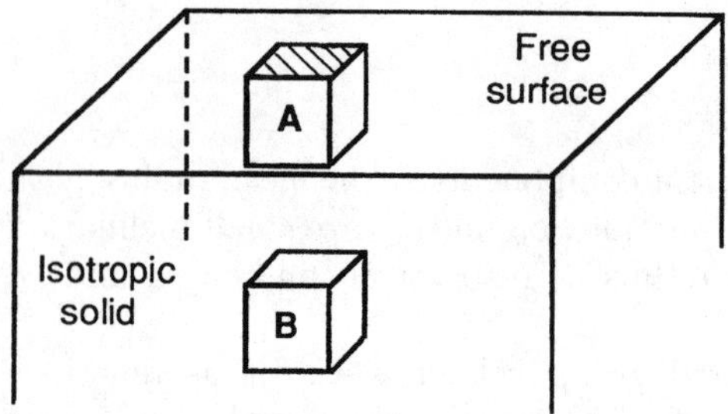

Fig. 5.47. The stiffness for cubes A and B is different

Solution. From (5.45), the normalized components can be written as

$$\begin{cases} U_1(X_2) = \dfrac{A}{1-\sqrt{\chi_1\chi_2}}\left(\mathrm{e}^{-\chi_1 X_2} - \sqrt{\chi_1\chi_2}\,\mathrm{e}^{-\chi_2 X_2}\right) \\ U_2(X_2) = \mathrm{i}A\dfrac{\sqrt{\chi_1/\chi_2}}{1-\sqrt{\chi_1\chi_2}}\left(\mathrm{e}^{-\chi_2 X_2} - \sqrt{\chi_1\chi_2}\,\mathrm{e}^{-\chi_1 X_2}\right) . \end{cases}$$

From (5.47) we have

$$\int_0^\infty \left[|U_1(X_2)|^2 + |U_2(X_2)|^2\right] \mathrm{d}X_2 = 1$$

and this gives

$$A^2\left(\chi_1 + \frac{1}{2\chi_1} + \frac{\chi_1}{2\chi_2^2} - 2\frac{\sqrt{\chi_1\chi_2}}{\chi_2}\right) = \left(1-\sqrt{\chi_1\chi_2}\right)^2 .$$

Making use of (5.43) this becomes

$$A^2\left(1 + \frac{\chi_1-\chi_2}{2\chi_1\chi_2^2}\right) = \chi_2 .$$

In terms of the parameter $R = (V_\mathrm{R}/V_\mathrm{T})^2$ we find

$$A^2 = \frac{(1-R)^{1/2}}{1+R^2/[2(2-R)^2(1-R)]} \quad \text{and} \quad U_2(0) = A\sqrt{\frac{\chi_1}{\chi_2}} .$$

5.4. The table of Fig. 5.10 indicates that, for class 622, a Bleustein–Gulyaev wave can exist for any sagittal plane that includes the A_6 axis, even if this plane is not normal to one of the six dyad axes. Confirm this statement.

Solution. It is sufficient to show that the tensor e_{ijk} is invariant for any rotation about the Ox_3 axis, which is parallel to the A_6 axis. For this class, (3.118) shows that only $e_{123} = e_{14}$ and $e_{213} = -e_{14}$ are non-zero, giving $e'_{ijk} = (\alpha_i^1\alpha_j^2\alpha_k^3 - \alpha_i^2\alpha_j^1\alpha_k^3)e_{14}$. Considering the constants of (3.117), we have

$$\begin{aligned} &e'_{31} = 0 \quad \text{because} \quad k \neq 3; \quad e'_{113} = e'_{15} = 0; \\ &e'_{333} = e'_{33} = 0 \quad \text{because} \quad i = j \quad \text{and} \quad k = 3; \\ &e'_{14} = e'_{123} = (\cos^2\varphi + \sin^2\varphi)e_{14} = e_{14} \quad \text{so} \quad e_{ijk} \quad \text{is unchanged}. \end{aligned}$$

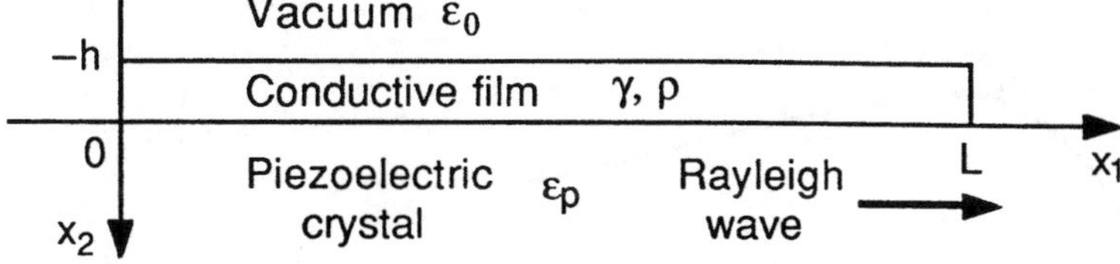

Fig. 5.48. Conductive film on a piezoelectric solid

5.5. A piezoelectric solid is covered with a conducting metal film whose thickness h is much less than the wavelength of the Rayleigh wave propagating along x_1, as in Fig. 5.48. Define ρ, $\boldsymbol{j}$, Φ and $\boldsymbol{E}$ as the charge and current densities, the potential, and the electric field associated with the wave.

(a) Using conservation of charge, given by (3.22), show that j_1 is proportional to the phase velocity $V = \omega/k$ of the wave.

(b) Calculate the potential Φ, assumed to be constant throughout the film thickness h, by applying Ohm's law $\boldsymbol{j} = \gamma\boldsymbol{E}$, where γ is the conductivity of the metal.

(c) Regarding the film as a layer of charge, with area density $\sigma = \rho h$, express the normal component of electric displacement $D_2(x_2 = 0)$ in terms of $D_2(x_2 = -h)$, and show that the surface permittivity of the medium at $x_2 < 0$ has the complex form $\varepsilon_{\mathrm{d}} = \varepsilon_0 - \mathrm{i}\varepsilon$, with ε positive.

(d) Write down the electrical boundary conditions at the surface $x_2 = 0$, taking account of the surface permittivity of the piezoelectric solid, as given by (5.80b). Show that the wavenumber k is complex, so that $k = \beta - \mathrm{i}\alpha$, and derive expressions for α and β. State whether the Rayleigh wave is amplified or attenuated as it propagates.

(e) *Example.* Consider an aluminium film of length $L = 1\,\mathrm{cm}$ and conductivity $\gamma = 200\,\Omega^{-1}\mathrm{m}^{-1}$ on a Y-cut crystal of lithium niobate with propagation along Z, which gives $V_0 = 3\,500\,\mathrm{m/s}$, $\Delta V/V = 2.5 \times 10^{-2}$ and $\varepsilon_{\mathrm{p}} = 4.1 \times 10^{-10}\,\mathrm{F/m}$. Calculate h_{m}, the thickness for which the attenuation is maximized, and the corresponding attenuation coefficient α_{m}, for a frequency $f = 20\,\mathrm{MHz}$. Derive the maximal attenuation A_{m} corresponding to a length L.

Solution.

(a) We have $-\mathrm{i}kj_1 + \mathrm{i}\omega\rho = 0$, so $j_1 = \rho V$.

(b) Here $j_1 = \gamma E_1 = \mathrm{i}k\gamma\Phi$, giving $\Phi = \rho V/\mathrm{i}k\gamma$.

(c) Since $D_2(0) - D_2(-h) = \sigma = \rho h$, we have

$$\varepsilon_{\mathrm{d}} = -\frac{D_2(0)}{k\Phi(0)} = -\frac{D_2(-h)}{k\Phi(-h)} - \frac{\rho h}{k\Phi(0)} = \varepsilon_0 - \mathrm{i}\varepsilon\,, \quad \text{where} \quad \varepsilon = \frac{\gamma h}{V}\,.$$

(d) The electrical boundary conditions give

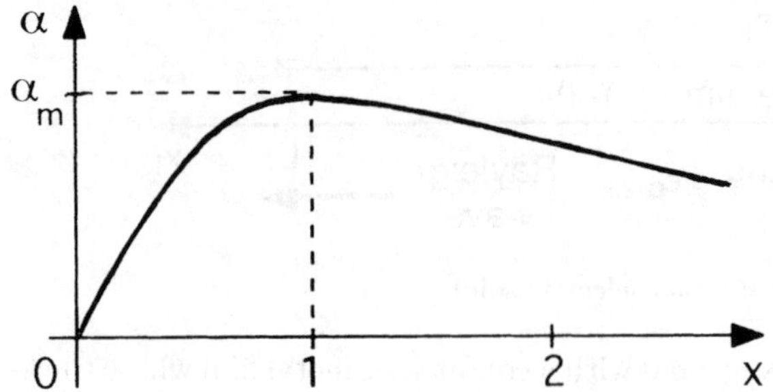

Fig. 5.49. Attenuation coefficient as a function of the normalized layer thickness

$$\varepsilon_{\mathrm{p}}\frac{\omega - kV_0}{\omega - kV_\infty} + \varepsilon_0 - \mathrm{i}\varepsilon = 0 \quad \text{and hence, with} \quad v = V_\infty/V_0\,,$$

$$k = \frac{\omega}{V_0}\frac{(\varepsilon_{\mathrm{p}}+\varepsilon_0)(\varepsilon_{\mathrm{p}}+v\varepsilon_0) + v\varepsilon^2 + \mathrm{i}\,\varepsilon\varepsilon_{\mathrm{p}}(v-1)}{(\varepsilon_{\mathrm{p}}+v\varepsilon_0)^2 + v^2\varepsilon^2} = \beta - \mathrm{i}\alpha\,.$$

This gives

$$\beta = \frac{\omega}{V_0}\frac{(\varepsilon_{\mathrm{p}}+\varepsilon_0)(\varepsilon_{\mathrm{p}}+v\varepsilon_0) + v\varepsilon^2}{(\varepsilon_{\mathrm{p}}+v\varepsilon_0)^2 + v^2\varepsilon^2} \cong \frac{\omega}{V_0}\,,$$

and

$$\alpha = \frac{\omega}{V_0}\frac{(1-v)\varepsilon\varepsilon_{\mathrm{p}}}{(\varepsilon_{\mathrm{p}}+v\varepsilon_0)^2 + v^2\varepsilon^2} \cong \frac{\omega}{V_0}\frac{\Delta V}{V}\frac{\varepsilon\varepsilon_{\mathrm{p}}}{(\varepsilon_{\mathrm{p}})^2 + \varepsilon^2}\,.$$

Thus α is maximized when $\varepsilon = \varepsilon_{\mathrm{p}} = \gamma h/V$, giving

$$h_{\mathrm{m}} = \varepsilon_{\mathrm{p}}V/\gamma \quad \text{and} \quad \alpha_{\mathrm{m}} = \omega\Delta V/2VV_0\,.$$

Defining $x = h/h_{\mathrm{m}} = \varepsilon/\varepsilon_{\mathrm{p}}$, we have $\alpha = 2\alpha_{\mathrm{m}}x/(1+x^2)$. This function is shown in Fig. 5.49.

(e) *Example.* The quoted values give $h_{\mathrm{m}} = 7.2\,\mathrm{nm}$, $\alpha_{\mathrm{m}} = 448.8\,\mathrm{m}^{-1}$ and $A_{\mathrm{m}} = \exp(\alpha_{\mathrm{m}}L) = 89$, so that $20\log A_{\mathrm{m}} = 39\,\mathrm{dB}$.

5.6. A metallic film in the form of a strip of width d is deposited on a piezoelectric solid taken to be isotropic in the x_1x_3 plane, as in Fig. 5.50.

(a) Compare the Rayleigh-wave velocity V_{R}' under the film with the velocity V_{R} outside the film, and express the relative difference $(V_{\mathrm{R}} - V_{\mathrm{R}}')/V_{\mathrm{R}}$ in terms of K_{R}.

(b) What condition does the angle φ need to satisfy in order that an incident Rayleigh wave (with displacement u_{I}) is totally reflected at the film edges (giving a reflected wave with displacement u_{R})? Express the critical angle φ_{c} in terms of K_{R}.

(c) In the case of total reflection, Sect. 1.2.4.3, the reflected wave suffers a phase change by an amount $\delta > 0$, such that

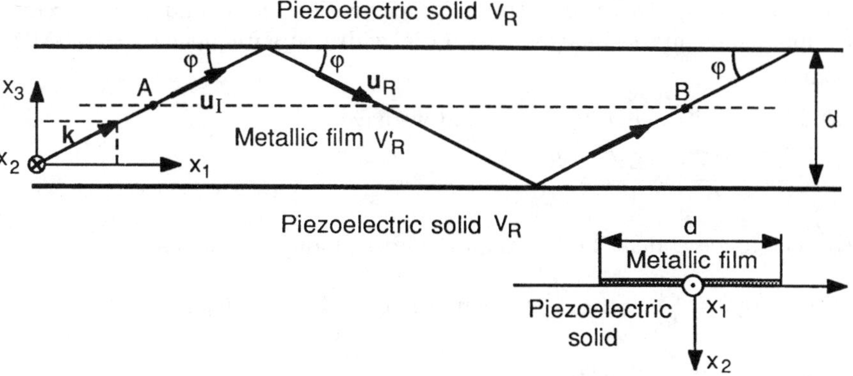

Fig. 5.50. Metallic film on a piezoelectric solid

$$\frac{u_R}{u_I} = \frac{\sin\varphi + \mathrm{i}\sqrt{\cos^2\varphi - \cos^2\varphi_c}}{\sin\varphi - \mathrm{i}\sqrt{\cos^2\varphi - \cos^2\varphi_c}} = \mathrm{e}^{\mathrm{i}\delta}\,.$$

Express $\cos\delta/2$ in terms of $\sin\varphi$ and $\sin\varphi_c$.

(d) For a given wavenumber k_1, an infinite number of modes with indices $m = 0, 1, 2, \ldots$ can propagate along the x_1 axis. Taking account of the transverse component k_3 of the wave vector $\boldsymbol{k}$, and the phase change δ, deduce the phase difference $\Phi_B - \Phi_A$ for two points A and B at the same x_3 value. This phase difference must be equal to $-k_1 l - 2\pi m$, where l is the distance between the two points and m is the mode index. Introducing the dimensionless parameter $p = (k_1 d/\pi)\varphi_c$, derive an equation relating $y = \varphi/\varphi_c$ to p and m, for the case $\varphi_c \ll 1$ rad. Sketch the curves for $y = f_m(p)$ with $m = 0, 1, 2, 3$, using the particular values $y = 1$, $\sqrt{3}/2$, $1/\sqrt{2}$, and $1/2$, and solving for p.

(e) *Application.* To compress a beam of Rayleigh waves, a film in the form of a horn can be used, as in Fig. 5.51.
What condition should the angle α satisfy for the Rayleigh-wave beam to be guided? Assuming the solid to be a $Y - Z$ lithium niobate crystal, with $K_R^2 = 0.048$ and $V_R = 3\,500$ m/s as in Fig. 5.27, calculate φ_c. For illustration, assume $\varphi = \varphi_c/2$. Deduce α and p, and the ratio d/λ_R for the fundamental

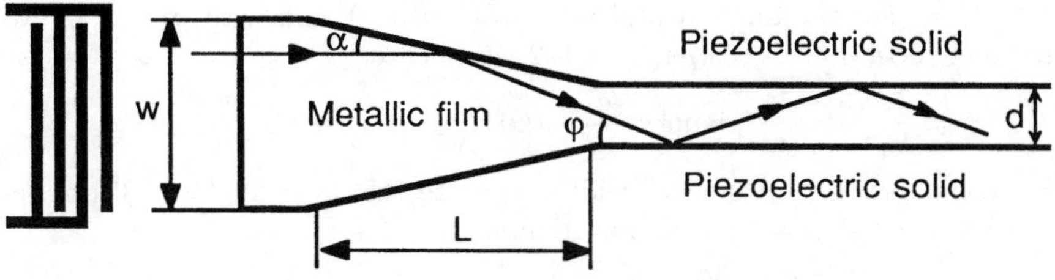

Fig. 5.51. A horn can focus a Rayleigh-wave beam from width w to width d

mode ($m = 0$). Determine the guide dimensions d, L and w required to compress a beam to a quarter of its initial width, at a frequency $f = 100\,\text{MHz}$.

Solution.

(a) The velocities give $V'_{\text{R}} < V_{\text{R}}$, and we have

$$\frac{V_{\text{R}} - V'_{\text{R}}}{V_{\text{R}}} = \frac{\Delta V}{V} = \frac{K_{\text{R}}^2}{2} .$$

(b) The angle φ must be less than the critical angle φ_{c}, given by

$$\cos\varphi_{\text{c}} = \frac{V'_{\text{R}}}{V_{\text{R}}} = 1 - \frac{K_{\text{R}}^2}{2} \quad \text{which gives} \quad \sin\varphi_{\text{c}} = K_{\text{R}} .$$

(c) Writing $\dfrac{u_{\text{R}}}{u_{\text{I}}} = \dfrac{a+ib}{a-ib} = \mathrm{e}^{i\delta}$, yields

$$\cos\frac{\delta}{2} = \frac{a}{\sqrt{a^2+b^2}} = \frac{\sin\varphi}{\sqrt{1-\cos^2\varphi_{\text{c}}}} = \frac{\sin\varphi}{\sin\varphi_{\text{c}}} .$$

(d) The phase change of B relative to A is due to an acoustic path, giving a phase change of the form $-k_i x_i$, and two total reflections, each with phase change δ. Thus, $\Phi_{\text{B}} - \Phi_{\text{A}} = -k_1 l - 2k_3 d + 2\delta$. With $\Phi_{\text{B}} - \Phi_{\text{A}} = -k_1 l - 2\pi m$, we find

$$\frac{\delta}{2} = \frac{k_3 d}{2} - \frac{\pi}{2} m = \frac{\pi}{2}\left(\frac{k_3 d}{\pi} - m\right) ,$$

and hence

$$\frac{\sin\varphi}{\sin\varphi_{\text{c}}} = \cos\frac{\pi}{2}\left(\frac{k_3 d}{\pi} - m\right) \quad \text{with} \quad k_3 = k_1 \tan\varphi .$$

When $\varphi_{\text{c}} \ll 1$ rad, we have

$$\frac{\varphi}{\varphi_{\text{c}}} \cong \cos\frac{\pi}{2}\left(p\frac{\varphi}{\varphi_{\text{c}}} - m\right) , \quad \text{where} \quad p = \frac{k_1 d}{\pi}\varphi_{\text{c}} = \frac{2d}{\lambda_{\text{R}}}\varphi_{\text{c}} . \tag{5.149}$$

Hence,

$$y \cong \cos\frac{\pi}{2}(py - m)$$

giving the curves shown in Fig. 5.52.

(e) *Application.* Here $\varphi = 2\alpha < \varphi_{\text{c}}$, so we need $\alpha < \varphi_{\text{c}}/2$. For lithium niobate we have $\sin\varphi_{\text{c}} = K_{\text{R}} = 0.219$, giving $\varphi_{\text{c}} = 12.6°$. Thus $\varphi = \varphi_{\text{c}}/2 = 6.3°$ and $\alpha = 3.15°$. For the fundamental mode with $m = 0$, and with $\varphi = \varphi_{\text{c}}/2$, we see that (5.149) gives $\cos(p\pi/4) = 1/2$, and hence

$$p = \frac{2d}{\lambda_{\text{R}}}\varphi_{\text{c}} = \frac{4}{3} \quad \text{giving} \quad \frac{d}{\lambda_{\text{R}}} = 3.04 .$$

For compression to a quarter width, we note that $\lambda = V_{\text{R}}/f = 35\,\mu\text{m}$, so $d = 106\,\mu\text{m}$ and $w = 4d = 424\mu\text{m}$. Hence

$$\tan\alpha = \frac{w-d}{2L} = \frac{3d}{2L} \cong \frac{\varphi_{\text{c}}}{4} , \quad \text{giving} \quad L \cong 6d/K_{\text{R}} = 2.9\,\text{mm} .$$

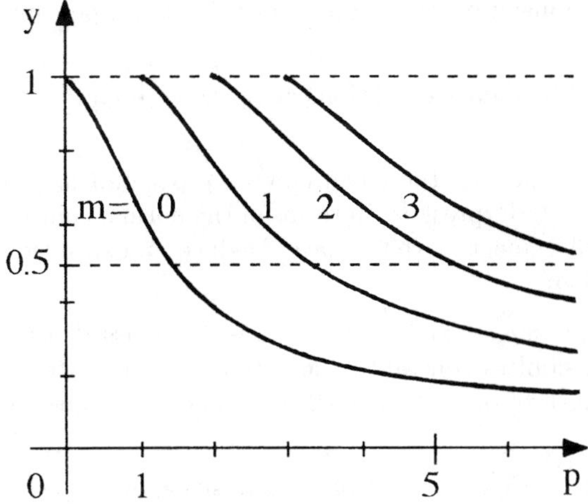

Fig. 5.52. Normalized angle of incidence as a function of normalized film width

5.7. A harmonic Rayleigh wave, with width w in the x_3 direction, propagates along x_1 with velocity V_R in an isotropic solid (Fig. 5.53). For $x_1 > 0$, the solid is covered by a non-viscous liquid with mass density ρ and acoustic impedance $Z = \rho c$, where c is the velocity of longitudinal waves in the liquid, and $V_R > c$.

(a) Calculate the angle θ at which the Rayleigh wave radiates a longitudinal wave into the liquid.

(b) For $x_1 > 0$, each variable associated with the Rayleigh wave decreases exponentially with x_1, with the form

$$a(x_1, x_2, t) = a_0(x_2) \exp(-\alpha x_1) \exp[i(\omega t - \beta x_1)] .$$

Express the mean power $P(x_1)$ transported by the Rayleigh wave in terms of its value P_0 at $x_1 = 0$. Express the attenuation coefficient α in terms of

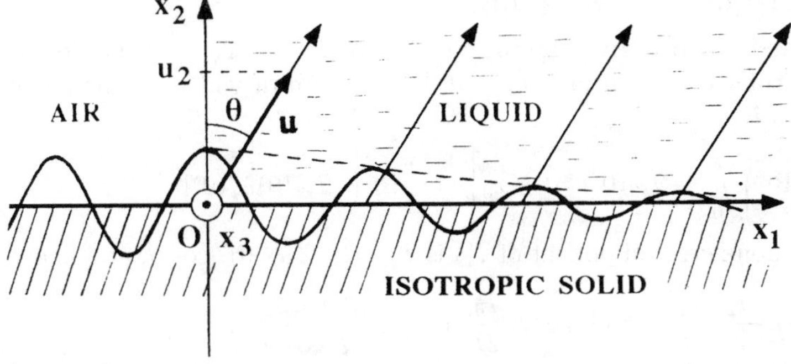

Fig. 5.53. Beyond the point $x_1 = 0$, the solid is covered by a non-viscous liquid

P and P_L, the latter being the mean power radiated into the liquid per unit length, so that $P_L = -dP/dx_1$.

(c) Write down the mechanical boundary conditions at the solid–liquid interface $x_2 = 0$.

(d) Using the Poynting vector, calculate the mean power P_L per unit length radiated across the surface $x_2 = 0$. Express P_L in terms of the normal component $u_2(0)$ of mechanical displacement at the surface. Deduce an expression for the attenuation coefficient α.

(e) Given that $|u_2(0)| = A(2P/w\rho_s\omega V_R^2)^{1/2}$, where ρ_S is the mass density of the solid and A is a dimensionless constant, calculate α and the attenuation $\gamma = \alpha\lambda_R$ for a path length of one wavelength. Examine the frequency dependence of γ.

(f) *Example.* Consider water, with $c = 1\,500\,\mathrm{m/s}$, and silica, with $\rho_S = 2\,200\,\mathrm{kg/m^3}$, $V_R = 3\,400\,\mathrm{m/s}$ and $A = 0.59$. Evaluate θ and γ. Calculate the attenuation of Rayleigh waves in dB, for a path length of $10\lambda_R$ under the liquid.

Solution.

(a) Longitudinal waves are emitted coherently into the liquid if

$$\sin\theta = \lambda_L/\lambda_R = c/V_R\,, \quad c < V_R\,.$$

(b) The mean power has the form $P = \mathrm{Re}[ab^*]/2$, and therefore varies as $P(x_1) = P_0\exp(-2\alpha x_1)$. Per unit length, we have

$$P_L = -dP/dx_1 = 2\alpha P_0\exp(-2\alpha x_1)\,,$$

so $\alpha = P_L/2P$.

(c) Since the fluid is non-viscous, the boundary conditions are

(i) the normal component of displacement is continuous, so

$$u_2(0) = u(0)\cos\theta\,;$$

(ii) the tangential stresses T_{12} and T_{32} are zero, and the normal stress is continuous, so that

$$T_{22}(0) = -\delta p = -Z\dot{u}(0)\,.$$

(d) The radiated power is given by the component $P_2 = -T_{i2}\dot{u}_i$ of the Poynting vector, normal to the surface $x_2 = 0$. For a beam of width w, the mean power radiated is

$$P_L = \frac{1}{2}\mathrm{Re}\left[\int_0^w -T_{i2}(0)\dot{u}_i^*(0)dx_3\right] = \frac{w}{2}\mathrm{Re}\left[-T_{i2}(0)\dot{u}_i^*(0)\right]\,.$$

Here the only non-zero component of T_{i2} is $T_{22} = -Z\dot{u}_2(0)/\cos\theta$, and hence

$$P_L = \frac{w}{2}Z\frac{|\dot{u}_2(0)|^2}{\cos\theta} \quad \text{and} \quad \alpha = \frac{P_L}{2P} = \frac{w}{4}Z\frac{\omega^2|u_2(0)|^2}{P\cos\theta}\,.$$

(e) Using the expression for $|u_2(0)|$, we find

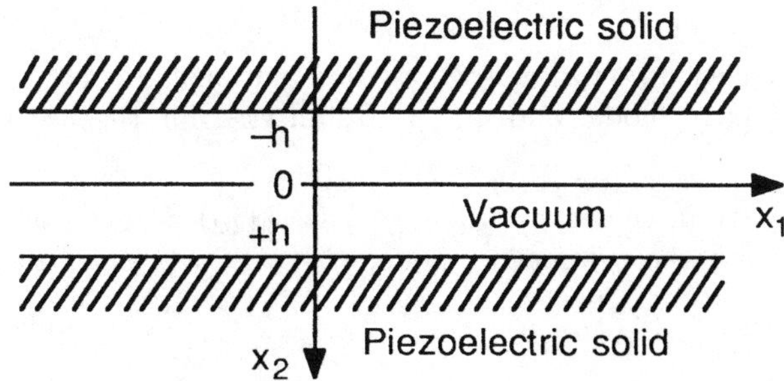

Fig. 5.54. Waveguide consisting of two semi-infinite identical solids separated by a vacuum of width $2h$

$$\alpha = \frac{1}{2}\frac{Z\omega A^2}{\rho_S V_R^2 \cos\theta} \quad \text{giving} \quad \gamma = \alpha\lambda_R = \frac{\pi\rho c}{\rho_S V_R \cos\theta} A^2 .$$

Thus γ is independent of frequency.

(f) *Example.* The values quoted give $\theta = 0.457\,\text{rad} = 26.2°$ and $\gamma = 0.245$. A path length of $10\lambda_R$ gives $a_1/a_0 = 0.0867$, that is, $-21\,\text{dB}$.

5.8. The waveguide of Fig. 5.54 consists of two semi-infinite solids, of the same piezoelectric material, separated by a gap of width $2h$ containing a vacuum. For the normal modes of this guide, the material displacements and the electric potential can be assumed to be symmetric, so that $\Phi_S(-x_2) = \Phi_S(x_2)$, or antisymmetric, so that $\Phi_A(-x_2) = -\Phi_A(x_2)$.

(a) Write down Laplace's equation for the vacuum, and give expressions for $\Phi_S(x_2)$ and $\Phi_A(x_2)$ for $|x_2| < h$. Sketch these as functions of kx_2.

(b) Derive the surface permittivity $\varepsilon_d(x_2) = -D_2/k\Phi$ in the vacuum, for the symmetric mode and for the antisymmetric mode. Evaluate it at $x_2 = 0$, and give a description of the surface $x_2 = 0$ for these two cases.

(c) Using (5.80b) for the surface permittivity of a piezoelectric material, derive the Rayleigh-wave velocity V_R when the half-space $x_2 < 0$ is a vacuum.

(d) Give expressions for the velocities V_S and V_A of the *symmetric* and *antisymmetric* modes, as functions of kh. Comment on their limiting values at $kh = 0$ and $kh = \infty$.

(e) For lithium niobate, $\varepsilon_p/\varepsilon_0 = 40 \gg 1$. Calculate V_S and V_A when $kh = \varepsilon_0/\varepsilon_p \ll 1$.

(f) A Rayleigh wave, denoted R_+, is launched on the surface $x_2 = +h$. Show that after a distance L_t the wave has been transferred to the surface $x_2 = -h$ (and is then called R_-). Calculate the coupling distance L_t when $kh = \varepsilon_0/\varepsilon_p$.

(g) *Example.* If $V_R = 4\,000\,\text{m/s}$, $\Delta V/V = 2.5\,\%$ and $f = 40\,\text{MHz}$, calculate the value of $2h$.

Solution.

(a) Laplace's equation gives $\mathrm{d}^2\Phi/\mathrm{d}x_2^2 - k^2\Phi = 0$. Solutions are $\Phi_{\mathrm{S}} = A\cosh kx_2$ for the *symmetric* mode, and $\Phi_{\mathrm{A}} = B\sinh kx_2$. for the *antisymmetric* mode.

(b) The surface permittivity in the vacuum $|x_2| < h$ is $\varepsilon_{\mathrm{d}}(x_2) = -D_2/k\Phi$, with $D_2 = -\varepsilon_0\mathrm{d}\Phi/\mathrm{d}x_2$. For the symmetric mode, $D_2 = -\varepsilon_0 Ak\sinh kx_2$, giving

$$\varepsilon_{\mathrm{d}}^{(\mathrm{S})} = \varepsilon_0 \tanh kx_2 \,.$$

Hence $\varepsilon_{\mathrm{d}}^{(\mathrm{S})}(0) = 0$, so the surface can be described as 'open circuit'. The antisymmetric mode has $D_2 = -\varepsilon_0 Bk\cosh kx_2$, so

$$\varepsilon_{\mathrm{d}}^{(\mathrm{A})} = \varepsilon_0/\tanh kx_2 \,,$$

and hence $\varepsilon_{\mathrm{d}}^{(\mathrm{A})}(0) = \infty$, so the surface is 'short-circuit'.

(c) When the half-space $x_2 < 0$ is a vacuum we have $\varepsilon_{\mathrm{d}} = \varepsilon_0$, and V_{R} is given by

$$\varepsilon_{\mathrm{p}}\frac{V - V_0}{V - V_\infty} + \varepsilon_0 = 0 \quad \text{and hence} \quad V_{\mathrm{R}} = \frac{V_0 + rV_\infty}{1 + r} \quad \text{where} \quad r = \frac{\varepsilon_0}{\varepsilon_{\mathrm{p}}} \,.$$

(d) For the symmetric mode $\overline{\varepsilon}(V) + \varepsilon_{\mathrm{d}}^{(\mathrm{S})}(h) = 0$, giving

$$V_{\mathrm{S}} = \frac{V_0 + V_\infty r \tanh kh}{1 + r \tanh kh} \,.$$

For $kh \to 0$ this gives $V_{\mathrm{S}} = V_0$, so the surface $x_2 = h = 0$ is 'open circuit'. For $kh \to \infty$ we have $V_{\mathrm{S}} = V_{\mathrm{R}}$, equivalent to a surface in contact with a vacuum.

For the antisymmetric mode,

$$V_{\mathrm{A}} = \frac{V_0 \tanh kh + rV_\infty}{\tanh kh + r} \,.$$

For $kh \to 0$, this gives $V_{\mathrm{A}} = V_\infty$, so the surface is 'short circuit'.
For $kh \to \infty$, we have $V_{\mathrm{A}} = V_{\mathrm{R}}$, corresponding to a surface in contact with a vacuum.

(e) For $kh = r = \varepsilon_0/\varepsilon_{\mathrm{p}} \ll 1$ we find

$$V_{\mathrm{S}} = \frac{V_0 + r^2 V_\infty}{1 + r^2} \cong V_0 \quad \text{and} \quad V_{\mathrm{A}} = \frac{V_0 + V_\infty}{2} \,.$$

(f) The Rayleigh wave R_+ on the surface $x_2 = +h$ is the sum of the symmetric and antisymmetric modes, and the R_- wave on the surface $x_2 = -h$ is the difference. The two modes of the structure propagate with slightly different velocities. The Rayleigh wave is transferred from one surface to the other in a distance L_{t} corresponding to a relative phase change of π. Hence L_{t} is given by

$$\left(\frac{\omega}{V_{\mathrm{A}}}-\frac{\omega}{V_{\mathrm{S}}}\right)L_{\mathrm{t}}=\pi=2\pi\frac{V_{\mathrm{S}}-V_{\mathrm{A}}}{V_{\mathrm{S}}V_{\mathrm{A}}}fL_{\mathrm{t}}\quad\text{so that}\quad L_{\mathrm{t}}=\frac{V_{\mathrm{S}}V_{\mathrm{A}}}{2(V_{\mathrm{S}}-V_{\mathrm{A}})f}\,.$$

For $\mathrm{LiNbO_3}$ and $kh=\varepsilon_0/\varepsilon_{\mathrm{p}}$ we find $V_{\mathrm{S}}\approx V_0$ and $V_{\mathrm{A}}=(V_0+V_\infty)/2$, so

$$V_{\mathrm{S}}-V_{\mathrm{A}}=\frac{V_0-V_\infty}{2}=\frac{\Delta V}{2}\,,$$

and hence

$$L_{\mathrm{t}}=\frac{V_{\mathrm{S}}V_{\mathrm{A}}}{f\Delta V}\cong\frac{V_{\mathrm{R}}}{f}\cdot\frac{V}{\Delta V}=\frac{\lambda_{\mathrm{R}}}{\Delta V/V}\,.$$

(g) *Example.* For $V_{\mathrm{R}}=4\,000\,\mathrm{m/s}$, $\Delta V/V=2.5\times10^{-2}$ and $f=40\,\mathrm{MHz}$, we find $\lambda_{\mathrm{R}}=100\,\mu\mathrm{m}$ and $L_{\mathrm{t}}=4\,\mathrm{mm}$. With $\varepsilon_{\mathrm{p}}/\varepsilon_0=40$, we have $2h=(\lambda_{\mathrm{R}}/\pi)(\varepsilon_0/\varepsilon_{\mathrm{p}})=0.8\,\mu\mathrm{m}$.

5.9. Consider a $\overline{\mathrm{TH}}$ wave polarized in the x_3 direction, with $u\equiv u_3$, propagating along x_1 in a piezoelectric plate, as in Fig. 5.55. The plate has symmetry $6mm$ with the A_6 axis parallel to x_3, and one face, at $x_2=0$, is firmly bonded to a metallic substrate with infinite stiffness. The other face, at $x_2=h$, is covered with a metallic film connected electrically to the substrate, so that $\Phi=0$ for both surfaces.

(a) For the harmonic regime, write down the propagation equations involving u and Φ.

(b) Eliminating the potential Φ, show that the equation for the displacement u can be written in the form $\mathrm{d}^2u/\mathrm{d}x_2^2+q^2u=0$. Express q^2 in terms of ω^2, k^2 and $\overline{V}_{\mathrm{T}}^2=(c_{44}+e_{15}^2/\varepsilon_{11})/\rho$. It is assumed that $q^2>0$. Give the solution satisfying the boundary conditions at $x_2=0$.

(c) Show that the right side of the equation obeyed by Φ is proportional to u. Find the general solution for Φ.

(d) Specify the electrical and mechanical boundary conditions at the surface $x_2=h$. Introduce the electromechanical coupling coefficient $K_{\mathrm{T}}^2=$

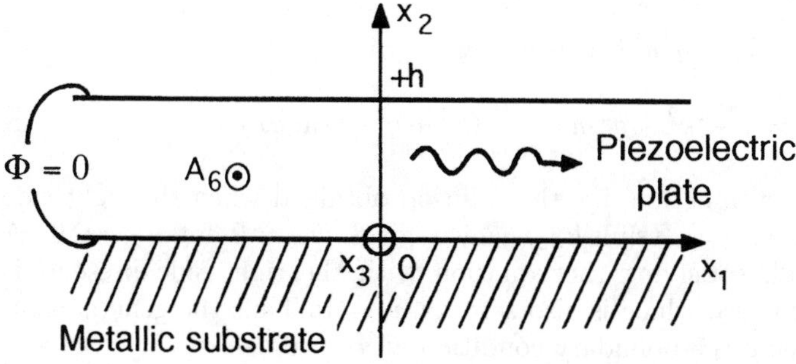

Fig. 5.55. A piezoelectric plate with thickness h on an infinitely rigid metal substrate

$e_{15}^2/(\varepsilon_{11}c_{44}+e_{15}^2)$. Establish two equations which implicitly yield the dispersion relation $\omega(k)$. Use the normalized variables $x = kh$, $y = qh$ and $\Omega = \omega h/\overline{V}_\mathrm{T}$.

(e) Assuming that $K_\mathrm{T}^2 = 0.5$, calculate the cut-off frequency Ω_c.

Solution.

(a) From (5.89), (5.90) and (5.102) we find

$$c_{44}\left(1-\frac{\partial^2}{\partial X_2^2}\right)u + e_{15}\left(1-\frac{\partial^2}{\partial X_2^2}\right)\Phi = \rho V^2 u\,, \quad X_2 = kx_2 \tag{5.150}$$

$$e_{15}\left(1-\frac{\partial^2}{\partial X_2^2}\right)u - \varepsilon_{11}\left(1-\frac{\partial^2}{\partial X_2^2}\right)\Phi = 0\,. \tag{5.151}$$

(b) Substituting Φ from (5.151) into (5.150) gives

$$\left(c_{44}+\frac{e_{15}^2}{\varepsilon_{11}}\right)\left(1-\frac{\partial^2}{\partial X_2^2}\right)u = \rho V^2 u$$

and hence

$$\overline{V}_\mathrm{T}^2\left(1-\frac{\partial^2}{k^2\partial x_2^2}\right)u = V^2 u\,.$$

This gives

$$\frac{\mathrm{d}^2 u}{\mathrm{d}x_2^2} + q^2 u = 0\,,$$

where

$$q^2 = \frac{\omega^2}{\overline{V}_\mathrm{T}^2} - k^2\,.$$

Noting that $u(x_2=0)=0$, the solution is $u(x_2) = A\sin qx_2$.

(c) Equation (5.151) can be written as

$$\left(k^2-\frac{\mathrm{d}^2}{\mathrm{d}x_2^2}\right)\Phi = \frac{e_{15}}{\varepsilon_{11}}\left(k^2-\frac{\mathrm{d}^2}{\mathrm{d}x_2^2}\right)u\,.$$

Replacing $\mathrm{d}^2u/\mathrm{d}x_2^2$ by $-q^2u$, this becomes

$$k^2\Phi - \frac{\mathrm{d}^2\Phi}{\mathrm{d}x_2^2} = \frac{e_{15}}{\varepsilon_{11}}(k^2+q^2)u = \frac{e_{15}}{\varepsilon_{11}}(k^2+q^2)A\sin qx_2\,.$$

The complementary function, i.e. the solution obtained when the right side is set to zero, is $\Phi = B\sinh kx_2$, which gives $\Phi = 0$ for $x_2 = 0$. A particular integral, satisfying the equation with the right side present, is $\Phi = (e_{15}/\varepsilon_{11})A\sin qx_2$, which is also zero for $x_2 = 0$. Thus the general solution for Φ, satisfying the boundary condition at $x_2 = 0$, is

$$\Phi = \frac{e_{15}}{\varepsilon_{11}}A\sin qx_2 + B\sinh kx_2\,.$$

(d) From Sect. 5.4.2 we have

$$T_{32} = \frac{\partial}{\partial x_2}(c_{44}u + e_{15}\Phi) = q\left(c_{44} + \frac{e_{15}^2}{\varepsilon_{11}}\right) A\cos qx_2 + e_{15}kB\cosh kx_2$$

and, introducing the coupling coefficient, this becomes

$$T_{32} = \left(c_{44} + \frac{e_{15}^2}{\varepsilon_{11}}\right)\left(qA\cos qx_2 + \frac{\varepsilon_{11}}{e_{15}}K_{\mathrm{T}}^2\, kB\cosh kx_2\right).$$

The boundary conditions are

$$T_{32}(x_2 = h) = 0 \quad \text{giving} \quad qA\cos qh = -\frac{\varepsilon_{11}}{e_{15}}K_{\mathrm{T}}^2\, kB\cosh kh$$

and

$$\Phi(x_2 = h) = 0 \quad \text{giving} \quad A\sin qh = -\frac{\varepsilon_{11}}{e_{15}}B\sinh kh\,.$$

In terms of normalized variables,

$$\frac{\tan y}{y} = \frac{1}{K_{\mathrm{T}}^2}\frac{\tanh x}{x} \quad \text{and} \quad y^2 + x^2 = \Omega^2\,. \tag{5.152}$$

(e) For $K_{\mathrm{T}}^2 = 0.5$ we have $(\tan y)/y = 2(\tanh x)/x$. To find the cut-off frequency, we note that $x = kh \approx 0$, so that $\tan y \approx 2y$, giving $y \approx 1.17$. Thus, from (5.152), $\Omega_{\mathrm{c}} = 1.17$, and so $\omega_{\mathrm{c}} = 1.17\, V_{\mathrm{T}}/h$.

5.10. Considering the Lamb-wave dispersion relation (5.119), show that, when $p = \mathrm{i}k\chi$ and $kh \to \infty$, the symmetric mode solutions ($\alpha = 0$) have $q_n h = n\pi$. Write down the corresponding expressions for the two displacement components u_1 and u_2, and sketch these for the S_1 mode.

Solution. From (5.119), with $\alpha = 0$, we have

$$(k^2 - q^2)^2 \tan qh = -4k^2 pq \tan ph\,,$$

and with $p = \mathrm{i}k\chi$ this gives

$$\left(1 - \frac{q^2}{k^2}\right)^2 \frac{\tan qh}{qh} = 4\chi\frac{\tanh \chi kh}{kh}\,.$$

For $kh \to \infty$ we have $\tanh(\chi kh) \to 1$, and hence

$$\left(1 - \frac{q^2}{k^2}\right)^2 \frac{\tan qh}{qh} \to 0\,, \quad \text{giving} \quad q_n h = np\,,\ n = 0, 1, 2, \ldots\,.$$

Thus, using (5.121), the S_n mode has displacements

$$\begin{cases} u_1 \cong \dfrac{n\pi}{h}A\left[\cos\left(n\pi\dfrac{x_2}{h}\right) - 2(-1)^n\dfrac{\cosh \chi kx_2}{\cosh \chi kh}\right] \\[2ex] u_2 \cong \mathrm{i}kA\left[\sin\left(n\pi\dfrac{x_2}{h}\right) - 2(-1)^n\dfrac{n\pi\chi}{kh}\dfrac{\sinh \chi kx_2}{\cosh \chi kh}\right]. \end{cases}$$

For the S_1 mode ($n = 1$), the displacements are sketched in Fig. 5.56.

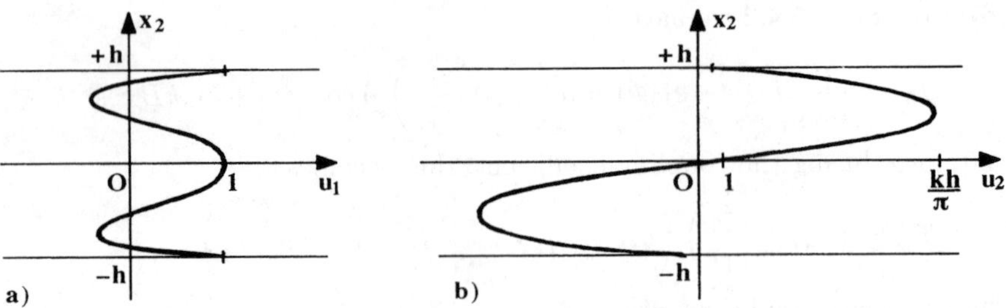

Fig. 5.56. Symmetric Lamb wave mode, S_1. **(a)** Longitudinal, u_1, and **(b)** transverse, u_2, components of mechanical displacement, as functions of the transverse coordinate, for $kh \gg 1$

The predominance of the component u_2 is not surprising in view of the position of the line ω/k in relation to the slowness circles when p is imaginary (Fig. 5.37b). The dispersion is weak, since the phase velocity is given by

$$V = V_{\mathrm{T}}(1 + n^2\pi^2/k^2h^2)^{1/2} .$$

5.11. Considering the Lamb-wave dispersion relation (5.119), show that, when $p = \mathrm{i}k\chi$ and $kh \to \infty$, the antisymmetric mode solutions ($\alpha = \pi/2$) have $q_n h = (2n + 1)\pi/2$. Write down the corresponding expressions for the two displacement components u_1 and u_2, and sketch these for the A_1 mode.

Solution. From (5.119) with $\alpha = \pi/2$, we find

$$(k^2 - q^2)^2 \tan ph = -4k^2 pq \tan qh ,$$

and with $p = \mathrm{i}k\chi$ this gives

$$k\left(1 - \frac{q^2}{k^2}\right)^2 \tanh \chi hk = -4\chi q \tan qh .$$

For $k \to \infty$ we have $\tan qh \to \infty$, and hence $q_n h \to (2n + 1)\pi/2$, with $n = 0, 1, 2, \ldots$. Using (5.121), the displacements for mode A_n are

$$\begin{cases} u_1 \cong \left(n + \dfrac{1}{2}\right) \dfrac{\pi}{h} A \left\{ -\sin\left[(2n+1)\pi \dfrac{x_2}{2h}\right] + 2(-1)^n \dfrac{\sinh \chi k x_2}{\sinh \chi kh} \right\} \\ u_2 \cong \mathrm{i}kA \left\{ \cos\left[(2n+1)\pi \dfrac{x_2}{2h}\right] + (2n+1)(-1)^n \dfrac{\pi\chi}{kh} \dfrac{\cosh \chi k x_2}{\sinh \chi kh} \right\} . \end{cases}$$

For mode $A_1 (n = 0)$, the displacements are shown in Fig. 5.57.

5.12. Taking p and q to be imaginary (Fig. 5.37), so that $p = \mathrm{i}\chi_1 k$ and $q = \mathrm{i}\chi_2 k$, show that the Rayleigh–Lamb equation (5.120) transforms into the Rayleigh equation (5.43) when $kh \to \infty$.

Solution. From (5.112b) we have

$$\frac{\omega^2}{V_{\mathrm{T}}^2} = \left(1 - \chi_2^2\right) k^2$$

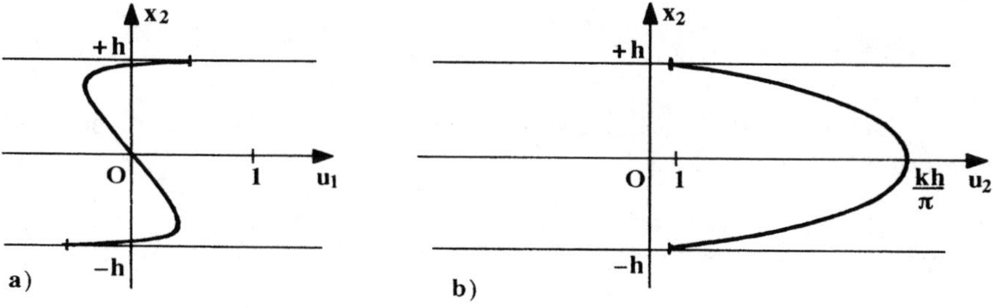

Fig. 5.57. Antisymmetric Lamb wave mode, A_1. (**a**) Longitudinal, u_1, and (**b**) transverse, u_2, components of mechanical displacement, as functions of the transverse coordinate, when $kh \gg 1$

and substitution into (5.120) gives

$$(1-\chi_2^2)^2 = -4\,\chi_2^2\left[1-\frac{\chi_1}{\chi_2}\left(\frac{\tanh\chi_1 kh}{\tanh\chi_2 kh}\right)^{\sigma}\right],$$

where $\sigma = 1$ for a symmetric mode and $\sigma = -1$ for an antisymmetric mode. Hence,

$$(1+\chi_2^2)^2 = 4\,\chi_1\chi_2\left(\frac{\tanh\chi_1 kh}{\tanh\chi_2 kh}\right)^{\sigma}$$

and for $kh \to \infty$ this becomes

$$(1+\chi_2^2)^2 = 4\,\chi_1\chi_2 \quad \text{with} \quad \chi_1 = \left(1-\frac{V^2}{V_\mathrm{L}^2}\right)^{1/2} \quad \text{and} \quad \chi_2 = \left(1-\frac{V^2}{V_\mathrm{T}^2}\right)^{1/2}.$$

This is the same as (5.43), so V approaches V_R , which is independent of k.

5.13. For an isotropic solid, show that the velocity of longitudinal waves propagating along the axis x_1 of a waveguide whose lateral dimensions are much smaller than the wavelength is $V_\mathrm{b} = 1/(\rho s_{11})^{1/2} = (E/\rho)^{1/2}$, where E is Young's modulus.

Solution. When the wavelength is much larger than the lateral dimensions of the waveguide, the stresses T_{12}, T_{13}, T_{22}, T_{23} and T_{33}, which are zero on the lateral surfaces (assumed to be free), must also be practically zero at all points in the waveguide section. Hooke's Law $S_{ij} = s_{ijkl}T_{kl}$, from 3.45, thus reduces to

$$S_{11} = \frac{\partial u_1}{\partial x_1} = s_{11}T_{11}\,.$$

The fundamental dynamic equation (3.15) becomes

$$\rho\frac{\partial^2 u_1}{\partial t^2} = \frac{\partial T_{11}}{\partial x_1} = \frac{1}{s_{11}}\frac{\partial^2 u_1}{\partial x_1^2}$$

showing that a longitudinal wave can propagate along the guide, with displacement u_1 and velocity

$$V_b = \frac{1}{\sqrt{\rho s_{11}}} = \sqrt{\frac{E}{\rho}}.$$

This velocity, called the *bar velocity*, is independent of both frequency and the form of the guide section.

Appendix A. Cylindrical and Spherical Coordinates

The use of cylindrical or spherical coordinates simplifies the solution of equations describing propagation in structures with axial or spherical symmetry. Thus, propagation in a homogeneous cylinder, or radiation from a source in an isotropic medium, call for cylindrical or spherical coordinates, respectively.

Let $\mathrm{d}s^2$ be the square (invariant) of the length of an element $\mathrm{d}\boldsymbol{x}$ with components $\mathrm{d}x_i$ referred to the orthonormal cartesian reference frame x_i with unit vector $\boldsymbol{e}_i$, so that

$$\mathrm{d}\boldsymbol{x} = \sum_{i=1}^{3} \boldsymbol{e}_i \mathrm{d}x_i \quad \text{giving} \quad \mathrm{d}s^2 = \mathrm{d}\boldsymbol{x} \cdot \mathrm{d}\boldsymbol{x} = \mathrm{d}x_1^2 + \mathrm{d}x_2^2 + \mathrm{d}x_3^2 \,.$$

The differential $\mathrm{d}\boldsymbol{v}$ of a vector $\boldsymbol{v} = \sum_{i=1}^{3} v_i \boldsymbol{e}_i$ is written as

$$\mathrm{d}\boldsymbol{v} = \sum_{i,j=1}^{3} \frac{\partial (v_i \boldsymbol{e}_i)}{\partial x_j} \mathrm{d}x_j = \sum_{i,j=1}^{3} \left(\boldsymbol{e}_i \frac{\partial v_i}{\partial x_j} + v_i \frac{\partial \boldsymbol{e}_i}{\partial x_j} \right) \mathrm{d}x_j \tag{A.1}$$

and, since the derivative in the second term is zero,

$$\mathrm{d}\boldsymbol{v} = \sum_{i,j=1}^{3} \boldsymbol{e}_i \frac{\partial v_i}{\partial x_j} \mathrm{d}x_j \quad \text{and hence} \quad \mathrm{d}v_i = \sum_{j=1}^{3} \frac{\partial v_i}{\partial x_j} \mathrm{d}x_j \,.$$

This relation is fundamental to the definition of strain in Sect. 3.1.1.

In another reference frame, with curvilinear coordinates q_i and with unit vectors $\boldsymbol{e}'_i$, any increase $\mathrm{d}q_k$ in the $\boldsymbol{e}'_k$ direction gives a change $\mathrm{d}s_k$ of the element length in this direction, so that

$$\mathrm{d}s_k = h_k \mathrm{d}q_k \quad \text{and hence} \quad \frac{\partial q_k}{\partial s_k} = \frac{1}{h_k} \,. \tag{A.2}$$

Here, h_k is often called the scaling factor. Bearing in mind the orthogonality of the axes, we can write $\mathrm{d}s^2$ as the formulae

$$\mathrm{d}s^2 = \mathrm{d}s_1^2 + \mathrm{d}s_2^2 + \mathrm{d}s_3^2 = \sum_{k=1}^{3} h_k^2 \, \mathrm{d}q_k^2 \,. \tag{A.3}$$

Expressing the curvilinear coordinates q_i as functions of the cartesian coordinates x_j, yields

$$q_j = q_j(x_i)\,, \quad \text{so that} \quad x_i = x_i(q_j)\,, \quad \text{and hence} \quad \mathrm{d}x_i = \sum_{j=1}^{3} \frac{\partial x_i}{\partial q_j} \mathrm{d}q_j\,.$$

The differential $\mathrm{d}s^2$ is written

$$\mathrm{d}s^2 = \sum_{i=1}^{3} \mathrm{d}x_i \mathrm{d}x_i = \sum_{i,j,k=1}^{3} \frac{\partial x_i}{\partial q_j} \frac{\partial x_i}{\partial q_k} \mathrm{d}q_j \mathrm{d}q_k\,. \tag{A.4}$$

Comparing (A.3) with (A.4) shows that

$$\mathrm{d}s^2 = \sum_{i,k=1}^{3} \left(\frac{\partial x_i}{\partial q_k}\right)^2 \mathrm{d}q_k^2 = \sum_{k=1}^{3} h_k^2 \mathrm{d}q_k^2\,,$$

where

$$h_k^2 = \sum_{i=1}^{3} \left(\frac{\partial x_i}{\partial q_k}\right)^2\,. \tag{A.5}$$

The *gradient* of a function $f(q_i)$ is

$$\nabla f = \sum_{k=1}^{3} \left(\frac{\partial f}{\partial s_k}\right) \boldsymbol{e}'_k\,, \tag{A.6}$$

and the components of this in the $\boldsymbol{e}'_k$ directions can be found using the scaling factors h_k of (A.2), giving

$$\frac{\partial f}{\partial s_k} = \frac{\partial f}{\partial q_k} \frac{\partial q_k}{\partial s_k} = \frac{1}{h_k} \frac{\partial f}{\partial q_k}\,. \tag{A.7}$$

The *divergence* of a vector $\boldsymbol{v}$ in the reference frame q_i is obtained by calculating the flux of the vector leaving an elementary cube with side $h_i \Delta q_i$, and then dividing by the volume. The contribution to the flux from each pair of parallel faces has the form $\partial(h_j h_k v_i)/\partial q_i$, with the indices in cyclic order. The result is

$$\mathrm{div}\boldsymbol{v} = \nabla \cdot \boldsymbol{v} = \frac{1}{h_1 h_2 h_3} \left[\frac{\partial}{\partial q_1}(h_2 h_3 v_1) + \frac{\partial}{\partial q_2}(h_3 h_1 v_2) + \frac{\partial}{\partial q_3}(h_1 h_2 v_3)\right]. \tag{A.8}$$

The *Laplacian* is $\Delta f = \nabla^2 f = \nabla \cdot (\nabla f) = \mathrm{div}(\mathrm{grad}\, f)$. From the above equations, this is given by

$$\Delta f = \nabla^2 f = \frac{1}{h_1 h_2 h_3} \times \left[\frac{\partial}{\partial q_1}\left(\frac{h_2 h_3}{h_1} \frac{\partial f}{\partial q_1}\right) + \frac{\partial}{\partial q_2}\left(\frac{h_3 h_1}{h_2} \frac{\partial f}{\partial q_2}\right) + \frac{\partial}{\partial q_3}\left(\frac{h_1 h_2}{h_3} \frac{\partial f}{\partial q_3}\right)\right]. \tag{A.9}$$

The curl (or rotation) of a vector is defined by

$$\operatorname{curl} \boldsymbol{v} = \nabla \wedge \boldsymbol{v} = \frac{1}{h_1 h_2 h_3} \begin{vmatrix} h_1 \boldsymbol{e}_1' & h_2 \boldsymbol{e}_2' & h_3 \boldsymbol{e}_3' \\ \dfrac{\partial}{\partial q_1} & \dfrac{\partial}{\partial q_2} & \dfrac{\partial}{\partial q_3} \\ h_1 v_1 & h_2 v_2 & h_3 v_3 \end{vmatrix} . \tag{A.10}$$

The Laplacian of a vector results from the formula

$$\Delta \boldsymbol{v} = \nabla^2 \boldsymbol{v} = \nabla(\nabla \cdot \boldsymbol{v}) - \nabla \wedge \nabla \wedge \boldsymbol{v} . \tag{A.11}$$

The derivation of this is rather long, so it is omitted here.

The above formulae are general. We now give the results for particular coordinate systems.

- *Cylindrical Coordinates*, Fig. A.1a. Here $q_1 = r$, $q_2 = \theta$ and $q_3 = z$.

$$\begin{cases} x_1 = r\cos\theta = q_1 \cos q_2 \,, & h_1^2 = \cos^2 q_2 + \sin^2 q_2 \,, \text{ hence } h_1 = 1 \\ x_2 = r\sin\theta = q_1 \sin q_2 \,, & h_2^2 = (-q_1 \sin q_2)^2 + (q_1 \cos q_2)^2 \,, \text{ hence } h_2 = r \\ x_3 = z = q_3 \,, & h_3^2 = 1 \,, \text{ hence } h_3 = 1 \,. \end{cases} \tag{A.12}$$

- *Spherical Coordinates*, Fig. A.1b. Here $q_1 = r$, $q_2 = \theta$ and $q_3 = \varphi$.

$$\begin{cases} x_1 = r\sin\theta\cos\varphi = q_1 \sin q_2 \cos q_3 \\ x_2 = r\sin\theta\sin\varphi = q_1 \sin q_2 \sin q_3 \\ x_3 = r\cos\theta = q_1 \cos q_2 \,. \end{cases}$$

From these we find

$$\begin{aligned} h_1^2 &= (\sin q_2 \cos q_3)^2 + (\sin q_2 \sin q_3)^2 + \cos^2 q_2 \,, \text{hence } h_1 = 1 \\ h_2^2 &= (q_1 \cos q_2 \cos q_3)^2 + (q_1 \cos q_2 \sin q_3)^2 + (-q_1 \sin q_2)^2 \,, \text{hence } h_2 = r \\ h_3^2 &= (-q_1 \sin q_2 \sin q_3)^2 + (q_1 \sin q_2 \cos q_3)^2 \,, \text{hence } h_3 = r\sin\theta \,. \end{aligned} \tag{A.13}$$

In *cylindrical* coordinates we have the formulae

$$\begin{aligned} \operatorname{grad} f &= \nabla f = \left[\frac{\partial f}{\partial r}, \; \frac{1}{r}\frac{\partial f}{\partial \theta}, \; \frac{\partial f}{\partial z}\right] \\ \operatorname{div} \boldsymbol{v} &= \nabla \cdot \boldsymbol{v} = \frac{1}{r}\frac{\partial (r v_r)}{\partial r} + \frac{1}{r}\frac{\partial v_\theta}{\partial \theta} + \frac{\partial v_z}{\partial z} \\ \Delta f &= \nabla^2 f = \frac{\partial^2 f}{\partial r^2} + \frac{1}{r}\frac{\partial f}{\partial r} + \frac{1}{r^2}\frac{\partial^2 f}{\partial \theta^2} + \frac{\partial^2 f}{\partial z^2} \\ \operatorname{curl} \boldsymbol{v} &= \nabla \wedge \boldsymbol{v} = \left[\frac{1}{r}\frac{\partial v_z}{\partial \theta} - \frac{\partial v_\theta}{\partial z}, \; \frac{\partial v_r}{\partial z} - \frac{\partial v_z}{\partial r}, \; \frac{1}{r}\left(\frac{\partial (r v_\theta)}{\partial r} - \frac{\partial v_r}{\partial \theta}\right)\right] . \end{aligned} \tag{A.14}$$

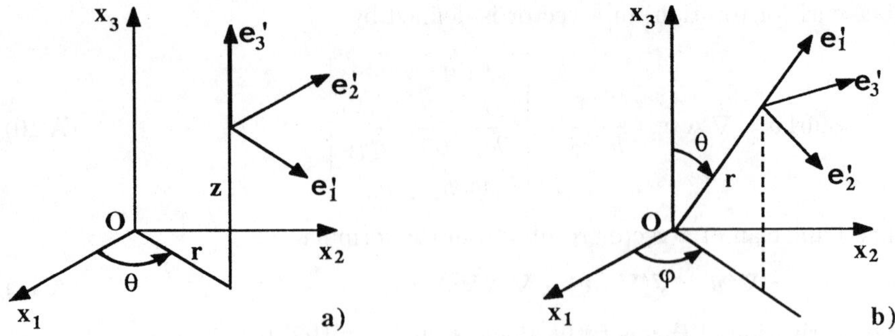

Fig. A.1. Coordinate systems. (**a**) Cylindrical, (**b**) Spherical

In *spherical* coordinates, we have

$$\operatorname{grad} f = \nabla f = \left[\frac{\partial f}{\partial r}, \quad \frac{1}{r}\frac{\partial f}{\partial \theta}, \quad \frac{1}{r\sin\theta}\frac{\partial f}{\partial \varphi}\right]$$

$$\operatorname{div} \boldsymbol{v} = \nabla \cdot \boldsymbol{v} = \frac{1}{r^2}\frac{\partial(r^2 v_r)}{\partial r} + \frac{1}{r\sin\theta}\left[\frac{\partial}{\partial\theta}(\sin\theta\, v_\theta) + \frac{\partial v_\varphi}{\partial\varphi}\right] \tag{A.15}$$

$$\Delta f = \nabla^2 f = \frac{1}{r^2}\left[\frac{\partial}{\partial r}\left(r^2\frac{\partial f}{\partial r}\right) + \frac{1}{\sin\theta}\frac{\partial}{\partial\theta}\left(\sin\theta\frac{\partial f}{\partial\theta}\right) + \frac{1}{\sin^2\theta}\frac{\partial^2 f}{\partial\varphi^2}\right].$$

These formulae, with the relation $\boldsymbol{u} = \nabla\phi + \nabla\wedge\Psi$, yield the displacement components. For example, in cylindrical coordinates we find

$$\begin{cases} u_r = \dfrac{\partial\phi}{\partial r} + \dfrac{1}{r}\dfrac{\partial\Psi_z}{\partial\theta} - \dfrac{\partial\Psi_\theta}{\partial z} \\[2ex] u_\theta = \dfrac{1}{r}\dfrac{\partial\phi}{\partial\theta} + \dfrac{\partial\Psi_r}{\partial z} - \dfrac{\partial\Psi_z}{\partial r} \\[2ex] u_z = \dfrac{\partial\phi}{\partial z} + \dfrac{1}{r}\dfrac{\partial(r\Psi_\theta)}{\partial r} - \dfrac{1}{r}\dfrac{\partial\Psi_r}{\partial\theta}. \end{cases} \tag{A.16}$$

To express the *strains* in terms of the displacements, it is necessary to revert to the basic relation (A.1), so that

$$\mathrm{d}\boldsymbol{u} = \sum_{k=1}^{3}\sum_{i=1}^{3}\left(\boldsymbol{e}_i'\frac{\partial u_i}{\partial s_k} + u_i\frac{\partial \boldsymbol{e}_i'}{\partial s_k}\right)\mathrm{d}s_k\,.$$

We consider this for the case of *cylindrical* coordinates. The second term contains only two non-zero derivatives, which are

$$\frac{\partial \boldsymbol{e}_1'}{\partial s_2} = \frac{1}{h_2}\frac{\partial \boldsymbol{e}_1'}{\partial q_2} = \frac{\boldsymbol{e}_2'}{r} \quad \text{and} \quad \frac{\partial \boldsymbol{e}_2'}{\partial s_2} = \frac{1}{h_2}\frac{\partial \boldsymbol{e}_2'}{\partial q_2} = -\frac{\boldsymbol{e}_1'}{r}\,.$$

This result, not proved here, expresses the change of direction of the vectors $\boldsymbol{e}_1'$ and $\boldsymbol{e}_2'$ with the observation point. Hence,

$$\mathrm{d}\boldsymbol{u} = \boldsymbol{e}_1' \left(\sum_{k=1}^{3} \frac{\partial u_1}{\partial s_k} \mathrm{d}s_k - \frac{u_2}{r} \mathrm{d}s_2 \right) + \boldsymbol{e}_2' \left(\sum_{k=1}^{3} \frac{\partial u_2}{\partial s_k} \mathrm{d}s_k + \frac{u_1}{r} \mathrm{d}s_2 \right)$$
$$+ \boldsymbol{e}_3' \left(\sum_{k=1}^{3} \frac{\partial u_3}{\partial s_k} \mathrm{d}s_k \right) .$$

Replacing u_1 by u_r, u_2 by u_θ and u_3 by u_z, and also $\mathrm{d}s_1$ by $\mathrm{d}r$, $\mathrm{d}s_2$ by $r\mathrm{d}\theta$ and $\mathrm{d}s_3$ by $\mathrm{d}z$, we obtain the gradient of displacements. This has the form of an array with three rows and three columns. As shown in Sect. 3.1.1.1, only the symmetric part represents the strain, and we thus find

$$S_{rr} = \frac{\partial u_r}{\partial r}, \qquad S_{\theta\theta} = \frac{u_r}{r} + \frac{1}{r}\frac{\partial u_\theta}{\partial \theta}, \qquad S_{zz} = \frac{\partial u_z}{\partial z},$$

$$S_{r\theta} = S_{\theta r} = \frac{1}{2}\left(\frac{\partial u_\theta}{\partial r} - \frac{u_\theta}{r} + \frac{1}{r}\frac{\partial u_r}{\partial \theta} \right), \qquad S_{\theta z} = S_{z\theta} = \frac{1}{2}\left(\frac{1}{r}\frac{\partial u_z}{\partial \theta} + \frac{\partial u_\theta}{\partial z} \right),$$

$$S_{rz} = S_{zr} = \frac{1}{2}\left(\frac{\partial u_r}{\partial z} + \frac{\partial u_z}{\partial r} \right). \tag{A.17}$$

The sum of the diagonal components of the gradient is the dilatation S, so

$$S = \frac{\partial u_r}{\partial r} + \frac{u_r}{r} + \frac{1}{r}\frac{\partial u_\theta}{\partial \theta} + \frac{\partial u_z}{\partial z}. \tag{A.18}$$

In the case of an isotropic solid, the expressions for the stresses in terms of the strains are similar to those for cartesian coordinates (Sect. 3.2.3.1). They are given by

$$\begin{aligned} T_{rr} &= \lambda S + 2\mu \frac{\partial u_r}{\partial r}, & T_{\theta\theta} &= \lambda S + 2\mu \left(\frac{u_r}{r} + \frac{1}{r}\frac{\partial u_\theta}{\partial \theta} \right), \\ T_{zz} &= \lambda S + 2\mu \frac{\partial u_z}{\partial z}, & T_{r\theta} &= \mu \left(\frac{\partial u_\theta}{\partial r} - \frac{u_\theta}{r} + \frac{1}{r}\frac{\partial u_r}{\partial \theta} \right), \\ T_{\theta z} &= \mu \left(\frac{1}{r}\frac{\partial u_z}{\partial \theta} + \frac{\partial u_\theta}{\partial z} \right), & T_{rz} &= \mu \left(\frac{\partial u_r}{\partial z} + \frac{\partial u_z}{\partial r} \right). \end{aligned} \tag{A.19}$$

The *Laplacian* of a vector $\boldsymbol{\Psi} = (\Psi_r, \Psi_\theta, \Psi_z)$ is, in cylindrical coordinates,

$$\nabla^2 \boldsymbol{\Psi} \begin{cases} \left(\nabla^2 \Psi\right)_r = \nabla^2 \Psi_r - \dfrac{\Psi_r}{r^2} - \dfrac{2}{r^2}\dfrac{\partial \Psi_\theta}{\partial \theta} \\ \left(\nabla^2 \Psi\right)_\theta = \nabla^2 \Psi_\theta - \dfrac{\Psi_\theta}{r^2} + \dfrac{2}{r^2}\dfrac{\partial \Psi_r}{\partial \theta} \\ \left(\nabla^2 \Psi\right)_z = \nabla^2 \Psi_z . \end{cases} \tag{A.20}$$

Appendix B. Eigenvectors and Eigenvalues of a Matrix

For many problems in physics, the application of the linear approximation leads to a system of three (or n) linear homogeneous equations in three (or n) unknowns u_1, u_2, u_3, with the form

$$C_{ij}u_j = 0\,, \quad i = 1, 2, 3\,, \tag{B.1}$$

where the coefficients C_{ij} have fixed values. For this system to have a solution other than $u_1 = u_2 = u_3 = 0$, the determinant of the coefficients C_{ij} must be zero, so that

$$|C_{ij}| = \begin{vmatrix} C_{11} & C_{12} & C_{13} \\ C_{21} & C_{22} & C_{23} \\ C_{31} & C_{32} & C_{33} \end{vmatrix} = 0\,.$$

This is the condition for the equations (B.1) to be compatible.

An important example of a linear homogeneous system is provided by the following equation for the eigenvectors u_j and eigenvalues λ of a matrix A_{ij}:

$$A_{ij}u_j = \lambda u_i\,. \tag{B.2}$$

Since $u_i = \delta_{ij}u_j$, this can be written as

$$(A_{ij} - \lambda\delta_{ij})u_j = 0 \tag{B.3}$$

and this has the same form as (B.1). The compatibility condition is

$$|A_{ij} - \lambda\delta_{ij}| = \begin{vmatrix} A_{11} - \lambda & A_{12} & A_{13} \\ A_{21} & A_{22} - \lambda & A_{23} \\ A_{31} & A_{32} & A_{33} - \lambda \end{vmatrix} = 0\,. \tag{B.4}$$

This equation, called the secular equation, determines the eigenvalues λ. It is of third degree, and can be expanded as

$$\begin{aligned}(A_{11} - \lambda)(A_{22} - \lambda)(A_{33} - \lambda) + A_{21}A_{32}A_{13} + A_{12}A_{23}A_{31} \\ - (A_{22} - \lambda)A_{31}A_{13} - (A_{11} - \lambda)A_{23}A_{32} - (A_{33} - \lambda)A_{12}A_{21} = 0\,.\end{aligned}$$

There are in general three distinct roots, $\lambda^{(1)}$, $\lambda^{(2)}$, $\lambda^{(3)}$, which are the eigenvalues, or principal values, of the matrix A_{ij}. For each eigenvalue $\lambda^{(k)}$ there is a corresponding eigenvector $u_j^{(k)}$ whose components satisfy the three equations

$$A_{ij}u_j^{(k)} = \lambda^{(k)}u_i^{(k)}, \tag{B.5}$$

which are not independent. Note that the Einstein summation convention does not apply to indices shown in brackets. In fact, a vector $\mu u_j^{(k)}$ is also an eigenvector with eigenvalue $\lambda^{(k)}$, for any value of the scalar μ. It is therefore preferable to speak of eigendirections, or principal directions, of a matrix.

Using the eigenvectors as basis vectors, and taking them to be normalized so that $u_i^{(k)} = \delta_{ik}$, the relation (B.5) becomes, with new components A'_{ij},

$$A'_{ij}\delta_{jk} = \lambda^{(k)}\delta_{ik},$$

so that

$$A'_{ik} = \lambda^{(k)}\delta_{ik}.$$

Thus the matrix has the diagonal form

$$A'_{ij} = \begin{pmatrix} \lambda^{(1)} & 0 & 0 \\ 0 & \lambda^{(2)} & 0 \\ 0 & 0 & \lambda^{(3)} \end{pmatrix}.$$

This form shows that the trace of the matrix, i.e. the invariant quantity A_{ii}, is given by

$$A_{ii} = \lambda^{(1)} + \lambda^{(2)} + \lambda^{(3)} \tag{B.6}$$

and is thus equal to the sum of the eigenvalues.

When two eigenvalues, such as $\lambda^{(2)}$ and $\lambda^{(3)}$, are equal, any linear combination of the eigenvectors $u_i^{(2)}$ and $u_i^{(3)}$, such as

$$u_i = \mu_2 u_i^{(2)} + \mu_3 u_i^{(3)}$$

is also an eigenvector. This form for u_i gives

$$A_{ij}u_j = \mu_2 A_{ij}u_j^{(2)} + \mu_3 A_{ij}u_j^{(3)}$$

and hence

$$A_{ij}u_j = \lambda^{(2)}(\mu_2 u_i^{(2)} + \mu_3 u_i^{(3)}) = \lambda^{(2)}u_i.$$

Hence u_i is an eigenvector. A double (or degenerate) eigenvalue no longer corresponds to a direction – it corresponds to the plane defined by the vectors $u_i^{(2)}$ and $u_i^{(3)}$.

Symmetric Matrices. We show here that symmetric matrices ($A_{ij} = A_{ji}$), with real coefficients, have two fundamental properties:

- their eigenvalues are real; and
- for distinct eigenvalues, the principal directions are orthogonal.

The first property follows from (B.2) on taking the scalar product with u_i^*, giving

$$u_i^* A_{ij}u_j = \lambda u_i^* u_i. \tag{B.7}$$

This becomes, on interchanging the dummy indices i and j on the left side,

$$u_j^* A_{ji} u_i = \lambda u_i^* u_i \,.$$

Since A_{ij} is symmetric ($A_{ij} = A_{ji}$) this gives

$$u_j^* A_{ij} u_i = \lambda u_i^* u_i \,.$$

Taking the conjugate, and noting that the matrix is real, gives

$$u_j A_{ij} u_i^* = \lambda^* u_i u_i^*$$

and comparison with (B.7) shows that the eigenvalues must be real, since $\lambda^* = \lambda$.

To demonstrate the second property, we write the following equations using the eigenvectors $u_i^{(1)}$ and $u_i^{(2)}$:

$$\begin{cases} A_{ij} u_j^{(1)} = \lambda^{(1)} u_i^{(1)} \\ A_{ij} u_j^{(2)} = \lambda^{(2)} u_i^{(2)} \end{cases}$$

and, for both cases, form the scalar product $u_i^{(1)} u_i^{(2)}$, giving

$$\begin{cases} A_{ij} u_j^{(1)} u_i^{(2)} = \lambda^{(1)} u_i^{(1)} u_i^{(2)} \\ A_{ij} u_j^{(2)} u_i^{(1)} = \lambda^{(2)} u_i^{(2)} u_i^{(1)} \,. \end{cases}$$

Subtracting these equations term by term gives

$$A_{ij} u_j^{(1)} u_i^{(2)} - A_{ij} u_j^{(2)} u_i^{(1)} = \left(\lambda^{(1)} - \lambda^{(2)}\right) u_i^{(1)} u_i^{(2)}$$

and, interchanging the dummy indices in the first term,

$$\left(A_{ji} - A_{ij}\right) u_i^{(1)} u_j^{(2)} = \left(\lambda^{(1)} - \lambda^{(2)}\right) u_i^{(1)} u_i^{(2)} \,.$$

Hence the scalar product $u_i^{(1)} u_i^{(2)}$ is zero if the matrix is symmetric ($A_{ij} = A_{ji}$) and if the eigenvalues are distinct, so that $\lambda^{(1)} - \lambda^{(2)} \neq 0$.

When two of the eigenvalues are equal, there is a principal plane, normal to the third eigenvector, in which it is always possible to choose two orthogonal vectors. In all cases, the principal directions of a matrix form an orthogonal triad.

The search for eigenvalues and eigenvectors is particularly simple if at least two of the non-diagonal elements of the matrix are zero, for example A_{13} and A_{23}. In this case, expansion of the determinant of (B.4) with respect to the third row and third column gives

$$(A_{33} - \lambda)[(A_{11} - \lambda)(A_{22} - \lambda) - (A_{12})^2] = 0 \tag{B.8}$$

showing that one root is $\lambda^{(3)} = A_{33}$. The other eigenvalues, $\lambda^{(1)}$ and $\lambda^{(2)}$, are solutions of a quadratic equation, giving

$$\lambda^{(k)} = \frac{A_{11} + A_{22}}{2} - \frac{(-1)^k}{2} \sqrt{(A_{22} - A_{11})^2 + 4(A_{12})^2} \quad \text{with} \quad k = 1, 2 \,. \tag{B.9}$$

Equations (B.5) become

$$\begin{cases} \left(A_{11} - \lambda^{(k)}\right) u_1^{(k)} + A_{12} u_2^{(k)} = 0 \\ A_{12} u_1^{(k)} + \left(A_{22} - \lambda^{(k)}\right) u_2^{(k)} = 0 \\ \left(A_{33} - \lambda^{(k)}\right) u_3^{(k)} = 0 \end{cases}$$

and this shows that, for the value $\lambda^{(3)} = A_{33}$, the Ox_3 axis is a principal axis, so that

$$u_1^{(3)} = u_2^{(3)} = 0 \quad \text{and} \quad u_3^{(3)} \quad \text{is arbitrary}\,.$$

For the other eigenvalues, $k = 1, 2$, we have $\lambda^{(k)} \neq A_{33}$. This gives $u_3^{(k)} = 0$ and

$$\frac{u_2^{(k)}}{u_1^{(k)}} = \frac{\lambda^{(k)} - A_{11}}{A_{12}}$$

leading to

$$\frac{u_2^{(k)}}{u_1^{(k)}} = \frac{A_{22} - A_{11}}{2A_{12}} - (-1)^k \sqrt{\left(\frac{A_{22} - A_{11}}{2A_{12}}\right)^2 + 1}\,. \tag{B.10}$$

The principal directions, contained in the Ox_1x_2 plane, are defined by the angles β_1 and $\beta_2 = \beta_1 + \pi/2$ which they make with Ox_1, such that

$$\tan\beta_k = \frac{u_2^{(k)}}{u_1^{(k)}} = a - (-1)^k \sqrt{a^2 + 1}\,,$$

where we have defined

$$a = \frac{A_{22} - A_{11}}{2A_{12}}\,.$$

Alternatively, the double angle is given by

$$\tan 2\beta_k = \frac{2\left[a - (-1)^k \sqrt{a^2+1}\right]}{1 - \left[a^2 + a^2 + 1 - 2a(-1)^k \sqrt{a^2+1}\right]} = -\frac{1}{a}\,. \tag{B.11}$$

Given the definition of a, and noting that changing $2\beta_k$ by π makes no essential difference, the formula

$$\tan 2\beta = \frac{2A_{12}}{A_{11} - A_{22}} \tag{B.12}$$

gives the two orthogonal eigendirections.

Appendix C. Tensor Representation of a Surface Element

This appendix refers to the integration of tensors along a curve, over a surface and in a volume.

In the first case, the integration variable is the vector $\mathrm{d}x_i$, and the integral of a tensor $A_{\dots i \dots}$ is written

$$\int_c A_{\dots i \dots} \mathrm{d}x_i \, .$$

For a surface integral, the infinitesimal element is a parallelogram constructed from the vectors $\mathrm{d}\boldsymbol{x}^{(1)}$ and $\mathrm{d}\boldsymbol{x}^{(2)}$. The projection of this parallelogram onto the coordinate plane $x_j x_k$ has area

$$\mathrm{d}s_{jk} = \mathrm{d}x_j^{(1)} \mathrm{d}x_k^{(2)} - \mathrm{d}x_k^{(1)} \mathrm{d}x_j^{(2)} \, .$$

These quantities form the following antisymmetric tensor of rank two:

$$(\mathrm{d}s_{jk}) = \begin{pmatrix} 0 & \mathrm{d}s_{12} & \mathrm{d}s_{13} \\ -\mathrm{d}s_{12} & 0 & \mathrm{d}s_{23} \\ -\mathrm{d}s_{13} & -\mathrm{d}s_{23} & 0 \end{pmatrix} .$$

The three independent components can be specified using a single index i, so that $\mathrm{d}s_i = \mathrm{d}s_{jk}$. We define i such that (ijk) has an even permutation, so $\mathrm{d}s_i$ is given by

$$\begin{cases} \mathrm{d}s_1 = \mathrm{d}s_{23} = \mathrm{d}x_2^{(1)} \mathrm{d}x_3^{(2)} - \mathrm{d}x_3^{(1)} \mathrm{d}x_2^{(2)} \\ \mathrm{d}s_2 = \mathrm{d}s_{31} = \mathrm{d}x_3^{(1)} \mathrm{d}x_1^{(2)} - \mathrm{d}x_1^{(1)} \mathrm{d}x_3^{(2)} \\ \mathrm{d}s_3 = \mathrm{d}s_{12} = \mathrm{d}x_1^{(1)} \mathrm{d}x_2^{(2)} - \mathrm{d}x_2^{(1)} \mathrm{d}x_1^{(2)} \, . \end{cases}$$

The transformation law (2.6) for rotation of axes can be applied to the components $\mathrm{d}s_i$ only if the axis transformation does not change the sense (handedness) of the axes. If the sense is changed (as in the case of symmetry with respect to a point or a plane), there must be a sign change, as shown in Problem 2.18. The vector $\mathrm{d}s_i$ is a pseudo-vector, or axial vector, and is simply the cross product of the vectors $\mathrm{d}\boldsymbol{x}^{(1)}$ and $\mathrm{d}\boldsymbol{x}^{(2)}$. From the properties of this product, $\mathrm{d}\boldsymbol{s} = \mathrm{d}\boldsymbol{x}^{(1)} \wedge \mathrm{d}\boldsymbol{x}^{(2)}$, it appears that $\mathrm{d}\boldsymbol{s}$ is normal to the parallelogram and its length is equal to the parallelogram area $\mathrm{d}s$. Hence, in place of the antisymmetric tensor $\mathrm{d}s_{jk}$, we can take the pseudo-vector $\mathrm{d}s_i$ as the integration variable. Taking $\mathrm{d}s_i = l_i \mathrm{d}s$, where l_i is the unit vector normal to the surface, the surface integral of the tensor $A_{\ldots i \ldots}$ can be written as

$$\int_s A_{\ldots i \ldots} l_i \mathrm{d}s \,.$$

If the surface is closed, this integral can be transformed into a volume integral, using Green's theorem (also called the divergence theorem). In terms of vectors, this theorem is

$$\int_s \boldsymbol{A} \cdot \mathrm{d}\boldsymbol{s} = \int_{\mathrm{V}} (\mathrm{div}\, \boldsymbol{A}) \mathrm{d}V \,,$$

and in tensor notation this becomes

$$\int_s A_i \mathrm{d}s_i = \int_s A_i l_i \mathrm{d}s = \int_{\mathrm{V}} \frac{\partial A_i}{\partial x_i} \mathrm{d}V \,,$$

where the limits of the volume V are given by the closed surface s. The two area integrals represent the flux of the vector $\boldsymbol{A}$. Generalizing to a tensor of arbitrary rank $A_{\ldots i \ldots}$, we have

$$\int_s A_{\ldots i \ldots} l_i \mathrm{d}s = \int_{\mathrm{V}} \frac{\partial A_{\ldots i \ldots}}{\partial x_i} \mathrm{d}V \,. \tag{C.1}$$

The reader is reminded of the expansion

$$\frac{\partial A_{\ldots i \ldots}}{\partial x_i} = \frac{\partial A_{\ldots 1 \ldots}}{\partial x_1} + \frac{\partial A_{\ldots 2 \ldots}}{\partial x_2} + \frac{\partial A_{\ldots 3 \ldots}}{\partial x_3} \,.$$

Note that, when the vector $\boldsymbol{A}$ represents the displacement $\boldsymbol{u}$ of the surface, the divergence is the local dilatation S of the elementary volume $\mathrm{d}V$, so that

$$S = \frac{\delta(\mathrm{d}V)}{\mathrm{d}V} = \mathrm{div}\, \boldsymbol{u} \,. \tag{C.2}$$

References

Historical Survey

1. P. and J. Curie, *Développement, par pression, de l'électricité polaire dans les cristaux hémièdres à faces inclinées*, C.R. Acad. Sc. Paris, 92, 1880.
2. P. and J. Curie, *Contractions et dilatations produites par des tensions électriques dans les cristaux*, C.R. Acad. Sc. Paris, 92, 1881.
3. Lord Rayleigh, *On waves propagating along the plane surface of an elastic solid*, Proc. London Math. Soc., **7**, p. 4–11, 1885.
4. P. Langevin and C. Chilowsky, *Procédés et appareils pour la production de signaux sous-marins dirigés et pour la localisation à distance d'obstacles sous-marins*, Brevet français n° 502.913 du 29 mai 1916.
5. P. Langevin, *Procédé et appareils d'émission et de réception des ondes élastiques sous-marines à l'aide des propriétés piézo-électriques du quartz*, Brevet français n° 505.703 du 17 septembre 1918.
6. C. Florisson and F. Vecchiacchi, *Les applications des ultrasons*, Congrès international d'électricité Paris (1932). Section 12, Rapport 8.
7. W.G. Cady, *Piezoelectricity*, McGraw-Hill, New York (1946)
8. R.A. Heising, *Quartz crystals for electrical circuits*, Van Nostrand, Princeton, New Jersey (1946).
9. P. Vigoureux and C. F. Booth, *Quartz vibrations and their applications*, H. M. Stationery Office, London (1950).
10. W.P. Mason, *Piezoelectric crystals and their application to ultrasonics*, Van Nostrand, Princeton, New Jersey (1950).
11. M. Tournier, *Cours de piézoélectricité*, Ecole Supérieure d'Electricité Paris (1953)
12. Don A. Berlincourt, D. Curran and H. Jaffe, *Piezoelectric and piezo-magnetic materials and their function in transducers*, Physical Acoustics. Principles and methods. Vol. 1, part A, p. 169–268, Academic Press. New York (1964).
13. K.F. Graff, *a – Ultrasonics : Historical aspects*, IEEE Ultrason. Symp. Proc. (1977) pp. 1–10. *b – A History of Ultrasonics*, Physical Acoustics. Principles and Methods. Vol. 15, p. 1–97, Academic Press, New York (1981).
14. C. Campbell, *Surface acoustic wave devices and their signal processing applications*, Academic Press. San Diego (1989).
15. Sadao Taki, *Improvement of growth process and characterization of quartz crystals*, Prog. Crystal Growth and Charact., **23**, p. 313–339, Pergamon, London (1992).

Chapter 1

1.1 E. Dieulesaint and D. Royer, *Automatique appliquée*, vol. 1, chap. 3, Masson, Paris (1987).
1.2 L.E. Kinsler, A.R. Frey, A.B. Coppens and J.V. Sanders, *Fundamentals of acoustics*, Chap. 7, J. Wiley, New York (1982).
1.3 G. Kino, *Acoustic waves: devices, imaging and analog signal processing*, Prentice Hall, Englewood Cliffs (1987).
1.4 P.R. Stephanishen, *Transient radiation from pistons in an infinite planar baffle*, J. Acoust. Soc. Amer., vol. **49**, p. 1629–1638 (1971).

Chapter 2

2.1 G. Binnig and H. Rohrer, *Scanning tunneling microscopy*, Helv. Phys. Acta, **55**, 726 (1982).
2.2 E.A. Ash and G. Nicholls, *Superresolution aperture scanning microscope*, Nature, **237**, 510 (1972).
2.3 D.M. Eigler and E.K. Schweiger, *Positioning single atoms with a scanning tunneling microscope*, Nature, **344**, 524 (1990).
2.4 D. Schechtman, I. Blech, D. Gratias and J.W. Cahn, *Metallic phase with long range orientational order and no translational symmetry*, Phys. Rev. Lett. **53**, 1951 (1984).
2.5 A. Katz and M. Duneau, *Quasiperiodic patterns*, Phys. Rev. Lett. **54**, 2688 (1985).
2.6 R. Penrose, *The role of aesthetics in pure and applied mathematical research*, Bull. Inst. Math. Appl. **10**, 266 (1974).
2.7 G.A.M. Reynolds, B. Golding, A.R. Kortan and J.M. Parsey Jr., *Isotropic elasticity of the Al-Cu-Li quasicristal*, Phys. Rev. B, **41**, 1194 (1990).
2.8 Y. Amazit, M. de Boissieu and A. Zarembowitch, *Evidences for elastic isotropy and ultrasonic attenuation anisotropy in Al-Mn-Pd quasicrystals*, Europhys. Lett., **20**, 703 (1992).
2.9 H.W. Kroto, J.R. Heath, S.C. O'Brien, R.F. Curl and R.E. Smalley, *C_{60}: Buckminsterfullerene*, Nature, **318**, 162 (1985).
2.10 L. Eyraud, *Diélectiques solides anisotropes et ferroélectricité*, Chap. 6. Paris, Gauthier-Villars (1967).
2.11 K. Schubert, *Kristallstrukturen zweikomponentiger Phasen*, p. 200, Fig.6. Berlin, Springer Verlag (1964).
2.12 H.D. Megaw, *Ferroelectricity in crystals*, Chap. 5, §. 5, Methuen and Co., London (1957).
2.13 M. Di Domenico, Jr. and S.H. Wemple, *Oxygen-octahedra ferroelectrics*, J. Appl. Phys., **40**, 720 (1969).
2.14 P.W. Krempl, *Quartz homeotypic gallium-orthophosphate, a new high tech piezoelectric material*, IEEE Ultrason. Symp. Proc., p. 949 (1994).
2.15 H.A.A. Sidek, G.A. Saunders, Wang Hong, Xu Bin and Han Jianru, *Elastic behavior under hydrostatic pressure and acoustic mode vibrational anharmonicity of single crystal berlinite*, Phys. Rev., **36**, 7612 (1987).
2.16 J. Detaint, E. Philippot, J.C. Jumas, J. Schwartzel, A. Zarka, B. Capelle and J.C. Doukhan, *Crystal growth, physical characterization and BAW devices applications of berlinite*, 39th Freq. Control Symp. Proceedings, (Philadelphia, May 1985).

Chapter 3

3.1 J. Vallin, M. Mongy, K. Salama and O. Beckman, J. Appl. Phys., **35**, 1825 (1964).
3.2 T.B. Bateman, H.J. McSkimin and J.M. Whelan, J. Appl. Phys., **30**, 544 (1959).
3.3 E.G. Spencer, R.T. Denton, T.B. Bateman, W.B. Snow and L.G. Van Uitert, J. Appl. Phys. **34**, 3059 (1963).
3.4 A.J. Slobodnik, Jr, and J.C. Sethares, J. Appl. Phys., **43**, 247 (1972).
3.5 Y.A. Chang and L. Himmel, J. Appl. Phys., **37**, 3567 (1966).
3.6 R.E. MacFarlane, J.A. Rayne and C.K. Jones, Phys. Letters, **18**, 91 (1965).
3.7 J.J. Hall, Phys. Rev., **161**, 756 (1967).
3.8 R. Lowrie and A.M. Gonas, J. Appl. Phys., **38**, 4505 (1967).
3.9 J.M. Smith and C.L. Arbogast, J. Appl. Phys., **31**, 99 (1960).
3.10 T.B. Bateman, J. Appl. Phys., **33**, 3309 (1962).
3.11 J.A. Corll, Phys. Rev., **157**, 623 (1967).
3.12 E.S. Fischer and C.J. Renken, Phys. Rev., **135A**, 482 (1964).
3.13 B.S. Chandrasekhar and J.A. Rayne, Phys. Rev., **124**, 1011 (1961).
3.14 G.A. Coquin, D.A. Pinnow and A.W. Warner, J. Appl. Phys., **42**, 2162 (1971).
3.15 W.J. Alton and A.J. Barlow, J. Appl. Phys., **38**, 3817 (1967).
3.16 Y. Ohmachi and N. Uchida, J. Appl. Phys., **41**, 2307 (1970).
3.17 R.K. Verma, J. Geophys. Res., **65**, 757 (1960)
3.18 D. Berlincourt and H. Jaffe, Phys. Rev., **111**, 143 (1958).
3.19 J.B. Watchman, W.E. Tefft, D.G. Lam and R.P. Stinchfield, J. Res. Natl. Bur. Std. **64A**, 213 (1960).
3.20 A.W. Warner, M. Onoe, and G.A. Coquin, J. Acoust. Soc. Am., **42**, 1223 (1967).
3.21 R. Bechman, Phys. Rev., **110**, 1060 (1958).
3.22 J.L. Malgrange, C. Quentin, and J.M. Thuillier, Phys. Status Solidi, **4**, 139 (1964).
3.23 S. Haussühl, Acta Cryst., **A24**, 697 (1968).
3.24 A.W. Warner, G.A. Coquin, and J.L. Fink, J. Appl. Phys., **40**, 4353 (1969).
3.25 H. Jaffe and D.A. Berlincourt, Proc. IEEE, **53**, 1372 (1965).
3.26 G. Arlt and P. Quadflieg, Phys. Status Solidi, **25**, 323 (1968).

Chapter 4

4.1 F.I. Fedorov, *Theory of elastic waves in crystals*, Plenum Press: New York (1968)
4.2 R.W. Damon, IEEE Spectrum, p. 87 (June 1967).
4.3 A.W. Warner, M. Onoe and G.A. Coquin, J. Acoust. Soc. Am., **42**, 1223 (1967).
4.4 K.A. Ingebrigtsen, *Surface waves in piezoelectrics*, J. Appl. Phys., **40**, p. 2681–2686 (1969).
4.5 M. Feldmann and J. Hénaff, *Surface acoustic waves for signal processing*, Artech House, Boston (1989) (1986).
4.6 J.L. Synge, Proc. R. Irish Acad., **A58**, p. 13–20 (1957).
4.7 M. Hayes, *Energy flux for trains of inhomogeneous plane waves*, Proc. R. Soc. Lond., **A370**, p. 417–429 (1980).
4.8 D.L. Arenberg, *Ultrasonic delay lines*, J. Acoust. Soc. Am., **20**, p. 1–25 (1948).

4.9 C. Potel and J.F. de Belleval, *Surface waves in an anisotropic periodically multilayered medium : influence of the absorption*, J. Appl. Phys., **77**, p. 6152-6161 (1995).

Chapter 5

5.1 G.W. Farnell and E.L. Adler, *Elastic wave propagation in thin layers*, In Physical Acoustics (W.P. Mason and R.N. Thurston, Eds), **9**, p. 35–127, Academic Press, New York (1972).

5.2 R.M. White and F.W. Voltmer, *Direct piezoelectric coupling to surface elastic waves*, Appl. Phys. Lett., **7**, 314 (1965).

5.3 Réf. 1, fig. 25.

5.4 M. Feldmann, *Théorie des réseaux et systèmes linéaires*, p. 123, Eyrolles, Paris (1981).

5.5 I.A. Viktorov, *Rayleigh and Lamb Waves*, p. 3, Plenum press, New York (1970)

5.6 D. Royer and E. Dieulesaint, *Rayleigh wave velocity and displacement in orthorhombic, tetragonal, hexagonal and cubic crystals*, J. Acoust. Soc. Am., **76**, p. 1438–1444 (1984).

5.7 R. Stoneley, *The propagation of surface elastic waves in cubic crystals*, Proc. Roy. Society, **232A**, p. 447–458 (1995).

5.8 R. Stoneley, *The seismological implications of aeolotropy in continental structure*, Mon. Not. Astr. Soc. Geophys. Suppl., **5**, p. 343–353 (1949).

5.9 G.W. Farnell, *Properties of elastic surface waves*, In Physical Acoustics (W.P. Mason and R.N. Thurston, Eds), **6**, p. 109–166, Academic Press, New York (1970).

5.10 M. Feldman and J. Hénaff, *Propagation des ondes élastiques de surface*, Rev. Phys. Appliquée, **12**, p. 1775–1789 (1977).

5.11 A.J. Slobodnik Jr., E.D. Conway and R.T. Delmonico, *Microwave acoustics handbook*, vol. 1A. *Surface wave velocities*, AFCRL Report n° 73–0597, Air Force Cambridge Research Lab. Bedford, Mass. (1973).

5.12 M. Moriamez, E. Bridoux, J.M. Desrumeaux, J.M. Rouvaen and M. Delannoy, *Propagation des ondes acoustiques superficielles dans les cristaux piézoélectriques*, Rev. Phys. Appliquée, **6**, p. 333 (1971).

5.13 G.A. Coquin and H.F. Tiersten, *Analysis of the excitation and detection of piezoelectric surface waves in quartz by means of surface electrodes*, J. Acoust. Soc. Am., **41**, 921 (1967).

5.14 C. Campbell, *Surface acoustic wave devices and their signal processing applications*, Academic Press, San Diego (1989).

5.15 R.G. Pratt, G. Simpson and W.A. Crossley, Electron. Letters, **8**, 127, (1972).

5.16 J.J. Campbell and W.R. Jones, J. Appl. Phys., **41**, 2796 (1970).

5.17 M. Feldmann and J. Hénaff, *Surface acoustic waves for signal processing*, Artech House, Boston (1989).

5.18 S. Fujishima, *Piezoelectric devices for frequency control and selection in Japan*, IEEE Ultrason. Symp. Proc., p. 87–94 (1990).

5.19 J.L. Bleustein, *A new surface wave in piezoelectric materials*, Appl. Phys. Lett., **13**, 412 (1968).

5.20 Yu. V. Gulyaev, *Electroacoustic surface waves in solids*, JETP Letters, **9**, p. 37–38 (1969).

5.21 L.P. Solie and B. Auld, *Elastic waves in free anisotropic plates*, J. Acoust. Soc. Am., **54**, p. 50–65 (1973).

5.22 J.D. Achenbach, *Wave propagation in elastic solids*, chap. 6, p. 236, North-Holland, Amsterdam (1975).

5.23 T.R. Meeker and A.H. Meitzler, *Guided wave propagation in elongated cylinders and plates*, in Physical Acoustics, Vol. **1A**, W.P. Mason Ed., Academic Press, New York (1964).
5.24 D.C. Gazis, *Three-dimensional investigation of the propagation of waves in hollow circular cylinders, I) Analytical foundation, II) Numerical results*, J. Acoust. Soc. Am. **31**, p. 568–78 (1959).
5.25 N.C. Nicholson and W.N. McDicken, *Mode propagation of ultrasound in hollow circular waveguides*, Ultrasonics, **29**, p. 411–416 (1991).
5.26 R.M. Cooper and P.M. Naghdi, *Propagation of nonaxially symmetric waves in elastic cylindrical shells*, J. Acoust. Soc. Am. **29**, p. 1365–73 (1957).
5.27 I. Mirsky and G. Hermann, *Nonaxially symmetric motions of cylindrical shells*, J. Acoust. Soc. Am. **29**, p. 1116–23 (1957).

Further Reading

J. Bok and P. Morel, *Mécanique - Ondes*, Hermann, Paris (1968)

W.C. Elmore and M.A. Head, *The physics of waves*, McGraw-Hill, New York (1969)

J.L. Davis, *Wave propagation in solids and fluids*, Springer-Verlag, New York (1988)

J.W. Goodman, *Introduction à l'optique de Fourier et à l'holographie*, Masson, Paris (1972)

M.A. Fink and J.F. Cardoso, *Diffraction effects in pulse echo measurement*, IEEE Trans. Sonics and Ultrasonics, vol. **SU-31**, p. 313-329 (1984)

J. J. Rousseau, *Cristallographie géométrique et radiocristallographie*, Masson, Paris (1995).

C. Janot, Quasicrystals, *A primer*, Clarendon Press, Oxford (1992).

F. Hippert and D. Gratias, *Lectures on quasicrystals*, les Editions de Physique, Orsay (1994).

J.C. Gualtieri, J.A. Kosinski and A. Ballato, *Piezoelectric materials for acoustic wave applications*, IEEE Trans. Ultrason. Ferroel. and Freq. Control, 41, p. 53 (1994).

F. C. Phillips, *An Introduction to Crystallography*, Longmans, London (1956).

J.F. Nye, *Physical properties of crystals*, chaps. V, VI, VIII, Clarendon Press, Oxford (1964).

F.I. Fedorov, *Theory of elastic waves in crystals*, chap. 1, Plenum Press, New York (1968).

B. Auld, *Acoustic fields and waves in solids*, vol. 1, 2nd edition, R. E. Krieger (1990).

V. M. Ristic, *Principles of acoustic devices*, Wiley, New York (1983).

J.F. Rosenbaum, *Bulk wave theory and devices*, Artech House, Boston (1988).

G.S. Kino, *Acoustic Waves, devices, imaging and analog signal processing*, Prentice-Hall, Englewood Cliffs (1987).

J.D. Achenbach, *Wave propagation in Elastic Solids*, North-Holland, Amsterdam (1975).

L.M. Brekhovskikh, *Waves in layered media*, Academic Press, New York (1980).

H. F. Pollard, *Sound waves in solids*, Pion, Londres (1977).

J. Miklowitz, *The theory of elastic waves and waveguides*, North-Holland, New York (1978).

E.A. Ash, G.W. Farnell, H.M. Gerard, A.A. Oliner, A.J. Slobodnik, Jr. and H.I. Smith, *Acoustic Surface Waves*, A.A. Oliner, Ed., Springer Verlag, Berlin (1978).

Symbols

c: elastic stiffness constant
e, d: piezoelectric constant
V: bulk wave velocity
K: bulk wave electromechanical coupling coefficient
V_{R}: Rayleigh wave velocity
K_{R}: Rayleigh wave electromechanical coupling coefficient
ρ: mass density
ε: dielectric constant

Index

Printing: Mercedes-Druck, Berlin
Binding: Buchbinderei Lüderitz & Bauer, Berlin